International Conference on Functional Materials and Nanotechnologies

(FM&NT 2012)

IOP Conference Series: Materials Science and Engineering Volume 38

Riga, Latvia
17 – 20 April 2012

ISBN: 978-1-63439-849-7
ISSN: 1757-8981

Printed from e-media with permission by:

Curran Associates, Inc.
57 Morehouse Lane
Red Hook, NY 12571

Some format issues inherent in the e-media version may also appear in this print version.

Printed by Curran Associates, Inc. (2015)

For permission requests, please contact the Institute of Physics at the address below.

Institute of Physics
Dirac House, Temple Back
Bristol BS1 6BE UK

Phone: 44 1 17 929 7481
Fax: 44 1 17 920 0979

techtracking@iop.org

Additional copies of this publication are available from:

Curran Associates, Inc.
57 Morehouse Lane
Red Hook, NY 12571 USA
Phone: 845-758-0400
Fax: 845-758-2634
Email: curran@proceedings.com
Web: www.proceedings.com

TABLE OF CONTENTS

MAGNESIUM NANOPARTICLES FOR HYDROGEN STORAGE: STRUCTURE, KINETICS AND THERMODYNAMICS ... 1
L Pasquini, M Brighi, A Montone, M V Antisari, B Dam, V Palmisano, E Bonetti

SCINTILLATING CERAMICS BASED ON ZINC OXIDE ... 9
P A Rodnyi, E I Gorokhova, K A Chernenko, I V Khoduk

SPIN-ORBIT GAPS IN ARMCHAIR NANOTUBES CALCULATED USING THE LINEAR AUGMENTED CYLINDRICAL WAVE METHOD ... 16
P D'Yachkov, D Kutlubaev

AB-INITIO STUDY OF METAL-ZIRCONIA INTERFACES ... 24
S Kulkova, A Bakulin, S Hocker, S Schmauder

AB INITIO SIMULATIONS ON RUTILE-BASED TITANIA NANOWIRES ... 31
Yu F Zhukovskii, R A Evarestov

FLUORESCENCE OF TIO_2:SM^{3+} COMPOSITE STIMULATED BY PLASMON WAVES ... 37
L Dolgov, V Kiisk, S Pikker, A Loot, I Sildos

LOW-ENERGY CHARGE TRANSFER EXCITATIONS IN NIO ... 43
V I Sokolov, V A Pustovarov, V N Churmanov, V Yu Ivanov, A Ye Yermakov, M A Uimin, N B Gruzdev, P S Sokolov, A N Baranov, A S Moskvin

VIBRATIONAL SPECTRA OF OPAL-BASED PHOTONIC CRYSTALS ... 49
G Dovbeshko, O Fesenko, V Boyko, V Romanyuk, V Moiseyenko, V Gorelik, L Dolgov, V Kiisk, I Sildos

PROCESSING OF AMORPHOUS SI BY PULSED LASER IRRADIATION AT DIFFERENT WAVELENGTHS ... 55
J Butikova, G Marcins, B Polyakov, A Muhins, A Voitkans, I Tale

MAGNETIC PROPERTIES OF CARBON NANOPARTICLES ... 61
E Lähderanta, A V Lashkul, K G Lisunov, D A Zherebtsov, D M Galimov, A N Titkov

HOLOGRAPHIC STUDIES OF PHOTOINDUCED ANISOTROPY IN MOLECULAR GLASSY FILMS CONTAINING DIPHENYLAMINE AZOCHROMOPHORES ... 68
A Ozols, V Kokars, P Augustovs, K Traskovskis, D Saharov

OPTICAL FIELD-INDUCED SURFACE RELIEF FORMATION ON CHALCOGENIDE AND AZO-BENZENE POLYMER FILMS ... 74
J Teteris, U Gertners

INFLUENCE OF SPECIFIC AND NON-SPECIFIC SOLVENT EFFECTS ON FRONTAL MO OF DMABI SYSTEM ... 80
V Kampars, P Pastors, G Ornicane

INVESTIGATION OF CHEMICAL BONDING AT METAL-CERAMIC INTERFACES ... 86
A Bakulin, S Kulkova, S Hocker, S Schmauder

THE EFFECT OF ZN VACANCIES AND GA DOPANTS ON THE ELECTRONIC STRUCTURE OF ZNO: AB INITIO SIMULATIONS ... 90
A V Sorokin, Y F Zhukovskii, J Purans, E A Kotomin

THE ROLE OF NANOPOWDER PARTICLE SURFACES AND GRAIN BOUNDARY DEFECTS IN THE SINTERING OF ZNO CERAMICS ... 94
F Muktepavela, R Zabels, V Sursajeva, L Grigorjeva, K Kundzins

SHEAR BANDING MECHANISM OF PLASTIC DEFORMATION IN LIF IRRADIATED WITH SWIFT HEAVY IONS ... 98
J Maniks, R Zabels, I Manika

DIMENSIONAL EFFECTS OF SAMPLE GEOMETRY AND MICROSTRUCTURE OF MNZN AND NIZN FERRITES ... 103
J Jankovskis, N Ponomarenko, N Mironova-Ulmane, D Jakovlevs

CHARACTERIZATION OF $LIFEPO_4$/C COMPOSITE THIN FILMS USING ELECTROCHEMICAL IMPEDANCE SPECTROSCOPY ... 107
G Bajars, G Kucinskis, J Smits, J Kleperis, A Lusis

STUDY THE EFFECTS OF MOISTURE CONTENT ON THE ELECTRICAL PROPERTIES OF TECHNICAL TEXTILES BY IMPEDANCE SPECTROSCOPY ... 112
A Lusis, E Pentjuss, G Bajars, J Gabrusenoks, R Janeliukštis, J Zandersons

ELECTROCHEMICAL IMPEDANCE AND MOISTURE CONTENT OF GLASS FABRIC ... 117
E Pentjuss, A Lusis, G Bajars, J Gabrusenoks, L Jekabsone

COTTON FABRIC SURFACE MODIFICATION BY SOL-GEL DEPOSITION OF ZNO THIN FILMS 121
S Vihodceva, S Kukle
PREPARATION AND CHARACTERIZATION OF CNT/TIO2 BASED TRANSPARENT FIBER ELECTRODES 126
M Paalo, T Tätte, M Hussainov, K Hanschmidt, M Lobjakas, A Lõhmus, U Mäeorg, I Kink
PSEUDOISOCHROMATIC TEST PLATE COLOUR REPRESENTATION DEPENDENCE ON PRINTING TECHNOLOGY 130
K Luse, S Fomins, M Ozolinsh
THE EXPRESSION OF CYTOKINES AND ß -DEFENSIN 2, - 3, -4 IN RABBIT BONE TISSUE AFTER HYDROXYAPATITE (HAP), α- TRICALCIUM PHOSPHATE (α -TCP) AND POLYMETHYLMETHACRYLATE (PMMA) IMPLANTATION 134
J Vamze, M Pilmane, A Skagers
THE IMPACT OF LIGHT POLARIZATION ON THE DIRECT RELIEF FORMING PROCESSES IN AS_2S_3 THIN FILMS 138
U Gertners, J Teteris
UV OPTICAL RECORD AND ELECTRON BEAM LITHOGRAPHY IN POLYMER FILMS 142
A Gerbreders, O Shimane, V Kolobjonoks, J Teteris
PHOTOINDUCED PHENOMENA IN AZO-DYED GELATINE FILMS 146
J Aleksejeva, J Teteris
IRON COMPLEXES EMBEDDING INFLUENCE ON PMAA HYDROGEL 150
L Khenkin, T Baluyan, A Novakova, I Rebrin, E Makhaeva
MECHANICAL, STRUCTURAL AND MAGNETIC PROPERTIES OF POLYPROPYLENE/IRON FERRITE MAGNETIC NANOCOMPOSITES 155
I Reinholds, V Kalkis, J Zicans, R Merijs Meri, A Grigalovica, M Maiorov
SYNTHESIS HIGHLY ACTIVE PLATINUM TRI-METALLIC ELECTROCATALYSTS USING "ONE-STEP" ORGANOMETALLIC CHEMICAL VAPOUR DEPOSITION TECHNIQUE FOR METHANOL OXIDATION PROCESS 159
Q-L Naidoo, S Naidoo, L Petrik, A Nechaev, P Ndungu, G Vaivars
FORMATION OF NANOSIZED STRONTIUM SUBSTITUTED HYDROXYAPATITES 164
K A Gross, A Jeršova, D Grossin, C Rey, A Viksna
ROLE OF PARAMAGNETIC POLYCONJUGATED CLUSTERS IN LIGNIN ANTIOXIDANT ACTIVITY (IN VITRO) 169
T Dizhbite, J Ponomarenko, A Andersone, G Dobele, M Lauberts, J Krasilnikova, N Mironova-Ulmane, G Telysheva
NONLINEAR OPTICAL PROPERTIES OF LOW MOLECULAR ORGANIC GLASSES FORMED BY TRIPHENYL MODIFIED CHROMOPHORES 173
A Tokmakovs, M Rutkis, K Traskovskis, E Zarins, L Laipniece, V Kokars, V Kampars
HYPER-RAYLEIGH SCATTERING AND TWO-PHOTON LUMINESCENCE OF PHENYLAMINE- INDANDIONE CHROMOPHORES 177
I Mihailovs, J Kreicberga, V Kampars, S Miasojedovas, S Juršenas, L Skuja, M Rutkis
MULTISCALE APPROACH TO THE STRUCTURE AND SPECTRA OF NILE RED ADSORBED ON POLYSTYRENE NANOPARTICLES 184
A Freidzon, V Tikhomirov, A Odinokov, A Bagaturyants
NEAR-BAND LUMINESCENCE OF CDZNTE DETECTOR CRYSTALS 190
L Grigorjeva, D Millers, V Ivanov, L Aleksejeva
SYNTHESIS OF CUBIC AND HEXAGONAL $NAYF_4:ER^{3+}$ 195
A Sarakovskis, M Voss, G Doke, D Jankovica, J Grube
STABILITY OF COLOR CENTERS IN $LIYF_4$ CRYSTALS AT LOW TEMPERATURES 199
V M Lisitsyn, Y V Bikhert, V I Korepanov, L A Lisitsyna, A K Dauletbekova, A T Akylbekov, V M Reiterov
COLOR CENTERS AND STRUCTURAL DAMAGE IN LIF INDUCED BY 150 MEV KR IONS 205
A Rusakova, J Maniks, K Schwartz, A Dauletbekova, A Akilbekov, V Lisitsin, M Zdorovets
AB INITIO CALCULATIONS OF THE H CENTERS IN MGF_2 CRYSTALS 210
F U Abuova, A T Akilbekov, E A Kotomin
FLUCTUATIONS IN AN INORGANIC GLASS FORMING SYSTEM CAPABLE OF LIQUID-LIQUID PHASE SEPARATION 214
V Bogdanov, L Maksimov, A Anan'Ev, S Nemilov, V Rusan
MEASUREMENTS OF SIO_2 GLASS SURFACE PARAMETERS BY METHODS OF MICROSCOPY 218
E Gavars, A Svagere, A Skudra, N Zorina, R Poplausks
OPTICAL PROPERTIES OF IRRADIATED YTTRIUM ALUMINUM GARNET 222
V Skvortsova, N Mironova-Ulmane

THE LUMINESCENT PROPERTIES OF PERSISTENT STRONTIUM ALUMINATE PHOSPHOR PREPARED BY SOLAR INDUCED SOLID STATE SYNTHESIS 226
V Liepina, K Smits, D Millers, L Grigorjeva, C Monty
EPR SPECTRA OF THE MN^{2+} ION IN THE OXYFLUORIDE GLASS CERAMICS CONTAINING BAF_2 NANOCRYSTALLINE PHASE 231
Dz Berzins, A Fedotovs, O Kiselova, A Sarakovskis
CHARACTERISTICS OF THE MN^{2+} EPR SPECTRA IN THE OXYFLUORIDE GLASS CERAMICS CONTAINING SRF_2 NANOCRYSTALS 235
A Fedotovs, Dz Berzins, O Kiselova, A Sarakovskis
WORK FUNCTION OF COLLOIDAL SEMICONDUCTING NANOCRYSTALS MEASURED BY KELVIN PROBE 239
P Kulis, J Butikova, B Polyakov, G Marcins, J Pervenecka, K Pudzs, I Tale
DEPOSITION OF CU NANOPARTICLES ON THE SURFACE OF METALLIC ALUMINUM 243
A Lescinskis, J Katkevics, D Erts, A Viksna
ZIRCONIA NANOCRYSTALS AS SUBMICRON LEVEL BIOLOGICAL LABEL 248
K Smits, J Liepins, M Gavare, A Patmalnieks, A Gruduls, D Jankovica
STRUCTURE OF NANOPOROUS CARBON MATERIALS FOR SUPERCAPACITORS 252
A Volperts, N Mironova-Ulmane, I Sildos, D Vervikishko, E Shkolnikov, G Dobele
SYNTHESIS OF SILICA-CARBON NANOTUBE COMPOSITE MATERIALS AND THEIR APPLICATION FOR LASER SYSTEMS 257
J Pilipavicius, D Sakalauskas, A Beganskiene, A Kareiva
THE EFFECT OF NANOSIZE ZNO ON THE PROPERTIES OF THE SELECTED POLYMER BLEND COMPOSITES 261
A Grigalovica, I Bochkov, R M Meri, J Zicans, J Grabis, R Kotsilkova, I Borovanska
CARRIER DYNAMICS IN EPILAYERS AND NANOCOLUMNS OF TERNARY ALGAN WITH A TUNABLE BANDGAP 265
R Aleksiejunas, P Ščajev, S Nargelas, S Miasojedovas, K Jarašiunas, L Trinkler, J Grigorjeva, B Berzina, K H Chen, Y T Chen, M W Chen, L C Chen
INVESTIGATION OF ELECTRIC PROPERTIES OF NANOCOMPOSITES IN FREQUENCY DOMAIN - COMPACT MEASUREMENT DATA REPRESENTATION 270
I Matiss, E Pentjušs, O Krievs, J Zicans
STRUCTURE ANALYSIS AND SIZE DISTRIBUTION OF PARTICULATE MATTER FROM CANDLES AND KEROSENE COMBUSTION IN BURNING CHAMBER 274
M Baitimirova, A Osite, J Katkevics, A Viksna
AB INITIO MODELING OF SULPHUR DOPED TIO_2 NANOTUBULAR PHOTOCATALYST FOR WATER-SPLITTING HYDROGEN GENERATION 278
O Lisovski, S Piskunov, Y F Zhukovskii, J Ozolins
AB INITIO MODELLING OF UN GRAIN BOUNDARY INTERFACES 283
E A Kotomin, Yu F Zhukovkii, D Bocharov, D Gryaznov
PREPARATION AND PHOTOACTIVITY OF ELECTROPHORETIC TIO_2 COATING FILM 287
I Liepina, G Bajars, J Gabrusenoks, L Grinberga, J Kleperis, A Lusis
HYDROGEN ADSORPTION IN ZEOLITE STUDIED WITH SIEVERT AND THERMOGRAVIMETRIC METHODS 291
P Lesnicenoks, A Sivars, L Grinberga, J Kleperis
THRESHOLD CHARACTER OF TEMPERATURES ON DEUTERIUM THERMAL DESORPTION IN MG-V COMPOSITE GROWN ATOM-BY-ATOM 297
I M Neklyudov, O M Morozov, V G Kulish, V I Zhurba, A G Galitskiy, N S Lomino, A S Kuprin, V D Ovcharenko, E N Reshetnyak
EPM FINE-DISPERSE PLATINUM COATING ON POWDER CARRIERS 301
V Serga, L Kulikova, A Cvetkov, A Krumina
THE MEASUREMENT OF IMPURITIES IN SILICON FOR SOLAR CELL PRODUCTION 305
G Kolosovska, A Viksna, G Chikvaidze, A Osite, A Opalais
PROPERTIES OF THE NAFION MEMBRANE IMPREGNATED WITH HYDROXYL AMMONIUM BASED IONIC LIQUIDS 309
V Garaev, J Kleperis, S Pavlovica, G Vaivars
Author Index

International Conference on Functional Materials and Nanotechnologies (FM&NT2012) IOP Publishing
IOP Conf. Series: Materials Science and Engineering **38** (2012) 012001 doi:10.1088/1757-899X/38/1/012001

Magnesium Nanoparticles for Hydrogen Storage: Structure, Kinetics and Thermodynamics

L Pasquini[1], M Brighi[1], A Montone[2], M Vittori Antisari[2], B Dam[3], V. Palmisano[3] and E Bonetti[1]

[1] Department of Physics, University of Bologna, viale Berti-Pichat 6/2, 40127 Bologna, Italy

[2] ENEA, Technical Unit "Materials Technology", Research Centre Casaccia, via Anguillarese 301, 00123 Rome, Italy

[3] Department of Chemical Engineering, Faculty of Applied Sciences, Delft University of Technology, Julianalaan 136, 2628BL Delft, The Netherlands

E-mail: luca.pasquini@unibo.it

Abstract: Magnesium nanoparticles coated by a native oxide shell and decorated by palladium clusters were synthesized by the inert gas condensation technique. The kinetics and thermodynamics of hydrogen sorption were investigated by Sieverts measurements at high temperature and by optical hydrogenography close to ambient temperature. The structure and morphology of the nanoparticles were studied by electron microscopy and X-ray diffraction both in the as-prepared state and after hydrogen sorption cycles.

1. Introduction

The strong interest towards efficient solid-state hydrogen (H) storage media has triggered the investigation of a quite impressive variety of candidate materials [1]. An alternative approach to the synthesis of new bulk compounds is currently represented by the engineering of many different nanostructures, from carbon nanotubes to metal organic frameworks [2] to metal-hydride based systems such as multilayers [3] and mechanically alloyed nanocomposites [4].

Along this route, in this work we report on the H-storage properties of magnesium (Mg) nanoparticles (NPs) coated by a native MgO oxide shell and decorated by ultrafine palladium (Pd) crystallites. Magnesium and its hydride MgH_2 display several attractive features like low cost, high H-storage capacity (7.6 wt%), non toxicity, but also a scarce ability to dissociate H_2 which is generally overcome by the addition of a suitable catalyst [3].

We address several issues relevant to the physics and chemistry of Mg-based nanostructures for H-storage. In particular, we examine in detail the effect of the Pd catalyst not only on the kinetics but also on the thermodynamics of the metal-hydride transformation. Furthermore, the instability problem related to the evolution of NPs' structure and morphology upon H-cycling at elevated temperature is discussed. Finally, we characterize the H-absorption behavior close to ambient temperature by means of optical techniques, studying both the pressure-dependence of the kinetics and the equilibrium thermodynamics.

2. Experimental

Mg/MgO core/shell NPs with Pd decoration were synthesized by Inert Gas Condensation (IGC) as described in [5]. Two different operating conditions and collection methods were employed. In the first case, the evaporation tungsten boat was kept at about 840 K and the helium (He) pressure in the chamber was 250 Pa. The collection surface consisted in a rotating stainless steel cylinder cooled by liquid nitrogen (LN), from which the NPs were scraped after the deposition of Pd. In a synthesis time of 1 hour, NPs batches of about 100 mg could be obtained. Two materials with different Pd contents were prepared for the present study and will be referred in the following as Mg-Pd_s1 and Mg-Pd_s2 ("s" stands for "scraped"). In the second case, the evaporation source was operated at about 700 K under a He pressure of 60 Pa. The collection surface was a LN-cooled copper plate on which the substrates suitable for subsequent measurements were mechanically fixed. The amount of material collected in this low-power condition is rather small (≈1 mg) and the samples are obtained in the form of nanoparticulate films resulting from the soft-landing of the NPs on the deposition substrates. For this reason they will be referred to as Mg-Pd_f in the following. After evaporation of Mg, pure oxygen was admitted into the chamber (O_2 partial pressure=100 Pa) with the aim to form a MgO protective shell around the NPs. Finally, Pd evaporation was conducted under high vacuum (10^{-5} Pa) and was monitored the by a quartz microbalance placed in proximity of the collection surface.

Transmission electron microscopy (TEM) on scraped NPs was carried out with a Tecnai G2 30F microscope, equipped with a Schottky type field emission source operated at 300 kV. A small amount of NPs was dispersed in hexane by sonication and a drop of the suspension was allowed to dry on holey carbon support grid. Scanning electron microscopy (SEM) on both scraped and film-like NPs was performed using a ZEISS EVO MA15 equipped with secondary (SE) and backscattered (BSE) electron detectors and energy dispersive x-ray microanalysis (EDS).

X-ray diffraction (XRD) was performed with a Rigaku DMAX-IIIC Bragg Brentano diffractometer employing Cu K-α radiation and a graphite monochromator on the diffracted beam.

H-sorption properties at high temperature (550-650 K) were studied on scraped NPs using a volumetric Sieverts' apparatus in which we loaded ≈ 30 mg of material. Low-temperature H-sorption (333 K) was investigated on Mg-Pd_f NPs films deposited onto 1 mm thick glass substrates by means of optical hydrogenography [6]. The purity of the H_2 gas was in both cases 99.9995 %.

3. Results and Discussion

3.1. Structural analysis

Figure 1 displays a TEM image of scraped NPs (Mg-Pd_s1), while figure 2 shows a SEM image of Mg-Pd_f on a glass substrate. In both cases, the NPs appear as single crystals with six-fold symmetry. In Mg-Pd_s1, a significant size dispersion is observed, with NPs sizes ranging from ~ 100 nm to ~ 1 μm. The average size is determined as 450 nm from the statistical analysis of TEM images. In Mg-Pd_f, the NPs tend to align the hexagonal (001) basal plane parallel to the deposition surface: the lateral extension -in directions parallel to the basal plane- spans the 200-300 nm range, whereas along the [001] direction a smaller size of about 60 nm was estimated by preliminary atomic force microscopy observations. The smaller average NP volume observed for Mg-Pd_f originates with the lower rate of Mg vapour production, which results in a slower NP growth by condensation and coagulation in proximity of the vapour source.

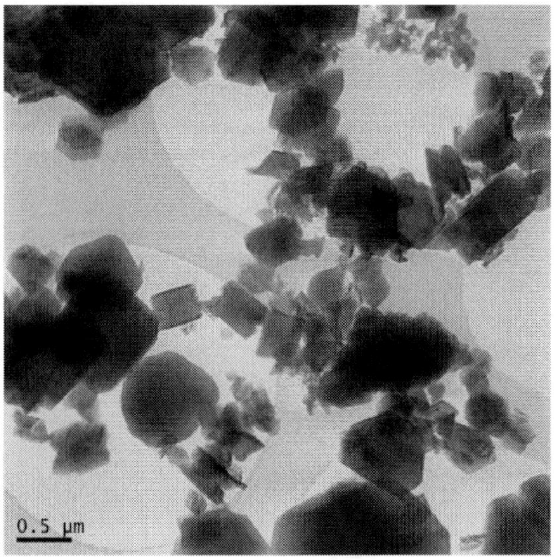

Figure 1. Bright field TEM image of as-prepared Mg-Pd_s1.

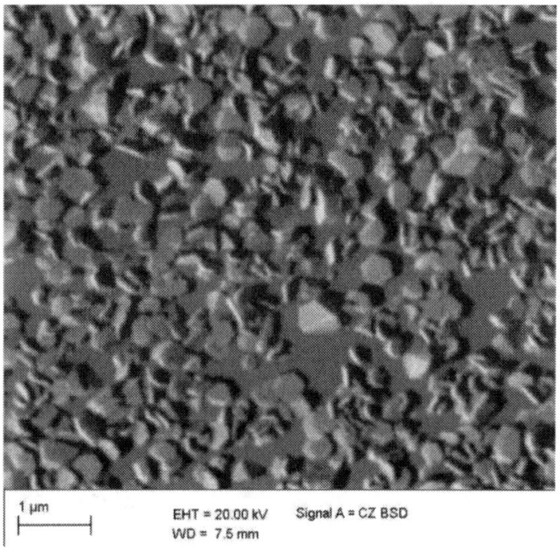

Figure 2. SEM image of Mg-Pd_f deposited onto a glass substrate.

The XRD profiles on the three samples under study are reported in figure 3. The preferential orientation of the NPs in Mg-Pd_f is confirmed by the strong (002) reflection of *hcp* Mg (the X-ray scattering vector is perpendicular to the deposition surface). The Pd content was evaluated by quantitative Rietveld analysis of XRD profiles, resulting in: 8(1) wt% (Mg-Pd_s1); 13(1) wt% (Mg-Pd_s2); 50(5) wt% (Mg-Pd_f). These data are consistent with the results of EDS microanalysis. From previous TEM observations, we could estimate a uniform MgO thickness of 4-5 nm around the NPs [5]. The structure of Pd in the scraped Mg-Pd NPs was studied in previous papers by a combination of different techniques [7], showing that a portion of the MgO shell is decorated by ultrafine metallic *fcc* Pd crystallites (3-4 nm) with the same interatomic distance as bulk Pd.

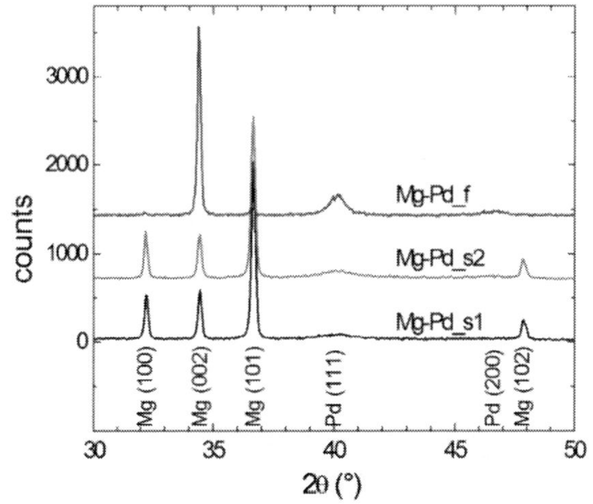

Figure 3. XRD profiles of as-prepared samples. The reflections of *hcp* Mg and *fcc* Pd are indicated. The data have been shifted vertically to allow for a better comparison.

3.2. Hydrogen sorption at high temperature

Figures 4 and 5 display hydrogen absorption and desorption kinetics at selected temperatures for the scraped NPs. Sample Mg-Pd_s1, with a lower Pd-content of 8 wt%, exhibits a reversible H-storage capacity ρ_m=5.0(1) wt%. The transformed fraction $\alpha(t)$ can be fitted quite well by a Johnson-Mehl-Avrami (JMA) equation for nucleation and growth $\alpha(t)=1-\exp[-(kt)^n]$ (see continuous lines in figures 4-5), where k is the rate constant. The reaction orders are determined as n=1.5 and n=1 for desorption and absorption, respectively, consistently with a decreasing nucleation rate and 1-dimensional growth controlled by diffusion (absorption) and interface motion (desorption) [5]. The effective activation energies, calculated from the temperature dependence of the rate constant under the assumption of Arrhenius-like behaviour, are E_{abs}=145±5 kJ/mol and E_{des}= 155±5 kJ/mol. Four different temperatures in the 550-610 K range were chosen for determining activation energies. In sample Mg-Pd_s2, with a higher Pd content of 13 wt%, we obtain ρ_m=4.0(1) wt%. The JMA analysis of H-absorption yields the same values n=1 and E_{abs} =145±10 kJ/mol obtained for Mg-Pd_s1. However, a modest deviation from the JMA behaviour during H-desorption at short times is observed. In fact, the double-logarithmic plot of $-\ln(1-\alpha)$ *vs* time clearly shows two distinct slopes at short and long times (inset of figure 5-(b)). Therefore a fit using a superposition of two JMA components with different k, n and relative amplitude was executed.

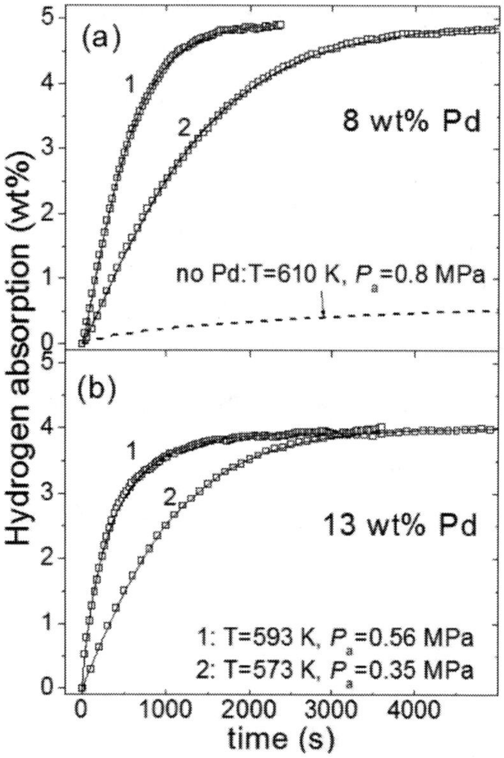

Figure 4. H-absorption in samples Mg-Pd_s1 (a) and Mg-Pd_s2 (b). The behavior of a similar NPs sample without Pd is shown for comparison.

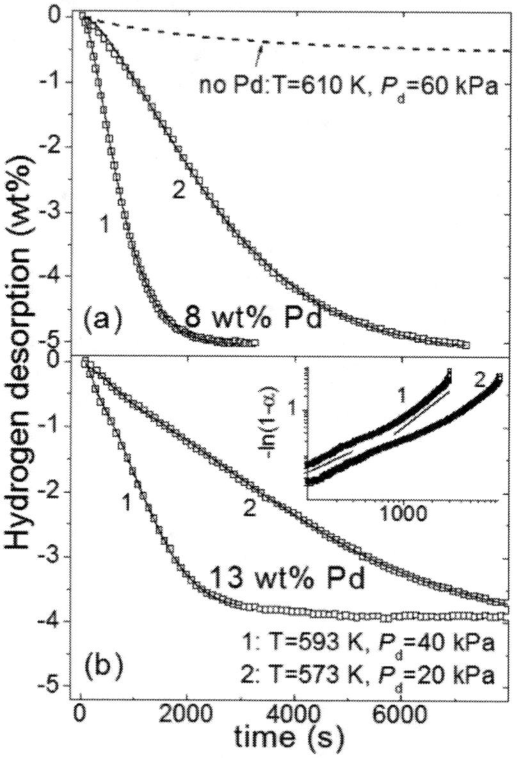

Figure 5. H-desorption in samples Mg-Pd_s1 (a) and Mg-Pd_s2 (b). The behavior of a similar NPs sample without Pd is shown for comparison. The inset in (b) shows the same kinetics in coordinates such that a single JMA equation would appear linear. The lines in the inset represent the slopes n=1 and n=1.7.

The slower, dominant component accounts for 3.4 wt.% H-desorption and has similar parameters to Mg-Pd_s1, i.e n=1.7 and E_{des}=170±10kJ/mol. The faster, minority component accounts for 0.6 wt.% H-desorption and is characterized by n=1, E_{des}=130±10kJ/mol. We suggest that it is associated with the transformations that involve Mg_xPd_y compounds, as pointed out recently by *in situ* powder diffraction using synchrotron radiation [8]. The presence of Mg-Pd phases together with MgO is also responsible for the reduced ρ_m values compared to 7.6 wt% of pure Mg.

The equilibrium properties of Mg-Pd_s2 were evaluated by a pressure-composition-isotherm (PCI) at 593 K (figure 6). The mean value between the absorption and desorption mid-plateau pressures is 0.28 MPa, very close to 0.31 MPa expected from the Mg-H phase diagram (dashed line in figure 6). The shoulder in proximity of 4 wt% H-content might again be associated with transformations involving Mg-Pd phases and characterized by a higher equilibrium pressure.

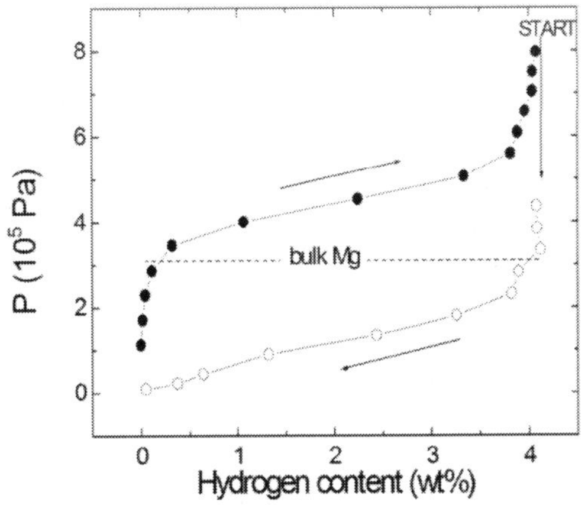

Figure 6. Pressure-composition isotherm measured at 593 K on the sample Mg-Pd_s2. The sample was previously hydrogenated for two hours at a pressure of 1 MPa, The equilibrium pressure for the metal-hydride transformation in bulk Mg is indicated.

3.3. Formation of hollow NPs

A significant modification of the NPs morphology, which eventually leads to the formation of hollow MgO shells, can be induced if several H-sorption cycles are carried out at sufficiently high temperature. This phenomenon is exemplified in figure 7, showing SEM images of NPs subjected to 6 complete cycles followed by one H-absorption step (total treatment time=28 hours) at 600 K (a) and 650 K (b). In (b), the transparency of almost all NPs indicates that they have actually been transformed into empty MgO shells. Furthermore, the Mg extracted from the MgO shells appears in form of larger structures with a smooth shape, which do not contain Pd according to EDS microanalysis. In (a), hollow NPs are seldom observed (one example is marked by the circle). Other experiments have demonstrated that the formation of hollow MgO shells does not occur if the NPs are simply kept at 650 K under low hydrogen pressure for the same total time (28 hours) [9].

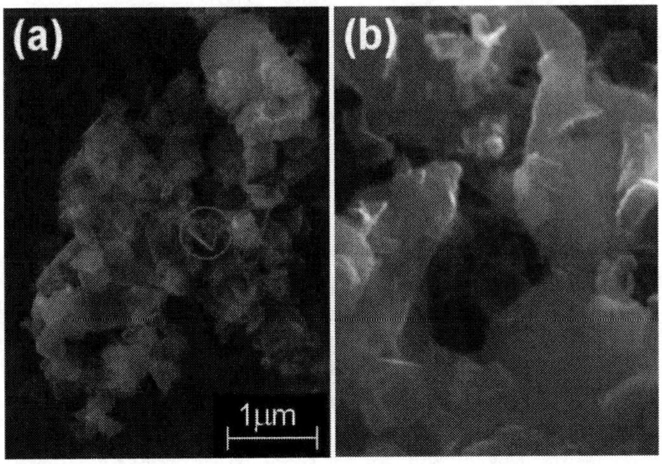

Figure 7. SEM images with SE taken on Mg-Pd_s1 NPs subjected to 6 complete H-sorption cycles plus one final H-absorption step (total time=28 hours) at the temperatures of 600 K (a) and 650 K (b). To be noticed that almost all NPs appear transparent in (b) and are actually hollow. The circle highlights a hollow NP in (a).

On these bases, we propose that in a first step, H-absorption in the Mg/MgO NPs drives the outward diffusion of Mg through the MgO shell. Two factors set the driving force: i) the tendency of Mg to react with H (lower chemical potential μ_{Mg} in the hydride phase) and ii) the stress which develops due to the 33% volume increase due to hydride formation. In the second step, following H-desorption, the Mg extracted from the MgO shell evaporates and condenses forming the larger structures without Pd. The relevant timescales are plotted in figure 8: the evaporation time τ_{ev} given by the Langmuir equation, the metal-hydride (τ_{MH}) and hydride-metal (τ_{HM}) transformation times derived from the JMA analysis, and the diffusion times for hydrogen ($\tau_{D,H}$) and Mg ($\tau_{D,Mg}$) across a MgO layer with thickness l=5nm. The diffusion times are evaluated as $\tau_D = l^2/2D$, where D is the relevant diffusivity from literature studies of H and Mg diffusion in MgO, taking into account accelerated diffusion in nanocrystalline materials. Figure 8 reveals that $\tau_{D,H}$ is about 5 orders of magnitude shorter than τ_{MH} and τ_{HM}, which span the 10^2-10^4 s range. Therefore the transport of H through the MgO shell is always much faster than the rate limiting steps of the transformations. Conversely, $\tau_{D,Mg}$ is longer than the transformation times and exhibits a larger slope in the Arrhenius plot of figure 8, owing to its larger activation energy. However, with increasing temperature, $\tau_{D,Mg}$ gets closer to τ_{MH}, explaining the observed temperature dependence in the yield of hollow NPs. In fact, at $T \approx 650$ K where $\tau_{D,Mg} \approx \tau_{MH}$, complete formation of hollow NPs takes place in about 6 cycles, while at $T \approx 600$ K, where $\tau_{D,Mg} >> \tau_{MH}$, hollow NPs are seldom detected. The Mg evaporation time τ_{ev} is 1-2 orders of magnitude shorter than τ_{MH} and τ_{HM}, suggesting that the extracted core material is removed rather quickly from the NPs' surface after H-desorption.

Differently from the classic and nanoscale Kirkendall Effect, the reaction that controls the developing porosity here does not simultaneously contribute to the growth of the shell, which acts more like a marker of the differential diffusion fluxes and potentially as a selective filter for particular atomic species. This feature offers more flexibility in the structural and chemical design of materials: in the present case the shell thickness is independent on the particle size, and perfectly hollow particles with geometrically faceted morphology are obtained in a wide size range.

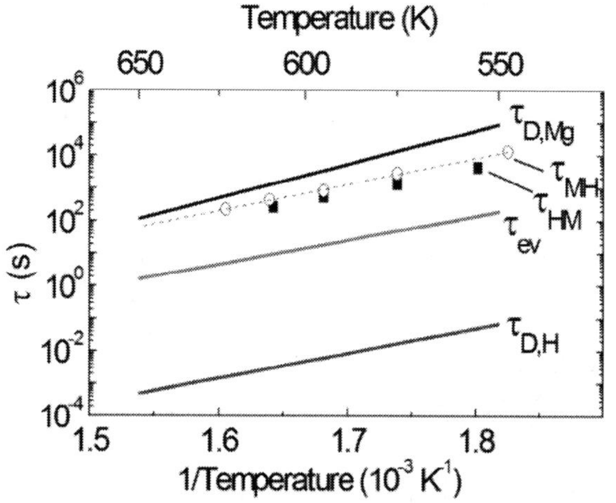

Figure 8. Characteristic timescales *vs* temperature of the processes involved in the formation of hollow NPs. The dashed line represents the Arrhenius-like fit to the experimental τ_{MH} data.

3.4. Hydrogen absorption at low temperature

Mg-Pd_f samples on glass substrates were employed for the investigation of H-sorption close to ambient temperature (333 K) by means of hydrogenography. In this technique, the logarithm $\ln(T/T_M)$, where T_M and T are the optical transmissions in the initial metallic state and during the experiment, is measured isothermally either as a function of time at fixed hydrogen pressure (kinetics) or as a function of hydrogen pressure in equilibrium (thermodynamics) [6]. Hydrogen desorption was extremely slow at this temperature and was therefore carried out under oxygen atmosphere (10^3 Pa O_2

pressure) which induces a significant kinetics speed up. For this reason only the H-absorption experiments reported below are relevant to the H-sorption properties of the Mg-H system.

The H-absorption kinetics during the first 5 cycles are reported in Figure 9, where a strong speed up after the first cycles is worth noticing. The time dependence is almost linear in cycles 1 and 2, suggesting a surface-limited transformation with constant rate. In cycles 4 and 5 the transformation exhibits instead a sigmoidal shape which can be described by a JMA curve with reaction order n=2.

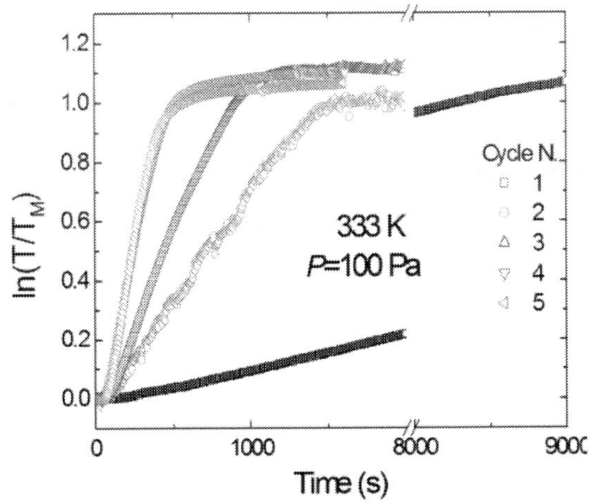

Figure 9. H-absorption kinetics at 333 K measured by hydrogenography during successive H-sorption cycles on Mg-Pd_f. H-desorption between the measurements was carried out under 1000 Pa O_2.

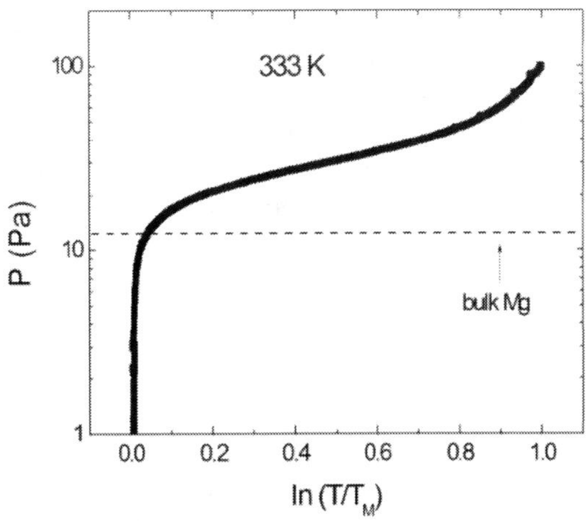

Figure 10. Pressure-transmission isotherm at 333 K showing equilibrium H-absorption in Mg-Pd_f. The pressure at which metal-hydride coexistence begins to appear in a bulk Mg film is marked by the dashed line [11].

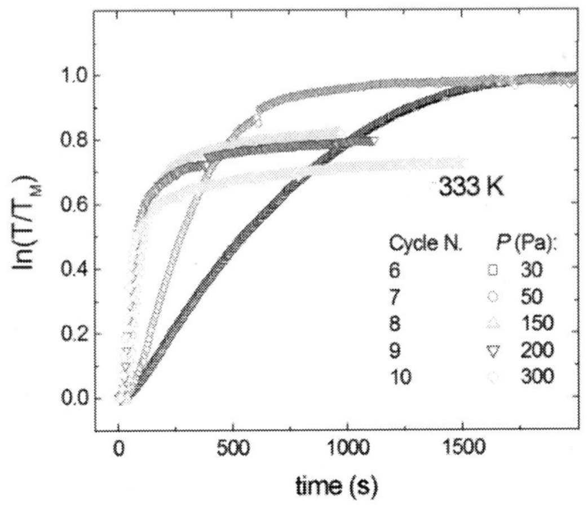

Figure 11. Pressure dependence of H-absorption kinetics in Mg-Pd_f measured after the cycles reported in figure 9.

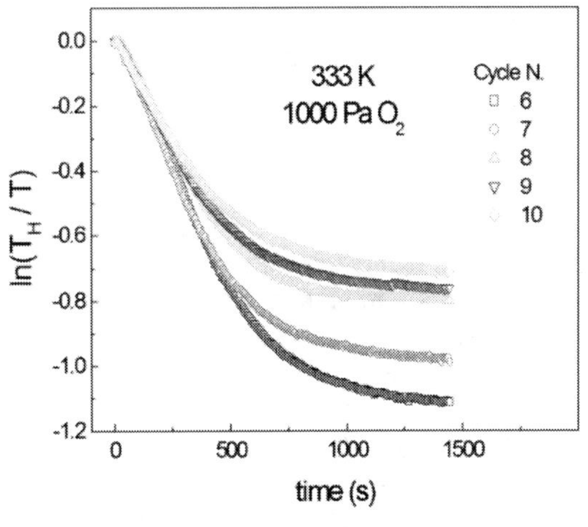

Figure 12. H-desorption kinetics measured after the H-absorption curves of figure 12. T_H is the transmission in the initial hydride state.

This finding might originate with a surface activation occurring during the very first H-sorption cycles and resulting in a change of the rate-limiting step from surface dissociation of the H_2 molecule to nucleation and growth of the hydride phase. The determination of the actual storage capacity from optical transmission change is more complicated for NPs than for thin films due to the very inhomogeneous thickness and to the non complete substrate coverage. However, assuming a 80% coverage and an effective thickness of 60 nm (see figure 2), we have estimated that $\alpha(t_{\max}) \geq 0.9$, which means that the metal-hydride transformation is almost complete.

Figure 10 reports a pressure-transmission isotherm which proves that the plateau pressure for H-absorption is close to the one measured in a bulk Mg film [11], as expected for NPs in this size range. Finally, figure 11 illustrates the pressure dependence of H-absorption. At the early times (t=100 s), $\ln(T/T_M)$ increases by a factor of ten with increasing H_2 pressure from 30 to 300 Pa. This dependence suggests that the driving force for the metal-hydride transformation increases linearly with pressure in this range. Differently, the maximum transformed fraction decreases with increasing pressure. An explanation for this behavior might be searched in a "blocking effect" exerted by the growing hydride phase, as reported recently for Mg thin films [12]. In fact, the increasing pressure favors the quick formation of a continuous hydride layer all around the Mg core. This layer strongly slows down the transformation rate because H diffusion through the hydride is rather sluggish at low temperature. The corresponding H-desorption curves shown in figure 12 confirm that the maximum transformed fraction decreases with increasing pressure.

4. Conclusions

The formation of intermetallic compounds at the interface between Mg and catalyst, driven by a negative heat of mixing and predictable on the basis of the binary phase diagram, is shown to have a profound influence not only on the kinetics of H-sorption but also on the thermodynamics, because new reaction paths with different reaction enthalpies come into play. Noteworthy, this phenomenon occurs also in the presence of a separation layer –MgO in this case.

An important issue set forth by this investigation relates to the possible morphological instability of nanostructures for H-storage due to irreversible transport phenomena at the nanoscale. In fact, although the MgO shell seems effective in avoiding agglomeration and coarsening of the NPs at moderate temperatures, repeated H-sorption cycles at high temperature progressively induce a dramatic change of the NPs' morphology, resulting in an increasing porosity which eventually leads to hollow MgO shells and large Mg condensates depleted of catalyst. A stable confinement of H-storage nanostructures is then possible only in low diffusivity conditions.

In this respects, the possibility to operate NPs at low temperatures where significant diffusion does not occur appears extremely important. Here, we showed that complete metal-hydride transformation of the Mg/MgO core/shell NPs can be achieved in about 10 minutes at 333 K and 100 Pa H_2 pressure.

References

[1] U. Eberle, M. Felderhoff, F. Schüth, Angew. Chem. Int. Ed. **48**, 6608 (2009)
[2] M. Hirscher, B. Panella, Scripta Mater. **56**, 809 (2009)
[3] A. Baldi et al., Phys. Rev. B **81**, 224203 (2011)
[4] G. Barkhordarian, T. Klassen, R. Bormann, J. Phys. Chem. B **110**, 11020 (2006)
[5] E. Callini, L. Pasquini, E. Piscopiello, A. Montone, M. Vittori Antisari, E. Bonetti, Appl. Phys. Lett. **94**, 221905 (2009)
[6] R. Gremaud et al., Adv. Mater. **19**, 2813 (2007)
[7] L. Pasquini, F. Boscherini, E. Callini, C. Maurizio, L. Pasquali, M. Montecchi, E. Bonetti, Phys. Rev. B. **83**, 184111 (2011)
[8] E. Callini, L. Pasquini, L.H. Rude, T.K. Nielsen, T.R. Jensen, E. Bonetti, J. Appl. Phys. **108**, 073513 (2010)
[9] L. Pasquini, A. Montone, E. Callini, M. Vittori Antisari, E. Bonetti, Appl. Phys. Lett. **99**, 021911 (2011)
[10] G. Krishnan, B.J. Kooi, G. Palasantzas, Y. Pivak, B. Dam, J. Appl. Phys. **107**, 053504 (2010)
[11] A. Baldi, M. Gonzalez-Silveira, V. Palmisano, B. Dam, R. Griessen, Phys. Rev. Lett. **102**, 226102 (2009)
[12] H. Takahiro Uchida, R. Kircheim, A. Pundt, Scripta Mater. **64**, 935 (2011)

IOP Conf. Series: Materials Science and Engineering **38** (2012) 012002 doi:10.1088/1757-899X/38/1/012002

Scintillating Ceramics Based on Zinc Oxide

P A Rodnyi, E I Gorokhova, K A Chernenko and I V Khoduk

Saint-Petersburg State Polytechnical University, 29, Polytekhnicheskaya, 195251, Saint-Petersburg, Russia

E-mail: Rodnyi@tuexph.stu.neva.ru

Abstract. Optical transparent ceramics were manufactured from different ZnO powders by uniaxial hot pressing method. Structural, optical and scintillating characteristics of the ceramics have been studied. For all samples, an increase in grain size by one or two decimal order in ceramics as compared to that of the initial powder was observed. Undoped ZnO ceramics shows a high light output 28 500 ± 800 photons/MeV, decay constant ~ 1 μs, a very low afterglow, optical transparency closed to that for single crystal and a linearity of the relative light output in range 35-652 keV. Introduction of Li in ZnO decreases decay time constant of the ceramics. ZnO:Ga ceramics show an ultrafast decay, 0.7 ns, not high light output: ~ 1000 photons/MeV, and a low optical transparency.

1. Introduction

Zinc oxide (ZnO) has a unique combination of properties of a semiconductor material with a significant fraction of ionic bonding. This material attracts much attention in current phosphor research, due to its superior optical and luminescent properties [1]. As a rule, various forms of ZnO, such as single crystals, thin films and threads, nanocrystals, needles, etc., exhibit two luminescence bands. One of them is a narrow band located near the absorption edge of the crystal; therefore it is called near-band-edge (NBE) emission. Another one is a broad band, its maximum usually lies in the green spectral range, this luminescence is associated with deep-level (DL) emission. The NBE band with a maximum at 3.35 eV and decay time of ~ 0.7 ns has an excitonic nature [2]. The origin of the DL emission is not fully understood, despite a vast number of investigations. In the initial stage, the researches offer a wide variety models of luminescence centers: impurity Cu^{2+} ions [3], zinc vacancies V_{Zn} [4,5], oxygen vacancies V_O [6,7], interstitial zinc ions Zn_i [8], oxygen antisites Zn_O [9,10], and transitions $Zn_i \rightarrow V_{Zn}$ [1]. This situation is obtained due to the imperfection of zinc oxide crystals, the instability of certain point defects, and the variety of their forms. Now we can consider it established that, in samples with an excess of oxygen, the maximum of the DL emission band is located at 2.35 ± 0.05 eV, and zinc vacancies V_{Zn} are responsible for the luminescence [10, 11]. In samples with an excess of zinc, the maximum of the DL emission band is located at 2.53 ± 0.05 eV and the luminescence centers are oxygen vacancies V_O [11]. Clearly, in O excess samples, interstitial oxygen centers O_i are also formed, while, in Zn excess samples, interstitial zinc centers Zn_i are formed; both of these centers form shallow donor levels in the crystal [1, 11].

Basic characteristics of ZnO are: the gap width is E_g = 3.437 eV (1.6 K) and 3.37 eV (300 K), the exciton binding energy is 60 meV, the energy of longitudinal optical phonons is 72 meV, the melting point is 1975 °C , the density is 5.67 g/cm^3.

Zinc oxide is a multifunctional material that can be applied in varistors, anti-reflection coatings, and transparent electrodes in solar cells, gas sensors, and surface acoustic wave devices [1]. This material has a high efficiency as a low-voltage phosphor and has been used in vacuum fluorescent and field emission displays. Much interest has been shown in ZnO as a promising candidate for ultraviolet optical applications such as light-emitting diodes or laser diodes. Using of zinc oxide as a scintillator is an innovative application. In this case, the following characteristics of ZnO are important: a short decay time of the NBE band, a high mechanical and chemical stability, a high radiation hardness and sufficient stopping power. A ZnO:Ga scintillator emitting an ultrafast NBE band is applied as an efficient detector of a particles in deuterium–tritium (D–T) neutron generators, which are used in customs inspection of containers [12]. Undoped ZnO films also show short times of the NBE emission (τ = 550 –700 ps at 300 K and τ = 60 – 130 ps at 4 K) upon fast electron excitation [13]. The best characteristics (high intensity of the NBE and absence of the DL emission) belong to the ZnO:Ga powder with grain size 8.2 μm that is produced at the Lawrence Berkeley Laboratory [14].

Commonly, ZnO and ZnO:Ga are used in the form of powder [14,15] or thin film [16] coatings. However, powdered and thin film materials can be used for detecting only neutrons and alpha particles, whereas the detection of gamma quanta and X ray photons requires scintillators with a large volume (single crystals and/or optical ceramics) and high transparency in the spectral range of emission. Growing ZnO single crystals is a complicated, long and expensive technological process [17]. Thus, optical ceramics, representing a pressed polycrystalline material with grain dimensions variable within broad limits, offer a promising alternative to traditional single crystals in both scintillator and laser technology.

This paper describes the preparation and presents the results of investigation of the optical properties and scintillation characteristics of ZnO based ceramics. A problem when sintering ZnO ceramics is the hexagonal symmetry of its crystal structure. Anisotropic grains in the ceramics cause considerable light scattering in the grain boundaries resulting in light absorption in the bulk. Nevertheless, the ZnO ceramics with reasonable optical transparency have been obtained. To create a high quality optical ZnO ceramics, we have used our preceding experience in the ceramics sintering from non-cubic material Gd_2O_2S [18].

2. Experimental results and discussion

Densification of different ZnO powders into ceramics was made by uniaxial hot pressing methods in Research and Technological Institute of Optical Materials All-Russia Scientific Center "S.I. Vavilov State Optical Institute" St. Petersburg [19]. The sample preparation conditions (temperature, pressure, gaseous medium, etc.) were varied so as to obtain the optical ceramics with maximum transparency and light yield. First, the initial ZnO powder was subjected to cold pressing in air at P = 12 MPa for 15 - 20 minutes. Then, a hot pressing was produced in vacuum at temperature 900 - 1100 °C and P = 100-200 MPa for 60 minutes. As a result, the transparent ceramics have been obtained in the form of discs with a diameter of 20 - 25 mm and a thickness of 1.0 - 1.5 mm.

Studied ceramics were obtained from six kinds of ZnO powders, differing in purity composition, size and shape of grains. The experiment shows that the transparency of the ceramics is primarily determined by the purity and grain size of the starting powder. Grain size effect on the NBE band: the smaller the grains, the greater NBE emission intensity (the minimum grain size was 60 - 150 nm). For brevity, we present data for the two most typical samples of ceramics which possess the best characteristics. ZnOI sample was prepared from a commercial zinc oxide powder of Alfa Aesar Company with initial particle sizes within 120 – 160 nm. ZnOII sample was prepared from zinc oxide powder of special purity produced in Russia with initial particle sizes within 300 – 600 nm.

The lattice parameters and phase composition of the samples were determined with the X-ray diffractometer DRON-2. Interpretation of the results was based on the database of the International

Centre for Diffraction Data (ICDD), which contains the following parameters for the ZnO: a = 3.2497 Å, c = 5.2069 Å. Comparative analysis of the parameters "a" and "c" of the crystal lattice and the initial powders made from these ceramics shows that the uniaxial hot pressing leads to compression of the lattice in ceramics in the direction parallel to the axis c for all the investigated samples. In particular, in the transition from powder to ceramics a parameter a increased from 3.2479 to 3.2490 Å, as in ZnOI, and in ZnOII. Parameter c in ZnOI ceramics decreased from 5.2040 Å in the powder to 5.1986 Å , and in ZnOII ceramics: from 5.2027 to 5.2012 Å.

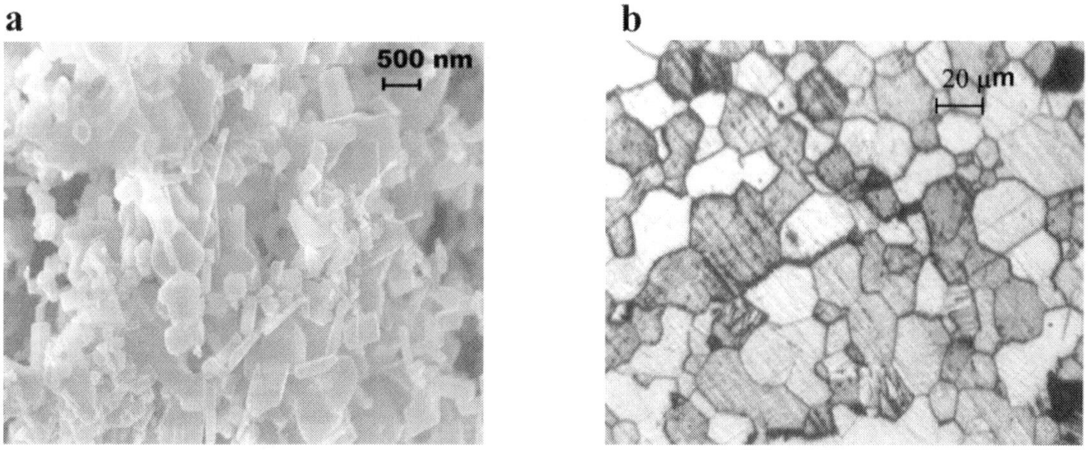

Fig. 1. a) Scanning electron microscopy image of ZnOII powder;
b) optical-microscope image of ZnOII ceramic produced from this powder.

The grain sizes of powders were determined with a scanning electron microscope SEM LEO-420. As an illustration, fig. 1a shows a SEM image of the ZnOII initial powder. Studies of the microstructure of the ceramic samples were carried out with an optical microscope POLAM R-312, and the exposed surface of the ceramic pre-etched with hydrochloric acid. Fig. 1b shows an image of ZnOII ceramics. For the ZnOII sample we have got: grain size of the initial powder is 300 - 600 nm, while grain size of the ceramics is 5 - 30 µm. For all samples, an increase in grain size by one or two decimal order in ceramics as compared to that of the initial powder was observed.

The total transmittance spectra of the samples were measured using a Hitachi-330 spectrophotometer with an integral sphere of Ø 60 mm. ZnOI sample with thickness 0.4 mm has total transmittance of 60% at 500-nm wavelength (fig. 2), which is a rather high value for ceramics based on a material possessing a hexagonal structure. The total transmittance of 1.0-mm thick ZnOI and ZnOII ceramics was 45% at 520 nm. The inset of fig. 2 shows a photograph of ZnOI ceramic of 1-mm thickness. It should be noted that the transparency of the obtained ceramics is close to that of ZnO single crystal [17].

The radioluminescence (RL) spectra were measured under steady state X-ray excitation (40 kV, 10 mA) with a photomultiplier tube FEU-106 and a MDR-2 grating monochromator with 1200 grooves/mm. The spectra were corrected for the quantum efficiency of the PMT and for the monochromator transmission. Figure 3 shows the RL spectra of ZnO I and ZnO II ceramics in comparison to the spectrum of a standard $YAlO_3$:Ce (YAP) scintillator crystal. Both ceramics exhibit intense luminescence in a long wavelength spectral range, that is the DL emission dominates. The maximum of the RL band, 2.37 eV, observed in our ZnO ceramics allows this emission band to be assigned to V_{Zn} centers. Conditions of the ceramic synthesis are such that zinc vacancies are dominating luminescence centers. However, the full width at half maximum ($\Delta E_{1/2} \approx 500$ meV) of the luminescence band in ceramics more than that of the crystals ($\Delta E_{1/2} \approx 400$ meV) [1], which suggests that in the emission of ceramics can participate other centers of luminescence (not only V_{Zn}). The integral light yield determined as the area under the RL curves for ZnOI and ZnOII ceramics was 1.88 and 1.61 times, respectively, greater than the same value for the YAP crystal scintillator.

The decay time profiles were recorded with the experimental system described in detail in [20]. A pulsed X-ray source with 1-ns pulse duration operates at 30 kV and maximum current amplitude of 500 mA. ZnOII ceramics shows a long decay time ~1.6 μs (fig. 4, curve 1). The hyperbolic shape of the decay curve indicates presence of the shallow traps, which can be associated with interstitial zinc Zn_i. Decay time constant of ZnOI ceramics is about 1.0 μs. In order to improve the decay-time characteristics of the ceramics Li^+ ions were introduced in the samples. Among the elements of the first group, Li is considered the best acceptor impurity, since the bond length $d_{Li\text{-}O}$ = 2.03 Å is the closest to average the bond length Zn–O 1.93 Å [1]. The decay profile of ZnO:Li ceramics contains two main time components: fast 17 ns and slow ~ 1 μs (fig. 4, curve 2). The fast decay component contains about 60-80% of the total light yield (depends on the dopant concentration).

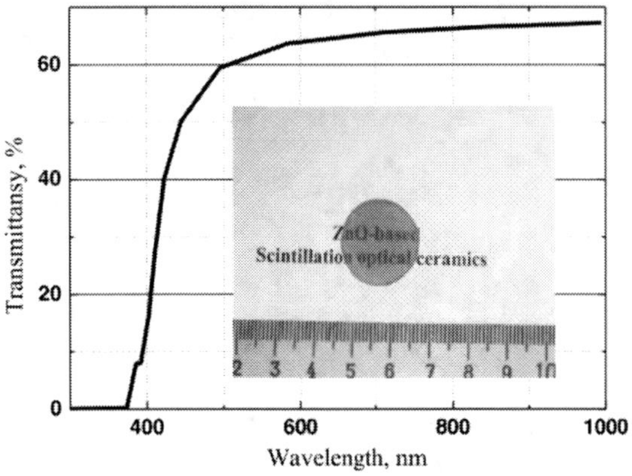

Fig. 2. Total transmittance spectrum of ZnOI ceramic of 0.4-mm thickness. Inset: Photograph of ZnOI ceramic of 1-mm thickness.

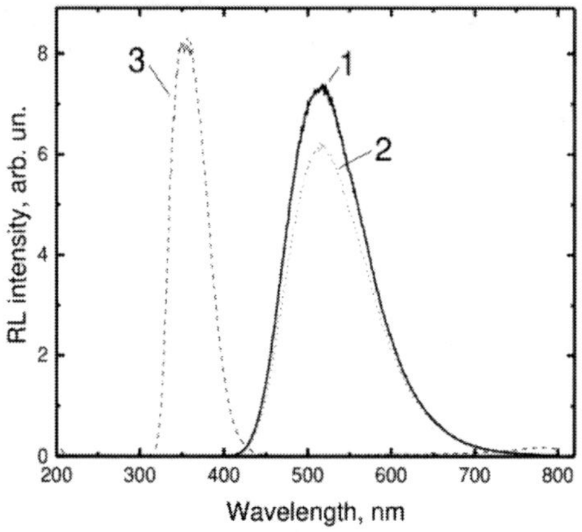

Fig. 3. Radioluminescence spectra: 1- ZnO I and 2 - ZnO II ceramics; in comparison with 3 - $YAlO_3$:Ce (YAP) standard scintillator crystal.

Scintillation characteristics of ZnO:Li ceramics have allowed us to consider it as a promising material for neutral-particle analyzers on the International Thermonuclear Experimental Reactor (ITER) [21]. Neutral particles, that leave the tokamak chamber, become ionized by falling through the stripping foil. After that the bunch of ions goes through magnetic and electron-optical systems. The main feature of these devices is a necessity to work for a long time without maintenance in the powerful background flux of heavy particles and electrons. A scintillation spectrometer with a multianod photomultiplier tube (PMT) is used as a position sensitive detector. There are the following requirements for the scintillator: high light yield (for detecting 100 keV particles), short decay time (in order to avoid pileup of signals of the electronic background, which leads to signal amplitude increase), small decrease in scintillation efficiency (α/β ratio) when excited by α-particles, nonhygroscopicity, possibility to manufacture the scintillator in the form of a plate with thickness ≤ 20 μm (electrons are almost not absorbed in the thin sample). ZnO-based ceramic scintillators can meet these requirements.

ZnO:Li ceramic samples with a thickness of 20 μm were prepared and investigated under alpha-particle excitation by ^{241}Am source, 5.5 MeV [22]. The alpha-particle range in ZnO (8.61 μm) is much shorter than the sample thickness so the energy of an alpha-particle is completely released in the scintillator. ZnO:Li ceramic shows good characteristics under alpha-particle excitation: a shot decay constant ~ 0.5 μs, a sufficient light yield (of about 25% of that for CsI:Tl) and it can be applied for the heavy-particle analyzers on ITER.

The absolute light yield (ALY) of the ceramic scintillators was determined using a Hamamatsu R6231-100 photoelectron multiplier. The detected signal was fed to an Ortec 114 16k analog to digital converter via a self-made preamplifier and an Ortec 672 spectrometric amplifier [23]. The absolute light yield was determined by comparing the position of a photoelectron peak in the spectrum of ^{137}Cs (662 keV) radiation to that of the center of a weighed one electron spectrum. The highest light yield, 28 500 ± 800 photons/MeV, has been obtained for the ZnOI ceramic.

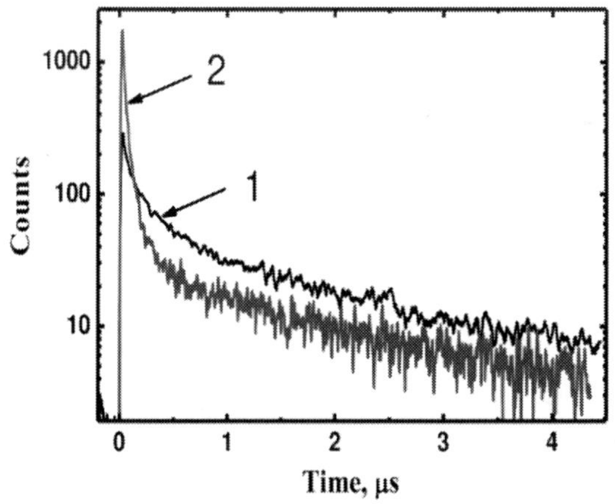

Fig. 4. Radioluminescence decay time curves: 1- undoped ZnO and 2 – ZnO:Li ceramics.

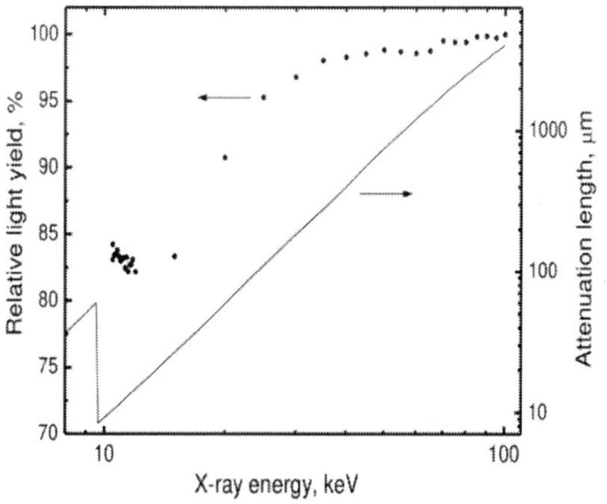

Fig. 5. Relative light yield (points) of ZnOII ceramic versus incident synchrotron radiation energy; the solid line shows the energy dependence of the X-ray-quntum relaxation length in ZnO.

Fig. 5 shows a plot of the relative light yield (RLY) in ZnO II ceramics versus incident X-ray energy. These measurements were performed using a scheme similar to that for determining the ALY, except that the luminescence was excited by synchrotron radiation [23]. The RLY was defined as the ratio of the ALY at certain energy of the incident X-ray quanta to the quantum energy. In fig. 5, the RLY is plotted as percentage of the ALY at 662 keV divided by 662 keV. The RLY varies within 2% in the energy range of 35–662 eV (fig. 5), it is less than the same deviation in the known NaI:Tl and $LaCl_3$:Ce scintillators [24]. However, as the incident radiation energy was decreased, the RLY exhibits a local minimum (82%) at ~12 keV. In the range12–35 keV, the RLY variation amounts to 16%. This decrease correlates with the energy dependence of the X-ray-quantum relaxation length in ZnO ceramics, which is depicted by the solid line in fig. 5. The dependence of the RLY on the energy of incident quanta is determined by a more basic dependence of the light yield on the energy of secondary electrons [24]. As the incident photon energy approaches the *K*-shell electron binding energy (9.659 keV) in Zn atom, the energy spectrum of secondary electrons shifts down to lower energies. This shift leads to an increase in the ionization rate that, in turn, leads to a decrease in the efficiency of the scintillator material.

A very important parameter for the scintillators operating in X-ray computed tomography (CT) is a low level of afterglow. At measuring of the afterglow the duration of X-ray pulse, formed by a mechanical shutter was 2 s, and the shutter time was ~ 0.5 ms [19]. Fig. 6 shows the afterglow curves of our ZnOII sample in comparison with Gd_2O_2S:Pr,Ce ceramics, which are widely used in computer tomography. It is seen that ZnOII ceramic has a fairly low level of afterglow and putting it quite suitable for use in X-ray CT.

It is known that the introduction of gallium in the zinc oxide suppresses the slow DL emission and the fast NBE luminescence increases [14]. So, by varying dopant concentration and synthesis parameters it is possible to produce samples with only a short-wavelength emission band, with a maximum at 3.21 eV (fig. 7). For ZnO:Ga ceramics optimal concentration of Ga_2O_3, giving complete suppression of the DL band was found to be 0.075 wt. %. The 3.21-eV (385 nm) NBE band is associated with emission of the donor bound D^0X excitons [10]. Gallium ions replace zinc ions, as well as fill zinc vacancies, in ZnO which results in suppression of intensity of the DL emission. The decay time constant of the NBE emission is 0.7 ns (inset fig. 7), such a short decay time is very important for the Positron Emission Tomography (PET) which operates in the time-of-flight mode [25]. The position of the luminescence band of ZnO:Ga coincides with the maximum sensitivity of the new ultrafast photodetector SiPM [26], which opens up the prospect of ZnO:Ga scintillation ceramic using. However, the light yield of ZnO:Ga ceramics is only ~ 3 % of that of the undoped ZnO ceramics. The main reason of that is a low transparency of the ZnO:Ga ceramics. At present we are carrying out work to improve the optical transparency of the ceramics.

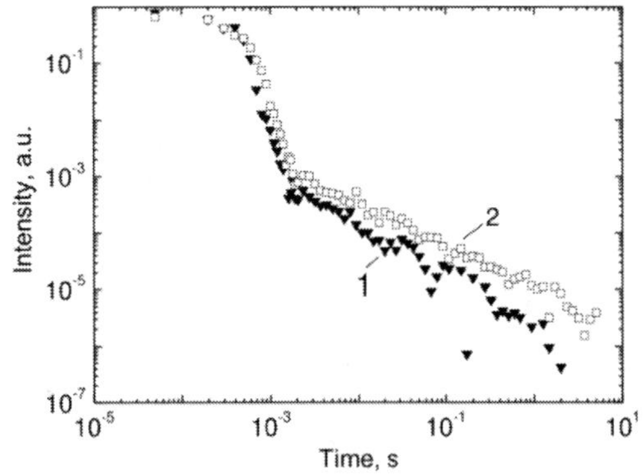

Fig. 6. Radioluminescence after the termination of the X-ray irradiation (afterglow): 1 – ZnOI and 2 - Gd_2O_2S:Pr,Ce ceramics.

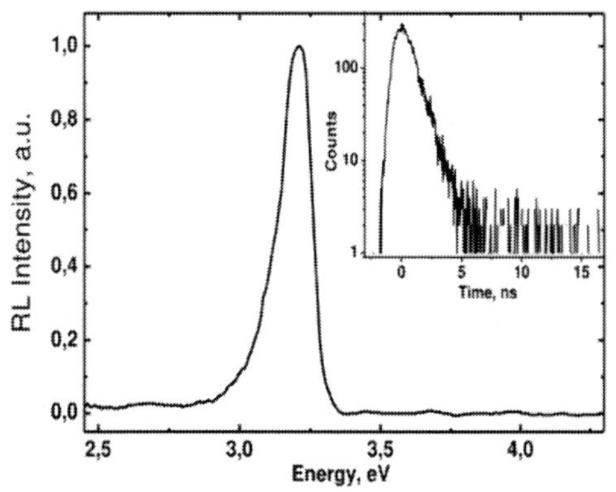

Fig. 7. Radioluminescence spectrum of ZnO:Ga ceramics. Inset: decay time curve of the ZnO:Ga ceramics.

Conclusions

The obtained ZnO ceramics are characterized by high transparency in the visible spectral range and exhibit high light yields, low afterglow, and linearity of the relative light yield in the energy range of 35–652 keV. These characteristics make ZnO ceramics promising scintillator materials for X-ray and gamma radiation detectors in the medium energy range.

The RL decay time of ZnO:Li ceramics is shorter than that of undoped ZnO. ZnO:Li ceramics shows good characteristics under alpha-particle excitation and can be applied for the heavy-particle analyzers on ITER.

The emission wavelength of ZnO:Ga ceramics (385 nm) is in a good agreement with the spectral sensitivity of silicon photomultipliers SiPM. ZnO:Ga has ultra fast decay time, 0.7 ns, and SiPMs have very high time resolution. It seems to be very attractive to use SiPMs in conjunction with ZnO:Ga ceramics for applications where fast timing is essential.

Acknowledgments

This work was supported in part by the TECHNOEXAN LTD, under Contract No. 11061-9.

References

[1] Ozgur U et al 2005 *J. Appl. Phys.* **98** 041301
[2] Meyer B K et al 2004 *Phys. Stat. Sol.(b)* **241** 231
[3] Alivov Ya I, Chukichev M V and Nikitenko V A 2004 *Semiconductors* **38** 31
[4] Kohan A F et al 2000 *Phys. Rev.* **61** ,15019
[5] Guo B, Qiu Z R and Wong K S 2003 *Appl. Phys. Lett.* **82** 2290
[6] Leiter F H et al 2001 *Phys. Status Solidi* **226** R4
[7] Shan F et al 2005 *Appl. Phys. Lett.* **86** 221910
[8] Liu M, Kitai A H and Mascher P 1992 *J. Lumin* **54** 35
[9] Reynolds D C et al 2000 *J. Appl. Phys.* **88** 2152
[10] Zhao Q X et al 2005 *Appl. Phys. Lett.* **87** 211912
[11] Berseth Moe T et al 2006 *Appl. Phys. Lett.* **89** 262112
[12] Beyerle A, Hurley J P and Tunnell L 1990 *Nucl. Instrum. Methods Phys. Res.* A **299** 458
[13] Lorenz M et al 2006 *Appl. Phys. Lett.* **89** 244510
[14] Moses W W 2002 *Nucl. Instrum. Methods Phys. Res.* A **487** 123–8
[15] Neal J S et al 2006 *Nucl. Instrum. Methods Phys. Res.* A **568** 803
[16] Yanagida T et al 2010 *IEEE Trans. Nucl. Sci.* **57** 1325
[17] Kortunova E V et al 2008 *J. Mater. Sci.* **43** 2336
[18] Gorokhova E I et al 2005 *IEEE Trans. Nucl. Sci.* **52** 3129
[19] Gorokhova E I et al 2011 *J. Opt. Technology* **78** 753
[20] Rodnyi P A et al 2001 *IEEE Trans. Nucl. Sci.* **48** 2340
[21] Afanasyev V I et al 2010 *Nucl. Instrum. Methods Phys. Res.* A **568** 456
[22] Rodnyi P A at al 2011 *11th Int. Conf. Inorg. Scint. Applic. "SCINT 2011" Giessen Germany Abstracts* P 1.25
[23] Rodnyi P A, Khodyuk I V and Gorokhova E I 2010 *Tech. Phys. Lett.* **36** 714
[24] Khodyuk I V, de Haas J T M and Dorenbos P 2010 *IEEE Trans. Nucl. Sci.* **57** 1175
[25] Fujimoto Y et al 2011 *Jap. J. Appl. Phys.* **50** 01BG04
[26] Yamaya T et al 2011 *Phys. Med. Biol.* **56** 6793

International Conference on Functional Materials and Nanotechnologies (FM&NT2012) IOP Publishing
IOP Conf. Series: Materials Science and Engineering **38** (2012) 012003 doi:10.1088/1757-899X/38/1/012003

Spin-Orbit Gaps in Armchair Nanotubes Calculated Using the Linear Augmented Cylindrical Wave Method

P D'yachkov[1] and D Kutlubaev
Institute of General and Inorganic Chemistry, Russian Academy of Sciences, Leninskii pr. 31, 119991 Moscow, Russia

E-mail: p_dyachkov@rambler.ru

Abstract. In the terms of the muffin-tin and local density functional approximations, a relativistic version of linear augmented cylindrical wave method is developed and applied to calculating the gaps in the Fermi energy region induced by the spin-orbit coupling in the armchair tubules (n, n) (where $4 \leq n \leq 12$). Due to the spin–orbit interaction, the energy gap opens and the metallic nature for armchair nanotubes is broken; however, each energy subband still has the two-fold degeneracy. The spin-orbit splitting decreases with growth of n from 0.537 meV for the (4, 4) tubule to 0.086 meV for the (12, 12) nanotube. A dependence of spin-orbit gap on the inverse diameter is not linear as different from a simple tight binding model.

1. Introduction

Carbon nanotubes are the most famous nanomaterials. The single-walled carbon nanotubes (CNTs) are the cylinders composed of the rolled graphene ribbons on which the carbon hexagons are arranged in a helical fashion about the axis. The large amount of theoretical and experimental investigations of nanotubes has brought to light many interesting properties of these materials. Particularly, the interaction of the spin of electrons with their orbital motion has become a focus of great attention in the nanotubes research. Due to the cylindrical structure of the CNTs, the electron states near gap correspond to the semiclassical clockwise and anticlockwise electron orbits encircling the tubule. In addition to the unique two-fold degree of freedom leading to the doubly degenerate orbits, the electrons can have different spins. In the absence of a spin-orbit coupling, the two-fold orbital degeneracy and two-fold spin degeneracy should yield a four-fold degenerate electronic spectrum in the CNT.

Based on low atomic number $Z = 6$ of carbon, the spin-orbit effects were commonly assumed to be negligible in carbon materials; in carbon atom, the spin-orbit splitting $\Delta_{S\text{-}O}$ equals to 6 meV and for flat graphene $\Delta_{S\text{-}O} = 10^{-2}\text{-}10^{-3}$ meV. However, already in 2000, the first theoretical study of spin-orbit interaction in a carbon nanotube was performed using the *kp*-method and simple tight-binding model with a single π orbital at each carbon atom, the spin-orbit effect being taken into account to the lowest order [1]. It was shown that the spin-orbit term gives rise to a splitting of energy levels and formation of gaps about 0.1-1 meV. In a more recent work [2], the spin-orbit effects were studied in the curved graphene, fullerene, and nanotubes also in the terms of simple π-electron model and perturbation theory, the importance of curvature of a carbon material surface for lager spin-orbital coupling being

[1] To whom any correspondence should be addressed.

indicated again. In Refs [2-8], the calculations of the curvature induced spin-orbit splitting in the Fermi level region were performed using an empirical tight-binding Hamiltonian.

Remarkably, the electrical measurements of the individual tubules in the presence of axial magnetic field show that the values of orbital magnetic moment μ_{orb} are 10-20 times larger than the Bohr magneton and scale with the tubule diameter d [9]. The reason of this large moment is the large size of electron orbits encircling the CNT compared to the radii of atomic orbitals. Later, the ultra-low-temperature (30 mK) measurements of the intrinsic electronic spectrum of the ultraclean nanotube quantum dot with $d = 5$ nm have shown that the expected four-fold symmetry is broken by spin-orbit splitting $\Delta_{S\text{-}O}$ equal to 0.37 and 0.21 meV for the electrons and holes, respectively [10,11]. Recently, the spin-orbit coupling was revealed by direct currant magnetoconductance and spectroscopy for the complementary situations of the open CNT-quantum wire and of the CNT-quantum dot in the general multielectron regime and in the presence of finite disorder [12,13].

In the theoretical works [1-8], the atomic spin-orbit interaction term included in Hamiltonian was given by $H_{S\text{-}O} = \lambda \mathbf{LS}$, where λ is the atomic spin-orbit coupling constant and **L** and **S** stands for the orbital and spin orbital angular momentum of electron, respectively. The parameters describing the strength of the spin-orbit interaction can not be determined in these calculations, because the simple tight-binding model used can describe the band structure only qualitatively [4,5]. The exact value of the spin-orbit coupling parameter is still under discussion [4,5]. For this reason, it is felt that the band structure calculation done with account of spin-orbit coupling from first principles would be of value in understanding these effects in the CNTs.

It is the aim of this work to calculate the gaps in the Fermi energy region induced by the spin-orbit coupling in the armchair tubules in the terms of a linear augmented cylindrical wave (LACW) method. The case of armchair CNTs is very interesting because formation of gap between the valence and conduction bands is totally relativistic effect here. The LACW method is just a reformulation of the augmented plane wave (APW) method well-known in the band structure theory of solids for cylindrical multiatomic systems. Thus, the relativistic version of LACW method can be readily obtained on the grounds of the relativistic APW techniques [14-17].

2. Theory

Following Ref. [14], in which the relativistic APW method was developed, let us use the one-electron approximation and start from the two-component relativistic Hamiltonian

$$H = \frac{p^2}{2m} + V + \frac{\hbar}{4m^2c^2}\boldsymbol{\sigma}\cdot\left[(\nabla V)\times\mathbf{p}\right] + \frac{\hbar}{8m^2c^2}\nabla^2 V - \frac{1}{8m^3c^2}p^4 \qquad (1)$$

which is obtained from the Dirac Hamiltonian by application of the Foldy-Wouthuysen transformation. Here, c is the velocity of light; V is the electron potential, σ are the Pauli matrices:

$$\sigma_1 = \begin{pmatrix} 0 & 1 \\ 1 & 0 \end{pmatrix},\ \sigma_2 = \begin{pmatrix} 0 & -i \\ i & 0 \end{pmatrix},\ \sigma_3 = \begin{pmatrix} 1 & 0 \\ 0 & -1 \end{pmatrix}. \qquad (2)$$

The sum $H_0 = p^2/(2m) + V$ is the nonrelativistic Hamiltonian which was previously used for the LACW band structure studies [18-23]. The third term of the Hamiltonian (1) is the familiar spin-orbit coupling term, which may split the degenerate levels and cause mixing of levels thus altering the band picture. The fourth and fifth terms (the Darwin and mass-velocity corrections) are invariant under the operations of the single group and therefore do not split the degenerate states at Fermi level under consideration [14]. As this work is restricted with states in the Fermi region, we can neglect both the Darwin and mass-velocity corrections in calculations of the levels splitting.

In order to take into account the spin-orbit term in the energy-band calculation for the nanotube, it is necessary to extend the formalism of the LACW method to include the matrix elements of the spin-orbit term between the spinor wave functions. Because the nonrelativistic terms make the largest contribution to the energy, the following procedure is used: First, the eigenfunctions $\Psi^0_{n,k}(\mathbf{r})$ and the eigenvalues $E_n^0(k)$ of the nonrelativistic Hamiltonian are found by the LACW method described in the references [18-23]. The spinor basis partners $\Psi^0_{n,k}(\mathbf{r})\alpha$ and $\Psi^0_{n,k}(\mathbf{r})\beta$ are formed, where α and β are the pure spin functions. The matrix elements of the spin-orbit operator between the spinor basis functions are determined, and the resulting Hamiltonian matrix is diagonalized to give the relativistic energies and eigenfunctions.

2.1. Cylindrical muffin-tin potential

In common with the standard and most simple both nonrelativistic and relativistic augmented plane wave techniques for bulk materials [14-17], in the LACW method, we apply the muffin-tin (MT) and local density functional approximations for electronic potentials. However, the potential of a nanotube differs drastically from that of bulk material. Indeed, infinite motion of an electron is possible in any direction in a crystal, but it is obviously limited in the case of nanotubes by their size and cylindrical shape. In terms of the LACW method, the atoms of nanotube are considered to be enclosed between two impenetrable cylinder-shaped potential barriers Ω_a and Ω_b, because there are two vacuum regions Ω_v on the outside and on the inside of the tubule. The radii a and b of these barriers are chosen so that the region confined by barriers accommodates a significant portion of the electron density of the tubule. Based on our previous nonrelativistic calculations of the CNTs, we take $a = R_{NT} + 2.3$ a.u., $b = R_{NT} + 2.3$ a.u., where, R_{NT} is radius of the tubule. Finally, the electronic potential is spherically symmetrical in the $\Omega_{MT\alpha}$ regions of MT-spheres of atoms α and constant in the interspherical region Ω_{II}. Inside these spheres, we calculate the electron potential by means of the local density approximation with Slater exchange. As usually, the radii of the MT spheres $r_{MT\alpha}$ were chosen so that the atomic spheres touch, but do not overlap. Outside the MT spheres up to the impenetrable potential barriers Ω_a and Ω_b, the potential is approximated by a constant value taken to be the energy zero point.

2.2. Eigenfunctions of nonrelativistic LACW method

For a further calculation of the spin-orbit interaction we need to present explicit form of the nonrelativistic solutions $\Psi^0_{n,k}(\mathbf{r})$. In the LACW method, the functions $\Psi^0_{n,k}(\mathbf{r})$ are written as the linear combinations of the basis functions Ψ^k_{PMN} called the linearized augmented cylindrical waves (LACWs). The basis LACWs are the solutions of the wave equation for the interspherical and MT regions of tubule sewn together so that the resulting LACWs are continuous and differentiable anywhere in the system. In the interspherical region, the LACWs Ψ^k_{PMN} are the solutions of the Schrödinger equation for a free electron movement in the cylindrical potential well. In the terms of the cylindrical coordinates (Z,Φ,R), this equation takes the form (in atomic Rydberg units)

$$\left\{-\left[\frac{1}{R}\frac{\partial}{\partial R}\left(R\frac{\partial}{\partial R}\right)+\frac{1}{R^2}\frac{\partial^2}{\partial\Phi^2}+\frac{\partial^2}{\partial Z^2}\right]+U(R)\right\}\Psi(Z,\Phi,R)=E\Psi(Z,\Phi,R), \tag{3}$$

where

$$U(R)=\begin{cases}0, & b\le R\le a\\ \infty, & R<b,\ R>a\end{cases}. \tag{4}$$

The solutions of Eq. (3) have the form $\Psi(Z,\Phi,R) = \Psi^k_{PMN}(Z,\Phi,R) = \Psi^k_P(Z)\Psi_M(\Phi)\Psi_{MN}(R)$, where $\Psi^k_P(Z) = (1/\sqrt{d})\exp[i(k+k_P)Z]$, $k_P = (2\pi/d)P$, $P = 0, \pm1, \pm2, \ldots$ The function $\Psi^k_P(Z)$ corresponds to the movement of an electron along the Z axis in a one-dimensional system with periodic boundary conditions; d is a lattice constant for the Z-direction; k is a one-dimensional crystal momentum, $-\pi/d \le k \le \pi/d$. The function

$$\Psi_M(\Phi)=\frac{1}{\sqrt{2\pi}}e^{iM\Phi},\ M = 0, \pm1, \pm2, \ldots \tag{5}$$

corresponds to the rotation of electron about the Z axis. The function $\Psi_{MN}(R)$, corresponds to the radial movement of an electron in the interspherical regions Ω_{II} of the tubule; it is a solution of the equation

$$\left[-\frac{1}{R}\frac{d}{dR}R\frac{d}{dR}+\frac{M^2}{R^2}\right]\Psi_{MN}(R)+U(R)\Psi_{MN}(R)=E_{|M|,N}\Psi_{MN}(R). \tag{6}$$

Here, $E_{|M|,N}$ is the energy spectrum, and N is the radial quantum number. The energy $E=(k+k_P)^2+E_{|M|,N}$, corresponds to the wave function Ψ^k_{PMN}(Z,Φ,R). At $b\le R\le a$, this equation is written as

$$\left[\frac{d^2}{dR^2}+\frac{1}{R}\frac{d}{dR}+\kappa^2_{|M|,N}-\frac{M^2}{R^2}\right]\Psi_{MN}(R)=0, \tag{7}$$

where $\kappa_{|M|,N}=\left\{E_{|M|,N}\right\}^{1/2}$. Any solution of the Bessel equation can be represented as a linear combination of cylindrical Bessel functions of the first J_M and second Y_M kinds of the Mth order:

$$\Psi_{MN}(R)=C^J_{MN}J_M\left(\kappa_{|M|,N}R\right)+C^Y_{MN}Y_M\left(\kappa_{|M|,N}R\right). \tag{8}$$

The C^J_{MN} and C^Y_{MN} are constants chosen so as to ensure the normalization of the wave function and its vanishing at the interior and exterior potential barriers Ω_a and Ω_b [18-23].
As in APW theory of solids, inside the MT sphere α_{MT} in the local spherical coordinate system {r,θ,φ,}, the basis function is expanded in spherical harmonics

$$\Psi^{PMN,k}_{\alpha_{MT}}(\rho,\theta,\varphi)=\sum_{l=0}^{\infty}\sum_{m=-l}^{l}\left[A^{PMN,k}_{lm,\alpha_{MT}}u_{l,\alpha_{MT}}\left(\rho,E_{l,\alpha_{MT}}\right)+B^{PMN,k}_{lm,\alpha_{MT}}\dot{u}_{l,\alpha_{MT}}\left(\rho,E_{l,\alpha_{MT}}\right)\right]Y_{lm}(\theta,\varphi). \tag{9}$$

Here, $u_{l,\alpha_{MT}}$ is the solution of radial Schrödinger equation in the MT sphere α_{MT} for the energy $E_{l,\alpha_{MT}}$:

$$\frac{1}{\rho}\frac{d^2\rho u_{l,\alpha_{MT}}(\rho)}{d\rho^2}+\left(E_{l,\alpha_{MT}}-V_{\alpha_{MT}}(\rho)-\frac{l(l+1)}{\rho^2}\right)u_{l,\alpha_{MT}}(\rho)=0, \tag{10}$$

$\dot{u}_{l,\alpha_{MT}}=\left[\partial u_{l,\alpha_{MT}}/\partial E\right]_{E_{l,\alpha_{MT}}}$, and $V_{\alpha_{MT}}$ is the local density spherically symmetric potential in the region of the MT sphere α_{MT}. Coefficients $A^{PMN,k}_{lm,\alpha_{MT}}$ and $B^{PMN,k}_{lm,\alpha_{MT}}$ are selected so that both the LACW $\Psi^k_{PMN}(Z,\Phi,R)$ and its derivative have no discontinuities at the boundaries of the MT spheres. The analytical expressions for coefficients $A^{PMN,k}_{lm,\alpha_{MT}}$ and $B^{PMN,k}_{lm,\alpha_{MT}}$ were obtained previously [18-20]:

$$A^{PMN,k}_{lm,\alpha_{MT}}=r^2_{\alpha_{MT}}D^{PMN,k}_{lm,\alpha_{MT}}a^{PMN,k}_{lm,\alpha_{MT}},\quad B^{PMN,k}_{lm,\alpha_{MT}}=r^2_{\alpha_{MT}}D^{PMN,k}_{lm,\alpha_{MT}}b^{PMN,k}_{lm,\alpha_{MT}}. \tag{11}$$

Here,

$$D^{PMN,k}_{lm,\alpha_{MT}}=\frac{1}{\sqrt{2d}}(-1)^{\frac{m+|m|}{2}+l}i^l\left[\frac{(2l+1)(l-|m|)!}{(l+|m|)!}\right]^{1/2}\exp\left\{i\left(K_PZ_{\alpha_{MT}}+M\Phi_{\alpha_{MT}}\right)\right\}\times$$
$$\times(-1)^M\left[C^J_{MN}J_{m-M}\left(\kappa_{|M|,N}R_{\alpha_{MT}}\right)+C^Y_{MN}Y_{m-M}\left(\kappa_{|M|,N}R_{\alpha_{MT}}\right)\right] \tag{12}$$

$$a^{PMN,k}_{lm,\alpha_{MT}}(r_{\alpha_{MT}})=I^{PMN,k}_{2,lm,\alpha_{MT}}\dot{u}_{l,\alpha_{MT}}(r_{\alpha_{MT}})-I^{PMN,k}_{1,lm,\alpha_{MT}}\dot{u}'_{l,\alpha_{MT}}(r_{\alpha_{MT}}), \tag{13}$$

$$b^{PMN,k}_{lm,\alpha_{MT}}(r_{\alpha_{MT}})=I^{PMN,k}_{1,lm,\alpha_{MT}}u'_{l,\alpha_{MT}}(r_{\alpha_{MT}})-I^{PMN,k}_{2,lm,\alpha_{MT}}u_{l,\alpha_{MT}}(r_{\alpha_{MT}}), \tag{14}$$

$Z_{\alpha_{MT}}$, $\Phi_{\alpha_{MT}}$, $R_{\alpha_{MT}}$ are cylindrical coordinates of atom α_{MT}. The prime on the functions u means derivative with respect to the radius r. Finally

$$I^{PMN,k}_{1,lm,\alpha_{MT}} = 2\int_0^{\pi/2} \exp\{iK_P r_{\alpha_{MT}} \cos\theta\} J_m\left(\kappa_{|M|,N} r_{\alpha_{MT}} \sin\theta\right) P_l^{|m|}(\cos\theta) \sin\theta d\theta , \tag{15}$$

$$I^{PMN,k}_{2,lm,\alpha_{MT}} = 2\int_0^{\pi/2} \exp\{i(K_P r_{\alpha_{MT}} \cos\theta)\} \{iK_P \cos\theta\, J_m(\kappa_{|M|,N} r_{\alpha_{MT}} \sin\theta) + (1/2)\kappa_{|M|,N} \sin\theta\}\times \tag{16}$$

$$[J_{m-1}(\kappa_{|M|,N} r_{\alpha_{MT}} \sin\theta) - J_{m+1}(\kappa_{|M|,N} r_{\alpha_{MT}} \sin\theta)]\}\ P_l^{|m|}(\cos\theta) \sin\theta\ d.$$

are integrals of the augmented Legendre polynomials $P_l^{|m|}$.

Thus, the final form of the basis LACWs Ψ^k_{PMN} is obtained. The techniques for calculating the overlap and nonrelativistic Hamiltonian matrix elements and solving the Schrödinger equation by means of a secular equation in the LACW basis have been discussed in references [18-23] and are not repeated here. The results of such the nonrelativistic calculation are the energies $E_n^0(k)$ for various bands and the eigenfunctions $\Psi^0_{n,k}(\mathbf{r})$ of the nanotube expressed as a linear combination of the basis functions

$$\Psi^0_{n,k}(\mathbf{r}) = \sum_{PMN} a^{kn}_{PMN} \Psi^k_{PMN}(\mathbf{r}). \tag{17}$$

2.3. Matrix elements of spin-orbit interaction

Let us use the functions $\Psi^0_{n,k}(\mathbf{r})\alpha$ and $\Psi^0_{n,k}(\mathbf{r})\beta$ as the spin-dependant basis for calculating the matrix elements of the spin-orbit coupling. Taking into account the Eq. (17), the calculation of the spin-orbit coupling in the basis $\Psi^0_{n,k}(\mathbf{r})\alpha$ and $\Psi^0_{n,k}(\mathbf{r})\beta$ is reduced to calculation of the $H_{S\text{-}O}$ matrix elements in the terms of the basis functions $\Psi^k_{PMN}(\mathbf{r})\alpha$ and $\Psi^k_{PMN}(\mathbf{r})\beta$.

Because the approximate potential is constant outside the MT spheres, this region gives no contribution to the spin-orbit matrix elements. The spherical symmetry of the MT potential $V_{\alpha_{MT}}(\rho)$ within each MT sphere allows the spin-orbit operator to be written in terms of the operator for angular momentum about the center of the sphere [14]:

$$H_{S\text{-}O} = \frac{1}{c^2}\boldsymbol{\sigma}\cdot\left[\left(\nabla V_{\alpha_{MT}}\right)\times\mathbf{p}\right] = \frac{1}{c^2}\frac{1}{\rho}\frac{dV_{\alpha_{MT}}}{d\rho}\boldsymbol{\sigma}\mathbf{L} = \frac{1}{c^2}\frac{1}{\rho}\frac{dV_{\alpha_{MT}}}{d\rho}\left(\frac{1}{2}\sigma_+ L_- + \frac{1}{2}\sigma_- L_+ + \sigma_z L_z\right). \tag{18}$$

The effect of spin operator on the functions α and β is the following

$$\sigma_+\alpha = 0,\ \sigma_-\alpha = 2\beta,\ \sigma_z\alpha = \alpha,\ \sigma_+\beta = 2\alpha,\ \sigma_-\beta = 0,\ \sigma_z\beta = -\beta. \tag{19}$$

The effect of the angular momentum operators on the spherical harmonics is

$$L_z Y_{lm}(\theta\varphi) = mY_{lm}(\theta\varphi),\ \ L_\pm Y_{lm}(\theta\varphi) = \left[l(l+1) - m(m\pm 1)\right]^{1/2} Y_{lm\pm 1}(\theta\varphi), \tag{20}$$

they do not effect the radial part of the wave function. Thus, the integral $\left\langle \Psi^k_{P_2M_2N_2}(\mathbf{r})\chi_2 \middle| H_{S-O} \middle| \Psi^k_{P_1M_1N_1}(\mathbf{r})\chi_1 \right\rangle$ is factored into the product of the radial and angular integrals. Use of the relations (18-20) together with the ortogonality and normalization of the spin functions α and β and of the spherical harmonics, allows the angular integration to be performed analytically to give finally

$$\left\langle \Psi^k_{P_2M_2N_2}(\mathbf{r})\beta \middle| H_{S-O} \middle| \Psi^k_{P_1M_1N_1}(\mathbf{r})\beta \right\rangle = -\left\langle \Psi^k_{P_2M_2N_2}(\mathbf{r})\alpha \middle| H_{S-O} \middle| \Psi^k_{P_1M_1N_1}(\mathbf{r})\alpha \right\rangle, \tag{21}$$

$$\left\langle \Psi^k_{P_2M_2N_2}(\mathbf{r})\alpha \middle| H_{S-O} \middle| \Psi^k_{P_1M_1N_1}(\mathbf{r})\beta \right\rangle = \left\langle \Psi^k_{P_2M_2N_2}(\mathbf{r})\beta \middle| H_{S-O} \middle| \Psi^k_{P_1M_1N_1}(\mathbf{r})\alpha \right\rangle^*, \tag{22}$$

$$
\begin{aligned}
&\left\langle \Psi^{k}_{P_2M_2N_2}(\mathbf{r})\alpha \middle| H_{S-O} \middle| \Psi^{k}_{P_1M_1N_1}(\mathbf{r})\alpha \right\rangle = \frac{1}{2c^2d}(-1)^{M_1+M_2}\sum_{\alpha_{MT}} r^4_{\alpha_{MT}} \sum_{l=0}^{\infty}(2l+1)\sum_{m=-l}^{l} m\frac{(l-|m|)!}{(l+|m|)!}\times \\
&\left\{\varsigma_{l,\alpha_{MT}}\left(a^{P_2M_2N_2,k}_{lm,\alpha_{MT}}\right)^*\left(a^{P_1M_1N_1,k}_{lm,\alpha_{MT}}\right)+\ddot{\varsigma}_{l,\alpha_{MT}}\left(b^{P_2M_2N_2,k}_{lm,\alpha_{MT}}\right)^*\left(b^{P_1M_1N_1,k}_{lm,\alpha_{MT}}\right)+\dot{\varsigma}_{l,\alpha_{MT}}\left[\left(a^{P_2M_2N_2,k}_{lm,\alpha_{MT}}\right)^*\left(b^{P_1M_1N_1,k}_{lm,\alpha_{MT}}\right)+\left(b^{P_2M_2N_2,k}_{lm,\alpha_{MT}}\right)^*\left(a^{P_1M_1N_1,k}_{lm,\alpha_{MT}}\right)\right]\right\}\times \\
&\exp\left\{i\left[\left(K_{p_2}-K_{p_1}\right)Z_{\alpha_{MT}}+\left(M_1-M_2\right)\Phi_{\alpha_{MT}}\right]\right\}\left[C^J_{M_2N_2}J_{m-M_2}\left(\kappa_{|M_2|,N_2}R_{\alpha_{MT}}\right)+C^Y_{M_2N_2}Y_{m-M_2}\left(\kappa_{|M_2|,N_2}R_{\alpha_{MT}}\right)\right]\times \\
&\left[C^J_{M_1N_1}J_{m-M_1}\left(\kappa_{|M_1|,N_1}R_{\alpha_{MT}}\right)+C^Y_{M_1N_1}Y_{m-M_1}\left(\kappa_{|M_1|,N_1}R_{\alpha_{MT}}\right)\right]
\end{aligned}
\tag{23}
$$

$$
\begin{aligned}
&\left\langle \Psi^{k}_{P_2M_2N_2}(\mathbf{r})\beta \middle| H_{S-O} \middle| \Psi^{k}_{P_1M_1N_1}(\mathbf{r})\alpha \right\rangle = \frac{1}{2c^2d}(-1)^{M_1+M_2}\sum_{\alpha_{MT}} r^4_{\alpha_{MT}} \sum_{l=0}^{\infty}(2l+1)\times \\
&\sum_{m=-l}^{l}\left[l(l+1)-m(m+1)\right]^{1/2}\left[\frac{(l-|m|)!(l-|m+1|)!}{(l+|m|)!(l+|m+1|)!}\right]^{1/2}(-1)^{(m+|m|)/2+[m+1+|m+1|]/2}\times \\
&\left\{\varsigma_{l,\alpha_{MT}}\left(a^{P_2M_2N_2,k}_{lm+1,\alpha_{MT}}\right)^*\left(a^{P_1M_1N_1,k}_{lm,\alpha_{MT}}\right)+\ddot{\varsigma}_{l,\alpha_{MT}}\left(b^{P_2M_2N_2,k}_{lm+1,\alpha_{MT}}\right)^*\left(b^{P_1M_1N_1,k}_{lm,\alpha_{MT}}\right)+\dot{\varsigma}_{l,\alpha_{MT}}\left[\left(a^{P_2M_2N_2,k}_{lm+1,\alpha_{MT}}\right)^*\left(b^{P_1M_1N_1,k}_{lm,\alpha_{MT}}\right)+\left(b^{P_2M_2N_2,k}_{lm+1,\alpha_{MT}}\right)^*\left(a^{P_1M_1N_1,k}_{lm,\alpha_{MT}}\right)\right]\right\}\times \\
&\exp\left\{i\left[\left(K_{p_2}-K_{p_1}\right)Z_{\alpha_{MT}}+\left(M_1-M_2\right)\Phi_{\alpha_{MT}}\right]\right\}\left[C^J_{M_2N_2}J_{m+1-M_2}\left(\kappa_{|M_2|,N_2}R_{\alpha_{MT}}\right)+C^Y_{M_2N_2}Y_{m+1-M_2}\left(\kappa_{|M_2|,N_2}R_{\alpha_{MT}}\right)\right]\times \\
&\left[C^J_{M_1N_1}J_{m-M_1}\left(\kappa_{|M_1|,N_1}R_{\alpha_{MT}}\right)+C^Y_{M_1N_1}Y_{m-M_1}\left(\kappa_{|M_1|,N_1}R_{\alpha_{MT}}\right)\right]
\end{aligned}
\tag{24}
$$

where

$$\varsigma_{l,\alpha_{MT}} = \int_0^{r_{\alpha_{MT}}}\left[u_{l,\alpha_{MT}}(\rho)\right]^2\frac{dV_{\alpha_{MT}}(\rho)}{d\rho}\rho d\rho, \tag{25}$$

$$\dot{\varsigma}_{l,\alpha_{MT}} = \int_0^{r_{\alpha_{MT}}} u_{l,\alpha_{MT}}(\rho)\dot{u}_{l,\alpha_{MT}}(\rho)\frac{dV_{\alpha_{MT}}(\rho)}{d\rho}\rho d\rho, \tag{26}$$

$$\varsigma_{l,\alpha_{MT}} = \int_0^{r_{\alpha_{MT}}}\left[\dot{u}_{l,\alpha_{MT}}(\rho)\right]^2\frac{dV_{\alpha_{MT}}(\rho)}{d\rho}\rho d\rho. \tag{27}$$

3. Results of calculations

Energy bands for the single-wall carbon armchair nanotubes (n, n) (where $4 \le n \le 12$) with spin–orbit interaction are calculated by solving the secular equation. As a typical example, figure 1 shows the energy bands of the (5, 5) tubule for the region near the Fermi level. For comparison, the energy bands calculated without spin–orbit interaction are shown in this figure too. The bands are plotted over only half of the zone, because the bands at $-k$ are the same as those at k.

For the absence of spin–orbit interaction, the two linear bonding (π) and antibonding (π^*) bands cross at the Fermi energy near $k = 2\pi/(3d)$. Due to the spin–orbit interaction, the energy gap opens between these bands, and the metallic nature for armchair nanotubes is broken. However, each energy subband still has the two-fold degeneracy, because the spin–orbit interaction does not break the time-reversal symmetry and the achiral armchair nanotubes have the inversion symmetry. These properties are consistent with the previous theories according to which the spin degeneracy survives for the achiral nanotubes [1-5].

Table. Spin–orbit energy gap for armchair nanotubes versus the roll-up index n.

Nanotube	Spin-orbit gap	Nanotube	Spin-orbit gap	Nanotube	Spin-orbit gap
(4, 4)	0.537 meV	(7, 7)	0.104 meV	(10, 10)	0.083 meV
(5, 5)	0.223 meV	(8, 8)	0.086 meV	(11, 11)	0.076 meV
(6, 6)	0.153 meV	(9, 9)	0.078 meV	(12, 12)	0.086 meV

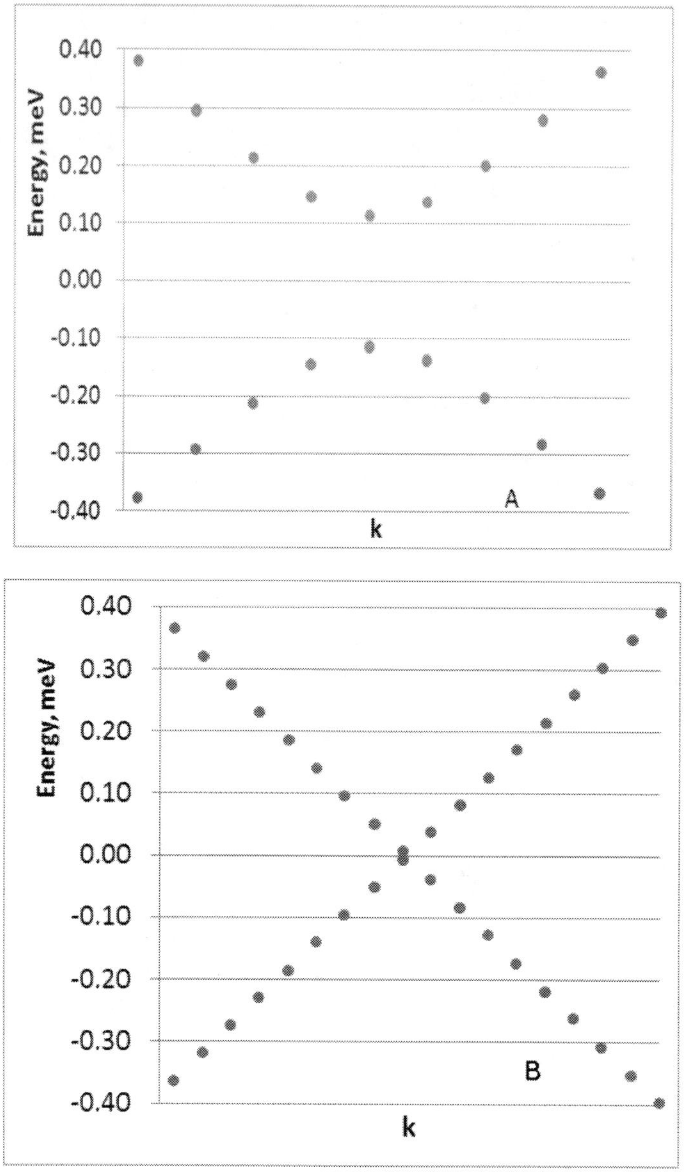

Figure 1. LACW bands of (5, 5) CNT calculated with (A) and without (B) spin-orbit coupling.

The present numerical LACW calculation shows a gap equal to 0.22 meV at the Fermi energy for the (5, 5) CNT (table). It is difficult to quantitatively compare the LACW spin-orbit gaps with those of the tight binding model because of ambiguity of the tight binding empirical parameters. In works [1-4], a formation of the spin-orbit gaps in the CNTs was analyzed qualitatively only. In Ref. [5], the energy gaps of nanotubes are presented as the functions of spin–orbit coupling constant λ varied between 0 and 15 meV; thus, in the (5, 5) CNT, the gap can take any value from zero up to about 2.2 meV. Application of the purely atomic $\lambda = 6$ meV gives the gap equal to 0.85 meV which is four times larger one than in this work. The gap equal to 1.3 meV is published in [7]; however, the authors of work [5] point that an unrealistic large spin–orbit coupling parameter has been assumed in the calculation [7]. According to the tight binding model developed in [8], the overlap of valence and conduction bands and the zero gaps are possible in some cases in the armchair tubules in spite of the spin-orbit coupling.

As expected, the spin-orbit splitting decreases with growth of n or, equivalently, of the tubule diameter $d = n(3/\pi)d_{C\text{-}C}$, where $d_{C\text{-}C} = 1.42$ Å is a C–C bond length, a dependence of gap on the inverse diameter is not linear as different from the tight binding models [17].

4. Conclusion

The relativistic version of the linear augmented cylindrical wave method is developed and applied to calculating the energy bands of the armchair (n, n) carbon nanotubes. It is shown that the energy gap opens in the Fermi energy region, and the metal nature of the band structure of the (n, n) tubules is broken due to the spin-orbit coupling. According to these first-principle calculations, the spin-orbit splitting varies between 0.537 and 0.086 meV meV for the (n, n) tubules with $4 \leq n \leq 12$.

Acknowledgments

The work was performed in frames of Russian Scientific and Research Program "Scientific and pedagogic personal for innovations in Russia in 2009–2013 years". The research was partially supported by the Russian Basic Research Foundation (grant 11-03-00691) and EU under Project No. FP7-247007 CACOMEL.

References

[1] Ando T 2000 *J. Phys. Soc. Jpn.* **69** 1757
[2] Huertas-Hernando D, Guinea F and Brataas A 2006 *Phys. Rev.* B **74** 155426
[3] Chiko L, López-Sancho M P and Miñoz M C 2004 *Phys. Rev. Lett.* **93** 176402
[4] Chiko L, López-Sancho M P and Miñoz M C 2009 *Phys. Rev.* B **79** 235423
[5] Izumida W, Sato K and Saito R 2009 *J. Phys. Soc. Jpn.* **78** 074707
[6] Wunsch B 2009 *Phys. Rev.* B **79** 235408
[7] Zhou J, Liang Q and Dong J *Phys. Rev.* B 2009 **79** 195427
[8] Liu H *Physica* B 2011 **406** 104
[9] Minot E D, Yaish Y, Sazonova V and McEuen P L 2004 *Nature (London)* **428** 536
[10] Kuemmeth F, Ilani S, Ralph D C and McEuen P L 2008 *Nature (London)* **452** 448
[11] Ilani S and McEuen P L 2010 *Annu. Rev. Condens. Matter Phys.* **1** 1
[12] Jhang S H, Marganska M, Skuorsky Y, Preusche D, Witkamp B, Grifony M, van der Zant H, Wosnitza J and Strank C 2010 *Phys. Rev.* B **82** 041404
[13] Jaspersen T S, Grove-Rusmussen K, Paaske J, Muraki K, Fujisawa T, Nigård J and Flensberg K 2011 *Nature Physics* **7** 348
[14] Conklin J B, Johnson L E and Pratt G W 1965 *Phys. Rev.* **137** 4A 1283
[15] Loucks T L 1965 *Phys. Rev.* **139** 4A 1333
[16] MacDonald A H, Pickettll W E and Koelling D D 1980 *J. Phys. C: Solid St. Phys.* **13** 2675
[17] Koelling D D and Harmon B N 1977 *J. Phys. C: Solid State Phys.* **10** 3107
[18] D'yachkov P N, Kepp O M and Nikolaev A V 1999 *Dokl. Chem.* **365** 67
[19] D'yachkov P N and Kirin D V 1999 *Dokl. Phys. Chem.* **369** 326
[20] D'yachkov P N 2004 In *Encyclopedia of Nanoscience and Nanotechnology* vol 1, ed H. S. Nalwa (American Scientific, Valencia CA) p 192
[21] D'yachkov P N and Makaev D V 2007 *Phys. Rev.* B **76** 195411
[22] D'yachkov P N and Makaev D V 2006 *Phys. Rev.* B **74** 155442
[23] D'yachkov P N and Makaev D V 2010 *Phys. Rev.* B **82** 035426

Ab-initio study of metal-zirconia interfaces

S Kulkova[1,2,4], A Bakulin[2], S Hocker[3], S Schmauder[3]
[1] Tomsk State University, pr. Lenina 36, Tomsk 634050, Russia
[2] Institute of Strength Physics and Materials Science of SB RAS, pr. Akademichesky 2/4, Tomsk 634021, Russia
[3] Institute of Materials Testing, Materials Science and Strength of Materials, University of Stuttgart, Germany

E-mail: kulkova@ms.tsc.ru

Abstract. A comparative theoretical study of metal-zirconia interfaces with BCC and FCC metals was performed using pseudopotential approach with LDA and GGA approximation for exchange-correlation functional. It was shown that the high adhesion can be achieved at the O-terminated *Me*/ZrO_2(001) interface with BCC metals that is related to large charge transfer from metal film to substrate and increase of an ionic contribution in the chemical bonding. The structural and electronic factors which are responsible for decrease of adhesion at differently oriented metal-zirconia interfaces are discussed. The influence of CaO, MgO and Y_2O_3 doping on the work of separation (W_{sep}) at *Me*(001)/c-ZrO_2(001) is analyzed.

1. Introduction

Phenomena occurring at metal–ceramic interfaces are very important in many technological processes and applications. The functional properties of corrosion-resistant coatings, high temperature composites and electronic materials depend strongly on the adhesion at the metal–ceramic interface [1]. Layered composite materials consisting of magnetic metal foils and insulating oxide layers can be used for the cores of motors, generators, transformers, etc. [2]. There are several reviews concerning the relationship between chemical bonding at the metal/ceramic interfaces and mechanical properties of materials [3,4]. In most studies the metal-alumina interfaces were considered [5-8]. Less attention was paid to metal-zirconia (*Me*/ZrO_2) interfaces. *Ab-initio* methods within the density functional theory (DFT) were applied to study the electronic structure of metal–ceramic interfaces in order to understand metal-ceramic cohesion at the microscopic level. In most papers concerning metal-zirconia interfaces the bonding at the Ni/ZrO_2 interface was discussed in detail [9-11]. In [9] authors predicted very strong adhesion at the Ni/ZrO_2(111) interface due to the formation of localized interface bonds. The existence of strong chemical bonding at the Ni(100)/ZrO_2(100) interfaces was also demonstrated in the work of Beltran et al. [10]. Two different types of interactions can provide this strong interfacial bonding: Ni–Zr electron hybridization and Ni–O ionic bonding. In our previous work [11] an even stronger interaction at the Ni/(ZrO_2)O(001) interface was shown for the less stable configuration (with a Ni interfacial atom located in the next-Zr or Zr2 (figure 1) position than that considered in [10] (Ni in the O-bridge position). It is necessary to point out that generalized gradient approximation (GGA) was used in [10] whereas we applied local density approximation (LDA) in [11]. Both obtained results

[4] To whom any correspondence should be addressed

are in disagreement with experiments which suggest a weak interaction between Ni films and zirconia [12,13]. This discrepancy between theoretical and experimental results can be attributed to the presence of surface vacancies. It should be noted that the cubic phase of zirconia exists above 2370 °C [14]. It can be stabilized at lower temperatures by addition of other oxides (Y_2O_3, MgO, CaO, etc.), which leads to the appearance of oxygen vacancies. The appearance of metal impurities leads to a decrease of the electrostatic interaction and the presence of oxygen vacancies, especially near an interface, can provide lower adhesion [1,15]. The influence of Fe additions on adhesion was also considered in our work [11]. It was shown that in the case of Ni–Fe ordered alloys the adhesion depends strongly on the composition of the interfacial layers and their magnetic states. The adhesion at metal/zirconia interfaces was considered in details by Mucoz et al. in review [1,16].

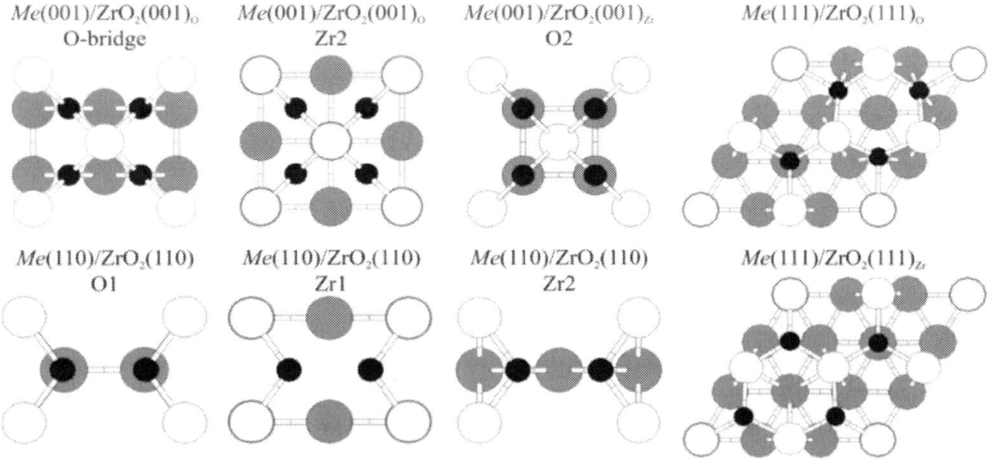

Figure 1. Atomic structure of the interfacial layers of *Me*(001)/ZrO_2(001), *Me*(110)/ZrO_2(110) and *Me*(111)/ZrO_2(111) interfaces. Atoms of *Me* and Zr are depicted by large grey and white circles and O atoms are shown by small black circles.

In this work we present the results of a comparative study of adhesion at *Me*/ZrO_2 interfaces with BCC and FCC metals and examine the influence of alloying oxides (CaO, MgO, Y_2O_3) on the electronic bonding properties at the metal-zirconia interfaces.

2. Computational details

Atomic and electronic structure of the *Me*/ZrO_2 interfaces were calculated using the pseudopotential approach with plane wave basis implemented in the Vienna ab-initio simulation package (VASP) [17-19]) and ultrasoft pseudopotentials [20]. Both LDA [21] and GGA-PBE [22] approximations for exchange-correlation functional were used. The equilibrium lattice parameter of cubic zirconia was found to be 5.082 Å using LDA which is in good agreement with the experimental value of 5.088 Å. The application of GGA-PBE resulted in 5.126 Å. An energy cutoff of 500 eV for plane wave basis set, and a 8×8×1 Monkhorst-Pack *k*-points mesh for integration over the Brillouin zone were used in calculations. The total energies were converged up to 10^{-4} eV. Metal-zirconia interfaces were modeled using a supercell approach with periodic boundary conditions. A sandwich configuration of the interface with a zirconia slab between two metal slabs with a vacuum region not less than 8 Å between them was used. It is known that the (001) and (111) surfaces of cubic ZrO_2 are characterized by alternating layers of O and Zr atoms and can be terminated by either Zr or O atomic planes whereas (110) surface has stoichiometric composition as bulk. In order to describe the cubic-ZrO_2(001) surface it is sufficient to use six atomic layers of O and five or seven atomic layers of Zr depending on surface termination. There are two different types of interfaces (*Me*-Zr and *Me*-O), depending on the termination of the (001) surface (either Zr or O atomic planes). One can see in figure 1 *Me* atoms are twofold coordinated by O atoms in the case of the O-bridge position but they are fourfold coordinated by O atoms in the case of Zr2 position at the Me-O interface. In the case of the Zr-terminated oxide surface *Me* atoms can occupy the next O atoms or O2-position (figure 1). Five atomic layer metal films were considered at the Me/ZrO_2 interfaces. The atoms of the three central atomic layers in the oxide film were fixed with the spacing between the layers being equal to its value in the bulk. The atoms of other layers including those in metal films were relaxed until the forces acting on the atoms were less than 0.025 eV/Å. Considered geometries for the case of (110) and (111) interfaces are shown in figure 1 as well.

The work of separation is a fundamental quantity, which controls the mechanical strength of the metal-ceramic interface. It is the work needed to separate the interface into two free surfaces and can be defined as

$$W_{sep} = (E_{me} + E_{ox} - E_{me-ox})/2S,$$

where E_{me-ox} is the total energy of supercell containing the multilayered slabs; E_{me} and E_{ox} are the total energies of the same supercell containing a single slab of the metal or oxide, respectively; S is the interface area and the factor 2 takes into account the presence of the two equivalent interfaces within a unit cell.

3. Results and discussion

The calculations of W_{sep} at the O-terminated $Me(001)/ZrO_2(001)$ interfaces with FCC metals (*Me*=Fe, Ni, Cu, Rh, Pd, Pt, etc.) with respect to the *Me*-O cleavage plane indicate much stronger chemical bonding at the interface with the metal in the O-bridge position than with the metal in the Zr2-position. Weaker adhesion of the *Me* films to zirconia in this position is in accordance with equilibrium *Me*–O bond length which is slightly longer in the former case. The bonding mechanism is primary ionic interaction between *Me* and O ions as can be clearly seen in figure 2 where the charge-density difference for Cu/ZrO_2 is presented.

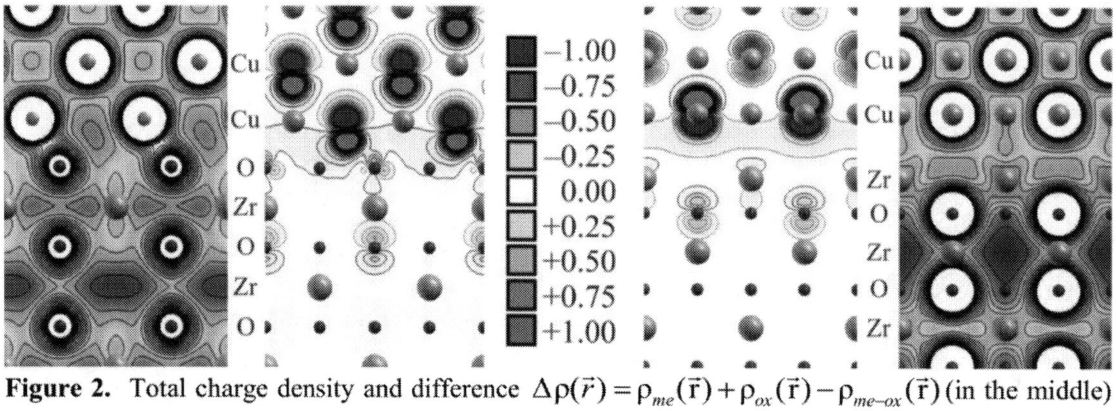

Figure 2. Total charge density and difference $\Delta\rho(\vec{r}) = \rho_{me}(\vec{r}) + \rho_{ox}(\vec{r}) - \rho_{me-ox}(\vec{r})$ (in the middle) at the $Cu/ZrO_2(001)$ interface: $Cu/(ZrO_2)_O$ (left) and $Cu/(ZrO_2)_{Zr}$ (right).

One can see the charge depletion region near Cu atoms and the corresponding charge accumulation region near O atoms. The charge-density distributions indicate that the Zr- and O-terminated interfaces differ in terms of the mechanism of bonding. Figure 2 reveals a metal bonding character between *Me* and Zr atoms at the *Me*-Zr interface, which arises from *s-d* and *d-d* hybridizations. Charge redistribution leads to the polarization of *Me* atoms (especially Ni, Fe atoms), therefore, the chemical bonding at the Zr-terminated interface is a combination of both metallic interaction and electron polarization. The estimation of the work of separation for example in the case of the $Cu/ZrO_2(001)_O$ interface (table 1) corresponds to a rather strong bonding than that calculated in [10].

As can be seen from table 1 the scatter in the values of W_{sep} at the Zr-terminated interfaces is less than 0.5 J/m^2 but it is larger at the O-terminated interface. Similar trends were obtained for other considered metals with FCC structure. Moreover the conclusions obtained for Cu are valid in the case of the HfO_2(001) substrate (table 1). For the Ni(001)/HfO_2(001) interfaces the values of W_{sep} are equal to 5.85 and 7.45 J/m^2 for Hf- and O-terminated interface respectively, but these values are 5.13 and 7.35 J/m^2 for Zr- and O-terminations. Our calculations show that the usage of a nonlocal correction in the exchange-correlation functional in the form [22] leads to lower values of W_{sep} (table 1). The difference in W_{sep} is significant less at the Zr-terminated interface due to more effective screening of Zr-states in oxide.

Table 1. Work of separation at Cu-zirconia interfaces with respect to their termination, configuration of the metal film as well as approximation for the exchange correlation potential (V_{xc}).

Interfaces	V_{xc}	Configuration[a]	W_{sep} (J/m^2)
Cu(001)/ZrO_2(001)$_{Zr}$	LDA	Zr1	3.48
		O2	3.92
		O-bridge	3.69
Cu(001)/ZrO_2(001)$_{Zr}$	GGA	O2	3.25 (3.66 [16])
Cu(001)/ZrO_2(001)$_O$	LDA	O-bridge	6.32
		O1	5.55
		Zr2	5.84
Cu(001)/ZrO_2(001)$_O$	GGA	O-bridge	4.90 (5.08 [16])
Cu(001)/HfO_2(001)$_{Hf}$	LDA	O2	4.07
Cu(001)/HfO_2(001)$_O$	LDA	O-bridge	6.10
Cu(110)/ZrO_2(110)	LDA	O1	1.96
		Zr1	0.45
		Zr2	1.50
Cu(111)/ZrO_2(111)$_O$	LDA		0.65
Cu(111)/ZrO_2(111)$_{Zr}$	LDA		3.70

[a] Symbols Zr1, O1, etc. denote interfacial atoms of oxide above which the metal atoms are located.

The investigation of the influence of the metal electronic structure on the adhesion at the *Me*(001)/ZrO_2(001)$_O$ interface shows that the work of separation decreases in the row of isoelectronic metals (5.96 J/m^2 (Ni $3d^8 4s^2$), 3.53 J/m^2 (Pd $4d^{10} 5s^0$) and 3.26 J/m^2 (Pt $5d^{10} 6s^0$), here GGA-PBE results are presented) due to filling of the *d*-band and decrease of charge transfer from the metal to the substrate. The structural factors caused by the increase of the metal and oxide lattices mismatch at the interface (2.06%, 9.11% and 9.74% for Ni, Pd and Pt, respectively) influence the work of separation also. It should be noted that for metal with not fully occupied *d*-band (Rh(001)/ZrO_2(001)$_O$) the value of the W_{sep} is still high (5.17 J/m^2) which is due to an increase of ionic contribution in the chemical bonding. The used projected augmented wave method (PAW-LDA) or PP LDA results in an even higher value of W_{sep} (6.72 J/m^2) than obtained with the pseudopotential GGA-PBE approach. The work of separation in the case of Pd film is 4.99 J/m^2 (PAW-LDA) and 5.04 J/m^2 (PP-LDA). The calculated values of W_{sep} in the case of Pt(111) and Pt(001) films are 3.37 J/m^2 and 3.26 J/m^2, respectively. In the case of Pt film W_{sep} depends slightly on the film orientation. It should be noted that adhesion of Cu in the case of non-polar stoichiometric ZrO_2(110) surface is significant less pronounced than at the interfaces with polar non stoichiometric Zr-terminated surfaces (001) or (111). The lower values of W_{sep} were obtained in the case of *Me* configuration in Zr1 i.e. above Zr of the first interfacial layer. Low W_{sep} (0.63 J/m^2) at the Cu/ZrO_2(111)$_O$ interface in comparison with that for (001) interfaces can be explained by the following factors: decrease of charge transfer to oxygen, increase of interfacial distance and decrease of *Me*-O bonds. Note that three from four oxygen atoms at the (111) interface are located in the position which is twofold coordinated by metals whereas the fourth atom is located in the *Me*-top position. Some metal atoms form the bond with Zr of the second layer from interface (bond lengths are varied from 2.50 up to 2.86 Å) whereas all metal atoms are bonded to oxygen atoms of interfacial layer in the case of (001).

At the Ni(111)/$ZrO_2(111)_O$ interface the W_{sep} reaches up to 1.45 J/m^2 due to an increase of the charge transfer. Electronegative oxygen can much easier get electrons for filling its *p*-band from Ni since its *d*-band is not fully occupied as that from Cu. Spin-polarized calculations of differently oriented Ni-zirconia interfaces demonstrate the decrease of W_{sep} by ~0.05-0.3 J/m^2 and an increase of the interfacial distance of 0.05-0.1 Å in dependence of the orientation and termination interface if the magnetism of Ni is taken into account. Note that the adhesion of the metal monolayer is higher than that for the metal film. The increase of W_{sep} up to 1.65 J/m^2 in the case of a Ni-monolayer correlates with the decrease of interfacial distance from 1.66 Å up to 1.55 Å. In general, our calculations demonstrate that high values of adhesion energy can be achieved at the metal-zirconia interface with BCC metals such as Nb, Ta, Mo, W (table 2) in the case of *Me*-$Al_2O_3(0001)$. The chemical interaction at the interface is mainly ionic (figure 3) with large charge transfer (~ 0.25-0.30 el) from the metal film to each oxygen atom.

Table 2. Work of separation at the *Me*(001)/$ZrO_2(001)_O$ where *Me*=Nb, Ta, Mo, W. The values in brackets were obtained using PP LDA

Interface	Method	W_{sep}(J/m^2)
Nb(001)/$ZrO_2(001)_O$	PAW-LDA	10.26 (10.29)
	PAW-PBE	8.82
Ta(001)/$ZrO_2(001)_O$	PAW-LDA	10.68
	PAW-PBE	9.30
Mo(001)/$ZrO_2(001)_O$	PAW-LDA	9.80 (9.94)
	PAW-PBE	9.65
W(001)/$ZrO_2(001)_O$	PAW-LDA	(10.03)
	PAW-PBE	9.35

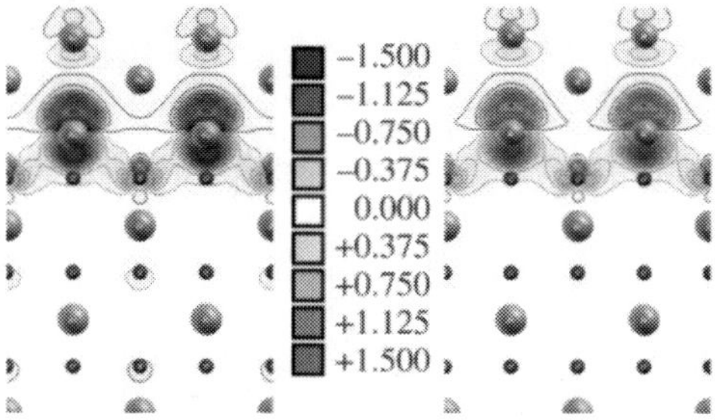

Figure 3. Charge density difference $\Delta\rho$ at Mo/$ZrO_2(001)_O$ (left) and W/$ZrO_2(001)_O$ (right) interfaces.

In order to study the influence of alloying oxide CaO, MgO and Y_2O_3 on the work of separation at the Cu/ZrO_2(001) interface we increased the size of computational cell. This allows us to decrease the concentration of defects. The positions of substitution impurities and oxygen vacancies are shown in figure 4. An O-vacancy was introduced removing one out four O atoms.

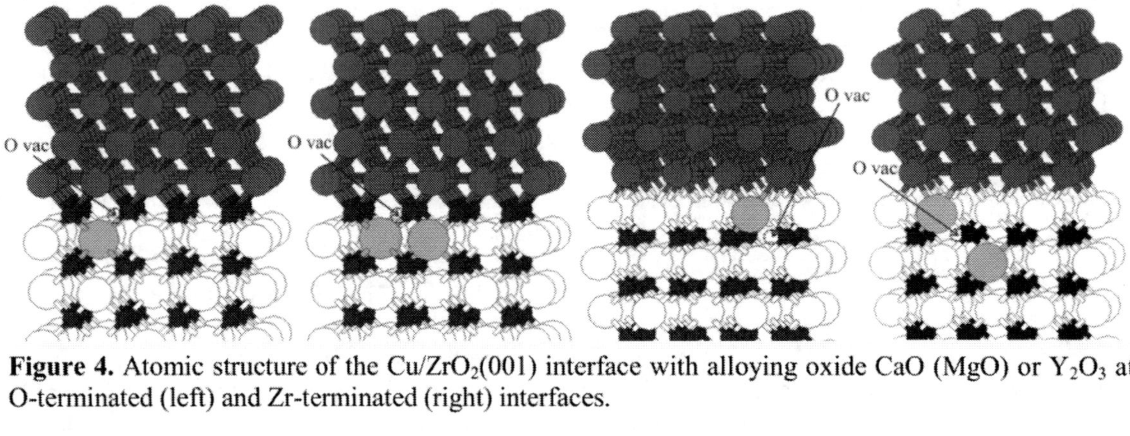

Figure 4. Atomic structure of the Cu/ZrO_2(001) interface with alloying oxide CaO (MgO) or Y_2O_3 at O-terminated (left) and Zr-terminated (right) interfaces.

Table 3. Work of separation (in J/m^2) at ideal and defective $Cu/ZrO_2(001)$ interface

Interface	Ideal	O-vac	CaO	MgO	Y_2O_3
$Cu/ZrO_2(001)_{Zr}$	3.25	3.03	2.26	2.44	2.64
$Cu/ZrO_2(001)_O$	4.90	2.96	2.98	2.93	4.63

As can be seen from table 3 the adhesion energy is reduced significantly on both interfaces in the case of CaO alloying. Appearance of Ca or Mg at the Zr-sites leads to a decrease of the hybridization of *Me*-Zr states which provides more substantial lowering of adhesion at the Zr-terminated interface than in the case of Y substitution impurity.

In the O-terminated interface the effect of the O-vacancies on the adhesion is pronounced more significantly and can even lead to instability of the cubic lattice and its transformation into the monoclinic one. The decrease of W_{sep} at the O-terminated interface is caused by the reduced number of interfacial bonds induced by the O-vacancies [11]. Substitution impurities stabilize the cubic structure due to keeping bond strengths as it was shown in [1]. The estimation of interatomic bonds and interfacial distances were performed in the presence of both defects (O-vacancy and *Me*-impurity). In the case of YSZ the restore of interatomic bonding occurs which decreases the critical influence of the O-vacancies on W_{sep}.

4. Conclusion

A systematic investigation of the adhesion properties at metal-zirconia interfaces with FCC and BCC metals has been performed by *ab-initio* density functional theory methods. General trends in the change of W_{sep} with respect to structure, composition of interfacial layers, and electronic structure of metal as well as defects are established. High adhesion is found for BCC transition metals at the O-terminated zirconia polar (001) surface that can be connected with their partially occupied *d*-band and a larger ionic contribution to the chemical bonding in comparison with FCC metals. It was shown that in zirconia stabilized by other oxides as CaO, MgO, Y_2O_3 critical influence of the O-vacancies in the interfacial oxide layers on W_{sep} can be partly compensated by substitutional impurities which lead to large rearrangements of the O and Zr atoms and restoration of the interfacial distance.

References

[1] Muñoz M C, Gallego S, Beltrán J I and Cerdá J 2006 *Surf. Sci. Rep.* **61** 303
[2] Moorhead A J and Kim H-E 1999 *J. Am. Ceram. Soc.* **82** 1349
[3] Klomp J T 1989 *Surfaces and interfaces of ceramic materials* ed L C Dufour (Norwell, MA: Kluwer)
[4] Finnis MW 1996 *J. Phys: Condens. Matter.* **8** 5811
[5] Kruse C, Finnis M W, Lin J S, Paine M C, Milman V, de Vita A and Gilan J 1996 *Phil. Mag.* **73** 377
[6] Batirev I, Alavi A and Finnis M 1999 *Phys. Rev. Lett.* **82** 1510
[7] Zhukovskii Y F, Kotomin E A, Herschend B, Hermansson K and Jacobs P W M 2002 *Surf. Sci.* **513** 343
[8] Feng J, Zhang W and Jiang W 2005 *Phys. Rev.* B **72** 115423
[9] Christensen A and Carter E A 2001 *J. Chem. Phys.* **114** 5816
[10] Beltrán J I, Gallego S, Cerdá J and Muñoz M C 2003 *Phys. Rev.* B **68** 075401
[11] Eremeev S V, Schmauder S, Hocker S and Kulkova S E 2009 *Surf. Sci.* **603** 2218
[12] Ladas S and Sotiropoulou D 1998 *Surf. Sci.* **408** 182
[13] Harel S, Mariot J M and Hague C F 1992 *Surf. Sci.* **269–270** 1167
[14] Ruff O, Ebert F and Anorg Z 1929 *Allg. Chem.* **180** 19
[15] Beltrán J I, Gallego S, Cerdá J and Muñoz M C 2003 *J. Eur. Ceram. Soc.* **23** 2737
[16] Beltrán J I and Muñoz M C 2008 *Phys. Rev.* B **78** 245417
[17] Kresse G and Hafner J 1993 *Phys. Rev.* B **47** 558
[18] Kresse G and Furthmüller J 1996 *Comput. Mater. Sci.* **6** 15

IOP Conf. Series: Materials Science and Engineering **38** (2012) 012004 doi:10.1088/1757-899X/38/1/012004

[19] Kresse G and Furthmüller J 1996 *Phys. Rev.* B **54** 11169
[20] Kresse G and Hafner J 1994 *J. Phys.: Condens. Matter* **6** 8245
[21] Ceperley D M and Alder B J 1980 *Phys. Rev. Lett.* **45** 566
[22] Perdew J P, Burke K and Ernzerhof M 1996 *Phys. Rev. Lett.* **77** 3865

IOP Conf. Series: Materials Science and Engineering **38** (2012) 012005 doi:10.1088/1757-899X/38/1/012005

Ab initio simulations on rutile-based titania nanowires

Yu F Zhukovskii[1,3], **R A Evarestov**[2]
[1]Institute of Solid State Physics, University of Latvia, 8 Kengaraga Str., LV-1063, Riga, Latvia.
[2]Department of Quantum Chemistry, St. Petersburg State University, 26 Universitetsky Ave., 198504, Petrodvorets, Russia.

E-mail: quantzh@latnet.lv

Abstract. The rod symmetry groups for monoperiodic (1D) nanostructures have been applied for construction of models for bulk-like TiO_2 nanowires (NWs) cut from a rutile-based 3D crystal along the chosen [001] and [110] directions of crystallographic axes. In this study, we have considered nanowires described by both the Ti-atom centered rotation axes as well as the hollow site centered axes passing through the interstitial positions between the Ti and O atoms closest to the axes. The most stable [001]-oriented TiO_2 NWs with rhombic cross sections are found to display the energetically preferable {110} facets only while the nanowires with quasi-square sections across the [110] axis are formed by the alternating $\{1\bar{1}0\}$ and {001} facets. For simulations on rutile-based nanowires possessing different diameters for each NW type, we have performed large-scale *ab initio* Density Functional Theory (DFT) and hybrid DFT-Hartree Fock (DFT-HF) calculations with total geometry optimization within the Generalized Gradient Approximation (GGA) in the form of the Perdew-Becke-Ernzenhof (PBE) exchange-correlation functionals (*PBE* and *PBE0*, respectively), using the formalism of linear combination of localized atomic functions (LCAO). We have simulated both structural and electronic properties of TiO_2 NWs depending both on orientation and position of symmetry axes as well as on diameter and morphology of nanowires.

1. Introduction

Due to a variety of promising applications, single-crystalline titania nanowires (TiO_2 NWs) attract enhanced attention in modern nanotechnology [1]. A number of theoretical studies on atomic-scale TiO_2 nanowires were published so far [2-5]. Large-scale *ab initio* calculations on thermodynamically preferable TiO_2 NWs in the rutile phase [2] clearly display only the energetically preferable {110} facets in the case of their [001] orientation as well as the alternating {110} and {001} facets in the case of [110] NW orientation, in accordance with the Wulff construction formalism [6]. The structure and electronic properties of TiO_2 anatase-based nanowires were calculated as well, with different orientations and surface terminations varied as a function of diameter [3]. In recent study, the stability and electronic structure of both anatase-based and rutile-based [001] nanowires were compared in a series of first-principles calculations [4]. Up to diameters of 2.1 nm in anatase-based and 3.7 nm in rutile-based nanowires, those {100}-walled anatase and {110}-walled rutile NWs were found to be the most stable, which retain the non-symmorphic character of the bulk space group. As to construction of monoperiodic (1D) atomistic models for titania nanowires, both the Ti atom- and hollow site-centered

[3] To whom any correspondence should be addressed.

symmetry axes (*i.e.*, translation two-fold rotation axis and four-fold screw axis, respectively) were considered in recent studies [4,5].

In this study, we apply the rod group symmetry for comparative Density Functional Theory (DFT) calculations on the rutile-based [001] and [110]-oriented TiO_2 nanowires with Ti atom- and hollow site-centered symmetry axes performed within the formalism of localized atomic orbitals (LCAO) using the Generalized Gradient Approximation (GGA) in the form of both the Perdew-Becke-Ernzenhof (*PBE*) and hybrid (DFT-Hartree-Fock) exchange-correlation functional *PBE0* as implemented in *CRYSTAL-09* code [7]. We construct four types of nanowires for energetically stable rutile-based TiO_2 NWs with rhombic square and rectangular quasi-square shapes. Let us underscore that the hollow site-centered [110]-oriented titania nanowires were not considered and simulated so far. Large scale *PBE* an *PBE0* calculations have allowed us to compare parameters of rutile-based titania bulk with those calculated on four types of models for [001] (rhombic shape) and [110] (quasi-square shape) rutile-based TiO_2 nanowires verifying them by analogous results calculated earlier.

2. Models and symmetry of TiO_2 nanowires

The symmetry of the rutile-based TiO_2 bulk structure is described by a 3D tetragonal lattice and space group $P4_2/mnm$ [5] with the two titania formula units *per* the primitive cell. Nanowires are constructed as 1D systems cut from the 3D crystal along the direction of one of the bulk crystal symmetry axes. The translational periodicity is maintained along this direction. In the case of a rutile-based [001] TiO_2 nanowire (Figs. 1,2) a direction of the translation axis is orthogonal to a pair of [110] and [1$\bar{1}$0] vectors while both titania formula units of primitive cell lie in the two cross-sectional planes. If the translational two-fold rotation axis goes through a Ti atom (Fig. 1), one can see that the reflection in the horizontal (*h*) plane and rotations around the two second order axes in this plane are the symmetry operations for [001]-oriented nanowires. The rod symmetry group of this system is *Pmmm* while its point symmetry group is D_{2h} [4]. As to the translation axis for a [001] rutile-based nanowire with the 4-fold screw axis going through the interstitials in the middle of the squares formed by Ti atoms (Fig. 2), its rod symmetry group is non-symmorphic $4_2/m$ while point symmetry group is C_{4h} [4]. It means that a number of symmetry operations used for description of [001]-oriented hollow site-centered nanowire (8) equals to that of Ti-atom centered nanowire.

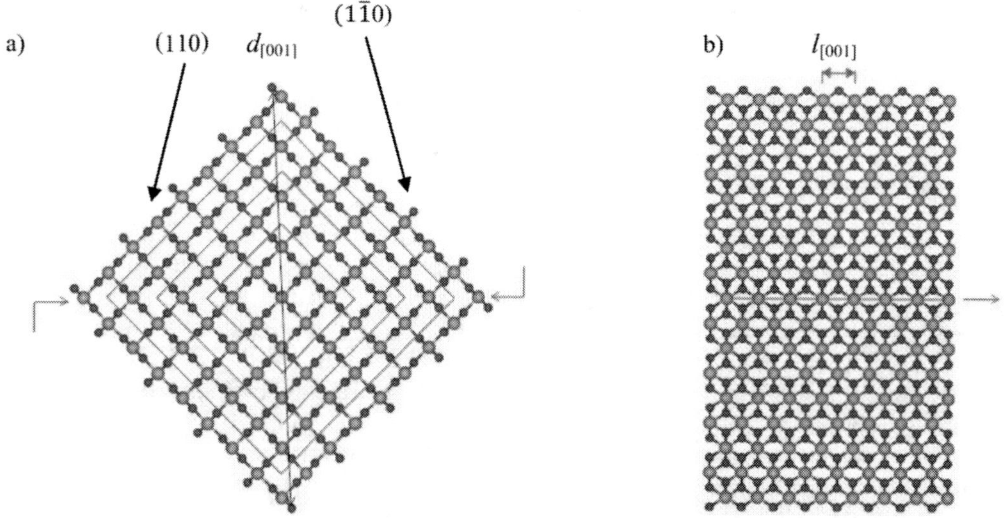

Figure 1. Cross sectional (a) and lateral (b) images of non-optimized large rutile-based [001]-oriented Ti atom-centered nanowire possessing the D_{2h} symmetry and containing 81 formula units *per* NW unit cell with aside (110) and (1$\bar{1}$0) facets shown in Fig. 1b. Red rhombs in Fig. 1a show borders for prism models of middle, small and smallest TiO_2 NWs (49, 25 and 9 formula units per UC, respectively). Diameter of a nanowire is shown by the twice-terminated arrow ($d_{[001]}$) while its period (length of UC) is shown in Fig. 1b as $l_{[001]}$.

Fig. 3 clearly shows how nanowires with hollow site-centered symmetry axes can be transformed to NWs with Ti atom-centered axes and *vice versa* by removal of external stoichiometric shells from the two adjacent facets.

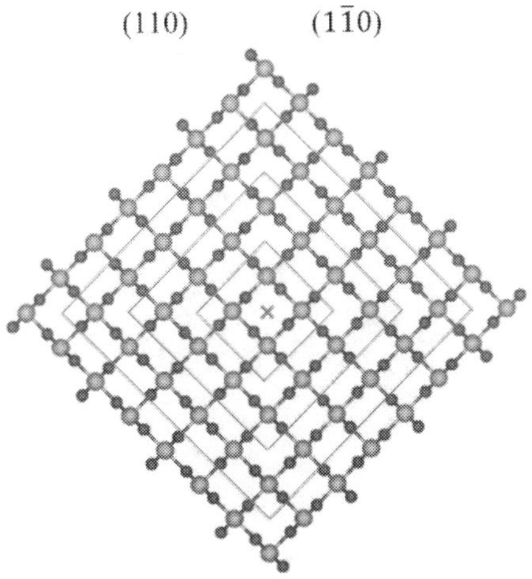

Figure 2. Cross-sectional view of non-optimized [001]-oriented hollow site-centered titania nanowire possessing the C_{4h} point symmetry and containing 64 formula units *per* NW unit cell. Red rhombs show borders for prism models of middle, small and smallest TiO_2 NWs (with 36, 16 and 4 formula units *per* unit cell, respectively). Indices correspond to the upper lateral facets while the center of nanowire is shown as a cross.

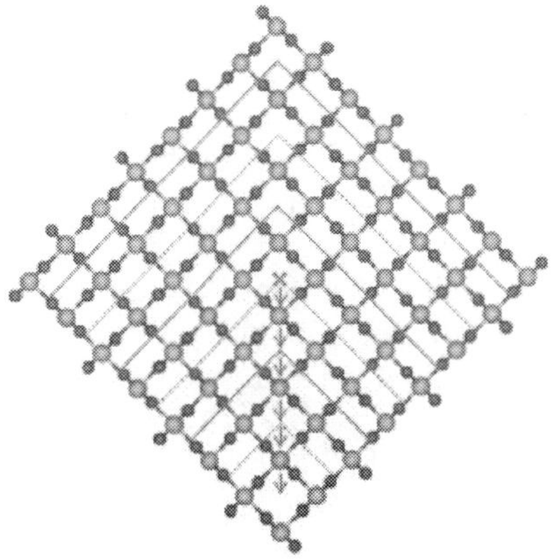

Figure 3. Cross-sectional image of non-optimized hollow site-centered TiO_2 [001] NW possessing four-fold screw axis (its track is shown by cross) and containing 64 formula units *per* NW UC. Small arrows show shifts of the nanowire axis during alternating NW transformations from Ti atom-centered to hollow site-centered structures and *vice versa*. Two adjacent facets of rhombs imaged by solid lines show borders for prism models of Ti atom-centered TiO_2 NWs while those limited by dashed lines correspond to hollow site-centered TiO_2 NWs. Two other lateral facets described here are remained fixed coinciding with lower sides of the cross-sectional image.

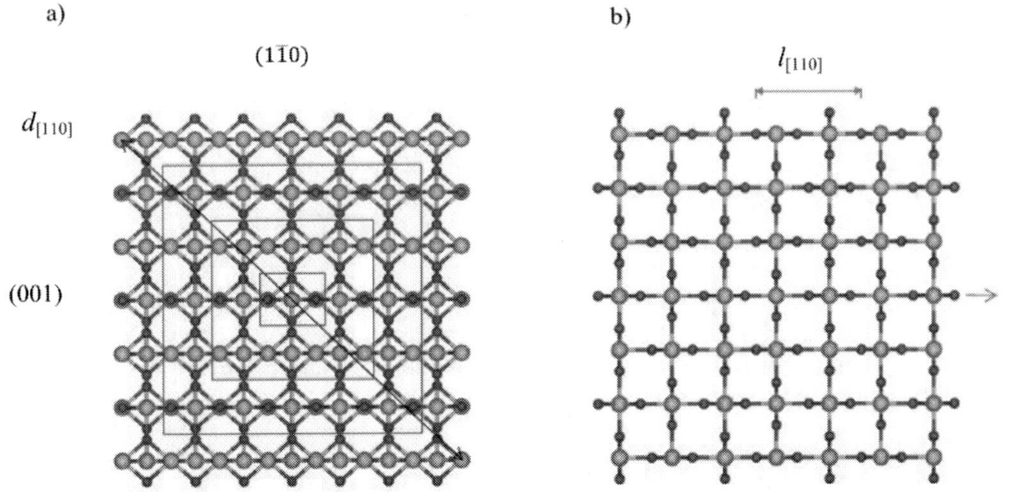

Figure 4. Cross sectional (a) and lateral (b) images of the non-optimized large rutile-based [110]-oriented Ti atom-centered NW possessing the D_{2h} symmetry and containing 105 formula units *per* NW UC, with aside (001) and (1$\bar{1}$0) facets (the former is shown in Fig. 4b). Rectangles in Fig. 4a show borders for prism models of middle, small and smallest TiO_2 NWs (55, 21 and 3 formula units *per* UC, respectively). NW diameter is shown by the twice-terminated arrow ($d_{[110]}$) while its period (length of UC) is shown in Fig. 4b as $l_{[110]}$.

For a rutile-based [110] nanowire (Fig. 4), the translation axis is orthogonal to [001] and [1$\bar{1}$0] vectors while four formula units of primitive cell lie in the six cross-sectional planes. The symmetry of a rutile-based [110] nanowire coincides with the symmetry of analogous type of [001] NW if the translation axis with rotation by π goes through Ti atoms. Indeed, Fig. 4, shows that the symmetry operations for [110] NWs are reflection in the vertical (v) plane (containing the translation axis) and rotation around the 2-fold translation axis in this plane. The symmetry operations include also rotations around the 2-fold axis in the horizontal (h) plane (orthogonal to the translation axis). Thus, the rod symmetry group of the system is again *Pmmm* (the point symmetry group is D_{2h}). As to symmetry of [110] NWs with hollow site centered symmetry axes (Fig. 5) it can be described by rather poor *Pm11* rod group characterized by presence of mirror plane only (C_s point symmetry) [7]. However, since lateral {1$\bar{1}$0} facets of this NW configuration are described by *Pm2m* rod group (C_{2v} point symmetry) the latter can be used to characterize the whole nanowire.

(1$\bar{1}$0)

(001)

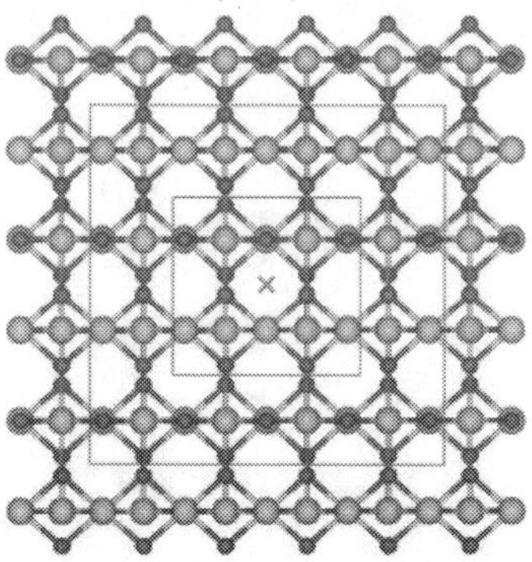

Figure 5. Cross-sectional view of non-optimized the [110]-oriented hollow site-centered TiO_2 nanowire possessing C_{2v} symmetry and containing 78 formula units *per* NW unit cell (its track is shown as a cross). Red rectangles show borders for prism models of middle and small TiO_2 NWs (with 36 and 10 formula units *per* unit cell, respectively). Indices correspond to the lateral facets of nanowire. This model is described by us for the first time.

3. Computational details

The first principles DFT-LCAO method, as implemented in the *CRYSTAL-09* code [7] describes 1D nanowires in their original space form. Calculations on DW NTs have been performed using both the DFT-*PBE* [8] and hybrid Hartree-Fock/Kohn-Sham *PBE0* [9] exchange-correlation Hamiltonians within the Generalized Gradient Approximation (GGA) as implemented in the *CRYSTAL-09* code. All-valence basis set (BS) in the form of 6*s*-311*sp*-1*d* Gaussian-type functions (GTFs) used for O atoms and a small-core pseudopotential of Ti atoms (3*s*, 3*p*, 3*d* and 4*s*-electrons were taken as valence electrons) have been recently adopted by us for TiO_2 nanoube calculations [10].

Calculations of titania nanowires possessing large unit cells have been performed using the full exploitation of the rod group symmetry. All the atomic coordinates in the monoperiodic unit cell have been generated from supercells of the corresponding 2D slabs. These coordinates have been allowed to relax when performing optimization procedure. Stability of nanowires after optimization of geometry depends on the surface energy of their lateral facets:

$$E_{surf} = (E_{NW\_UC} - N_{UC} E_{bulk\_FU})/S_{NW\_UC}, \qquad (1)$$

where E_{NW_UC} and E_{bulk_FU} are the calculated total energies of both nanowire *per* its unit cell and bulk *per* formula unit, respectively, N_{UC} the number of formula units *per* NW UC while S_{NW_UC} the total surface area of lateral NW facets *per* unit cell. The smaller is a surface energy, the more stable is a corresponding nanowire.

4. Results and discussion

The structural and electronic properties of the rutile TiO_2 bulk calculated using both the DFT-PBE and hybrid PBE0 methods as well as simulated earlier and measured experimentally are present in Table 1. Comparison of calculated parameters for TiO_2 bulk clearly shows qualitative correlation between themselves and their correspondence to the experimental values although certain dispersion of results obtained using different methods can be observed. It is also true for Ti-O bond lengths inside the first (I) and second (II) coordination shells consisting of four and two O atoms, respectively, around each

Ti atom. Calculated values of the effective charges q_O and q_{Ti} look also quite reasonable, which are found to be qualitatively comparable with the corresponding data obtained earlier [11].

Table 1. The structural and electronic properties of optimized TiO_2 bulk (rutile pjhase).

Calculation method and reference source	Lattice parameters			Parameters of bonds and atoms				Band gap
	a, Å	*c*, Å	*u*	$r_{Ti\text{-}O(I)}$, Å	$r_{Ti\text{-}O(II)}$, Å	q_O, *e*	q_{Ti}, *e*	$\Delta\varepsilon_g$, eV
PBE	4.647	2.980	0.306	1.964	2.006	-1.097	+2.194	1.74
PBE0	4.586	2.963	0.306	1.945	1.983	-1.172	+2.344	3.83
Results of other calculations	4.54-4.651[11]	2.93-3.00[11]	0.303-0.307[11]	1.91-1.95[11]	1.97-2.04[11]	-1.00--1.36[11]	+2.00-+2.72[11]	1.8-2.1[11]
Experimental data	4.593[11]	2.958[11]	0.305[11]	1.949[11]	1.980[11]	–	–	3.03[11]

As to a large difference between the experimental and calculated values of band gaps ($\Delta\varepsilon_g$), it is well known that standard DFT calculations essentially underestimate $\Delta\varepsilon_g$ for semiconductors and insulators. Only application of the hybrid exchange-correlation functionals, *e.g.*, *PBE0*, allows one to reproduce the band gaps of nanostructures with relatively high accuracy.

Results of large-scale *PBE* and *PBE0* calculations on all four types of rutile-based titania nanowires (Figs. 1, 2, 4, 5) with sets of different diameters for each type are systematized in Tables 2-5. They clearly demonstrate that [001]-oriented NWs are noticeably more stable than [110] ones, stability of all types of nanowires increases with diameter growth.

Table 2. The structural and electronic properties of optimized Ti atom-centered [001]-oriented NWs (Fig. 1).

Nanowire size (number N_{UC} of TiO_2 formula units *per* UC)	Initial diameter of NW d_{NW}, Å	Optimized NW diameter (its change) d_{NT}, Å (δd_{NT}, %)	Initial length of NW UC l_{NW}, Å	Optimized NW period (its change) l_{NW}, Å (δl_{NT}, %)	Surface energy of nanowire E_{surf}, J/m^2	Band gap (its change from that of bulk, Table 1) $\Delta\varepsilon_g$, eV ($\delta\varepsilon_g$, %)
			PBE			
9 (smallest)	11.122	11.493 (+3.33)	2.980	2.911 (-2.37)	1.023	1.04 (-40.38)
25 (small)	20.240	20.711 (+2.32)		2.944 (-1.22)	0.906	1.51 (-13.25)
49 (middle)	29.400	29.941 (+1.84)		2.957 (-0.78)	0.842	1.57 (-9.67)
81 (large)	38.670	39.270 (+1.55)		2.963 (-0.57)	0.789	1.60 (-8.18)
			PBE0			
9 (smallest)	10.979	11.268 (+2.63)	2.964	2.881 (-2.88)	1.072	3.15 (-21.59)
25 (small)	19.980	20.476 (+2.48)		2.927 (-1.26)	0.971	3.61 (-6.09)
49 (middle)	29.023	29.594 (+1.97)		2.940 (-0.82)	0.914	3.70 (-3.51)
81 (large)	38.174	38.712 (+1.41)		2.946 (-0.61)	0.861	3.74 (-2.41)

Table 3. The structural and electronic properties of optimized hollow site-centered [001]-oriented NWs (Fig. 2).

Nanowire size (number N_{UC} of TiO_2 formula units *per* UC)	Initial diameter of NW d_{NW}, Å	Optimized NW diameter (its change) d_{NT}, Å (δd_{NT}, %)	Initial length of NW UC l_{NW}, Å	Optimized NW period (its change) l_{NW}, Å (δl_{NT}, %)	Surface energy of nanowire E_{surf}, J/m^2	Band gap (its change from that of bulk, Table 1) $\Delta\varepsilon_g$, eV ($\delta\varepsilon_g$, %)
			PBE			
4 (smallest)	6.632	6.872 (+3.61)	2.980	2.893 (-3.01)	1.057	3.40 (95.40)
16 (small)	15.672	16.094 (+2.69)		2.937 (-1.46)	0.854	2.34 (34.48)
36 (middle)	24.818	25.327 (+2.05)		2.954 (-0.88)	0.773	2.24 (28.74)
64 (large)	33.984	34.567 (+1.72)		2.962 (-0.61)	0.746	2.03 (16.67)
			PBE0			
4 (smallest)	6.545	6.789 (+3.73)	2.964	2.876 (-3.06)	1.106	4.53 (18.28)
16 (small)	15.466	15.892 (+2.75)		2.919 (-1.54)	0.873	4.09 (6.79)
36 (middle)	24.492	25.007 (+2.10)		2.934 (-1.02)	0.842	3.98 (3.92)
64 (large)	33.538	34.140 (+1.80)		2.945 (-0.65)	0.829	3.93 (2.61)

Table 4. The structural and electronic properties of optimized Ti atom-centered [110]-oriented NWs (Fig. 4).

Nanowire size (number N_{UC} of TiO_2 formula units *per* UC)	Initial diameter of NW d_{NW}, Å	Optimized NW diameter (its change) d_{NT}, Å (δd_{NT}, %)	Initial length of NW UC l_{NW}, Å	Optimized NW period (its change) l_{NW}, Å (δl_{NT}, %)	Surface energy of nanowire E_{surf}, J/m^2	Band gap (its change from that of bulk, Table 1) $\Delta\varepsilon_g$, eV ($\delta\varepsilon_g$, %)
			PBE			
3 (smallest)	2.958	2.882 (-2.64)	6.572	6.251 (-5.14)	2.771	1.60 (-8.12)
21 (small)	10.996	11.237 (+2.19)		6.377 (-3.06)	1.453	1.46 (-16.47)
55 (middle)	19.683	20.094 (+2.09)		6.424 (-2.30)	1.341	1.61 (-7.52)
105 (large)	28.431	28.966 (+1.88)		6.452 (-1.86)	1.287	1.64 (-5.88)
			PBE0			
3 (smallest)	2.920	2.846 (-2.60)	6.487	6.210 (-4.46)	3.117	3.67 (-4.36)
21 (small)	10.852	11.125 (+2.52)		6.315 (-2.72)	1.520	3.65 (-4.93)
55 (middle)	19.430	19.907 (+2.45)		6.356 (-2.06)	1.462	3.76 (-1.86)
105 (large)	28.066	28.689 (+2.22)		6.382 (-1.65)	1.418	3.79 (-1.06)

Table 5. The structural and electronic properties of optimized hollow site-centered [110]-oriented NWs (Fig. 5).

Nanowire size (number N_{UC} of TiO_2 formula units *per* UC)	Initial diameter of NW d_{NW}, Å	Optimized NW diameter (its change) d_{NT}, Å (δd_{NT}, %)	Initial length of NW UC l_{NW}, Å	Optimized NW period (its change) l_{NW}, Å (δl_{NT}, %)	Surface energy of nanowire E_{surf}, J/m^2	Band gap (its change from that of bulk, Table 1) $\Delta\varepsilon_g$, eV ($\delta\varepsilon_g$, %)
			PBE			
10 (small)	6.748	6.954 (+3.06)	6.572	6.264 (-4.92)	1.651	3.40 (95.40)
36 (middle)	15.326	15.483 (+1.02)		6.371 (-3.15)	1.362	2.34 (34.48)
78 (large)	24.054	24.261 (+0.86)		6.424 (-2.30)	1.283	2.24 (28.74)
			PBE0			
10 (small)	6.659	6.861 (+3.03)	6.487	6.208 (-4.49)	1.886	4.53 (18.28)
36 (middle)	15.125	15,472 (+2.30)		6.311 (-2.79)	1.479	4.09 (6.79)
78 (large)	23.738	24.224 (+2.05)		6.358 (-2.03)	1.391	3.98 (3.92)

5. Conclusions

The rod group formalism allows us to construct models of [001]- and [110]-oriented titania nanowires cut from the rutile-based 3D crystal along the direction of a chosen crystallographic symmetry axis. To estimate stability of NWs, we have chosen the surface energies as a criterion. The most stable [001]-oriented TiO_2 NWs with rhombic cross sections are found to display the energetically preferable {110} facets only. Stability of all types of nanowires increases with diameter growth.

Acknowledgements. R.E. is grateful to Saint Petersburg State University for the financial support (grant Nr 12.0.108.2010). The authors thank B. Aradi, A.V. Bandura, P. Deák and D.B. Migas for stimulating discussions.

References

[1] Liu B, Boercker J E and Aydil E S 2008 *Nanotechnology* **19** 505604
[2] Migas D B, Shaposhnikov V L, Borisenko V E and D'Avitaya F A 2010 *J Phys Chem C* **114** 21013
[3] Iacomino A, Giovanni C, Train F and Ninno D 2010 *J Phys Chem C* **114** 12389
[4] Aradi B, Deák P, Huy H A, Rosenauer A and Frauenheim T 2011 *J Phys Chem C* **115** 18494
[5] Peng H and Li J 2008 *J Phys Chem C* **112** 20241
[6] Herring C 1951 *Phys Rev* **82** 87
[7] Dovesi R, Saunders V R, Roetti C, Orlando R, *et al* 2009 *CRYSTAL User Manual*: University of Torino
[8] Perdew J P, Burke K and Ernzerhof M 1996 *Phys Rev Lett* **77** 3865
[9] Ernzerhof M and Scuseria G E 1999 *J Chem Phys* **110** 5029
[10] Evarestov R A, Zhukovskii Yu F, Bandura A B and Piskunov S 2010 *J. Phys. Chem. C* **114** 21061
[11] Labat F, Baranek Ph, Domain Ch, Minot Ch and Adamo C 2007 *J. Chem. Phys.* **126** 154703

IOP Conf. Series: Materials Science and Engineering **38** (2012) 012006 doi:10.1088/1757-899X/38/1/012006

Fluorescence of TiO_2:Sm^{3+} composite stimulated by plasmon waves

L Dolgov[1], **V Kiisk, S Pikker, A Loot, I Sildos**
Institute of Physics, University of Tartu, Estonia

E-mail: dolgov@fi.tartu.ee

Abstract. Plasmon-coupled fluorescence of rare earth Sm^{3+} ions doped into the TiO_2 film placed on thin gold layer is investigated. It was demonstrated that plasmon waves induced in the gold layer strongly influence rare earth emission. We revealed that each spectral peak of Sm^{3+} emission can be directionally separated, polarized and enhanced by intensity due to coupling with plasmon waves propagating in the gold film. It is noticeable that we obtained about 10 % of non-linear growth in Sm^{3+} emission intensity over the pumping strength.

1. Introduction

Applications of plasmon-coupled fluorescence deal with organic fluorophores usually [1]. As a rule such fluorophore is placed near the noble metal nanoparticle or metal nano-coating. Light induced surface plasmon-polaritons in the metal can share their energy with fluorophores. Such plasmonic support of fluorescence is used for enhancement of weak optical signals in case of low amount of analyte down to single molecules [2, 3].

Rare earth (RE) ions can be considered as an alternative to organic fluorophores. Oxides of metals and semiconductors doped with RE ions are chemically stable and do not suffer from photobleaching. Enhancement of RE fluorescence by means of surface plasmons localized in the noble metal nanoparticles was demonstrated recently [4, 5]. But surface plasmon waves propagating in metal nano-coatings were not used for regulation of RE fluorescence widely, as far as is known.

Here we incorporated fluorescent Sm^{3+} ions into TiO_2 films deposited on the gold nanolayer. We demonstrated that plasmonic waves in such dielectric-metal films can adjust direction, polarization and intensity of RE fluorescence. This effect is connected with reciprocal fluorescence-plasmon stimulation described below. It is revealed that plasmons provide about 10 % faster growth of fluorescence intensity vs. excitation in comparison with non-plasmonic case. This effect can be promising for inducing of coherent plasmons with low damping.

2. Experimental

Commercial gold-covered (50 nm) glass substrates (PHASIS, Switzerland) were used for the preparation of the samples. Titanium alkoxide precursor containing 2.5 mol % of samarium salt was spin-coated on their aureate side. The thickness of deposited sol-gel film was controlled through the number of spin coating cycles. Solid TiO_2:Sm^{3+} films (thickness 40-200 nm) on the gold-covered glass were obtained after annealing at 500°C. To achieve the matching of wave vectors for light and

[1] To whom any correspondence should be addressed

plasmon waves, the glass side of the $TiO_2:Sm^{3+}$-Au sample was attached to a semicylindrical BK7 prism (Figure 1). Fluorescence was excited with the UV line of continuous argon laser (363.8 nm, 0.7-50 mW, beam diameter 1.5 mm) impinging along the normal onto the topmost layer in contact with air. Such arrangement is known as the reverse Kretschmann geometry [1].

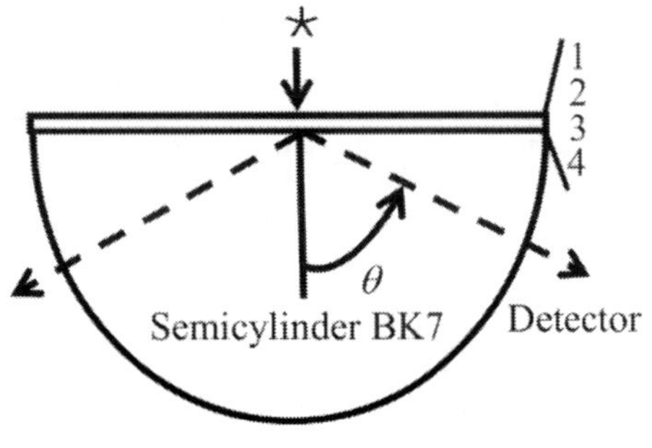

Figure 1. Scheme of experiment for detection of directed Sm^{3+} emission. The sample's layers 1-4 are marked: 1) $TiO_2:Sm^{3+}$; 2) gold; 3) glass; 4) immersion oil. Solid line with arrow indicates excitation and dashed lines represent emission light beams respectively. The drawing is not in scale for clarity.

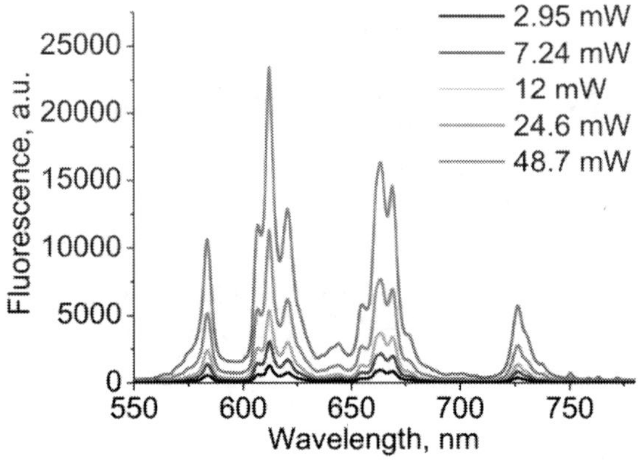

Figure 2. Fluorescence of $TiO_2:Sm^{3+}$-Au sample at different excitation powers (λ_{exc}=363.8 nm). Angle of detection θ=75°.

The fluorescent spectra were measured by using a fiber-coupled CCD spectrometer Ocean Optics 2000+ at the angles θ counted from the normal to the sample on the convex side of the prism (Figure 1). All measurements were done at room temperature. Laser beam did not cause any visible defects at the measured samples. So we did not take into consideration possible small heating of samples by the laser beam. The dependences of the spectra on the observation direction over the angular range 0-90° were recorded.

3. Results and discussion

Fluorescent spectra of $TiO_2:Sm^{3+}$-Au films have several spectral peaks at 583 nm, 616 nm, 665 nm, 728 nm, which are typical for split electron transitions of Sm^{3+} ions in the TiO_2 host: $^4G_{5/2}\rightarrow{}^6H_{5/2}$, $^4G_{5/2}\rightarrow{}^6H_{7/2}$, $^4G_{5/2}\rightarrow{}^6H_{9/2}$, $^4G_{5/2}\rightarrow{}^6H_{11/2}$ (Figure 2) [6, 7]. Fluorophore incorporated in continuous medium emits light uniformly in all directions. But incorporation of Sm^{3+} fluorophore into layered TiO_2-Au structure results in stronger fluorescence detected at certain angles θ. We guess that this effect has plasmonic nature. Namely projection of light wave vector on the dielectric-metal interface becomes equal to the surface plasmon wave vector at certain resonance angle θ_r [8, 9]:

$$(\varepsilon_p(\lambda))^{1/2}\sin\theta_r = \left(\frac{\varepsilon'(\lambda)\cdot\varepsilon_{TiO_2}(\lambda)}{\varepsilon'(\lambda)+\varepsilon_{TiO_2}(\lambda)}\right)^{1/2} + \beta \qquad (1),$$

where λ is the wavelength of fluorescent light, $\varepsilon_p(\lambda)$ is dielectric permittivity of the prism, $\varepsilon'(\lambda)$ is the real part of gold permittivity, $\varepsilon_{TiO_2}(\lambda)$ is permittivity of TiO_2 film and β is a small correction value depending on the imaginary part of metal permittivity and the thickness of the metal film. Thus light emission at the angle θ_r results in inducing of surface plasmon wave with wave vector corresponding to the right part of equation (1). Strong local field of induced plasmon stimulates following more effective emission at the angle θ_r. Therefore directional plasmon-coupled fluorescence appears at this angle.

According to equation (1) resonance angles for fluorescent bands detected at different λ must be different. Indeed, angular dependence of Sm^{3+} fluorescence proves that various spectral peaks become more pronounced at different resonance angles θ_r (Figure 3). One can see that maxima of spectral peaks with bigger wavelengths are observed at smaller resonant angles θ_r. Such regularity correlates with equation (1) and proves plasmonic nature of detected directional Sm^{3+} emission.

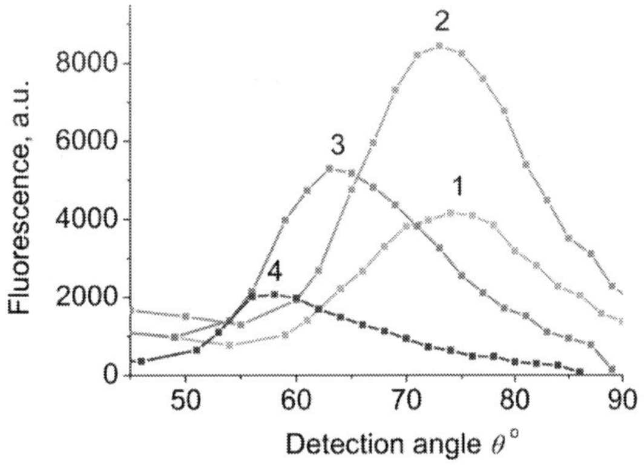

Figure 3. Measured angular dependences of fluorescence of the TiO_2:Sm^{3+}-Au structure. Data for different spectral peaks of Sm^{3+} are plotted: (1) 583 nm; (2) 616 nm; (3) 665 nm; (4) 728 nm. Excitation power is 24.6 mW, λ_{exc}=363.8 nm.

Directional fluorescence of TiO_2:Sm^{3+}-Au films is described in our paper [9] in more detail. It deserves to mention here only that the directional fluorescence is strongly (9:1) polarized. Type of polarization depends on the thickness of TiO_2:Sm^{3+} layer. Directional fluorescence is p-polarized in case of thin (<50 nm) films and s-polarized for films, which are thicker than 60 nm. Plasmon-coupled fluorescence is usually p-polarized. S-polarization can appear only in thick films as a result of coupling with waveguide modes. To accentuate attention on the coupling of fluorescence with plasmons, we shall consider the sample with thin TiO_2:Sm^{3+} layer further.

Light can interact with plasmons in several different ways. One possibility is radiative energy transfer from excited Sm^{3+} ions to plasmon waves. Locally enhanced electric field of plasmons increases the probability of radiative transitions of the Sm^{3+} ions [1, 10]. As a result plasmon waves support directional emission (Figure 3 and Figure 4a).

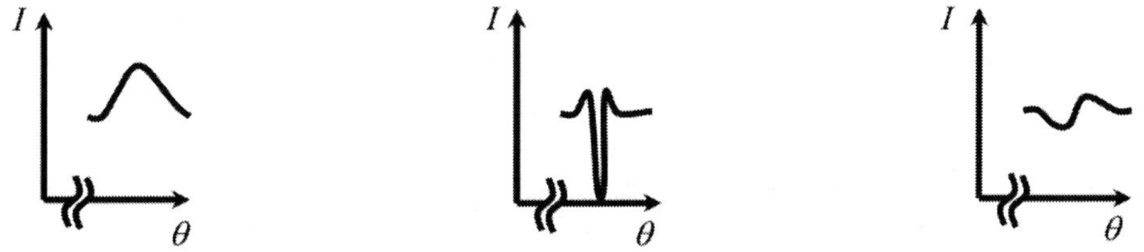

a b c

Figure 4. Scheme of emission intensity I vs. detection angle θ at various plasmon-fluorescence interactions: a) directional emission; b) absorption dip of coherent plasmons; c) competition between directional emission and absorption, caused by coherent plasmons. Scheme is prepared on the basis of data from Refs. [11-13].

Another possibility is non-radiative energy transfer from excited Sm^{3+} ions to plasmon waves [11]. Initially such non-radiative pumping of plasmons results in the compensation of losses for plasmon waves. Then system behaves as a generator of coherent plasmons. At high efficiency of such process, there would be no emission at plasmon resonance angle. Plasmons become coherent and very narrow dip in light intensity similar to laser line appears in the angular dependence (Figure 4b).

In practice there is a competition between two mentioned possibilities. Maxima of directional emission in Figure 3 and Figure 4a refer to the case where plasmons mainly support radiative process. But maximum of directional emission becomes less pronounced or even disappears when sufficient part of energy from excited fluorophore goes to plasmons non-radiatively (Figure 4c). That is, the emission remains plasmon-coupled but becomes non-directional [12, 13]. At the same time the lines of the spectrum become narrower, which testifies about induction of coherent light and coherent plasmons respectively. Such coherent processes work on the scheme similar to three-level lasers. That is plasmons are pumped from the overpopulated electronic level of fluorophore.

In our case the emitting energy level of Sm^{3+} ions, situated inside the bandgap of TiO_2 host, has relatively long lifetime (>100 μs). It is easy to excite optical electrons of Sm^{3+} through an energy transfer from the host, which itself can be excited by ultraviolet radiation. This can be associated with four-level system and creates a possibility to achieve inverted population of the emitting Sm^{3+} level under strong pumping. Farries et al. proved that lasing can be achieved by means of Sm^{3+} ions pumping [14]. We increased the energy of exciting light from 0.7 to 50 mW to estimate the contribution of coherent light and plasmons, which can be induced in $TiO_2:Sm^{3+}$-Au sample. The narrowing of Sm^{3+} spectral peaks was very small and can not be identified with coherent emission. Nevertheless the growth of fluorescence vs. excitation power demonstrates slight super-linearity (Figure 5).

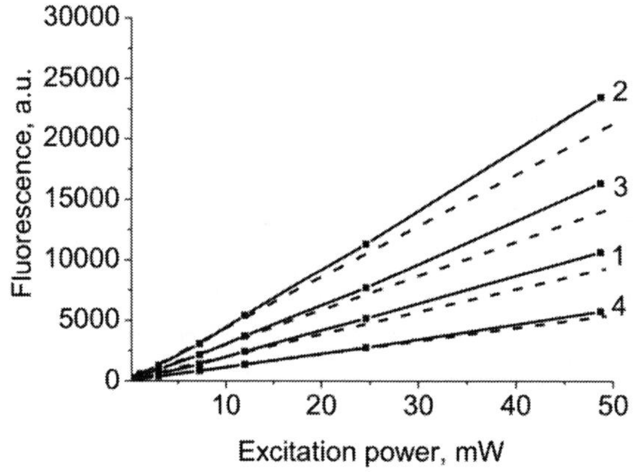

Figure 5. Intensity of fluorescence vs. excitation power for the $TiO_2:Sm^{3+}$-Au structure. Data for different spectral peaks of Sm^{3+} are plotted: (1) 583 nm; (2) 616 nm; (3) 665 nm; (4) 728 nm. λ_{exc}=363.8 nm, detection angle θ=75°. Dashed lines are corresponded to the linear dependence observed at low pumping powers.

Plotted graph consists of two linear parts having different slope to the horizontal axis. One can see that fluorescence grows more rapidly at the exciting energies higher than 12 mW (Figure 5, solid line) compared to weak excitations (Figure 5, dashed line). Deviation from initial growth is stronger for spectral peaks satisfying the plasmon resonance angle. Particularly fluorescence intensity deviates ~10% from the initial linear dependence (dashed line) for the spectral peaks at 616 nm and 665 nm detected near resonance angle 75°. Thus it testifies that about 10% of the pumping energy can be used for generation of the coherent plasmons.

An indirect proof of such possibility can be obtained also from the angular dependences of fluorescence measured at different excitation powers. Local minimum in the emission intensity detected at θ=55° (Figure 6, curve 3 under dashed line) corresponds to the similar local minimum in Figure 4c situated before the maximum of directed emission (see also Figure 2 in Ref. [13]). The local minimum in Figure 6 becomes more pronounced at higher excitation energies when the contribution of coherent plasmons is expected to grow.

As a next step it can be perspective to induce stimulated emission and coherent plasmons in the $TiO_2:Sm^{3+}$-Au samples more efficiently using additional feedback system and pulsed laser excitation.

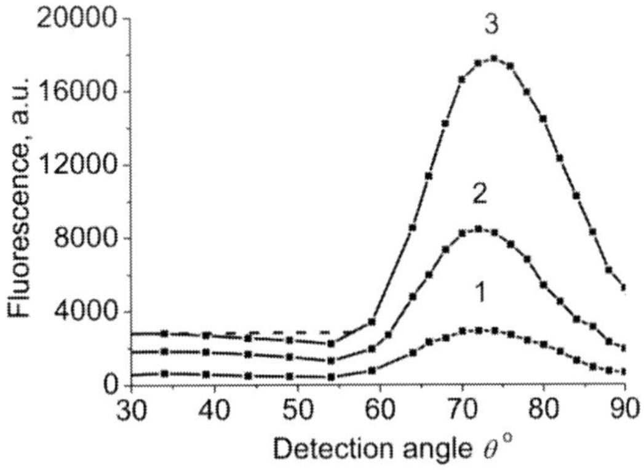

Figure 6. Angular dependence of fluorescence for the $TiO_2:Sm^{3+}$-Au structure measured for spectral peak situated near 616 nm in case of different intensities of exciting light (λ_{exc}=363.8 nm): (1) 7.24 mW; (2) 24.6 mW; (3) 48.7 mW.

4. Conclusions

Fluorescence of Sm^{3+} ions incorporated into TiO_2-Au nano-coatings was detected through the gold layer. Maxima in the angular dependence of rare earth emission are corresponded to its coupling with surface plasmon waves in the gold. Different peaks of Sm^{3+} fluorescence can be plasmonically supported at different detection angles. Thus adjustment of detection angle allowed separation of certain Sm^{3+} spectral peak from the others. Surface plasmon waves are induced mainly by fluorescent light propagating from $TiO_2:Sm^{3+}$ film through the gold at appropriate angles. At the same time about 10 % growth of fluorescence intensity vs. excitation detected at high excitation powers can testify about partially non-linear and non-radiative energy transfer to the coherent plasmons.

Acknowledgements
This work was supported by ERDF Centre of Excellence TK114 "Mesosystems: Theory and Applications"and partially by Nanotwinning project FP7-INCO-2011-6.

References
[1] Gryczynski Z, Matveeva E G, Calander N, Zhang J, Lakowicz J R, Gryczynski I 2007 Surface plasmon coupled emission *Surface Plasmon Nanophotonics (Springer Series in Optical Sciences* Vol 131*)* ed M L Brongersma and P G Kik (Springer) chapter 17 pp 247-265
[2] Stefani F D, Vasilev K, Bocchio N, Stoyanova N and Kreiter M 2005 *Phys. Rev. Lett.* **94** 023005
[3] Hoff M, Sars V, Oheim M 2008 *Opt. Express* **16(22)** 18495
[4] Zhang J, Fu Yi, Lakowicz J R 2009 *J. Phys. Chem.* **113** 19404-10
[5] Dolgov L, Reedo V, Kiisk V, Pikker S, Sildos I, Kikas J 2010 *Opt. Mater.* **32** 1540-44

[6] Kiisk V, Reedo V, Karbowiak M, Brik M G, Sildos I 2009 *J. Phys. D: Appl. Phys.* **42** 125107
[7] Kiisk V, Šavel M, Reedo V, Lukner A, Sildos I 2009 *Phys. Procedia* **2** 527–538
[8] Homola J 2006 Electromagnetic theory of surface plasmons *Surface Plasmon Resonance Based Sensors (Springer Series of Chemical Sensors and Biosensors* Vol 4*)* ed J Homola (Berlin Heidelberg: Springer-Verlag) pp 3-44
[9] Dolgov L, Kiisk V, Matt R, Pikker S, Sildos I 2012 *Appl. Phys. B* DOI 10.1007/s00340-012-5025-6
[10] Lakowicz J R 2005 *Anal. Biochem.* **337** 171-194
[11] Stockman M I 2008 *Nat. Photonics* **2** 327-329
[12] Noginov M A, Zhu G, Mayy M, Ritzo B A, Noginova N, Podolskiy V A 2008 *Phys. Rev. Lett.* **101** 226806
[13] Seidel J, Grafström S, Eng L 2005 *Phys. Rev. Lett.* **94** 177401
[14] Farries M C, Morkel P R, Townsend J E 1988 *Electron Lett* **24(11)** 709-710

IOP Conf. Series: Materials Science and Engineering **38** (2012) 012007 doi:10.1088/1757-899X/38/1/012007

Low-energy charge transfer excitations in NiO

V I Sokolov[1], V A Pustovarov[2], V N Churmanov[2], V Yu Ivanov[2], A Ye Yermakov[1], M A Uimin[1], N B Gruzdev[1], P S Sokolov[3], A N Baranov[3] and A S Moskvin[4]

[1]Institute of Metal Physics UD RAS, S. Kovalevskaya Str. 18, 620990, Ekaterinburg, Russia
[2]Ural Federal University, Mira Str., 19, 620002 Ekaterinburg, Russia
[3]Lomonosov Moscow State University, 119991, Moscow, Russia
[4]Department of Theoretical Physics, Institute of Natural Science, Ural Federal University, Lenin Str., 51, 620083 Ekaterinburg, Russia

E-mail: `alexandr.moskvin@usu.ru`

Abstract. Comparative analysis of photoluminescence (PL) and photoluminescence excitation (PLE) spectra of NiO poly- and nanocrystals in the spectral range 2-5.5 eV reveals two PLE bands peaked near 3.7 and 4.6 eV with a dramatic rise in the low-temperature PLE spectral weight of the 3.7 eV PLE band in the nanocrystalline NiO as compared with its polycrystalline counterpart. In frames of a cluster model approach we assign the 3.7 eV PLE band to the low-energy bulk-forbidden *p-d* ($t_{1g}(\pi)$-e_g) charge transfer (CT) transition which becomes the allowed one in the nanocrystalline state while the 4.6 eV PLE band is related to a bulk allowed *d-d* (e_g-e_g) CT transition scarcely susceptible to the nanocrystallization. The PLE spectroscopy of the nanocrystalline materials appears to be a novel informative technique for inspection of different CT transitions.

1. Introduction

Despite several decades of studies there is still no literature consensus on the detailed electronic structure of the prototype 3d oxide NiO, in particular, the character of the low-energy charge transfer excitations. NiO has long been viewed as a prototype "Mott insulator" [1] with a gap formed by intersite *d-d* charge transfer (CT) transitions, however, this view was later replaced by that of a "CT insulator" [2] with gap formed by *p-d* CT transitions. At present we have no comprehensive assignment of intensive spectral features in NiO to different *p-d* or *d-d* CT transitions. The main reason for this is that the absorption coefficient rises steeply above 3.5 eV and reaches a giant value of $10^6\,cm^{-1}$ at and above 4 eV [3] that complicates the detailed absorption and reflection measurements particularly near the fundamental edge. This difficulty can be avoided in part by making use of specific techniques such as electroreflectance measurements [4] or nonlinear absorption measurements [5] that are particularly effective for location of the forbidden transitions.

Main goal of our paper is to show with NiO as an example that the photoluminescence excitation (PLE) spectroscopy of the nanocrystalline materials appears to be a novel informative technique for inspection of different charge transfer transitions. Indeed, on the one hand, nanocrystalline state gives rise to a noticeable modification of the optical response of materials, in particular, due to an enhanced role of the surface induced local structural distortions resulting in a shift and splitting of the energy levels, and, mainly, in a liberalization of the selection rules. It means that optical response of nanocrystalline materials can reveal electron-hole excitations unresolved in bulk spectra. On the other hand, the photoluminescence (PL) spectroscopy can give a profound insight into the electronic structure, electron-hole excitations and their relaxation in the lattice. As regards nanocrystalline NiO, its optical properties both in absorption and emission have been scarcely investigated. To the best of our knowledge the photoluminescence studies of nanocrystalline NiO are restricted by just two papers [6, 7]. Volkov *et al.* [6] have performed the steady-state absorption and time-resolved luminescence measurements for nanoclustered NiO samples with the particle size of 0.6 nm. The authors of Ref. [7] have presented certain controversial results of absorption and photoluminescence measurements for NiO nanocrystals with the particle size of 40 nm. However, in the both researches there is no inspection of the PL photoexcitation over the CT band, though such a study together with the comparative analysis of the poly- and nanocrystalline NiO PL and PLE spectra can be an instructive tool to elucidate the mechanism of the CT transitions and the spectral selectivity of the photoluminescence. These issues did motivate our studies of the PL and PLE spectra in nanocrystalline NiO samples as compared with polycrystalline samples.

2. Experimental results

In order to make clear the effect of nanocrystallization the PL measurements were made both on conventional polycrystalline and nanocrystalline samples of NiO. As a starting material for polycrystalline NiO we have used the commercially available powder of NiO (99%; Prolabo) which has been pressed into pellets under pressure of about 0.125 GPa and placed into gold capsules. Quenching experiments at 7.7 GPa and 1000-1100 K have been performed using a toroid-type high-pressure apparatus. Detailes of experimental technique and calibration are described elsewhere [8]. Electron microscopy analysis shows the samples to be dense poreless oxide ceramics with rock salt cubic structure and grain size of about 10-20 μm. Nanocrystalline samples of NiO with a particle size of 25 nm were produced by the gas-condensation synthesis method [9] and annealed in air at a temperature of 350° C for an 24 hour. The samples for our measurements were produced by tablet pressing at a pressure of 0.5 GPa. The PL and PLE spectra were measured in the range (2-5.5) eV using two double DMP-4 prism monochromators (a reciprocal linear dispersion of 10 $\AA$/mm near 5 eV), an R6358-10 (Hamamatsu) photomultiplier tube, and a photon-counting system [10, 11]. A deuterium DDS-400 lamp was used for excitation. The PLE spectra were normalized to the same number of photons incident on the sample using yellow phosphor, which has an energy-independent PL quantum yield in the energy range under investigation. The PL spectra are presented without any normalization to the spectral sensitivity of the optical section. Typical PL spectra of poly- and nanocrystalline NiO samples measured at room temperature and low temperature T = 90 K under excitation energies suggestive the CT band excitation ($E_{exc} > E_g$) are presented in upper and lower panels of Fig. 1, respectively. Such a broad luminescence band in the region 2-3.5 eV was observed in earlier PL measurements performed with UV excitation below and near optical gap [12, 6, 13]. The low-energy excitation ($E_{exc} \approx 2.4$ eV) below the charge-transfer gap stimulates a photoemission in single-crystal NiO with two maxima, one at 1.5-1.6 eV and a larger one at 2.2-2.3 eV [14]. Green luminescence band with a maximum around 2.25 eV has been observed in nanoclustered NiO at $E_{exc} \leq 2.95$ eV [6]. The observed emission bands in the visible and near infrared spectral ranges are usually attributed to Ni^{2+} intrasite, or crystal field *d-d* transitions. In particular, the main green luminescence band peaked near 2.3 eV is attributed to a Stokes-shifted ${}^1T_{2g}(D) \rightarrow {}^3A_{2g}(F)$ transition while the low-energy band peaked near 1.5 eV is related to ${}^1E_g(D) \rightarrow {}^3A_{2g}(F)$ transition [14] (see Ref. [3] for the spectrum of the crystal field *d-d* transitions). However, the photoluminescence spectra of bulk single crystals and ceramics of NiO under 3.81 eV photoexcitation [13] suggestive the band gap excitation have revealed in addition to the green band a more intensive broad violet PL band with a maximum around

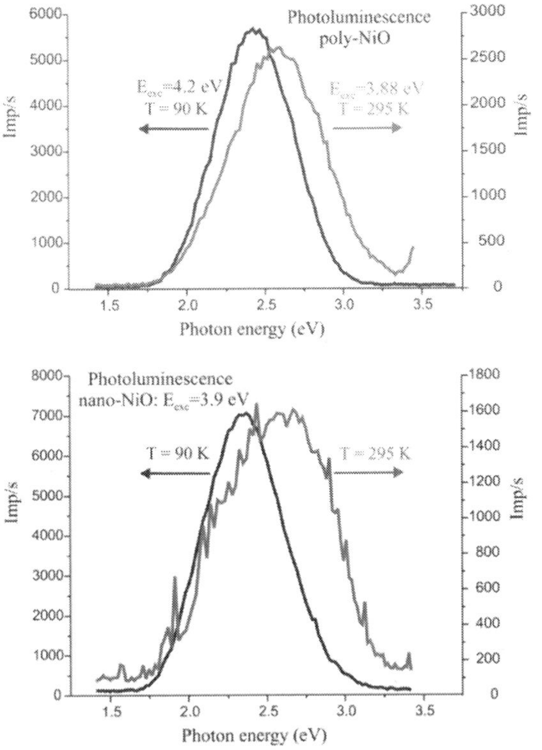

Figure 1. (Color online) Photoluminescence spectra of polycrystalline (upper panel) and nanocrystalline (lower panel) NiO samples at two temperatures and $E_{exc} > E_g$.

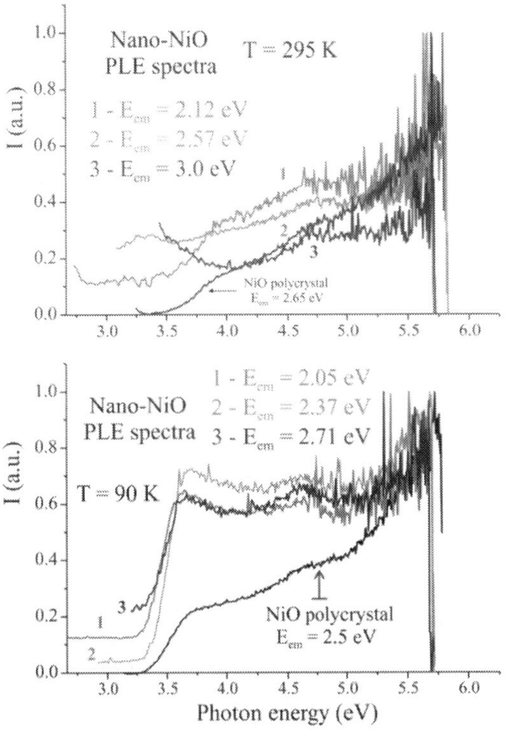

Figure 2. (Color online) The PLE spectra of nanocrystalline NiO at two temperatures (T = 295 K, upper panel, T = 90 K, lower panel) and different emission energies. For comparison in the both panels shown are the PLE spectra of polycrystalline NiO at corresponding temperatures.

3 eV. The band was related to a *p*-*d* charge transfer. Radiative recombination of carriers in powdered pellets of NiO under UV excitation with E_{exc} = 4.43 eV (280 nm) higher than the CT gap consists at 10 K of a broad intensive band peaked at 2.8 eV with a shoulder centered at about 3.2 eV [12]. At variance with Ref. [13] the PL emission bands observed in the work [12] were attributed to the crystal field *d*-*d* transitions. Fig. 1 clearly shows the blue shift of the PL band with increasing the temperature both for polycrystalline and nanocrystalline samples of NiO. Such a behavior was noticed in earlier measurements for polycrystalline samples of NiO [12]. Furthermore, the detailed temperature measurements [12] indicate that the broad PL spectra of NiO consist of several components with relative intensity dependent on the temperature and excitation energy.

Different kinetic properties, seemingly different temperature behavior [12] point to different relaxation channels governing the green and violet luminescence. Despite the PL spectra for nanocrystalline and polycrystalline NiO have similar lineshape, it is worth noting a clear red shift of the low-temperature PL spectra and a visible assymetry of the high-temperature PL spectra for nano-NiO.

Main results of our paper are related with the PLE spectral measurements in range 2-5.5 eV presented in Fig. 2. The polycrystalline NiO sample reveals a rather structureless PL excitation spectrum with a steep spectral weight rise from 3.3 towards 6 eV and scarcely marked weak features near 3.7 and 4.6 eV. The low-energy feature becomes more pronounced at low temperatures T = 90 K. The room temperature PLE spectra for nanocrystalline NiO corroborates the presence of at least the 4.6 eV feature while the obscure lower energy spectral behavior strongly depends on the emission energy. Just another lower energy spectral behavior in PLE spectra for nanocrystalline NiO is observed at low temperature T = 90 K (see lower panel in Fig. 2). Here we see a dramatic rise in the PLE spectral weight with a peak at excitation energy $E_{exc} \approx 3.7$ eV that points to the energy $E_g^{PLE} \approx 3.5$ eV as a specific PLE gap for the both poly- and nanocrystalline NiO. Interestingly that the spectral weight and position of the 4.6 eV PLE feature seem to weakly alter in between poly- and nanocrystalline NiO samples. Thus, we are drawn to the conclusion that the 3.7 eV band can be most likely related with the bulk-forbidden CT transition which becomes the allowed one in the nanocrystalline state while the 4.6 eV PLE band is believed to be related with a bulk allowed CT transition scarcely susceptible to the nanocrystallization. Such an attribution is supported by several arguments:

1. Electroreflectance spectra for bulk NiO [4] which detect the dipole-forbidden CT transitions clearly point to a forbidden transition peaked near 3.7 eV missed in reflectance and absorption spectra [3, 15, 16].

2. At variance with the bulk NiO a clearly visible intensive peak near 3.6-3.7 eV has been observed in absorption spectra of NiO nanoparticles [6, 7, 17]. Its intensity points to the CT origin.

3. Peak near 3.8 eV has been observed in nonlinear absorption spectra of NiO [5].

4. Photoconductivity threshold in bulk NiO has been observed at 3.7 eV [18].

5. A rather strong CT band peaked at 4.5 eV is clearly visible in the absorption of thin NiO films [19], RIXS spectra [20, 21], the reflectance spectra (4.3 eV) [15], the nonlinear absorption spectra [5]. All this points to dipole-allowed CT transition.

Below we will address the microscopic origin of the CT transitions defining the PLE bands at 3.7 and 4.6 eV.

3. Discussion

Despite the nature of radiative and nonradiative transitions in strongly correlated $3d$ oxides is far from full understanding some reliable semiquantitative predictions can be made in frames of a simple cluster approach (see, e.g. Refs. [22] and references therein). The method provides a clear physical picture of the complex electronic structure and the energy spectrum, as well as the possibility of a quantitative modelling. In a certain sense the cluster calculations might provide a better description of the overall electronic structure of insulating 3d oxides than the band structure calculations, mainly due to a better account for correlation effects. Starting with octahedral NiO_6 complex with the point symmetry group O_h we deal with five Ni 3d and eighteen oxygen O 2p atomic orbitals forming both hybrid Ni 3d-O 2p bonding and antibonding e_g and t_{2g} molecular orbitals (MO), and purely oxygen nonbonding $a_{1g}(\sigma)$, $t_{1g}(\pi)$, $t_{1u}(\sigma)$, $t_{1u}(\pi)$, $t_{2u}(\pi)$ orbitals. Typically for the octahedral MeO_6 clusters the non-bonding $t_{1g}(\pi)$ oxygen orbital has the highest energy and forms the first electron removal oxygen state while other nonbonding oxygen π-orbitals, $t_{2u}(\pi)$, $t_{1u}(\pi)$, and σ-orbital $t_{1u}(\sigma)$ have lower energy with the energy separation ~ 1 eV in between[22].

The p-d CT transitions in NiO_6^{10-} center are related with the transfer of the O 2p electron to the partially filled 3de_g-subshell with formation on the Ni-site of the $(t_{2g}^6 e_g^3)$ configuration of nominal Ni^+ ion isoelectronic to the well-known Jahn-Teller Cu^{2+} ion. Yet actually instead of a single p-d CT transition we arrive at a series of O 2p$\gamma \rightarrow$ Ni 3de_g CT transitions forming a complex p-d CT band. It should be noted that each single electron transition gives rise to two many-electron transitions. The band starts with the dipole-forbidden $t_{1g}(\pi) \rightarrow e_g$, or ${}^3A_{2g} \rightarrow {}^3T_{1g}, {}^3T_{2g}$ transitions, then includes two formally dipole-allowed so-called $\pi \rightarrow \sigma$ p-d CT transitions, weak $t_{2u}(\pi) \rightarrow e_g$, and relatively strong $t_{1u}(\pi) \rightarrow e_g$ CT transitions, respectively, each giving rise to $^3A_{2g} \rightarrow {}^3T_{1u}, {}^3T_{2u}$ transitions. Finally main p-d CT band is ended by the strongest dipole-allowed $\sigma \rightarrow \sigma$ $t_{1u}(\sigma) \rightarrow e_g$ ($^3A_{2g} \rightarrow {}^3T_{1u}, {}^3T_{2u}$) CT transition. Above estimates predict the separation between partial p-d bands to be ~ 1 eV. Thus, if the most intensive CT band with a maximum around 7 eV observed in RIXS spectra [20, 21] to attribute to the strongest dipole-allowed O 2p$t_{1u}(\sigma) \rightarrow$ Ni 3de_g CT transition then one should expect the low-energy p-d CT counterparts with maximuma around 4, 5, and 6 eV respectively, which are related to dipole-forbidden $t_{1g}(\pi) \rightarrow e_g$, weak dipole-allowed $t_{2u}(\pi) \rightarrow e_g$, and relatively strong dipole-allowed $t_{1u}(\pi) \rightarrow e_g$ CT transitions, respectively. It is worth noting that the $\pi \rightarrow \sigma$ p-d CT $t_{1u}(\pi) - e_g$ transition borrows a portion of intensity from the strongest dipole-allowed $\sigma \rightarrow \sigma$ $t_{1u}(\sigma) - e_g$ CT transition because the $t_{1u}(\pi)$ and $t_{1u}(\sigma)$ states of the same symmetry are partly hybridized due to p-p covalency and overlap. Interestingly that this assignement finds a strong support in the reflectance (4.9, 6.1, and 7.2 eV for allowed p-d CT transitions) [15] and, particularly, in electroreflectance spectra [15, 4] which detect dipole-forbidden transitions. Indeed, the spectra clearly point to a forbidden transition peaked near 3.7 eV (missed in reflectance spectra) which thus defines a p-d character of the optical CT gap and can be related with the on-set transition for the whole complex p-d CT band. It should be noted that a peak near 3.8 eV has been observed in nonlinear absorption spectra of NiO [5]. A rather strong $p(\pi)$-d CT band peaked at 6.3 eV is clearly visible in the absorption spectra of MgO:Ni [16].

As a result of the p-d CT transition, a photo-generated electron localizes on a Ni^{2+} ion forming Jahn-Teller 3d^9, or Ni^{1+} configuration, while a photo-generated hole can move more or less itinerantly in the O 2p valence band determining the hole-like photoconductivity [18]. The most effective channel of the recombinational relaxation for the spin-triplet p-d CT states $t_{2g}^6e_g^3\underline{\gamma}(\pi)$ ($\gamma(\pi) = t_{1g}(\pi), t_{2u}(\pi), t_{1u}(\pi)$) implies the $\pi \rightarrow \pi$ transfer $t_{2g} \rightarrow \gamma(\pi)$ with formation of excited spin-triplet $^3T_{1g}$ or $^3T_{2g}$ states of the $t_{2g}^5e_g^3$ configuration of Ni^{2+} ion followed by final relaxation to lowest singlet terms $^1T_{2g}$ and ^1E_g producing green and red luminescence, respectively. Obviously, this relaxation is strongly enhanced by any symmetry breaking effects lifting or weakening the selection rules. It means the three $\pi \rightarrow \sigma$ p-d CT transitions $t_{1g}(\pi) \rightarrow e_g$, $t_{2u}(\pi) \rightarrow e_g$, and $t_{1u}(\pi) \rightarrow e_g$ are expected to effectively stimulate the green luminescence due to large effective mass of π-oxygen holes and a well-localized and long-lived character of the p-d excitation. On the other hand, the most intensive $\sigma \rightarrow \sigma$ p-d CT transition $t_{1g}(\sigma) \rightarrow e_g$ appears to be significantly less effective due to small effective mass of the σ-oxygen holes and relatively short lifetime of the respective unstable p-d CT state.

Along with the p-d CT transitions an important contribution to the optical response of strongly correlated 3d oxides can be related with strong d-d CT, or Mott transitions [22]. In NiO one expects a strong d-d CT transition related with the $\sigma - \sigma$-type $e_g - e_g$ charge transfer $t_{2g}^6e_g^2 + t_{2g}^6e_g^2 \rightarrow t_{2g}^6e_g^3 + t_{2g}^6e_g^1$ between nnn Ni sites with the creation of electron $[NiO_6]^{11-}$ and hole $[NiO_6]^{9-}$ centers (electron-hole dimer), or nominally Ni^+ and Ni^{3+} ions. This unique anti-Jahn-Teller transition $^3A_{2g} + {}^3A_{2g} \rightarrow {}^2E_g + {}^2E_g$ creates a d-d CT exciton self-trapped in the lattice due to electron-hole attraction and strong "double" Jahn-Teller effect for the both electron and hole centers.

Strong dipole-allowed Franck-Condon $d(e_g)$-$d(e_g)$ CT transition in NiO manifests itself as a strong spectral feature near 4.5 eV clearly visible in the absorption of thin NiO films [19], the RIXS spectra [20, 21], the reflectance spectra (4.3 eV) [15], the nonlinear absorption spectra (4.1-4.3-4.5 eV) [5]. Such a strong absorption near 4 eV is beyond the predictions of the p-d CT model and indeed is lacking in absorption spectra of MgO:Ni [16]. Accordingly, the CT gap in NiO is believed to be formed by a superposition of the electro-dipole forbidden p-d ($t_{1g}(\pi) \rightarrow e_g$) and allowed $d(e_g)$-$d(e_g)$ CT transitions with close energies that is NiO is close to the intermediate regime of the Zaanen-Sawatzky-Allen scheme. Thus we see that simple cluster model unambiguously points to the low-energy bulk-forbidden p-d ($t_{1g}(\pi) \rightarrow e_g$) and allowed $d(e_g)$-$d(e_g)$ CT transitions to be transitions responsible for the 3.7 and 4.6 eV bands revealed by the PLE spectroscopy of nanocrystalline NiO.

4. Conclusion

In summary, we have measured the photoluminescence and photoluminescence excitation spectra of NiO poly- and nanocrystals in the spectral range 2-5.5 eV. The PL spectra for both types of samples have a very similar spectral shape with broad emission bands centered at about 2.5 eV. The PLE spectra reveal two PLE bands peaked near 3.7 and 4.6 eV with a dramatic rise in the low-temperature PLE spectral weight of the 3.7 eV PLE band in the nanocrystalline NiO as compared with its polycrystalline counterpart. We have performed a cluster model analysis of the CT transitions and assigned the 3.7 eV PLE band to the low-energy bulk-forbidden p-d ($t_{1g}(\pi)$-e_g) CT transition which becomes the allowed one in the nanocrystalline state while the 4.6 eV PLE band is related to the bulk allowed d-d (e_g-e_g) CT transition scarcely susceptible to the nanocrystallization. For the first time we succeded in distinct separating the contributions of the low-energy $p(\pi)$-d and d-d CT transitions to the fundamental absorption of NiO and show that NiO is close to the intermediate regime of the Zaanen-Sawatzky-Allen scheme. Thus, the PLE spectroscopy of the nanocrystalline materials is shown to be a novel informative technique for inspection of different CT transitions.

Acknowledgments

The authors are grateful to R.V. Pisarev for stimulating discussions. This work was partially supported by the Ural Branch RAS (Grant No. 12-U-2-1030) and RFBR (Grants No. 10-02-96032 and 12-02-01039).

References

[1] Brandow B 1977 *Adv. Phys.* **26** 651; 2008 *J. Phys.: Conf. Ser.* **108** 012009
[2] Zaanen J, Sawatzky G A and Allen J W 1985 *Phys. Rev. Lett.* **55** 418
[3] Newman R and Chrenko R M 1959 *Phys. Rev.* **114** 1507
[4] McNatt J L 1969 *Phys. Rev. Lett.* **23** 915; Glosser R. and Walker W C 1971 *Solid State Commun.* **9** 1599
[5] Shablaev S I and Pisarev R V 2003 *Physics of the Solid State* **45** 1742
[6] Volkov V V, Wang Z L and Zou B S 2001 *Chemical Phys. Lett.* **337** 117
[7] Chakrabarty S and Chatterjee K 2009 *Journal of Physical Sciences* **13** 245
[8] Baranov A N, Sokolov P S, Kurakevich O O, Tafeenko V A, Trots D. and Solozhenko V L 2008 *High Pressure Research* **28** 515
[9] Sokolov V I, Yermakov A E, Uimin M A, Mysik A A, Vykhodets V B, Kurennykh T E, Gaviko V S, Shchegoleva N N and Gruzdev N B 2007 *JETP* **105** 65
[10] Sokolov V I, Yermakov A E, Uimin M A, Mysik A A, Pustovarov V A, Chukichev M V and Gruzdev N B 2009 *J. Lumin.* **129** 1771
[11] Gruzdev N B, Sokolov V I, Yermakov A E, Uimin M A, Mysik A A and Pustovarov V A 2010 *JETP* **111** 231
[12] Díaz-Guerra C, Remón A, García J A and Piqueras J 1997 *Phys. Stat. Solidi (a)* **163** 497
[13] Mochizuki S and Saito T 2009 *Physica B* **404** 4850
[14] Kuzmin A, Mironova-Ulmane N and Rochin S 2003 *Proceedings SPIE* **5122** 61
[15] Powell R J and Spicer W E 1970 *Phys. Rev. B* **2** 2182
[16] Blazey K W 1977 *Physica B* **89** 47
[17] Absorbance spectrum in Ref. [7] distinctly points to an absorption peak near 3.7 eV, however, with overall view of the spectrum to be at odds with literature data
[18] Ksendzov Ja M and Drabkin I A 1965 *Sov. Phys. Solid State* **7** 1519
[19] Rossi C E and Paul W 1969 *J. Phys. Chem. Solids* **30** 2295
[20] Duda L-C, Schmitt T, Magnuson M, Forsberg J, Olsson A, Nordgren J, Okada K and Kotani A 2006 *Phys. Rev. Lett.* **96** 067402
[21] Hiraoka N, Okamura H, Ishii H, Jarrige I, Tsuei K D and Cai Y Q 2009 *Eur. Phys. J. B* **70** 157
[22] Moskvin A S 2002 *Phys. Rev. B* **65** 205113; *Optics and Spectroscopy* **111** 403;Pisarev R V, Moskvin A S, Kalashnikova A M and Rasing Th 2009 *Phys. Rev. B* **79** 235128;Moskvin A S and Pisarev R V 2010 *Low Temp. Phys.* **36** 613

IOP Conf. Series: Materials Science and Engineering **38** (2012) 012008 doi:10.1088/1757-899X/38/1/012008

Vibrational spectra of opal-based photonic crystals

G Dovbeshko[1,6], O Fesenko[1], V Boyko[1], V Romanyuk[2], V Moiseyenko[3], V Gorelik[4], L Dolgov[5], V Kiisk[5], I Sildos[5]
[1]Institute of Physics of NASU, Kyiv, Ukraine
[2]Institute for Physics of Semiconductors of NASU, Kyiv, Ukraine
[3]Dnipropetrovsk National University, Dnipropetrovsk, Ukraine
[4]P.N. Lebedev Physics Institute of RAS, Moscow, Russian Federation
[5]Institute of Physics, University of Tartu, Tartu, Estonia

Abstract. Synthetic silica opals were investigated by infrared and Raman spectroscopies. Vibrational modes associated with molecular groups of opal globules and admixtures were detected. Similarities in Raman and infrared spectra of synthetic opal with reference fused and α-quartz indicate the presence of amorphous phase in opal globules. Also some spectral bands designate on modified optical stretching vibrations at 1000-1200 cm^{-1} and bigger amount of Si–H defects in photonic crystal.

1. Introduction

Active search for new materials with controllable optical properties has resulted in the invention of new class of optical materials called photonic crystals. These materials consist of periodically arranged dielectric globules or cavities with sizes comparable to the wavelength of light. Therefore photonic crystals mediate between nano- and micro-structured composite materials [1]. Light interference on the mentioned globular superlattice results in forbidden photonic band gaps. Fabrication of photonic crystals with variable position of band gaps is one of the topical experimental tasks nowadays.

Natural photonic crystal – natural opal is named often opaline silica [2]. However, it could contain different admixtures, variety of SiO_2 phases and shape of the globules is far from regular sphere [3-5]. Growing of synthetic opal-like photonic structures is still not a simple task [1,6], therefore natural opals brought from different regions were investigated more frequently than synthetic ones [3-5].

The authors of [7] have pointed out that vibrational spectra of natural opals in terms of a simple superimposition of bands of cristobalite and tridymite phases is irrelevant to a great extent. In this paper we focus our efforts on the comparison studies of synthetic opal vibrational spectra with ones for natural opal, fused silica and α-quartz. Infrared and Raman scattering spectroscopies allow identify the framework of silicates in the globules via the position of the most intense signals arising from bridging oxygen breathing mode of the SiO_4 tetrahedrons organized in rings. Cristobalite, tridymite and amorphous phases of silica can be differently combined in globules and finally will result different

[6] G.Dovbeshko e-mail: matinelli@gmail.com, gd@iop.kiev.ua

optical behaviour of whole photonic structure. Therefore investigation of vibrational spectra for prepared synthetic opals in comparison with fused and crystalline quartz as well natural opal offers necessary.

2. Materials and methods

Nanodisperse silica globules were synthesized by the Stöber method. Growth of the photonic crystals was achieved by subsequent natural sedimentation. Preparation conditions were optimized for producing well-packed monodispersed 230-275 nm silica globules [8].The rests of organic components were removed from the samples by means of annealing in air at 600 °C. As a result different samples built from the 230-275 nm silica globules were obtained. The thickness of the samples was varied from 0.7 to 2.3 mm.

The infrared reflectance spectra from synthetic opal were registered by means of commercial Fourier spectrometer Bruker IFS-66 at the range from 4000 to 400 cm^{-1} with maximum resolution 0.2 cm^{-1}. Reflectance measurements were performed with IPO-22 reflectance unit (angle of light incidence 16.5°). Program Opus 5.5, which is a part of Bruker IFS-66 software, was used for registration and processing of the infrared spectra. Position of spectral bands and their intensity were determined with accuracy 0.05 cm^{-1} and ±0.0005 a.u. respectively.

Raman spectra were measured with Renishaw inVia micro-Raman spectrometer. Exciting (λ=514 nm) and scattering beams were propagated through the objective of the microscope. During the Raman measurements objective 20× was used, because it allows recognize the borders of domains in the samples and provide better signal to noise ratio than the 5× objective. The diameter of laser spot on the sample surface was about 5 μm. Laser power applied does not exceed 20 mW. All measurements were done at room temperature. We assume that testing laser beam does not cause essential heating of the samples.

X-ray diffraction was measured with Bruker 8D diffractometer in θ–2θ regime from 5° to 80° with Al Kα (1486.6 eV) source. Figure 1 indicates to the amorphous state of synthetic opals under study.

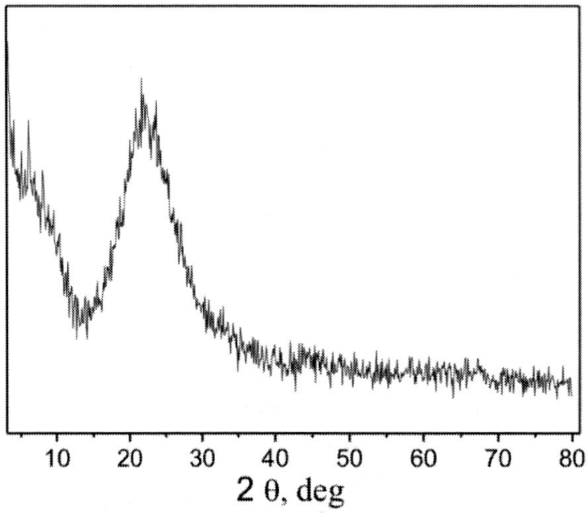

Figure 1. XRD pattern of synthetic opals.

3. Results and discussion

We have compared the vibrational spectra of synthetic silica opals with spectra got from quartz in crystalline and fused form. It is known that structure unit of any quartz is SiO_4, so the infrared spectra are associated with vibrations in SiO_x, which change dipole moments of molecules, whereas Raman spectra refer to the vibrations with changed polarizability. Therefore infrared and Raman spectra are complementary to each other. All defects and admixtures should give a contribution to both spectra also.

3.1. Raman scattering from synthetic silica opals

Periodic arrangement of the opal globules provides specific selective light reflection having interference nature and signals in the visible region for photonic crystals under study. The optical image of opal surface looks like a set of differently coloured domains (Figure 2a). This effect is caused by disorientation of domains in the sample. Size of the domains is in the range 40-250 µm depending on the perfection of the crystals [5].

a b

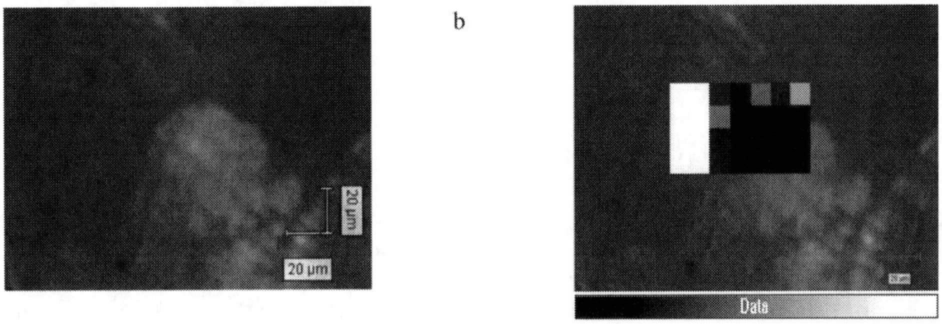

130000 a.u.
370000 a.u.

Figure 2. Optical microscope image of the opal surface (a) and map of Raman scattering intensity at 440 cm^{-1} measured from the mentioned domain (b).

It is necessary to point out that Raman spectra from darker areas of the samples are more intensive than from bright ones (Figure 2a, b). If one take into account the position of Bragg reflection for this photonic crystal (490-520 nm) and applied excitation laser wavelength (λ=514 nm) then one can suppose that larger penetration depth in photonic crystal is necessary for excitation of Raman scattering.

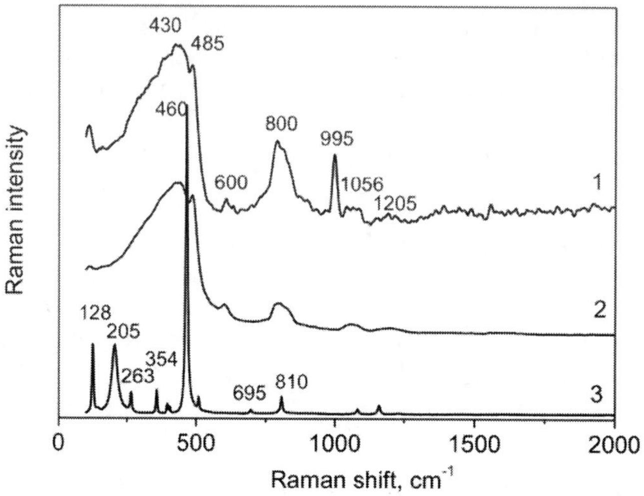

Figure 3. Raman spectra of investigated synthetic opal (1) compared to that of fused quartz (2) and α-quartz (3).

Raman spectrum of silica photonic crystal is determined mainly by SiO_2 vibrations (Figure 3, spectrum 1). The most pronounced band at 430 cm^{-1} is originated from oxygen atom vibrations with identical distortions of neighbouring Si-O bonds [9], i.e. symmetric stretching vibrations [10]. The band at 800 cm^{-1} is also associated with symmetric stretch vibrations of oxygen atoms, but these vibrations involve also substantial amount of surrounding Si atoms. The small peaks at 485 and 600 cm^{-1} are associated with defects in the silica lattice. These defects are due to the formation of 4-membered (4 oxygen atoms in the ring) and 3-membered (3 oxygen atoms in the ring) rings with vibrations at 485 and 600 cm^{-1} respectively [11]. The bands at 1056 and 1205 cm^{-1} (Figure 3, spectrum 1) are similar to those associated with Si-O transverse and longitudinal optical modes [10, 12].

RS spectrum of synthetic opal investigated is similar to the one of amorphous natural opal-A [2] or fused silica. But pronounced band at 995 cm^{-1} can be assigned with vibrations of small admixtures of crystobalite or trydimite phases [3] supported by finite-k vibrations of the corresponding bulk material [6].

3.2. Infrared reflectance of synthetic silica opals

The infrared spectra of opals consist of 2 mains bands: stretching vibration with peak position in the range 1114 cm^{-1} to 1122 cm^{-1} for different samples and deformational vibration at 468-470 cm^{-1} (Figure 4). Stretching vibrations are situated near 1122 cm^{-1} in fused quartz. Intensity of the stretching vibrations in opal is less than in quartz and comparable with intensity of deformational vibrations. A zone of localization for stretching vibration in opal became less than in both forms of quartz also. However, in some points of opals we have registered drastically increased intensity of stretching vibrations up to 60 %. The wide band registered in opal at the range 600-700 cm^{-1} could be assigned to Si-H vibration. Different environment and variable location of the Si-H group with respect to different cites of $Si\text{-}O_x$ network [13] can result in small shift and widening of this band. Data about O-H were presented in our early paper [5]. So, the IR spectrum of opal also suggests its similarity with fused silica rather than α-quartz.

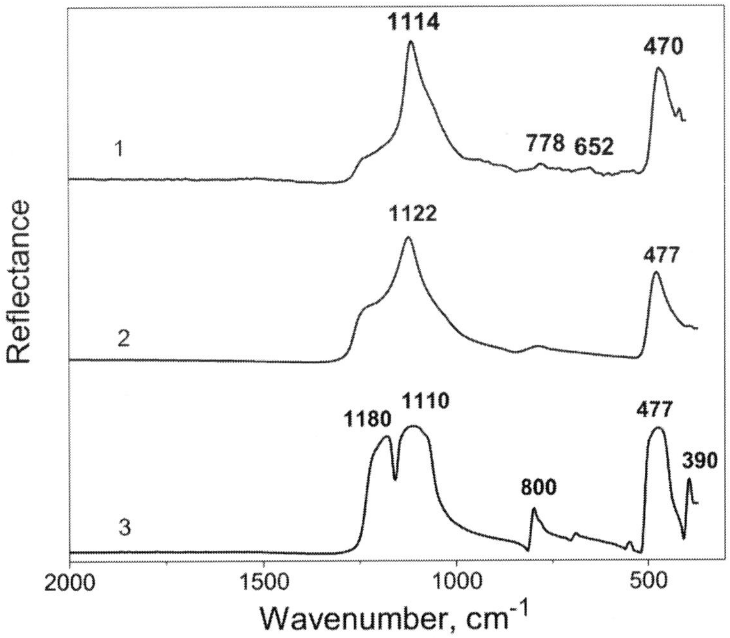

Figure 4. Infrared reflectance spectra of synthetic opal (1), fused quartz (2) and α-quartz (3).

4. Conclusions

Infrared and Raman spectra of the synthetic silica opals can be used for the determination of phase content of silica globules in the photonic crystal. Both spectra suggest that the structure of silica globules is similar to fused quartz in many aspects. Nevertheless, Raman band near the 995 cm^{-1} can be associated with the presence of small areas of globules in the cristobalite phase, which is absent in fused quartz. Infrared reflectance spectrum of photonic crystal differs from the fused quartz also by narrower band corresponding to the stretching vibrations analogous to transversal optical modes at 1000-1200 cm^{-1}. Structure of photonic crystal is characterized by higher amount of defects arising between globules, particularly Si-H defects with infrared absorption in the range 600-700 cm^{-1}.

Acknowledgment. This work was supported by Ukrainian-Russian project N60-02-12, Nanotwinning project N294952 and ERDF Centre of Excellence TK114 "Mesosystems: Theory and Applications".

References

[1] Sakoda K 2005 Optical properties of photonic crystals (*Springer Series in Optical Sciences* Vol 80) ed W. T. Rhodes (Berlin: Springer)
[2] Jones J B, Segnit E R 1971 *J. geol. Soc. Aus.*, **18(1)** 57
[3] Ivanov V, Reyes B, Fritsch E, Faulques E 2011 *J. Phys. Chem. C* **115** 11968
[4] Ilieva A, Mihailova B, Tsintsov Z, Petrov O 2007 Am. Mineralogist **92** 1325
[5] Fritsch E, Gaillou E, Rondeau B, Barreau A, Albertini D, Ostroumov M 2006 J. Non-Cryst. Solids **352** 3957
[6] Marlow F, Muldarisnur, Sharifi P, Brinkmann R, Mendive C 2009 *Angew. Chem. Int. Ed.* **48** 6212
[7] Kingma K J, Hemley R J 1994 *Am. Mineral.*, **79**, 269
[8] Boyko V, Dovbeshko G, Fesenko O, Gorelik V, Moiseenko V, Romanyuk V, Shvets T, Vodolazkyy P 2011 *Mol. Cryst. Liq. Cryst.* **535** 30
[9] Skuja L 1998 *Journal of Non-Crystalline Solids*, **239** 16-48
[10] Galeener F L 1979 *Phys. Rev. B* **19(8)** 4292–4297
[11] Galeener F L, Geissberger A E 1983 *Phys. Rev. B* **27**(10) 6199

[12] Galeener F L, Lucovsky G 1976 *Phys. Rev. Lett.* **37**(22) 1474
[13] Trukhin A N 2000 Excitons, localized states in silicon dioxide and related crystals and glasses *Defects in SiO_2 and Related Dielectrics: Science and Tecnology (NATO Science series II Mathematics, Physics and Chemistry, Vol. 2)* ed G Pacchioni et al. (Kluwer Academic Publishers) pp 235-283

IOP Conf. Series: Materials Science and Engineering **38** (2012) 012009 doi:10.1088/1757-899X/38/1/012009

Processing of amorphous Si by pulsed laser irradiation at different wavelengths

J Butikova[1], **G Marcins, B Polyakov, A Muhins, A Voitkans, I Tale**
Institute of Solid State Physics
8 Kengaraga str., Riga, LV-1063, Latvia

E-mail: but@latnet.lv

Abstract. Amorphous Si thin films deposited on thermally oxidized Si wafers have been processed by the 2nd and 3rd harmonics of Nd:YAG laser. Surface modification of amorphous silicon layers have been investigated by scanning electron microscopy before and after chemical etching of processed silicon films. The super-lateral crystal growth regime was achieved with laser fluence of 280 mJ/cm^2 for the 2nd harmonics and 155 mJ/cm^2 for the 3rd harmonics. The grain size in polycrystalline Si samples prepared by successive crystallization in the lateral growth regime is about 0.5 - 1 μm.

1. Introduction

Polycrystalline silicon (poly-Si) is a promising material for various thin film electronic devices. There have been the attempts of using laser crystallized polycrystalline silicon to increase the efficiency of thin film solar cells [1, 2]. One more field of interest is using laser-treated silicon as an active layer in thin-film transistors for large area liquid crystal displays [3, 4].

Excimer laser crystallization of amorphous silicon (a-Si) is a well established procedure to produce poly-Si [5-7]. The grain size of poly-Si strongly depends on the laser fluence. Large grains up to 4 μm can be fabricated [8].

Three distinct regimes can be distinguished in the process of laser crystallization of a-Si: partial melting regime, near complete melting regime and complete melting regime [9, 10]. In the case of lower laser fluence, the energy exceeds the threshold of Si crystallization, but the solid layer of Si still remains continuous. In a complete melting regime, the Si layer melts forming a small-sized grain structure as a result of the homogeneous nucleation. Between these two regimes, a super lateral growth (SLG) of poly-Si can be achieved. It is considered that non-melted solid clusters of silicon, separated by the completely molten silicon, act as solidification seeds for the laterally growing crystals [11].

In this paper, we investigate laser crystallization of amorphous silicon by the 2nd and 3rd harmonics of the pulsed Nd:YAG laser as an alternative method for the most common technique of obtaining laterally growing Si polycrystals when excimer laser processing is used.

2. Experimental

For crystallization experiments using the 2nd (532 nm) and the 3rd (355 nm) harmonics of the Q-switched Nd:YAG laser, a-Si samples deposited by electron beam evaporation were used, because this

[1] To whom any correspondence should be addressed.

method allows obtaining amorphous silicon films without undesired impurities [12]. The samples were deposited on thermally oxidized wafers. The thicknesses of SiO_2 and a-Si layers were 400 nm and 270 nm, respectively.

The samples were crystallized applying 1 to 3 consecutive pulses of the laser beam by varying laser energy until the a-Si layer was crystallized and super lateral growth regime was achieved. Pulse width was constant – 7.6 ns – in all the cases. Crystallization was performed at room temperature in air and vacuum (no differences found), applying an unfocused laser beam with diameter about 600 μm.

The topography of the samples was characterized by the scanning electron microscope before and after the chemical etching of the samples. The laser-processed samples were etched by Secco etching procedure described elsewhere [13]. The main advantage of this etchant is that it removes amorphous Si areas faster than crystallized ones. Secco solution was diluted with water in proportions 1:3 to remove the non-crystallized Si successfully and, at the same time, to prevent removing the crystallized layer. The etching time was 5 sec for all the samples.

The sizes of the grains of resulting poly-Si were determined on Secco-etched samples.

3. Results and discussion

3.1. Crystallization using 3rd harmonics of Nd:YAG (355 nm) laser beam

During the crystallization of a-Si samples using the 3rd harmonics of the Nd:YAG laser, the SLG regime was already reached with the single pulse processing. With this regime, when the a-Si layer is nearly completely melted, the relatively large poly-Si grains were formed. The size of the grains shown in the Fig. 1b is about 0.5 – 1 μm.

In the Fig. 1a for the non-etched crystallized sample, sphere-like structures can be observed. We believe these structures are of crystalline silicon, and they might be the result of the molten Si developed into spheres under the influence of surface diffusion and surface tension effects.

These sphere-like structures are not present in the etched part of the sample. The reason of their disappearance might be the fact that Secco etchant might flush under the gap between the spheres and the surface, and etch the spheres away slower than the a-Si, but faster than the rest of the crystallized silicon.

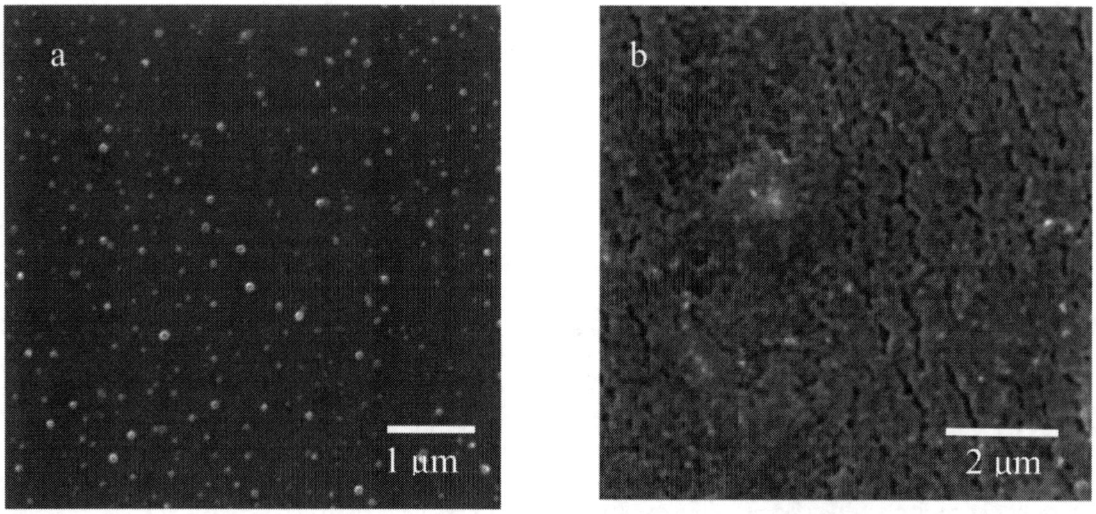

Figure 1. SEM images of the surface of crystallized a-Si before (a) and after (b) the Secco etching. The crystallization was performed by 1 pulse of the 3rd harmonics of Nd:YAG laser (λ = 355 nm) and energy density of 155 mJ/cm^2.

Onto the next sample, two consecutive laser pulses of the same fluence were applied. Fig.2 demonstrates the non-etched and etched processed surface of the sample exposed to twice as much amount of laser energy integrally as the previous one. Now we observe an increased amount of the sphere-like structures in the non-etched region. The diameter of the structures is increased as well. Some of the structures remained at the surface after the etching. In the etched samples, grain boundaries and small grains of about 0.3 – 0.5 μm can be distinguished.

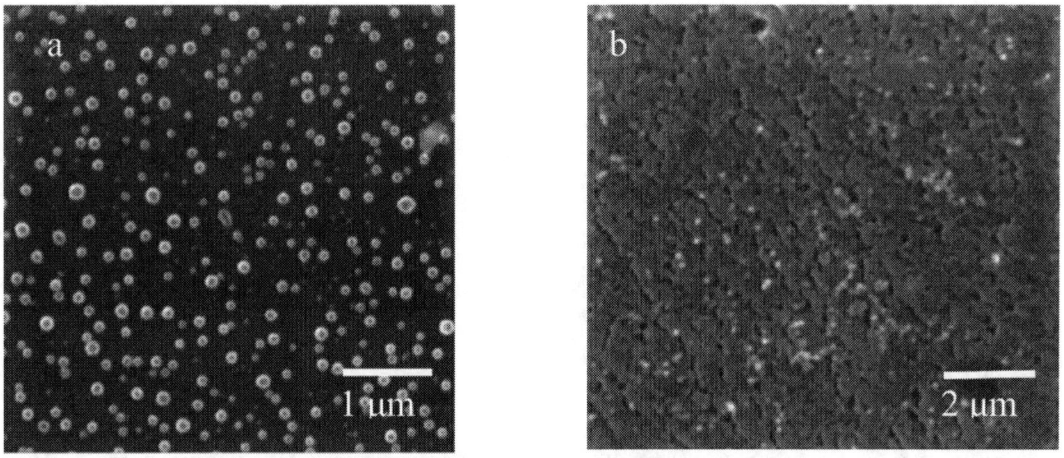

Figure 2. SEM images of the surface of crystallized a-Si before (a) and after (b) the Secco etching. The crystallization was performed by 2 consecutive pulses of the 3rd harmonics of Nd:YAG laser (λ = 355 nm) and energy density of 155 mJ/cm^2.

A Si sample crystallized by applying 3 consecutive laser pulses of the 3rd harmonics is shown in Fig.3. It is evident that the amount of the sphere-like structures is greatly increased both on the non-etched and etched part of the sample. This large number of the Si spheres cannot be etched away with the same etching time as in the previous cases.

The etched laser-processed silicon has much more porous structure than this processed with lower amount of energy. The etched area consists of the small grain polysilicon regions. This might be evidence that the full melting regime has been reached.

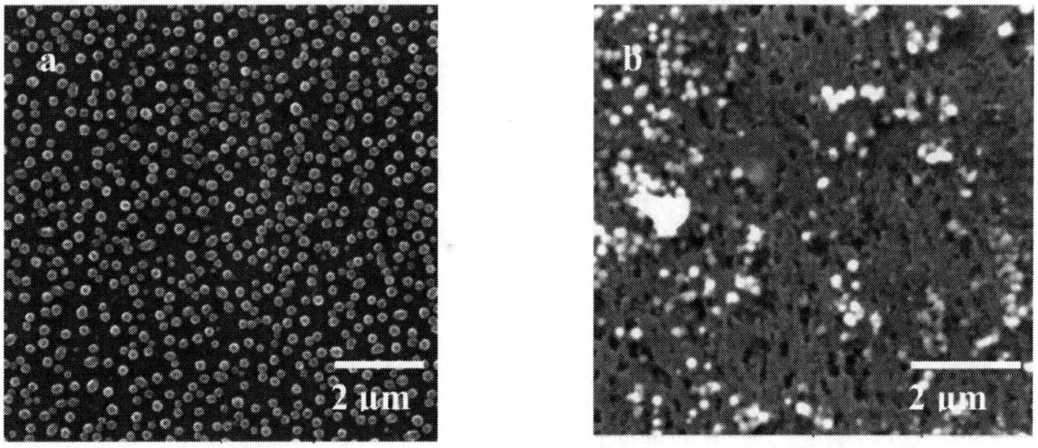

Figure 3. SEM images of the surface of the crystallized a-Si before (a) and after (b) the Secco etching. The crystallization was performed by 3 consecutive pulses of the 3rd harmonics of Nd:YAG laser (λ = 355 nm) and energy density of 155 mJ/cm^2.

3.2. Crystallization using 2nd harmonics of Nd:YAG (532 nm) laser beam

In the Fig. 4a-b, in the middle of the crystallized region, we can observe silicon layer which has been removed from the substrate after a single-pulse laser irradiation of the 2nd harmonics laser beam with energy density of 280 mJ/cm^2. The same situation is shown in the Fig. 5 a-d for the etched sample. This spot corresponds to the maximum intensity of the laser beam, and amorphous Si was melted completely.

In the crystallized region I (Fig. 4c), we can observe the same sphere-like structures as in the case of the 3rd harmonics crystallization. The occurrence of these structures is the evidence that in the middle of the laser spot the energy is much higher than the one needed for the SLG regime to obtain. After the etching, the severely textured Si layer in the region I is detached from the substrate completely.

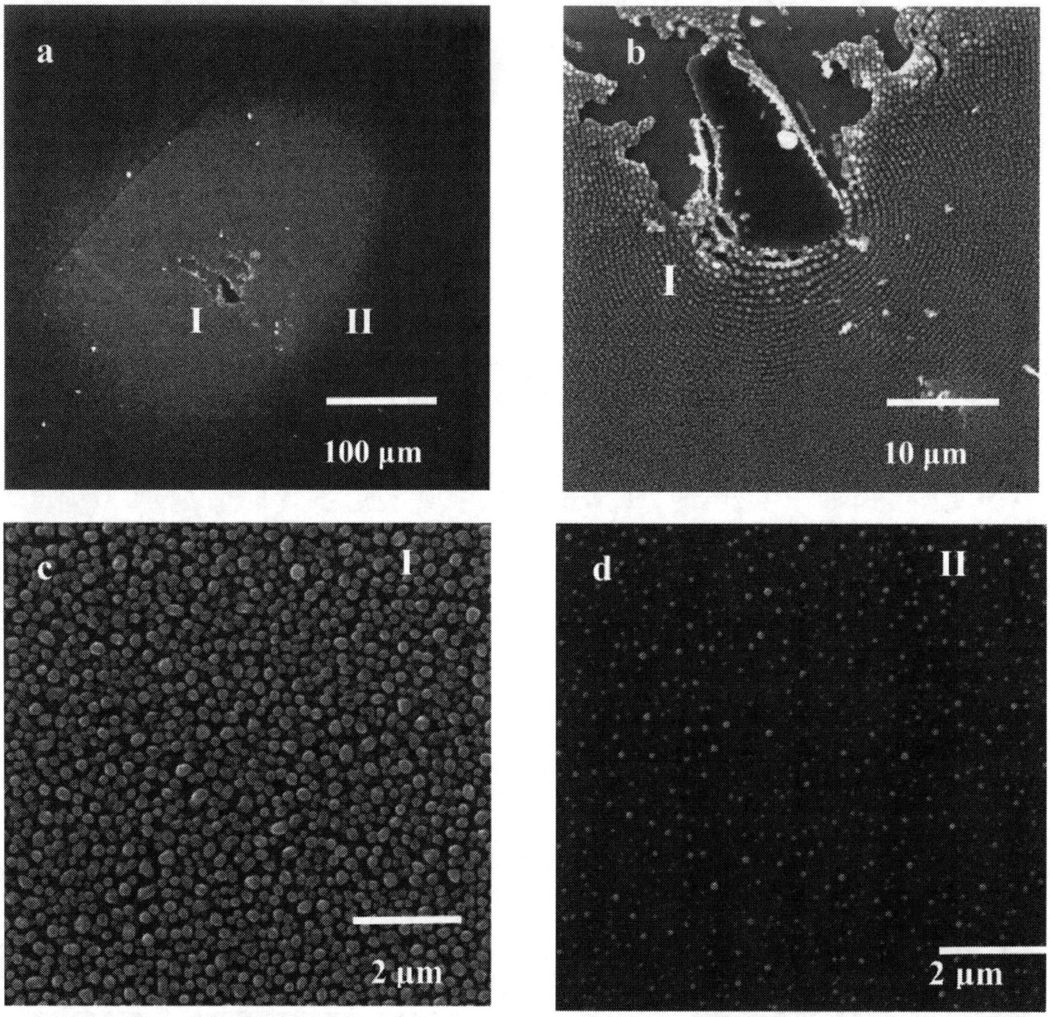

Figure 4. SEM images of the surface of the crystallized a-Si before the Secco etching. The crystallization was performed by the single pulse of the 2nd harmonics of Nd:YAG laser (λ = 532 nm) and energy density of 280 mJ/cm^2.

Figures 4b-d are larger magnifications of the corresponding areas (I and II) in Figure 4a.

In the crystallized region II (Fig. 4d), which is approximately equivalent to the region III (Fig. 5a, 5b), the amorphous silicon film is successfully crystallized, and the SLG regime was achieved. On the periphery of the region corresponding to the SLG regime, a fine-grain poly-Si structure is observed (Fig. 5c, 5d).

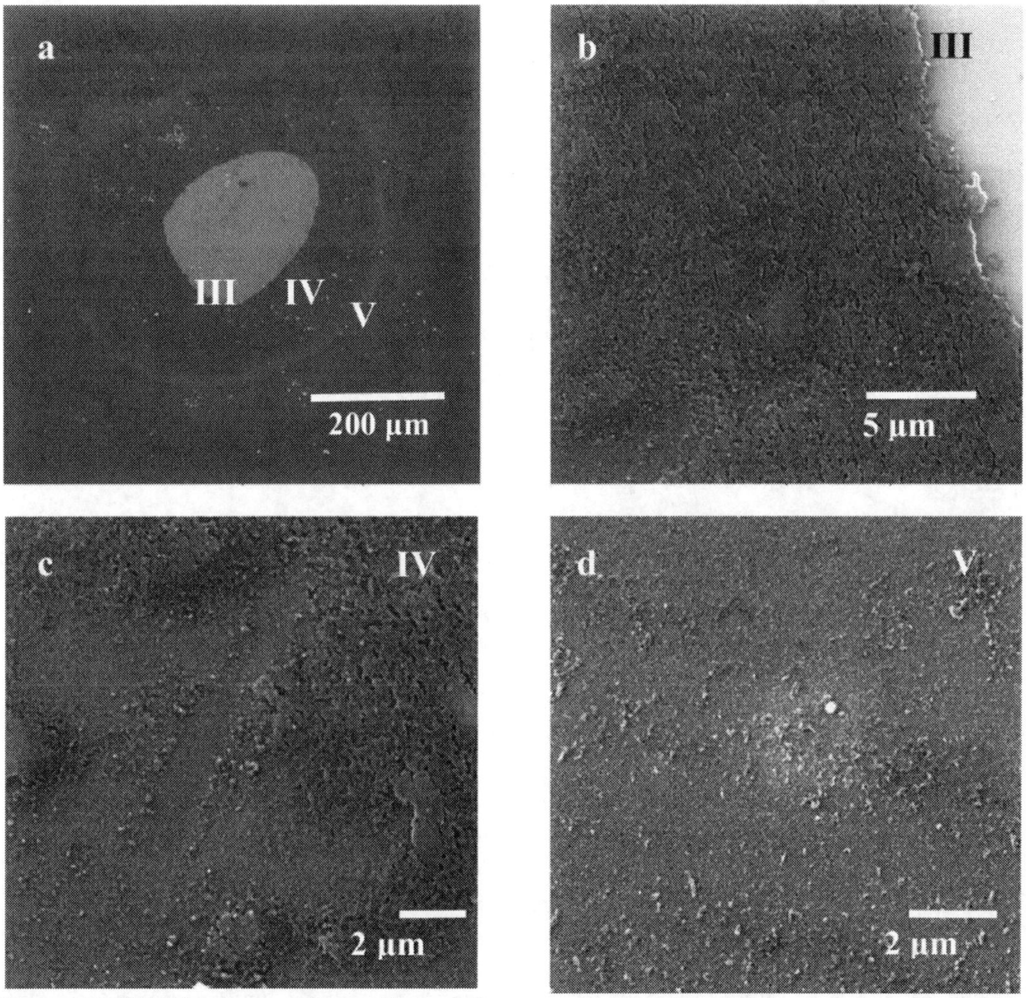

Figure 5. SEM images of the surface of the crystallized a-Si after the Secco etching. The crystallization was performed by the single pulse of the 2nd harmonics of Nd:YAG laser (λ = 532 nm) and energy density of 280 mJ/cm^2. Regions II (see Figs. 4a, 4d) and III are approximately equivalent; SLG regime has been reached.
Figures 5b-d are larger magnifications of the corresponding areas (III, IV and V) in Figure 5a.

Together with the crystallization by 532 nm laser pulses, it is possible to texture the surface of silicon to obtain, for instance, so-called black silicon (Fig. 4b). Black silicon is able to improve the efficiency of the solar sells significantly due to its low reflectivity and high absorption features [14 - 16].

The slight discrepancies in the laser intensities to reach SLG regime reported elsewhere [17, 18] can be explained by the different mechanisms of Si crystallization using the excimer laser and the 2nd harmonics of Nd:YAG laser, as well as the fact that the Si thin films used in this study was not hydrogenized.

4. Conclusions

Si crystallization experiments had demonstrated that to implement a successful crystallization procedure and to achieve the super lateral growth of polysilicon grains, the irradiation of 2nd and the 3rd harmonics of the Q-switched Nd:YAG laser can used with the same efficiency as that of excimer lasers.

By the single laser pulse of 532 nm and energy density of 280 mJ/cm^2, the SLG regime can be obtained resulting in about 0.5 μm large poly-Si grains. This regime can also be used for the texturing the surface of silicon to lower its reflectivity for further improving the efficiency of the Si-based solar cells.

The processing of the a-Si film by the single laser pulse with 355 nm wavelength and 155 mJ/cm^2 energy density is sufficient to obtain relatively large Si grains (0.5 - 1 μm) as a result of the super lateral growth.

Acknowledgments
The authors are grateful to Dr. L. Dimitrocenko for valuable discussions.

J. Butikova, B. Polyakov, A.Muhin, A.Voitkans and I.Tale are supported by the ERAF project Nr.2010/0252/2DP/2.1.1.1.0/10/APIA/VIAA/009. G. Marcins is supported by the European Social Fund within the project «Support for Doctoral Studies at the University of Latvia».

References

[1] Wang Q, Soderstrom T, Omaki K, Lennon A and Varlamov S 2012 *Energy Procedia* **15** 220
[2] Moschou D C, Vourdas N, Davazoglou D, Kouvatsos D N, Vamvakas V E and Voutsas A T 2012 *Microelectron Eng* **90** 69
[3] Uchikoga S and Ibaraki N 2001 *Thin Solid Films* **383** 19
[4] Exarchos M A, Moschou D C, Papaioannou G J, Kouvatsos D N, Arapoyanni A and Voutsas A T 2009 *Thin Solid Films* **517** 6375
[5] Loreti S, della Sala D and Garrozzo M 2000 *Micron* **31** 299
[6] Brotherton S D, Ayres J R, Edwards M J, Fisher C A, Glaister C, Gowers J P, McCulloch D J and Trainor M 1999 *Thin Solid Films* **337** 188
[7] Voogt F C and Ishihara R 2001 *Thin Solid Films* **383** 45
[8] Brendel K, Nickel N H, Lengsfeld P, Schöpke A, Siebera I, Nerding M, Strunk H P and Fuhs W 2003 *Thin Solid Films* **427** 86
[9] Chang C-H and Chao L-S 2008 *Int Commun Heat Mass* **35** 571
[10] Voutsas A T, Limanov A and Im J S 2003 *J Appl Phys* **94** 7445
[11] Lengsfeld P 2001 PhD thesis p.13
[12] Fang P H, Bai P, Kinnier J H, Huan Z and Schubert C C 1983 *J Non-Cryst Solids* **59-60** 819
[13] Mählißß J, Abbadie A and Kolbesen B O 2009 *Mat Sci Eng* B **159-160** 309
[14] Yoo J S, Parm I O, Gangopadhyay U, Kim K, Dhungel S K, Mangalaraj D and Yi J 2006 *Sol Energ Mat Sol C* **90** 3085
[15] Kumar D, Srivastava S K, Singh P K, Husain M and Kumar V 2011 *Sol Energ Mat Sol C* **95** 215
[16] Dobrzański L A, Drygała A 2007 *Mater Process Tech* **191** 288
[17] Lengsfeld P, Nickel N H and Fuchs W 2000 *J Non-Cryst Solids* **266-269** 659
[18] Mariucci L, Pecora A, Fortunato G, Spinella C and Bongiorno C 2003 *Thin Solid Films* **427** 91

IOP Conf. Series: Materials Science and Engineering **38** (2012) 012010 doi:10.1088/1757-899X/38/1/012010

Magnetic properties of carbon nanoparticles

E Lähderanta[1,5], A V Lashkul[1], K G Lisunov[1,2], D A Zherebtsov[3], D M Galimov[3] and A N Titkov[4]

[1] Department of Mathematics and Physics, Lappeenranta University of Technology, PO Box 20, FIN-53851 Lappeenranta, Finland
[2] Institute of Applied Physics, Academy of Sciences of Moldova, Academiei Str. 5, MD-2028 Kishinev, Moldova
[3] South Ural State University, 454080 Chelyabinsk, Russia
[4] A. F. Ioffe Physico-Technical Institute, 194021 St. Petersburg, Russia

E-mail: erkki.lahderanta@lut.fi

Abstract. Magnetization M (T, B) of powder and glassy samples containing carbon nanoparticles is investigated in the interval of temperatures T between ~ 3 – 300 K and magnetic fields B up to 5 T. Low-field magnetization, M (T), exhibits a strong magnetic irreversibility, which is suppressed above the field of ~ 1 T. The dependence of M (B) saturates at high temperatures above B ~ 2 T, magnetic hysteresis is observed already at 300 K. The values of the saturation magnetization, the coercivity field and the maximum blocking temperature are obtained. Analysis of the experimental data gives evidence for concentration of the magnetization close to the surface of the particles, which is consistent with the origin of magnetism in nanocarbon presumably due to intrinsic disorder and surface defects.

1. Introduction

Magnetism of carbon-based materials attracts considerable interest in recent years due to high-temperature ferromagnetic (FM) behavior, observed in various carbon structures [1]. This makes them attractive for applications not only in techniques, but in biology and medicine as well, being low-toxic due to vanishing concentrations of metallic elements [1, 2]. Generally, residual metallic impurities have been proved to play a negligible role in formation of the observed FM signal [1–3]. Therefore, the unconventional magnetism of carbon-based materials is associated with intrinsic defects or disorder introduced into an ordered host matrix [1, 2], which exhibits only a diamagnetic response as takes place in bulk ordered graphite and diamond [1–3]. In contrast, intentional incorporation of defects in graphite e. g. by proton irradiation leads to a pronounced ferromagnetism [2, 5, 6].

Experimental investigations establishing intrinsic magnetism of defect-rich carbon structures [1–6] have been supported by extensive theoretical work. Namely, FM behavior have been predicted in such structures as graphite surface with negative Gaussian curvature [7], a mixture of carbon atoms with alternation of $sp^2 - sp^3$ bonds [8], in presence of the graphene zigzag edges [9, 10] and in disordered graphite with random single-atom defects [11]. In turn, theoretical values of local magnetic moments $\mu_i \sim 1 - 2\ \mu_B$, connected to intrinsic defects or disorder [12, 13], have been supported by experimental

[5] To whom any correspondence should be addressed

investigations of highly-oriented pyrolitic graphite yielding μ_i ~ 0.2 – 1.5 μ_B per defect at a distance between defects ~ 0.5 – 4 nm [14].

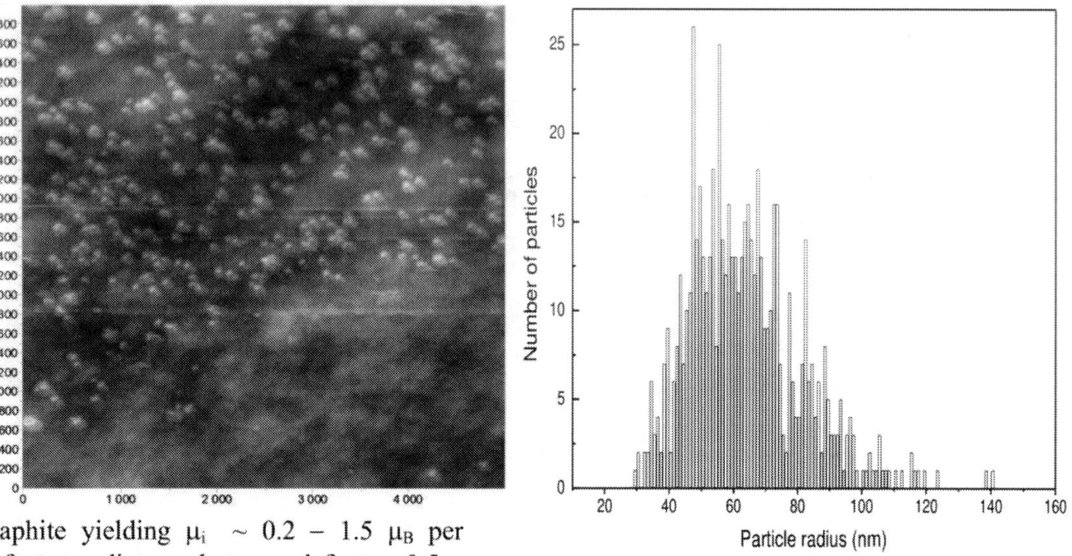

Figure 1. The AFM image of the surface of sample S2 with the area of 5000×5000 nm^2 (left panel) and distribution of the particle radius (right panel).

Carbon nanoparticles represent a promising case of defect-rich carbon media due to numerous surface defects, accompanied with a high surface-to-volume ratio. Such features suggest a possibility of interesting magnetic behavior, pertinent to an assembly of small magnetic particles, including large magnetic irreversibility, pronounced hysteresis and saturation of the magnetization [15–17], which are expected to be observed already near the room temperature.

In this work we investigate magnetic properties of the system of carbon nanoparticles to determine macroscopic magnetic behavior and to obtain microscopic information on magnetism of nanocarbon.

2. Experimental results

Not intentionally doped (undoped) powder nanocarbon sample (S1) and porous glassy carbon samples, doped with 0.00123 mass % of Ag (S2), 0.004 mass % of Au (S3) and 0.117 mass % of Co (S4) were investigated. The glassy samples were obtained with a method based on a triple component system, including furfural spirit (FS), dibutilphtalat (DBF) and polyethilen-10-glykol ether of isooktiphenol $C_8H_{17}C_6H_4O[C_2H_4O]_{10}H$ (OP-10), with the ratios of FS:DBF:OP = 20 % : 40 % : 40 %. After mixture it was added 2 – 15 drops of paratoluolsulfoacid dissolved in butanol (36.9 mass %) (TSA) for catalyses of reaction of FS polycondensation. Amount of TSA depends on the level of dissolving of FS and differs from 2 drops for pure FS up to 15 drops for solution with 20 mass % of FS. Such approach leads to the simultaneous transformation of FS to polymer and solution tinged to dark colours. Ready samples were kept for 20 – 30 days at room temperature in closed vessels for finishing of the polymerization process. After the polymerization was finished, samples were dried at T = 50, 100 and 150 °C with the drying time of 23 hours at every temperature. As a result, black samples were obtained and samples with 20% of FS were still plastic or were even paste. Samples dried at 150 °C were then heated in the oven with the speed of temperature increasing 50 °C/h up to T = 970 °C. They were placed into the crucible with the labyrinth bolt filled up by soot and were annealed for 1 hour in the recovery atmosphere created by soot. Finally, annealed samples were black glass-like porous nanocarbon materials having the initial shape but smaller in size.

As can be seen in Fig. 1 atomic force microscopy (AFM) measurements demonstrate a broad size distribution of carbon particles, characterized by the average radius $R_{av} \sim 60$ nm and the maximum radius $R_{max} \sim 110$ nm being close in all investigated glassy samples. In addition, the AFM images of the samples are quite similar, yielding only a small nonsphericity of the particles.

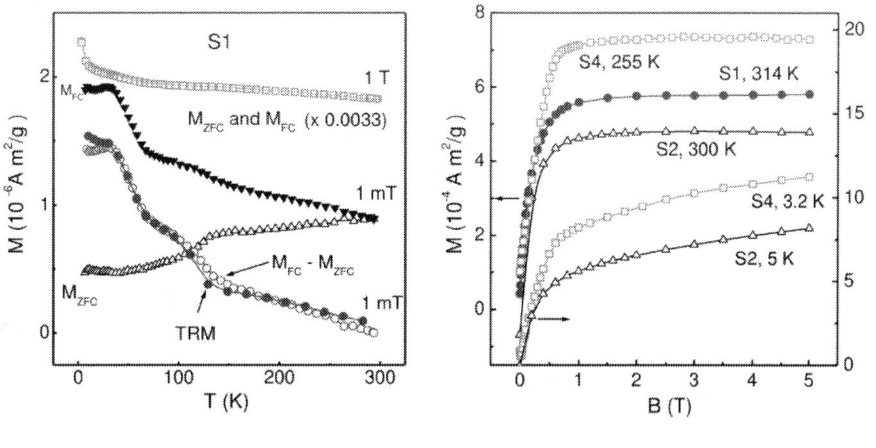

Figure 2. The plots of M_{ZFC}, M_{FC}, TRM and $M_{FC} - M_{ZFC}$ vs. T for sample S1. The plots at 1 T are multiplied by a factor 0.0033 for convenience (left panel). The dependence of M on B (right panel).

For investigations the sample was put into a plastic ampoule filled with helium. Magnetization M (B) was measured with a SQUID magnetometer in fields B up to 5 T by increasing and decreasing the field. The dependence of M (T) was measured in a constant magnetic field between 1 mT – 5 T, after cooling the sample from 300 K down to 3 K in zero field (zero-field cooled magnetization, M_{ZFC}) or in the applied field (field-cooled magnetization, M_{FC}). Thermoremanent magnetization (TRM) was investigated after cooling the sample from 300 K to 3 K and reducing the filed to zero. The magnetization data below are presented after subtraction of the diamagnetic contribution.

Temperature dependence of the magnetization in all the investigated samples in weak fields of 1 – 50 mT (an example for S1 is displayed in the left panel of Fig. 2) is characterized by a strong irreversibility or deviation of M_{ZFC} (T) from M_{FC} (T) and by TRM decreasing with T. Magnetic irreversibility is decreased with increasing the field and vanishes above $B \sim 1$ T. As can be seen in the right panel of Fig. 2, M (B) exhibits a saturation already above $B \sim 2$ T at high temperatures, whereas at low $T \sim 3 – 5$ K a deviation from such behavior is observed. Hysteresis is observed already at the room temperature. Hysteresis loops in S2 and S3 are similar to that of S1, shown in the left panel of Fig. 2, and differ substantially from the image observed in S4 (the right panel of Fig. 3).

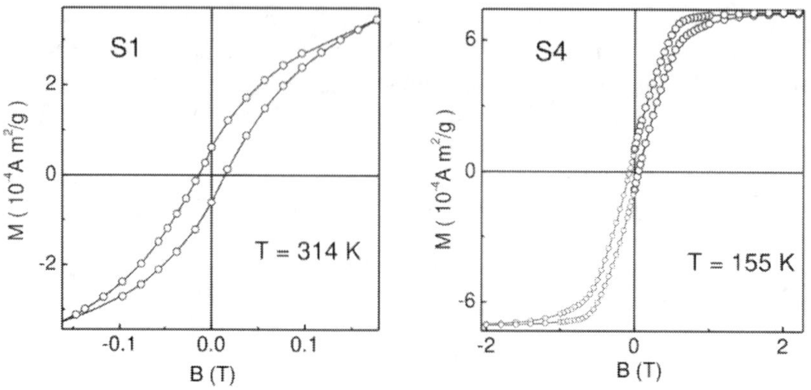

Figure 3. The hysteresis loops in samples S1 and S2.

3. Analysis and discussion

Strong magnetic irreversibility in the left panel of Fig. 2 is attributable to the large size distribution evident in the right panel of Fig. 1 and represents a typical property of an assembly of magnetic nanoparticles at $B < B_K$ and $T < T_b$, where B_K is the mean anisotropy field and $T_b \approx KV / (25\ k_B)$ is the blocking temperature, determined by the density of the anisotropy energy, K, and the particle volume, V [15, 16]. Above T_b thermal excitations cause a rapid chaotic rotation of particle moments and lead to superparamagnetic behaviour of the whole particle system, whereas below T_b the blocked or stable regime sets in, when the rotations of the particle moments are logarithmically slow.

The magnetic irreversibility is damped by the magnetic field due to diminution of the anisotropy barriers by the field and vanishes at $B > B_K$ [16, 18]. We obtained $B_K \approx 1$ T in S1 – S3 and it lies between ~ 1 – 5 T in S4, whereas the temperature dependence of the magnetization acquires an upturn below ~ 50 K. Only a small deviation of TRM (T) from the difference of M_{FC} (T) – M_{ZFC} (T) (left panel of Fig. 2) indicates an insignificant non-sphericity of magnetic particles [19], which is in agreement with the data in the left panel of Fig. 1.

Saturation of the magnetization at high temperatures (right panel of Fig. 2) characterizes the system of large magnetic particles with sizes ~ R_{av}, provided that T_b exceeds the room temperature. On the other hand, a substantial deviation from saturation at low T ~ 3 – 5 K, evident from the right panel of Fig. 2 as well, suggests presence of the magnetic particles even much smaller than those visible in the left panel of Fig. 1. Such particles can be unblocked down to the lowest temperatures. Therefore, they exhibit a paramagnetic or superparamagnetic response and make only a small contribution to the saturation magnetization of the material, M_s (0), which characterizes the blocked regime of the system of magnetic nanoparticles at $B > B_K$ and $T << T_b$. It can be shown that contribution of such small particles can be excluded by fitting the plots of M (T) at $B \geq 2$ T with the Curie-Weiss-like law, $M(T) \approx M_s(0) + C B / (T - \theta)$, yielding the values of M_s (0) ≈ 5.6, 4.9, 4.7 and 7.2 (in units of 10^{-4} Am^2/g).

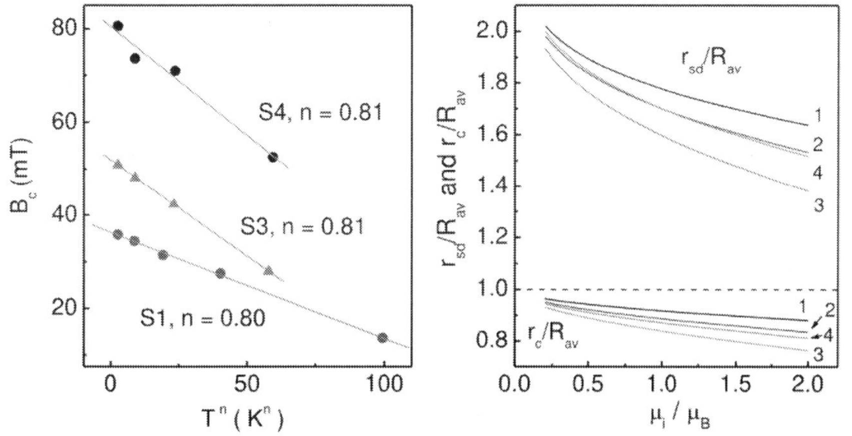

Figure 4. The plots of B_c vs. T^n for samples S1, S3 and S4 (left panel). The dependences of r_{sd}/R_{av} and r_c/R_{av} on μ_i/μ_B for samples S1 – S4 (lines 1 – 4, respectively).

Hysteresis loop in the left panel of Fig. 3 is typical of a system of blocked magnetic nanoparticles, whereas in the right panel it is broaden considerably and has a more rectangular shape, indicating a FM response from Co ions and/or clusters. In any case, we have observed a substantial decrease of the coercive field, B_c (T), with increasing T. In a system of magnetic nanoparticles in the blocked regime the temperature dependence of the coercivity is given by the expression $B_c(T) = B_c(0)\,[1 - (T/T_b)^n]$, where n and B_c (0) depend on the magnetization reversal mode and on the applicability of the single-domain (SD) regime. For SD particles n = 1/2 [20] and 2/3 [16] for coherent rotation and curling, respectively, otherwise the value of $n = 1$ is connected to domain wall effects [16, 21].

The analysis of B_c (T) with the expression above is done by minimizing the plots of the standard deviation vs. n, evaluated with a linear fit of the plots of B_c (T) vs. T^n by variation of n. We have found that the minimum of these plots yields $n \approx 0.80$ (S1) and 0.81 (S3 and S4), giving the best linearization of B_c vs. T^n displayed in the left panel of Fig. 5. The values of $T_b \approx 570$, 405 and 580 K and B_c (0) = 36, 52 and 80 mT for S1, S3 and S4, respectively, are also obtained from the fitting procedure above. The value of n lying between 2/3 and 1 suggests a crossover of a SD curling mode of the magnetization reversal and a multidomain regime, which is expectable due to the relatively large values of particle radii and the broad size distribution evident in the bottom panel of Fig. 1. On the other hand, the regime of curling is supported also by the relation of B_c (0) << B_K, whereas for coherent rotation these values should be comparable [16, 22].

Eventually, we present results of the microscopic analysis involving the macroscopic parameters M_s (0), B_c (0) and T_b found above. Calculations have been performed using expressions for a critical radius r_{sd} of the SD regime, taking place at $R << r_{sd}$ (where R is the particle radius), and for a critical radius r_c of curling, which sets in at $R >> r_c$ [16, 17]. A system of small magnetic particles has been treated with a conventional phenomenological approach, taking into account almost spherical shape of the particles, their additive contributions to the net magnetization of the sample and a supposition of strongly inhomogeneous distribution of the magnetization over the particle volume, concentrated within a thin near-surface layer with thickness h (i. e. $h/R << 1$).

At first we take $\mu_i \sim 0.2 - 2\ \mu_B$ (see Introduction) and evaluate the dependences of r_{sd}/R_{av} and r_c/R_{av} by putting $h = a$, where a is the average distance between the internal localized magnetic moments of the particle. As can be seen in the right panel of Fig. 4 the relations $R_{av} \leq r_{sd}$ and $R_{av} \geq r_c$ are satisfied reasonably for all investigated samples. Taking into account the broad size distribution, this supports the preliminary conclusion about the curling mode of the magnetization reversal in conditions of closeness to the limit of the SD regime made above. Eventually, the values of a (μ_i) are found to lie between ~ 0.3 – 1.2 nm which is consistent, according to the impurity models [8–13], with the strong FM interactions between the localized magnetic moments.

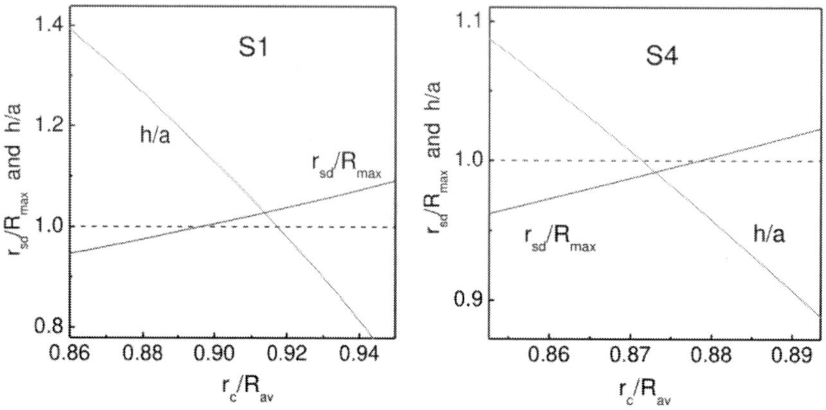

Figure 5. The plots of r_{sd}/R_{av} vs. r_c/R_{av} and h/a vs. r_c/R_{av}, evaluated for samples S1 and S4.

Another type of calculations is presented in Fig. 5, where we put $\mu_i = \mu_B$ and vary r_c. The plots of r_{sd}/R_{max} vs. r_c/R_{av} and h/a vs. r_c/R_{av} in Fig. 5 are compared with the horizontal line corresponding to the constant value of unity. Taking into account the evident relations, $r_{sd}/R_{max} \geq 1$ and $h/a \geq 1$, the interval of acceptable values of both ratios should be given by the plots of r_{sd}/R_{max} vs. r_c/R_{av} and h/a vs. r_c/R_{av} lying above unity, whereas the corresponding interval of r_c/R_{av} lies between the values of $X_{min} \equiv (r_c/R_{av})_{min}$ and $X_{max} \equiv (r_c/R_{av})_{max}$ given by intersections of these plots with the constant line of unity, respectively. Therefore, the values of a and h can be found within the error, determined by the difference of X_{min} and X_{max}, where $X_{min} \leq X_{max}$. As can be seen in the left panel of Fig. 5, in S1 the relation of $X_{min} \leq X_{max}$ is fulfilled with the difference of ~ 2 % between X_{min} and X_{max}. However, in S4, as well as in S2 and S3 (not shown) we have $X_{min} > X_{max}$, but the difference between them lying between ~ 1 % and 5 % is insignificant, taking into account uncertainty of determination of R_{max} as evident in Fig. 1. The calculations above demonstrate that the relation $h \approx a$ is observed within an acceptable error of ~ 15 % and 20 % in S1 and S3, respectively, and within the error as small as ~ 4 % and 3 % in S2 and S4, respectively, at least in the vicinity of μ_i to μ_B. The values of $a \approx h \approx 0.74$, 0.79, 0.81 and 0.65 nm are obtained in S1, S2, S3 and S4, respectively, in agreement with the relation of $h/R_{av} \ll 1$ laid down in the basis of the microscopic analysis above.

To conclude, magnetic properties of the system of carbon nanoparticles have been investigated. The values of the macroscopic parameters M_s (0), B_c (0) and T_b have been determined. Microscopic analysis involving these parameters demonstrates strongly inhomogeneous distribution of the magnetization, concentrating within a narrow near-surface layer of the particles. This is consistent with the origin of magnetism in nanocarbon presumably due to intrinsic near-surface defects.

References

[1] Makarova T and Palacio F (eds) 2006 Carbon-Based Magnetism (North-Holland: Elsevier)
[2] Makarova T 2005 Unconventional magnetism in carbon based materials, in: Frontiers in Magnetic Materials, Narlikar A V (ed) p. 5 (Berlin: Springer)

[3] Červenka J, Katsnelson M I and Flipse C F J 2009 *Nature Physics* **5** 840
[4] Stamenov P and Coey J M D 2005 *J. Magn. Magn. Mater.* **290-291** 279
[5] Esquinazi P, Spemann D, Schindler K, Höhne R, Ziese M, Setzer A, Han K-H, Petriconi S, Diaconu M, Schmidt H, Butz T and Wu Y H 2006 *Thin Soli Films* **505** 85
[6] Barzola-Quiquia J, Höhne R, Rothermel M, Setzer A, Esquinazi P and Heera V 2008 *Eur. Phys. J. B* **61** 127
[7] Park N, Yoon M, Berber S, Ihm J, Osawa E and Tomanek D 2003 *Phys. Rev. Lett.* **91** 237204
[8] Ovchinnikov A A and Spektor V N 1988 *Synth. Mather.* **27** B615
[9] Son Y-W, Cohen M L and Louie S G 2006 *Nature* **444** 347
[10] Yazyev O V and Katsnelson M I 2008 *Phys. Rev. Lett.* **100** 047209
[11] Yazyev O V 2008 *Phys. Rev. Lett.* **101** 37203
[12] Faccio R, Pardo H, Denis P A, Oeiras R Y, Araujo-Moreira F M, Verissimo-Alves M and Mombru A W 2008 *Phys. Rev. B* **77** 035416
[13] Yazyev O V and Helm L 2007 *Phys. Rev. B* **75** 125408
[14] Červenka J, Katsnelson M I and Flipse C F J 2009 *Nature Physics* **5** 840
[15] Bean C P and Livingston J D 1959 *J. Appl. Phys.* **30** 120S
[16] Sun L, Hao Y, Chien C L and Searson P 2005 *IBM J. Res. & Dev.* **49** 79
[17] Frei E H, Shtrikman S and Trevis D 1957 *Phys. Rev.* **106** 446
[18] Wolfarth E P 1980 *J. Phys. F: Metal. Phys.* **10** L241
[19] Belous N, Zorin I, Kulich N, Lezhnenko I and Tovstolytkin A 1990 *Sov. Phys.-Solid State* **32** 887
[20] Gullity B D 1972 Introduction to Magnetic Materials (London: Addison-Wesley)
[21] Gaunt P 1983 *Philos. Mag. B* **48** 261
[22] Bozorth R M Ferromagnetism 1956 (Princeton, New-Jersey: Van Nostrand)

Holographic studies of photoinduced anisotropy in molecular glassy films containing diphenylamine azochromophores

A Ozols[1], V Kokars, P Augustovs, K Traskovskis, and D Saharov

Faculty of Material Science and Applied Chemistry, Riga Technical University, Azenes iela 14/24, LV-1007, Riga, Latvia

E-mail: aozols@latnet.lv

Abstract. Photoinduced anisotropy (PA) is experimentally studied in three molecular glassy films synthesized in our Faculty and containing diphenylamine based azochromophores {4-((4-nitrophenyl)diazenyl)-N-(4-((4-nitrophenyl)diazenyl)phenyl)-N-(2-trityloxy)ethyl) benzeneamine (shortly K-D-24), 4-((2-chloro-4-tritylphenyl)diazenyl)-N-(4-((2-chloro-4-tritylphenyl)diazenyl)phenyl)-N-(2-(trityloxy)ethyl)benzenamine (shortly K-D-25), 4,4'-(2-(trityloxy)ethylazanediyl)bis(4,1-phenylene)bis(diazene-2,1-diyl)dibenzonitrile (shortly K-D-32) } and, for comparison, in an a-As_2S_3 film. Holographic method enabling simultaneous measurements of *s*- and *p*-polarized diffracted light powers was applied in both transmission and reflection modes. Holographic grating recording with the period of 2 μm was made by two equally strong 532 nm *p*- polarized laser beams with total light intensity of 0.81 W/cm^2. The readout was made by 633 nm circularly polarized beam, and the *s*- and *p*-polarized components of diffracted light were separated by polarization beam splitter prism. PA was characterized by diffracted power difference (DPD), $P_s - P_p$, and by anisotropy contrast $A = (\eta_s - \eta_p)/(\eta_s + \eta_p)$, η_s and η_p being the diffraction efficiencies for *s*- and *p*-polarized light. DPD exhibited markedly different kinetic behaviour in transmission and reflection modes. There was a negative minimum in transmission mode in all samples, and a growth with oscillations up to saturation in reflection mode. Small initial DPD maxima were specific to a-As_2S_3 film. The highest PA was found in K-D-24 film in transmission mode (A=-0.23) and in K-D-25 film in reflection mode (A=0.49). Holographic recording efficiency and PA do not correlate. In our opinion, PA in transmission mode of molecular glasses is due to the photoinduced volume processes including *trans-cis* photoisomerization , chromophore orientation and mechanical stress modulation. In the case of a-As_2S_3 film these photoinduced volume processes include photoinduced structural changes, D-centre orientation and, again, mechanical stress modulation. All these processes lead to a photoinduced birefringence. Surface relief grating recording and polarization-dependent reflection are responsible for diffractive anisotropy in reflection mode.

1. Introduction

Azobenzene-containing organic materials and chalcogenide inorganic materials have found numerous applications in optoelectronic devices, especially for holographic information recording [1,2,3]. They are also of large physical interest. In our previous studies we have investigated azobenzene and chalcogenide films with respect to holographic recording. We have experimentally found that quite efficient recording is possible in both classes of materials [4,5]. It is also found that recording efficiency strongly depends on both recording light and readout light polarization [6]. Thus the question of recording anisotropy arises. The aim of this study was to carry out direct real-time

---

[1] To whom any correspondence should be addressed

photoinduced anisotropy (PA) studies using novel holographic scheme in both azobenzene and chalcogenide films. The idea of this scheme is borrowed from the paper [7] and transformed for more convenient measurements in both transmission and reflection modes. The obtained results show that PA manifested by diffraction efficiency difference with linear *s*- and *p*-polarizations of readout light takes place in all the studied materials but PA properties differ for different materials and different readout modes.

2. Experiments and results

We have experimentally studied three molecular glassy films synthesized in our Faculty and containing diphenylamine based azochromophores{4-((4-nitrophenyl)diazenyl)-N-(4-((4-nitrophenyl)diazenyl)phenyl)-N-(2-trityloxy)ethyl) benzeneamine (shortly K-D-24), 4-((2-chloro-4-tritylphenyl)diazenyl)-N-(4-((2-chloro-4-tritylphenyl)diazenyl)phenyl)-N-(2-(trityloxy)ethyl) benzene amine (shortly K-D-25) , 4,4'-(2-(trityloxy)ethylazanediyl)bis(4,1-phenylene)bis(diazene-2,1-diyl)dibenzonitrile (shortly K-D-32) } and, for comparison, in an a-As_2S_3 film produced by J.Teteris at Institute of Solid State Physics, University of Latvia. Molecular glassy films with the thickness of about 1 μm were spin-coated onto the glass substrates. Their structural formulae are shown in figure 1. The a-As_2S_3 film with the thickness of 10 μm was thermally evapourated in vacuum also onto the glass substrate.

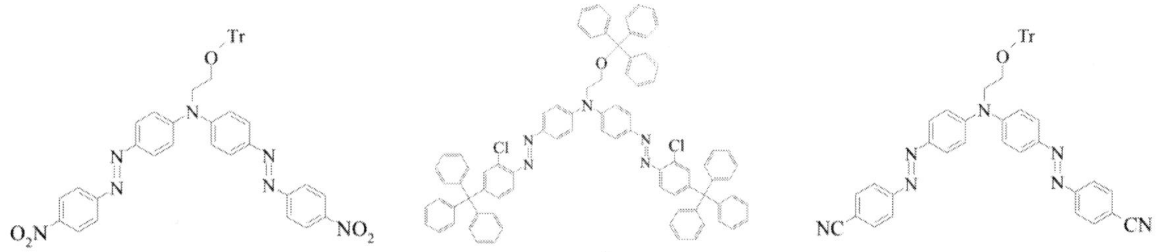

Fig.1. Chemical structural formulae of azobenzene derivatives K-D-24 (a), K-D-25 (b) and K-D-32 (c).They have the long wavelength absorption maxima at 514 nm, 498 nm and 485 nm, respectively. Tr stands for trityl group.

Experimental setups for PA measurements in transmission and reflection mode are shown in figures 2 and 3. The following notations are used: S-sample, λ_1=532nm is the recording wavelength, λ_2=632.8 nm≈633 nm is the readout wavelength, PBS – polarization beam splitter enabling *s*- and *p*-polarized diffracted light in the output, F-filter used to equal the transmitted *s*- and *p*-polarized light intensities without the sample, P-linear polarizer used to improve the *s*- polarization degree, PD1 and PD2-photodetectors, LPM – laser power meter.

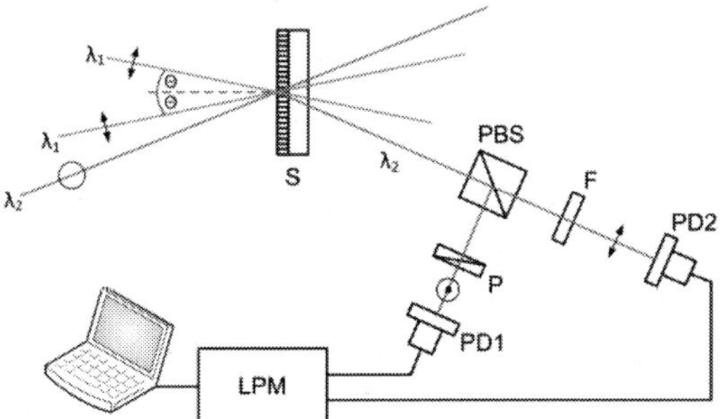

Fig.2. Experimental setup for transmission mode PA measurements. The notations are

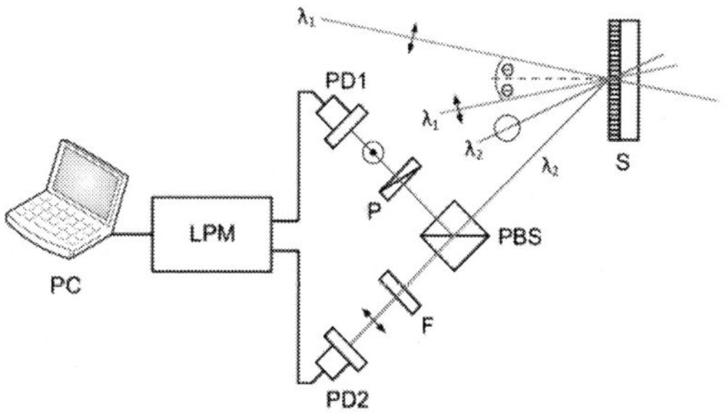

Fig.3. Experimental setup for reflection mode measurements. The notations are

explained in the text.

explained in the text.

As seen from figures 2 and 3, holographic grating (HG) recording is made by linearly *p*-polarized light, but its readout by circularly polarized light. Only central parts of experimental setups are shown. The part including lasers and optical elements for the adjustment of beam powers and polarizations is absent for simplicity.

Two symmetrically incident *p*-polarized 10.3 mW beams (with 0.81 W/cm^2 total intensity) from KLASTECH DENICAFC 532-300 diode pumped solid state laser (λ_1 = 532 nm) were used for HG recording with the period of 2 μm. One circularly polarized 1.1 mW beam from Melles Griot 25LH928-230 He-Ne gas laser (λ_2 = 632.8 nm) was used for HG readout. "Ophir Laserstar" laser power meter enabled the measurement and recording of diffracted light powers P_s and P_p as well as their ratio P_s/P_p and difference $P_s - P_p$. In our opinion, the difference $P_s - P_p$ is the best characteristic of photoinduced anisotropy to be measured directly, therefore, this was the main measured quantity. Diffraction efficiencies $\eta_s = P_s/P_{si}$ and $\eta_p = P_p/P_{pi}$ were also determined. The incident readout powers, P_{si} =0.313 mW and P_{pi}=0.334 mW, were very close to each other thus we assumed that the time dependence of difference $P_s - P_p$ qualitatively was the same as that of difference $\eta_s - \eta_p$. These time dependences were displayed on the PC screen, and they are presented in figures 4-11. All the measurements started from zero at 30 seconds. The estimated measurement errors did not exceed 5%.

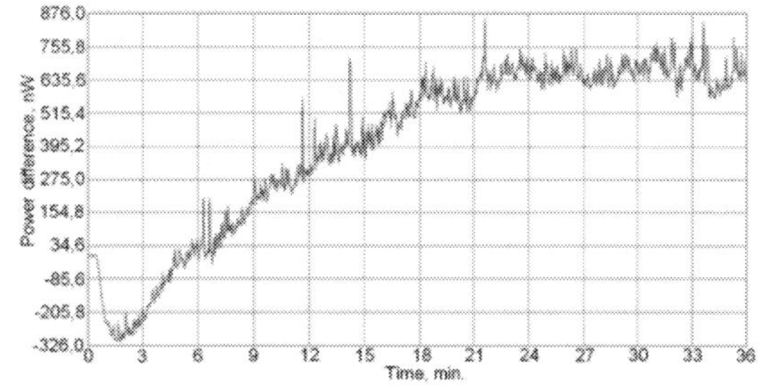

Fig. 4. Diffracted power difference $P_s - P_p$ (nW) versus time(minutes) in the case of K-D-24 sample. Transmission mode.

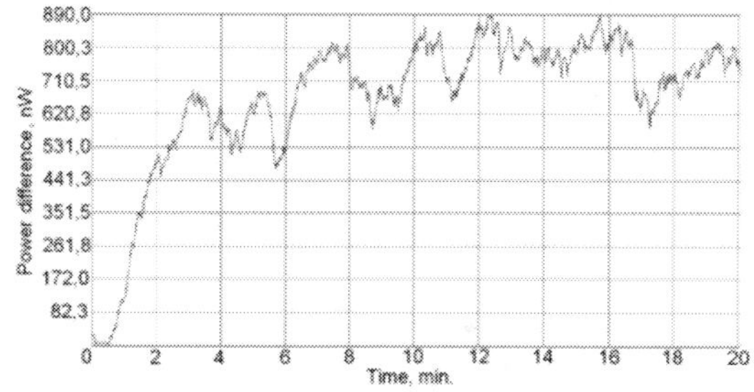

Fig. 5. Diffracted power difference $P_s - P_p$ (nW) versus time (minutes) in the case of K-D-24 sample. Reflection mode.

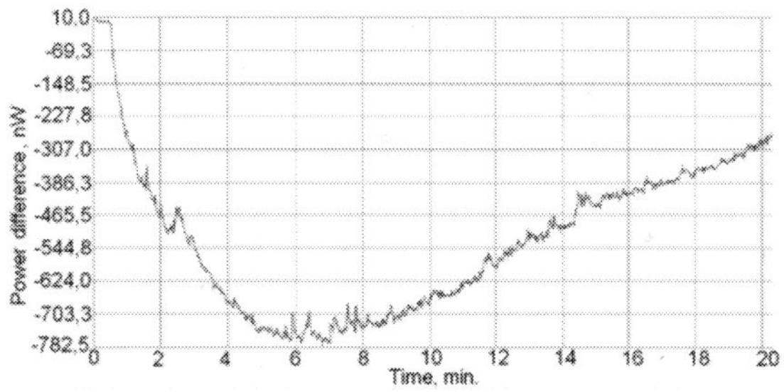

Fig. 6. Diffracted power difference $P_s - P_p$ (nW) versus time (minutes) in the case of K-D-25 sample. Transmission mode.

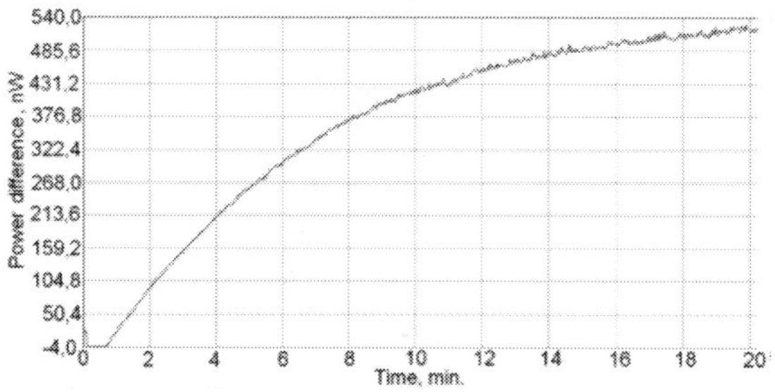

Fig.7. Diffracted power difference $P_s - P_p$ (nW) versus time (minutes) in the case of K-D-25 sample. Reflection mode.

Figures 4-11 show that time varying PA takes place in all the studied samples. It can be seen from the comparison of figures 4,6,8,9 with figures 5,7,9,11 that PA kinetic behaviour in transmission and reflection modes are markedly different. A rather deep negative minimum followed by the PA increase reaching positive values in the case of K-D-24 and K-D-32 films takes place in transmission mode whereas almost fully positive PA increasing up to saturation takes place in reflection mode.

To characterize the photoinduced anisotropy independently of particular diffracted powers and diffraction efficiencies we have introduced the quantity which we call photoinduced anisotropy contrast (or simply – anisotropy) $A = (\eta_s - \eta_p)/(\eta_s+\eta_p)$.This quantity changes from -1 to +1. In both these cases the anisotropy is maximal.

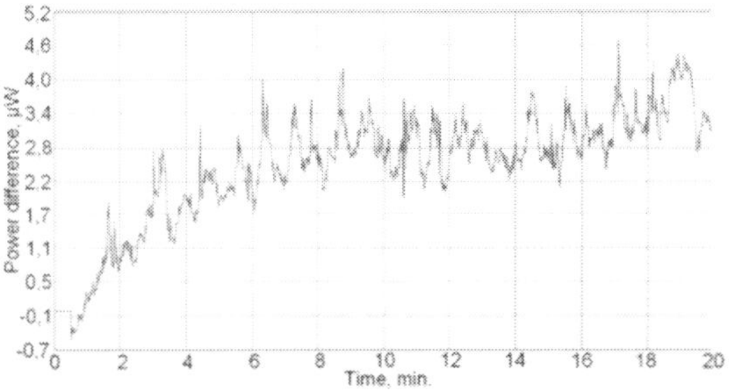

Fig.8. Diffracted power difference $P_s - P_p$ (μW) versus time (minutes) in the case of K-D-32 sample. Transmission mode.

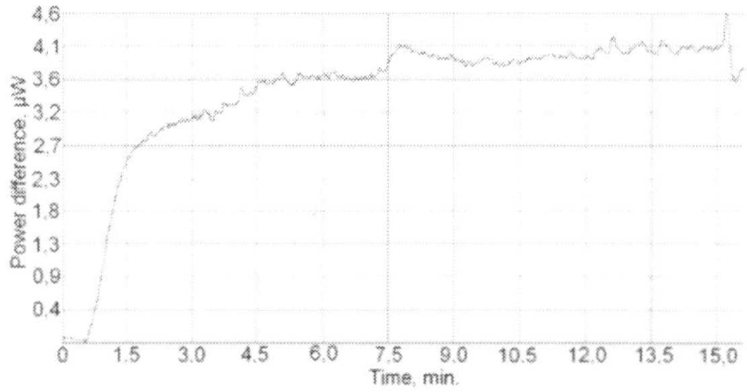

Fig.9. Diffracted power difference $P_s - P_p$ (μW) versus time (minutes) in the case of K-D-32 sample. Reflection mode.

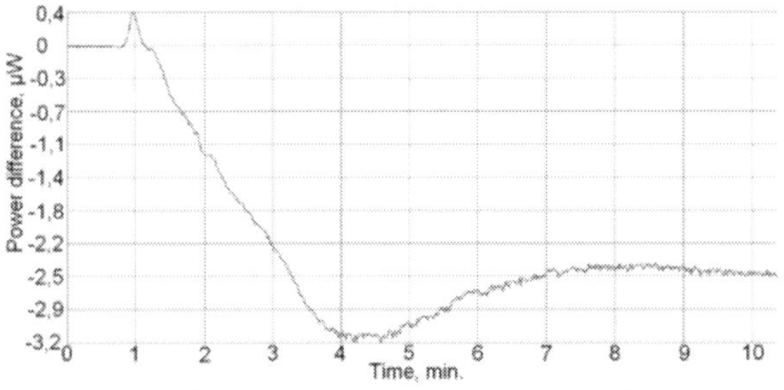

Fig.10. Diffracted power difference $P_s - P_p$ (μW) versus time (minutes) in the case of a-As_2S_3 sample. Transmission mode.

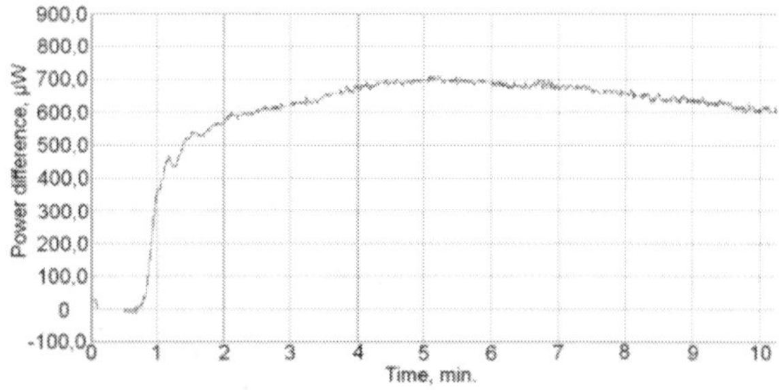

Fig.11. Diffracted power difference $P_s - P_p$ (μW) versus time (minutes) in the case of a-As_2S_3 sample. Reflection mode.

It can be seen from figures 10, 11 in comparison with figures 4-9 that PA kinetic dependences in the a-As_2S_3 film differ by the initial positive peak in transmission mode, and by PA decrease at large exposure times. Details of PA curves for molecular glasses are also different.

Experimental measurements are summarized in tables 1 and 2 where the values of diffraction efficiencies η_s and η_p, anisotropy contrast A are displayed corresponding to the extremal values of $P_s - P_p$ at exposure times t_{extr}, which are the the time moments corresponding to extremal (minimal or maximal) values of P_s - P_p .

Table 1. Transmission diffractive anisotropy

Sample	Measurement number	t_{extr}, min	$(P_s - P_p)_{extr}$, μW	η_s	η_p	A
a-As_2S_3	1	4.0	-3.8	0.0866	0.0925	- 0.033
	2	4.5	- 3.1	0.051	0.0575	- 0.060
K-D-24	1	7.2	- 3.9	0.0185	0.0290	- 0.22
	2	2.5	- 0.826	0.0096	0.0114	- 0.086
	2	19.5	0.409	0.0179	0.0157	0.065
	3	1.5	- 0.306	0.0150	0.0150	0.000
	3	33	0.796	0.0326	0.0284	0.069
K-D-25	1	7.0	- 0.773	0.0105	0.0123	- 0.079
K-D-32	1	21.5	4.9	0.173	0.147	0.081

In table 1 , there are several measurements in the case of a-As_2S_3 sample(measurement numbers 1,2) and K-D-24 sample (measurement numbers 1,2,3). Repeating numbers 2 and 3 correspond to two extremal values (one minimum and one maximum) of $P_s - P_p$ in one exposure. No correction of A values for Fresnel losses was made because these corrections were comparable to the measurement errors ($\delta\eta \approx 5\% \rightarrow \delta A \approx 26\%$).

Table 2. Reflection diffractive anisotropy

Sample	Measurement number	t_{extr}, min.	$(P_s - P_p)_{extr}$, μW	η_s	η_p	A
a-As_2S_3	1	3.2	0.850	0.0062	0.0033	0.33
	2	5.5	0.710	0.0051	0.0027	0.31
K-D-24	1	12.5	0.890	0.0061	0.0030	0.34
	2	9.7	0.935	0.0086	0.0054	0.23
K-D-25	1	18.0	0.458	0.0025	0.00093	0.46
	2	24.0	0.537	0.0027	0.00093	0.49
K-D-32	1	17.8	6.2	0.047	0.0254	0.30
	2	15.3	4.6	0.0345	0.0186	0.30

3. Discussion

The following observations can be made from the obtained results.

PA takes place in all the studied samples *($A \neq 0$)*. It is changing when the exposure time is changed. Kinetic PA behaviour is different in transmission and reflection modes as seen from different $P_s - P_p$ time dependences. The values of anisotropy contrast A are much higher in reflection mode and mainly positive whereas they are lower and often negative in transmission mode.

In transmission mode, evidently, the volume processes play the main role. In principle, volume processes mediated by multiple internal reflections can also explain the reflection mode results because reflection coefficients are higher for *s*-polarization than for *p*-polarization ($R_s > R_p$). This effect can compensate for higher refractive index changes at *p*-polarization. However, the quantitative evaluation of this effect by thin HG theory including multiple internal reflections [8] has shown that it can be neglected.

PA processes are qualitatively different in a-As_2S_3 film compared to molecular glasses since $P_s - P_p$ time dependences are different. The main features of processes in molecular glasses are similiar as judged by $P_s - P_p$ time dependences . However, the details are different. The most efficient recording is found in K-D-32: η_s =17.3% and η_p =14.7% in transmission mode, η_s =4.7% and η_p =2.5% in reflection mode. The largest transmission anisotropy is found in K-D-24 but the the largest reflection anisotropy takes place in K-D-25. Thus the most efficient recording does not mean the largest photoinduced anisotropy.

In the case of transmission mode for molecular glasses we propose the following PA explanation. HG recording in molecular glassy films is due to the *trans-cis* photoisomerization including photoorientation of chromophores perpendicularly to the *p*-polarized light electric field vector $\boldsymbol{E_p}$ because 532 nm recording wavelength is close to the *trans*-form absorption maxima (see figure 1) [2,3]. *Trans-cis* photoisomerization is followed by refractive index changes at 633 nm and by mass transfer leading to simultaneous surface relief HG formation [2]. Forces, initiating mass transfer (permittivity gradient force, electric gradient force, photoisomerization pressure [2]) create spatial mechanical stress modulation which elastooptically modulates refractive index mainly in $\boldsymbol{E_p}$ direction. This explains the first negative $P_s - P_p$ minimum in figures 4,6,8. After that the volume HG due to the chromophore orientation begins to prevail and $P_s - P_p$ increases.

Similiar processes take place in a-As_2S_3 film. Photoinduced structural changes (PSC) including (polarization-selective) As-As bond breaking and As-S bond formation are responsible for HG recording [3,4]. This again creates periodical mechanical stress modulation and larger refractive index changes in $\boldsymbol{E}_p$ direction responsible for the first $P_s - P_p$ minimum (figure 10). The small initial peak in figure 10 can be due to the D-centre (dangling bond) photoorientation [9]. Further on PSC HG prevails.

In reflection mode, surface relief HG play the main role and mainly A>0. If surface modulation depth is less than wavelength and HG period is sufficiently larger than wavelength surface relief HG can be treated as a scalar phase grating . Then diffraction efficiency polarization dependence is determined by R_s>R_p.condition as observed .

Transmission diffractive anisotropy can be treated as photoinduced birefringence. Due to the processes considered above optical axis arises parallel to $\boldsymbol{E}_p$. Basing on Kogelnik's theory [10] and our optical measurements we have calculated the photoinduced refractive index changes, Δn_s and Δn_p , corresponding to the measured diffraction efficiencies η_s and η_p. Maximal absolute values of Δn_s=0.094 and Δn_p=0.10 were found in the case of a-As_2S_3. In the case of molecular glasses the maximal absolute values Δn_s=0.124 and Δn_p=0.116 were found in the case of K-D-32. Thus the birefringence values $n_s - n_p = \Delta n_s - \Delta n_p$ as high as 0.194 and 0.24, respectively, are possible if the signs of Δn_s and Δn_p are opposite. We believe that this is the case because matter polarization increases in the direction perpendicular to $\boldsymbol{E}_p$ and decreases in $\boldsymbol{E}_p$ direction.

4. Conclusions

PA is found in all the studied samples by a novel holographic method in both transmission and reflection modes. PA is time dependent.

The diffractive anisotropy time dependences notably differ for a-As_2S_3 film and for molecular glassy films. Several differences can be found also between different molecular glasses. Holographic recording efficiency and photoinduced anisotropy do not correlate.

The PA time dependences measured by *s*- and *p*-polarized diffracted light power difference is different in transmission and reflection modes. In transmission mode, PA of molecular glasses is due to the photoinduced volume processes including *trans-cis* photoisomerization , chromophore orientation and mechanical stress modulation. In the case of a-As_2S_3 film these photoinduced volume processes include PSC, D-centre orientation and, again, mechanical stress modulation. All these processes lead to a photoinduced birefringence. Surface relief HG recording and polarization-dependent reflection are responsible for diffractive anisotropy in reflection mode.

Acknowledgment
The financial support from the Latvian State Program „Materials for photonics and nanoelectronics based on new functional low molecular and high molecular compounds „ is greatly acknowledged.

References

[1] 2008 *Introduction to Organic Electronic and Optoelectronic Materials and Devices* ed Sam-Shajing and Larry R Dalton (Boca Raton: CRC Press) 910 p
[2] 2009 *Smart Light-Responsive Materials Azobenzene-Containing Polymers and Liquid Crystals* ed Yue Zhao and Tomiki Ikeda (Hoboken , New Jersey: John Wiley & Sons, Inc) 514 p
[3] Schwartz K 1993 *The Physics of Optical Recording* (Berlin: Springer-Verlag) 190 p
[4] Ozols A Nordman N Salminen O Riihola P 1997 *Proc.SPIE* **2968** 282
[5] Ozols A Saharov D Kokars V Kampars V Maleckis A Mezinskis G Pludons A 2010 *J. of Phys.: Conf.Series* **249** 012055
[6] Ozols A Kokars V Augustovs P Uiska I Traskovskis K Saharov D 2011 *Centr. Eur. J.Phys.***9** 547
[7] Ogiwara A Kakiuchida H Tazawa M Ono H 2007 *Jap.J.Appl.Phys.* **46** 7341
[8] Ozols A Augustov P A Shvarts K 1978 *Opt.Spektrosk.* **44** 1171
[9] Mott N F Davis E A 1979 *Electron Processes in Non-Crystalline Materials* (Oxford: Clarendon Press) 663 p
[10] Kogelnik H 1969 *Bell Syst.Techn.J.***48** 2090

IOP Conf. Series: Materials Science and Engineering **38** (2012) 012012 doi:10.1088/1757-899X/38/1/012012

Optical field-induced surface relief formation on chalcogenide and azo-benzene polymer films

J Teteris[1] **and U Gertners**
Institute of Solid State Physics, University of Latvia, 8 Kengaraga Str., Riga, LV-1063, Latvia

E-mail: teteris@latnet.lv

Abstract. The dependence of the surface relief formation in amorphous As_2S_3 and Disperse Red 1dye grafted polyurethane polymer films on the polarization state of recording light was studied. It is shown that the direction of mass transport on the film surface is determined by the direction of light electric vector and photoinduced anisotropy in the film. We propose a photoinduced dielectropfhoretic model to explain the photoinduced mass transport in amorphous films. Model is based on the photoinduced softening of the matrix, formation of defects with enhanced or decreased polarizability, and their drift under the electrical field gradient of light.

1. Introduction

Chomat *et al* [1] already in 1976 showed that a thin film of an amorphous As_2Se_3 subjected to an intensity modulated red light interference pattern was significantly altered: a relief hologram was formed on the surface of the film. Almost twenty years later a similar phenomenon for thin layers of azobenzene-containing polymers was observed by Rochon *et al* [2]. A flat film can reach a corrugation whose amplitude may be up to fifty per cent of the initial thickness. This suggests that the material moieties are put in light-induced lateral transfer regarding the propagation direction of exciting light. The most commonly used method to induce mass motion is holographic recording with which a sinusoidal modulated intensity pattern can be obtained at the submicron scale. Recently, it has been demonstrated that surface relief structures can be formed also by a focused light spot [3] and illumination performed via a metal microgrid placed on the sample [4]. The surface patterns can be erased by homogeneous illumination or heating above glass transition temperature of recording media. The surface relief grating (SRG) direct formation by two coherent laser beam interference have been observed in different disordered materials: in amorphous inorganic (As-S, As-Se chalcogenides [1, 4, 5], Sb-P-O oxide-containing glass [6]) and organic (azobenzene-containing organic polymers [2, 7], acrylamide photopolymers [8], azocellulose [9], organic molecular glasses [10]) compounds.
The efficiency of SRG formation on the azobenzene containing polymer and chalcogenide semiconductor films strongly depends on the polarization state of the recording beams [5, 11-13]. The two *s*-polarized recording beams produce the largest light intensity modulation in interference pattern. But this configuration only produces a SRG with very low diffraction efficiency and small surface modulation. Under the *s-p* polarization recording conditions, the two writing beams possesses orthogonal polarization and the resultant electric vector of light on the film surface has the largest

[1] To whom any correspondence should be addressed.

variation but the light intensity is uniform over the entire exposed area. This purely polarization recording condition also produces very small surface modulation. The largest SR modulation was obtained under $+45^0$: -45^0 and RCP : LCP recording conditions. This indicates that the existence of both light intensity and resultant electric field variations are essential to the formation of SRG on amorphous chalcogenide and azobenzene polymer films.

A number of models have been proposed to explain the origin of the driving force responsible for SRG formation in the azo-polymers under the light illumination on the molecular level, including mean-field theory [14], permittivity gradient theory [15], gradient electric force model [16] and other. There is still not general agreement on the origin of the driving force for this process and no mechanism yet proposed accounts for all experimental observations.

The main goal of this paper is to explore the dependence of surface relief grating formation efficiency in amorphous As_2S_3 and azobenzene containing polymer films on recording parameters such as polarization state of recording beams and incoherent assisting light as well as to establish the spatial phase relationship between the surface features and the light intensity distribution.

2. Experimental details

Thin films of amorphous As_2S_3 were prepared by thermal deposition on glass substrates at room temperature. The pressure during deposition was around 10^{-6} Torr and the rate of evaporation was ~20 nm/s. A series of polyurethane polymers containing chemically bonded azobenzene dye DR1was synthesized [17]. The solution of polymer was cast on glass substrate and dried.

The surface relief (SR) formation experiments were performed using both a holographic recording system described in [18] and an illumination of amorphous films through optical slit (Fig.1). An additional illumination of grating performed during the holographic recording by an incoherent assisting beam (I_A) of *p*- or *s*-polarization regarding the recording beams comes from another laser. The SR grating recording efficiency was defined as a tangent of the slope angle α for the linear part of reflection diffraction efficiency (DE) growing curve [18].

The scheme for SR formation on the amorphous films by illumination through optical slit is shown in Fig.1. The sample (S) has been illuminated from the film side with a single recording light beam I_R (532 nm) through a slit with a width of ~10 μm. From the opposite side the film was illuminated through the substrate by assisting light beam I_A (532 nm) coming from another laser, i.e. incoherent to beam I_R. The intensities of the beams were I_R= 0.5 W/cm^2 and I_A=0.43 W/cm^2. The wave plates WP1 and WP2 were used for the combination of polarization states of the beams. The As_2S_3 and polyurethane samples with thickness 4.5 μm and 3.75 μm, respectively, were exposed 180 min. at room temperature.

The photo-induced surface relief of the films was measured by an atomic force microscope (AFM).

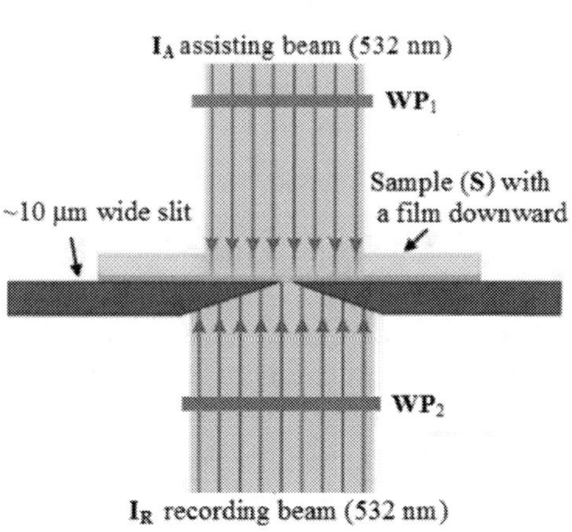

Figure 1. Experimental setup for formation of surface relief structures by illumination through optical slit.

3. Experimental results and discussion

3.1. Holographic recording of SRG.

It is known that the efficiency of SRG formation in amorphous chalcogenide and organic azobenzene polymer films essentially depends on the polarization state of the holographic recording beams [5, 11-13]. A large difference in SRG recording efficiencies has been observed. The dependence of SRG recording efficiency on polarization state combinations of recording beams and contrast of interference patterns at the sample surface during holographic recording are summarized in Table 1. The contrast of interference pattern (K) in amorphous As_2S_3 film for a period of 1 μm was calculated according to [19] using a value of refractive index of 2.6 at recording wavelength of 532 nm.

Under the intensity modulated holographic recording conditions, the two parallel linear (*s-s* and *p-p*) polarized beams produce the largest light intensity modulation, when the contrast of an interference pattern is equal to ~1. However, there is no spatial modulation in the direction of the resultant electric vector of light. These configurations only produce a very low DE and small SR modulation. Under the polarization modulated holographic recording, the two orthogonal linear (*s-p* and $+45^o$: -45°) and circular (*R-L*) polarized beams produce the largest variation in the resultant electric field direction on the film surface but the light intensity is uniform over the entire exposed area (K≈0). Very small DE and SR modulations were obtained for *s-p* recording configuration. The largest DE and SRG modulation were obtained for orthogonal linear ($+45^0$:-45^0) and circular (*R-L*) polarized beam configuration. For these configurations there are components of resultant electric vector of light parallel and perpendicular to the grating vector direction. This indicates that the resultant electric field and its direction variation are essential to the formation of SR gratings on the amorphous chalcogenide films.

Table 1. Dependence of SRG recording efficiency on amorphous As_2S_3 film with (column I_A=260) and without (column I_A=0) assisting illumination on polarization and contrast of interference patterns at the sample surface during holographic recording (532 nm, $I_1=I_2=0.25$ W/cm^2, Λ= 1 μm) using a variety of polarized recording beam combinations (x-axis coincides with the direction of grating vector). Assisting illumination was performed by 473 nm light with intensity I_A=260 mW/cm^2; polarization state (**S** or **P**) of an assisting beam is shown in cells.

Polarization of recording beams	Resulting electric field vector pattern (*x-y* plane)					Contrast of interf. pattern	SRG recording efficiency, a.u.	
x:	+Λ/2	+Λ/4	0	-Λ/4	−Λ/2	K	I_A=0	I_A=260
S:S ↕ ↕	•	↕	↕	↕	•	1	0.1	19 P
P:P ↔ ↔	◆	↔	↔	↔	◆	0.98	0.35	91 S
S:P ↕ ↔	↘	○	↗	○	↘	0	0.04	-
$+45^0$:-45^0 ↙ ↘	↔	○	↕	○	↔	0	44	88 P
RC:LC ○ ○	↔	↗	↕	↘	↔	0.01	90	60 P

3.2. Enhancement of SRG holographic recording by assisting light.

In 2001 Jager et al [20] showed that an efficiency of SRG formation on azobenzene polymer films can be essentially enhanced by an assisting light with an orthogonal polarization to s-s holographic recording light. This process was explained by "photo softening" the medium through enhancing the *trans-cis-trans* isomerization cycling. Similar enhancement of SRG formation by assisting light was observed also in amorphous chalcogenide As_2S_3 and $AsSe_4$ films [13, 21].

The enhancement efficiency strongly depends on the wavelength, intensity and polarization direction of an assisting light [18]. The results on dependence studies of SRG formation efficiency and enhancement in As-S film by assisting light on the polarization state combinations of recording and assisting beams are summarized in Table 1. We can see that even in *s-s* polarized writing conditions, a significant SR grating formation enhancement can be achieved if an incoherent assisting beam with orthogonal (p) polarization is used. An effective enhancement by an additional illumination has been observed for intensity modulated holographic recording conditions (*s-s* and *p-p*). However, a difference of an enhancement between *s-* and *p-* polarized assisting beams suggests the idea that the material photo-induced softening in the dark places of the grating is not the only reason for this phenomenon. Actually, the additional illumination of *s-s* or *p-p* holographic recording combination with orthogonal assisting beam creates the electric field distribution in the interference pattern that is similar with the *L-R* and $+45^{\circ}$: -45° recording situation without the assisting light when the largest SRG formation can be obtained.

3.3. SR relief formation by illumination through optical slit.

The phase relationship between the exciting light field and the resulting surface deformation is crucial in understanding the mechanism of SRG formation. Early investigations on azopolymer films showed that the light and surface relief are 180^0 out of phase [15]. It means that light intensity maxima correspond to valleys in the surface relief. In other words, material is moved out of the light intensity maxima into the dark regions.

In order to estimate the influence of polarization state of the recording beams on SR formation direction and efficiency we performed experiments with an illumination of amorphous As_2S_3 and azo-benzene containing films through optical slit. The experimental scheme is shown in Fig.1. After illumination the SR of the samples was studied by AFM.

Actually these experiments simulate the recording of single line grating enabling to evaluate the dependence of photoinduced mass transport direction on the polarization state of recording light. The single beam experiments with the writing beam (I_R) polarized parallel to the optical slit are similar to the holographic recording of SRG with *s:s* polarization, and the single beam experiments with the writing beam polarization perpendicular to the optical slit are similar to the *p-p* holographic recording. The illumination of the samples through slit performed with *s-*, *p-* recording beam (I_R) in combination

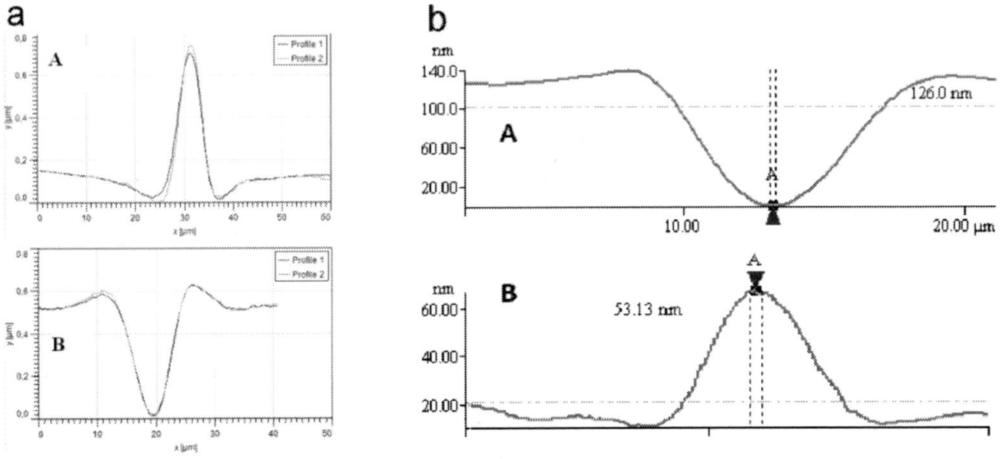

Figure 2. AFM measurements of SR on As_2S_3 (a) and polyurethane (b) films illuminated through optical slit. A – *p*-polarized recording beam (E vector perpendicular to slit) and *s*-polarized assisting beam (E vector parallel to slit); B – *s*-polarized recording beam and *p*-polarized assisting beam.

with, *p*-, *s*- polarized assisting beam (I_A) simulates the SRG formation with $+45^0$: -45^0 and RC:LC laser beams combination at holographic recording. In case of single beam illumination without assisting light very small SR changes (~10 nm) were obtained for *s*-and *p*- polarization recording. The largest SR changes in amorphous As_2S_3 films are observed for *s*- and *p*- polarized recording beam assisted with orthogonal *p*- and *s*- polarized additional illumination, respectively (Fig.2a). It is seen that the direction of the photo-induced matter transport in both cases is with opposite sign. Consequently, *s*-polarized recording beam assisted with *p*-polarized light creates a contraction of the film or a formation of valley but *p*-polarized recording beam assisted with *p*-polarized light creates an expansion of the film or a formation of hill. These results are consistent with a conclusion obtained by Trunov [21] that the direction of the material displacement in amorphous chalcogenide thin films for *p-p* holographic recording with *s*- assisting light (scheme *p-p*:+*s*) is in phase and for scheme *s-s*:+*p* out of phase to the maxima in the light interference pattern.

The similar experiments were performed on DR1 grafted polyurethane films. Fig.2b shows that in this case the direction of the photoinduced matter transport is opposite to that obtained with As_2S_3 films. The surface deformation goes down when the recording beam polarization is perpendicular to the slit and goes up when the recording beam polarization is parallel to the slit. Consequently, the photoinduced mass transport process depends not only on the polarization direction but also on the properties of material.

The interference of two orthogonal, circularly (*L:R*) or $+45^0$:-45^0 linearly polarized beams causes a space-varying polarization pattern on sample (Fig.3 and Fig.4). That space-varying polarization is linear changing periodically along the *x* direction. The peak of the SRG is formed at the position of p-polarization state for As_2S_3 films (Fig.3A) and s-polarization state for azobenzene polymer films (Fig.3B) in the polarization modulation pattern. This fact is crucial to evaluate the direction of the force to form the SRG. It is, since the SRG is formed by molecular migration, we can confirm that the force drives the migration from s-polarization state toward the position of p-polarization state in case of As_2S_3 films and in the opposite direction for azobenzene polymer films.

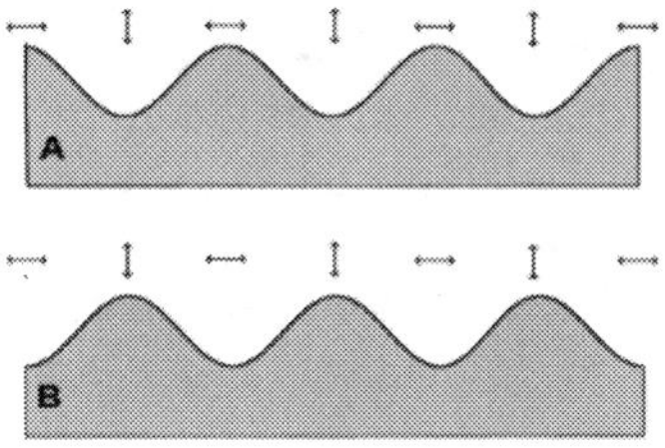

Figure 3. The position of the surface relief grating regarding an interference pattern using $+45^0$:-45^0 or LCP:RCP polarization combinations for recording: A) for As_2S_3 film; B) for DR1 dye grafted polyurethane polymer film.

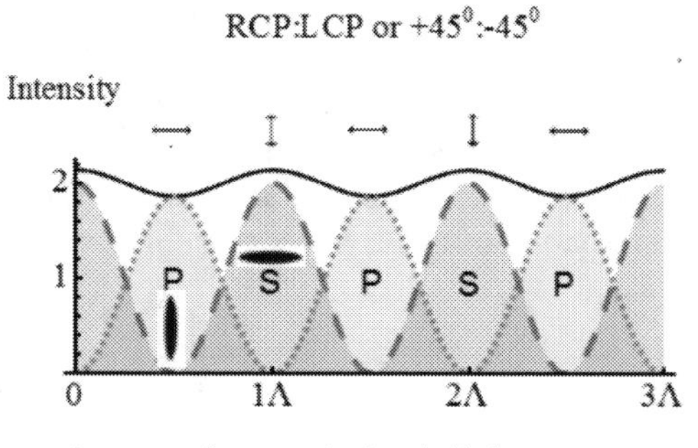

Figure 4. *s*- and *p*- polarized light components in space-varying polarization pattern on sample for $+45^0$:-45^0 or LCP:RCP recording. An orientation of azo-dye molecules due to photoinduced anisotropy is shown for *s*- and *p*- components of light electric field.

3.4. Photoinduced dielectrophoretic model of SRG formation.

We propose a photoinduced dielectrophoretic model to explain the photoinduced mass transport in amorphous films. Model is based on the photoinduced softening of the matrix, formation of defects with enhanced or decreased polarizability, and their drift under the electrical field gradient of light.

It is known that molecules and nano particles can be manipulated in fluidic systems using dielectrophoretic (DEP) forces [22]. A DEP force is generated when a neutral molecule or particle is suspended in a non-uniform electric field. This electric field induces electrical charges within the particle to establish a dipole. If a non-uniform electric field is applied, the ends of the dipole experience unequal Columbic forces, which result in a total non-zero, imposed force. If the particle is less polarizable than the suspending medium, it is repelled from the regions of higher electric field and the motion is called negative dielectrophoresis, while the opposite case is referred to as positive dielectrophoresis.

An interference pattern on sample surface at $+45^0$:-45^0 or LCP:RCP polarization recording can be expressed as a sum of two linearly polarized light components (see Fig.4): with electric vector parallel to grating vector (P) and with electric vector perpendicular to grating vector (S). For both components the electric field gradient vector is parallel to grating vector. Due to polarization modulation linear polarized light in azo dye containing polymers causes an arrangement of the photo active molecules (see Fig.4). As a result their palarizability regarding the light electric field gradient direction has been increased by S-electric field component and decreased by P-electric field component. Due to the dielectrophoretic forces azo-polymer material is moved out of the P-polarized light component region into S-polarized regions.

4. Conclusion

We have shown that the surface relief grating formation in amorphous As_2S_3 and azobenzene grafted polyurethane films strongly depends on the polarization state of recording beams. The surface relief grating formation efficiency of *s-s* and *p-p* recording beam combination can be essentially enhanced by additional illumination with orthogonal polarization. It is shown that the peak of the SRG is formed at the position of p-polarization state for As_2S_3 films and at s-polarization state for azobenzene polymer films in the polarization modulation pattern. The photoinduced dielectrophoretic model to explain the photoinduced mass transport in amorphous films has been proposed. Model is based on the photoinduced softening of the matrix, formation of defects with enhanced or decreased polarizability, and their drift under the electrical field gradient of light.

References

[1] Chomat H, Ležal D, Gregora I and Srb I 1976 *Journ. Non-Cryst. Sol.* **20** 427
[2] Rochon P, Batalla E and Natansohn A 1995 *Appl. Phys. Lett.* **66** 136
[3] Hisakuni H and Tanaka K 1994 *Appl. Phys. Lett.* **65** 2925
[4] Kokenyesi S, Ivan I, Takats V, Palinkas J, Biri S and Szabo I A 2000 *JNCS.* **353** 1470
[5] Asatryan K E, Galstian T and Vallee R 2005 *Phys. Rev. Lett.* **94** 087401
[6] De Vicente F S, Siu Li M and Messaddeq Y 2004 *Journ. Non-Cryst. Sol.* **348** 245
[7] Bellini B, Ackermann J, Klein H, Graves Ch, Dumas Ph and Safarov V 2006 *J. Phys.: Cond.Matter* **18** S1817
[8] Pavani K, Naydenova I, Martin S and Toal V 2007 *J. Opt. A: Pure Appl. Opt.* **9,** 43
[9] Yang K, Yang S and Kumar J 2006 *Phys. Rev. B,* **73** 165204
[10] Audorff H, Walker R, Kador L and Schmidt H W 2009 *Proc.SPIE* **7233** 72330O-1
[11] Kryshenik V M, Trunov M L and Ivanitsky V P 2007 *J. Optoelectron. Adv. Mater.* **9** 1949
[12] Jiang X L, Li L, Kumar J, Kim D Y, Shivshankar V and Tripathy S K 1996 *APL* **68** 2618
[13] Gertners U and Teteris J 2010 *Opt.Mat.* **32** 807
[14] Pederson T G, Johansen P M, Holme N C and Ramanujam P S 1998 *PRL* **80** 89
[15] Yager K G and Barret C J 2001 *Cur. Opinion Sol. St. Mat. Sc.* **5** 487
[16] Kumar J, Li L, Jiang X, Kim D, Lee T and Tripathy S 1998 *Appl. Phys. Lett.* **72** 2096
[17] Aleksejeva J, Gerbreders A, Gertners U, Reinfelde M and Teteris J 2011 *IOP Conf. Series: Mat.Sc.Eng.* **23** 012006
[18] Teteris J, Aleksejeva J and Gertners U 2011 *IOP Conf. Series: Mat.Sc.Eng.* **23** 012002
[19] Cai L Z and Yang X L 2002 *Optics&Laser Techn.***34** 671
[20] Jäger C, Bieringer T and Zilker S J 2001 *Appl. Opt.* **40** 1776
[21] Trunov M L, Lytvyn P M and Dyachyns`ka O M 2010 *Appl. Phys. Lett.* **97** 031905
[22] Wells G G, Sampara N, Kriezis E E, Fyson J and Brown C V 2011 *Opt.Lett.* **36** 4404

IOP Conf. Series: Materials Science and Engineering **38** (2012) 012013 doi:10.1088/1757-899X/38/1/012013

Influence of specific and non-specific solvent effects on frontal MO of DMABI system

V Kampars[1], P Pastors and G Ornicane

Institute of Applied chemistry, Riga Technical University, Azenes 14/24, Riga, LV-1048, Latvia

E-mail: Valdis.Kampars@rtu.lv

Abstract. Derivatives of 2-(4'-dimethylaminobenzylidene)-1,3-indandione (DMABI) are classical push-pull chromophores containing carbonyl groups like numerous other non-linear chromophores for electro-optical devices. The energy of the intramolecular charge transfer exciton of DMABI is determined by the difference of HOMO-LUMO energies and can be varied in width range by chemical modifications of the DMABI structure or by the non-specific and specific solvent effects if the chromophore is incorporated into condensed environment. The condensed medium in a thin film (polymer or molecular glass) exhibits a solid state solvation effect similar to the effect of the organic solvents in a solution. The value of the medium effect depends on specific and non-specific interactions between the chromophore and medium. The main purpose of this work is to evaluate the specific effects. The chromophores often contain few centres for specific interaction, and thus we have synthesised and investigated UV-Vis spectra of three derivatives of DMABI and also three derivatives of 4-aza-DMABI in solutions with strong H-bond donor properties. The experimentally obtained results show that the strong H-bond with carbonyl group lowers the intramolecular charge-transfer energy by 900 cm^{-1}, the strong H-bond with carbonyl group and heterocyclic nitrogen in 4-aza-DMABI by 1500 cm^{-1}, and the protonation of heterocyclic nitrogen in 4-aza-DMABI by 3250 cm^{-1} respectivelly.

1. Introduction

The carbonyl group containing molecules are popular nonlinear optical chromophores and components of the electro-optical devices [1], but derivatives of 2-(4'-dimethylaminobenzylidene)-1,3-indandione (DMABI) within this field are traditional objects of investigation [2, 3]. As a classical puch-pull chromophore DMABI has a different location of HOMO and LUMO and a characteristic intensive intramolecular charge transfer absorption band in the 20000 cm^{-1} region. The energy of this intramolecular charge transfer exciton is determined by the difference of HOMO-LUMO energies and can be varied in width range (approximately 10000 cm^{-1}) by introduction substituents in the benzylidene or phthaloyl part. Electro-optical device is build up from thin films, including thin film containing chromophore. The medium (polymer or molecular glass) alter the electronic energy levels of the individual, isolated chromophore molecule and all properties of the molecule incorporated in

[1] To whom any correspondence should be addressed

solid can be changed radically. The effect of the medium depends on the strength of the local electric fields inside the electro-optical material and also on the effects of specific interaction with medium (specific and non-specific effects). The solid state solvation effect like to the chemical modification can be used as a method for tuning the energy level structure of organic materials [4] and it is very important to evaluate how large this effect can be. Since the effect of the medium both in the films and in the solutions can be described by use the some models [4], the investigation of chromophore in different liquid media can give an important information about the quantity of the solid state solvation effects. In our previous work we have investigated the nonspecific and weak H-bond effect on the energy of intramolecular charge transfer (CT) exciton of DMABI [5] and we had to conclude, that the non-specific effects on DMABI is approximately 950 cm^{-1}, but weak H-bond with the solvent gives additionally 200 cm^{-1}. The investigation of DMABI analogues in solvents with or without weak H-donor properties has been also done by other researchers [3,4,6]. But it could be said that the investigation of 1,3-indandione dyes in strong H-donor solvents has not been done jet. For example, electronic spectra of derivatives of 2-aryl-1,3-indandiones were investigated only in trifluoroacetic acid and in its mixture with dichloromethane [7]. When the dye has one or more additionally heteroatoms, it is observed that the solvent interact not only with carbonyl groups, but also with these heteroatoms and so the strong acidic medium effect has been too complicated for further analysis [7]. In order to reinvestigate the problem of 1,3-indandione carbonyl group interactions with strong acidic media we have studied the electronic spectra of derivatives of 2-benzylidene-1,3-indandiones (DMABI) and 4-aza-2-benzylidene-1,3-indandiones (4-aza-DMABI) in an ethanol/hydrochloric acid media.

2. Experimental

2.1. Materials and measurements

Ethanol, 36% hydrochloric and 98% sulphuric acids - analytical grade quality commercial reagents, used as received. The elemental analyses were carried out on "Carlo Erba 1108" microelemental analyser. Melting point was determined by using the "Stuart SMP10" melting point apparatus. ^1H spectra were obtained on "Brucker "WH-90/DS"(compounds 1 a-c) and "Varian UNITY INOVA 600" (compounds 2 a-c) in $CDCl_3$ at 25 °C. Chemical shifts are given relative to the residual signal of the solvent (1H 7.24 ppm). IS-Spectra were recorded on "Thermo Electron Corporation Nicolet 5700" FT-IR spectrophotometer in Nujol. UV-Vis absorption spectroscopic studies were performed with LAMBDA 650 spectrometer. The acidity of ethanol solutions was raised by gradual addition of the hydrochloric acid solution.

2.2. Preparation

The investigated dyes 1 a-c and 2 a-c were synthesised accordingly to the figure 1.

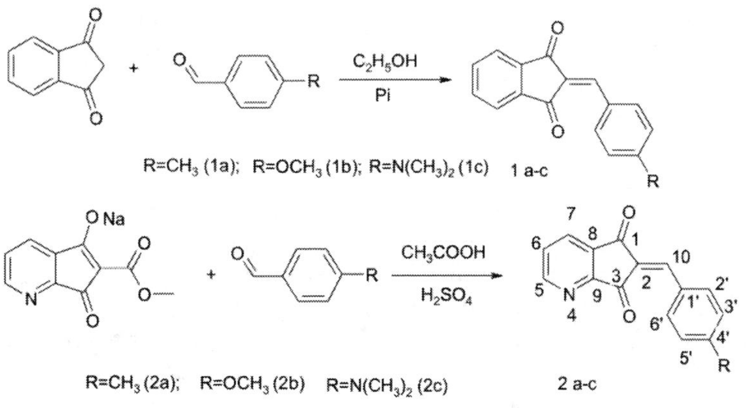

Figure 1. Synthesis of dyes 1 a-c and 2 a-c.

2.2.1. *The typical procedure:the preparation of 2-(4'-R-Benzylidene)- 1,3-indandiones 1a-c.* Indan-1,3-dione (0.5 g, 3.4 mmol) and 4-R-benzaldehyde (3.5 mmol) were dissolved in boiling ethanol (30 ml) and then the piperidine (0.3 ml) solution in ethanol (10 ml) was added. The solution becomes damson-coloured. Stirring for 10 minutes, then the acetic acid (1 ml) is added to the solution. After that the solution is cooled to room temperature, then placed in the refrigerator. After 12 hours the precipitate is filtered and crystallized from acetonitrile. Yield of compounds was 70-75 %.
1a. ^{1}H NMR (90 MHz, $CDCl_3$) δ (ppm): C- 4,7 7.89, C-5,6 7.70, C-10 7.78, C-2',6' 8.29, C-3',5' 6.67, R 2.36. FT-IR ν/cm^{-1}: 1719, 1679, 1586, 1509, 1351, 1247, 1183, 1079, 739. Anal. Calc. for $C_{17}H_{12}O_2$: C 82.24, H 4.87; found: C 82.30, H 4.76.
1b. ^{1}H NMR (90 MHz, $CDCl_3$) δ (ppm): C-4,7 7.89, C-5,6 7.72, C-10 7.78, C-2',6' 8.47, C-3',5' 6.96, R 3.84. FT-IR ν/cm^{-1}: 1717, 1690, 1586, 1580, 1312, 1266, 1174, 1082, 730. Anal. Calc. for $C_{17}H_{12}O_3$: C 77.26, H 4.58; found: C 76.92, H 4.64.
1c. ^{1}H NMR (90 MHz, $CDCl_3$) δ (ppm): C- 4,7 7.82, C-5,6 7.64, C-10 7.69, C-2',6' 8.44, C-3',5' 6.67, R 3.04. FT-IR ν/cm^{-1}: 1705, 1659, 1551, 1580, 1226, 1186, 1087, 723. Anal. Calc for $C_{18}H_{15}NO_2$: C 77.96, H 5.45, N 5.05; found: C 77.83, H 5.33, N 5.01.

2.2.2. *The typical procedure: the preparation of 2-(4'-R-Benzylidene)-4-aza-1,3-indandiones 2a-c.* The sodium salt of 4-aza-1,3-indandione-2-carboxylic acid methyl ester (0,45 g, 2 mmol) and 4-R-benzaldehyde (2 mmol) in glacial acetic acid (5 ml) were heated to 70-75 °C and then maintained at the same temperature for 1 h. The reaction mixture was left at the room temperature for 24 h and then filtered. The filtrate was evaporated in vacuum to get an oily residue, which was crystallized from isopropyl alcohol. Yield of the compounds was 40-60 %. M.p. > 200 °C (decomp.)
2a. ^{1}H NMR (600 MHz, $CDCl_3$) δ (ppm): C-5 9.15, C-6 7.74, C-7 8.37, C-10 7.98, C-2',6 '8.49, C-3',5' 7.80, R 2.45. FT-IR ν/cm^{-1}: 1728, 1688, 1595, 1580, 1312, 1289, 1189, 1001, 806. Anal. Calc. for $C_{16}H_{11}NO_2$: C 77.10, H 4.45, N 5.62; found: C 76.72, H 4.31, N 5.66.
2b. ^{1}H NMR (600 MHz, $CDCl_3$) δ (ppm): C-5 9.05, C-6 7.84, C-7 8.27, C-10 7.98, C-2',6' 8.59, C-3',5' 7.04, R 3.96. FT-IR ν/cm^{-1}: 1726, 1686, 1574, 1510, 1313, 1274, 1179, 1022, 836. Anal. Calc. for $C_{16}H_{11}NO_3$: C 72.45, H 4.18, N 5.28; found: C 72.26, H 4.02, N 5.31.
2c. ^{1}H NMR (600 MHz, $CDCl_3$) δ (ppm): C-5 8.98, C-6 7.56, C-7 8.18, C-10 7.91, C-2',6 '8.57, C-3',5' 6.75, R 3.16. FT-IR ν/cm^{-1}: 1719, 1677, 1555, 1516, 1306, 1271, 1189, 816. Anal. Calc. for $C_{17}H_{14}N_2O_2$: C 73.37, H 5.07, N 10.07; found: C 73.42, H 5.02, N 10.01.

3. Results and discussion

All the investigated compounds in ethanol solution have an intensive long wavelength absorption band with maxima between 19000 and 22000 cm^{-1} (table 1). Accordingly to the CT character of this band the increase of the electron donor properties of 4'-substituent in the donor part of dyes regular shifted this band to the long wavelength region. The replacement of the CH group with more electronegative nitrogen in the acceptor part gives the similar effect. If the acidity of the solvent is raised, the bathochromic shift of the long wavelength absorption band is observed, and this effect may be related to the solvent interaction with the only equal for all dyes 1,3-dicarbonyl part. However figure 2 shows, that we have two different groups of dyes (DMABI and 4-aza-DMABI groups) with specific behaviour of compounds 1c and 2c (dimethylamino derivatives). The DMABI derivatives have a considerable weaker effect from the medium variation than the 4-aza-DABI derivatives and the reason could be the N atom in the phthaloyl moiety which is an additional H-acceptor and protonation position. The simultaneous interaction of the acidic medium with the carbonyl group and heteroatom strengthen the acceptor properties of phthaloyl part more, than the interaction with carbonyl group and thus cause a greater bathochromic shift. The analysis of the variation of the intensities of the long wavelength absorption bands (figure 3) gives another classification of investigated dyes. In agreement with previous classification we have lines with and without curvature and can separate the compounds in two new groups: a group what unify dyes with small variations of absorption intensity (dyes 1a,

2a, and 2b) and a group with clearly defined decrease of the intensity with increase of hydrochloric acid concentration (dyes 1b, 1c and 2c). The several different group presence in dyes points to the important influence of all the heteroatoms. The only common part of all dyes is the carbonyl group and it is known, that the carbonyl group (the oxygen atom of the carbonyl group) is a proton-acceptor and can also be protonated in the highly acidic media [8,9].

Table 1. The maximum absorption wavelengths of the dyes in ethanol/hydrochloric acid mixture.

Concentration of HCl, mol/l	Maximum absorption wavelengths, cm^{-1}					
	1a	1b	1c	2a	2b	2c
0	28026	25853	20609	27215	24427	19762
1	27925	25712	20408	27046	24170	19292
2	27904	25612	20233	26730	23729	18328
3	27700	25447	20040	26013	22872	17934
4	27588	25334	19927	25155	22197	17770
5	27496	25245	-[a]	24714	21924	17700
6	27426	25197	-	24444	21750	17755
7	27371	25106	-	24271	21641	17958
8	27323	25073	-	24125	21519	-[a]
9	27266	25030	-	23987	21430	-
10	27219	24992	-	23883	21372	-
11	27207	24992	-	23828	21334	-

[a]Charge transfer band disappears due to the protonation of the dimethylamino group

The protonation of the carbonyl group remarkably lowers the energy of the lowest unoccupied molecular orbital (LUMO) and the bathochromic shift of long wavelength absorption band in UV-Vis spectra of such compounds as acetophenone and benzophenone is 5500-10000 cm^{-1} [7, 9, 10- 12].

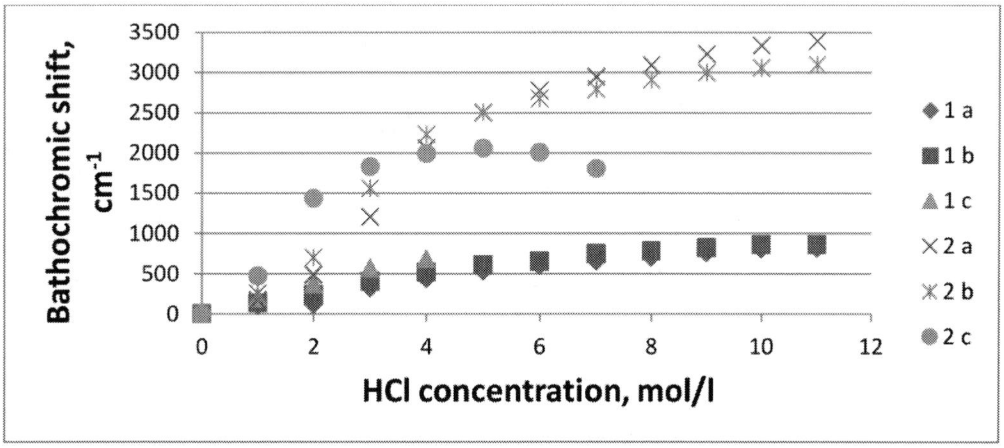

Figure 2. The bathochromic shift of the long wavelength absorption band with the increase of the hydrochloric acid concentration in ethanol/HCl mixture

Our investigation of dye 1a in concentrated sulphuric acid gives a bathochromic shift about 6500 cm^{-1}. So we can conclude, that the protonation of the carbonyl group in ethanol/HCl mixture is not observed. Likewise the protonation of 1,3-dicacarbonyl system in trifluoroacetic acid and in its mixture with dichloromethane also is not observed [7]. If the molecule of the dye has another kind of

IOP Conf. Series: Materials Science and Engineering **38** (2012) 012013 doi:10.1088/1757-899X/38/1/012013

heteroatom, the H-bond formation with the every atom containing electron pair could be possible and the situation would become more complicated. All the investigated compounds have the carbonyl groups of phthaloyl moiety, but each of them has a different number and a different type of other centres of H-bond formation and protonization.

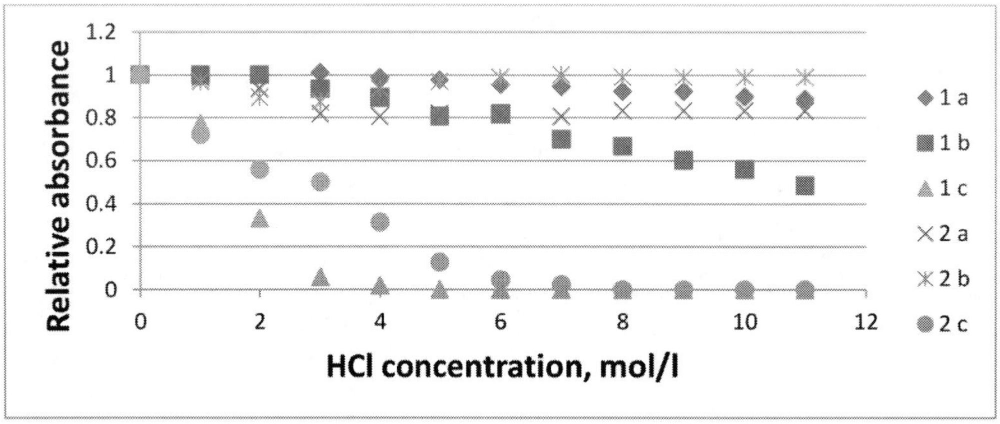

Figure 3. The relative intensity of the long wavelength absorption band with the increase of the hydrochloric acid concentration in ethanol/HCl mixture

The comparison of the values of the experimentally determined bathochromic shifts for the compounds I a-c and II a-c in ethanol/HCl mixtures together with the results of our previous investigation of solvatochromy in different organic solvents (non-specific interaction) and the full carbonyl group protonation in sulphuric acid gives the possibility to separate eight different effects that are shown detailed in table 2.

Table 2. The effect of the media interactions on CT exciton.

No	Weak H-bond	Strong H-bond			Protonation			Effect ±100, cm^{-1}
	=N-	CO	=N-	NR_2	CO	=N-	NR_2	
1.	-	-	-	-	-	-	-	950
2.	-	-	-	-	x	-	-	6500
3.	-	x	-	-	-	-	-	900
4.	-	x	x	-	-	-	-	1500
5.	-	x	-	x	-	x	-	2000
6.	-	x	-	-	-	x	-	3250
7.	-	x	-	-	-	-	x	Band vanish
8.	-	x	-	-	-	x	x	Band vanish

As it is shown in the Table 3 the effect of heterocyclic N protonation causes the most significant effect and can be achieved in the medium without an extreme acidity.

4. Conclusions

The strong acidic medium alter the electronic energy levels of 1,3-indandione dyes and can cause a remarkable bathochromic shift of long wave absorption band as a result of a strong H-bonds formation or the protonation of the heteroatoms. The carbonyl groups and the nitrogen atom in phthaloyl part is a place of localisation of LUMO. The strong H-bond formation and the protonation of these heteroatoms lower the LUMO and cause the bathochromic shift. The bathochromic shift as a result of the strong H-bond formation with the carbonyl groups is about 900 cm^{-1}, but as a result of the protonation of the 4-N atom 3250 cm^{-1}. The effect of the protonation of the carbonyl group is approximately 6500 cm^{-1}. H-bond formations with the heteratom of 4'-substituent lower the HOMO level and cause the hipsochromic shift of the long wavelength absorption band.

Acknowledgements
The authors gratefully acknowledge the financial support from the Lithuania-Latvia-Taiwan project.

References
[1] Raimundo J, Blanchard P, Gallego-Planas N, Mercier N, Ledoux-Rak I, Hierle R and Roncali J 2002 *J. Org. Chem.* **67** 205
[2] Rutkis M, Kreicberga J, Kokars V, Kampars V and Jecs E 2010 *Optical Materials* **32** 796
[3] Xiaochuan L, Sung-Hoon K, Son Y A 2009 *Dyes and Pigments* **82** 293
[4] Madigan C F, Bulovic V 2003 *Phys.Rev.Lett.* **91** 247403
[5] Kampars V and Pastors P 2008 *Scientific Proceedings of Riga Technical University, Material Science and Applied Chemistry* **18** 103
[6] Cui Y, Ren H, Yu J, Wang Z and Qian G. 2009 *Dyes and Pigments* **82** 293
[7] Chiba K, Aihara J, Araya K and Matsunaga Y 1980 *Bull.Chem.Soc.Japan* **43** 1703
[8] Sigalov M, Shainyan B, Chipanina N, Ushakov I and Shulunova A 2009 *J.Phys.Org.Chem.* **22**, 1178
[9] Stewart R, Granger M R, Moodie R B and Muenster J 1963 *Canadian J.Chem.* **41**, 1065
[10] Ghosh K K and Tamrakar P 2003 *Indian J.Chem.* A **42** 1081
[11] Alex S, Le Thanh H and Vocelle D 1992 *Can.J.Chem.* **70** 880
[12] Steward B R and Yatess K 1958 *J.Am.Chem.Soc.* **80** 6355

IOP Conf. Series: Materials Science and Engineering **38** (2012) 012014 doi:10.1088/1757-899X/38/1/012014

Investigation of chemical bonding at metal-ceramic interfaces

A Bakulin[1,4], S Kulkova[1,2], S Hocker[3], S Schmauder[3]
[1] Tomsk State University, pr. Lenina 36, Tomsk 634050, Russia
[2] Institute of Strength Physics and Materials Science of SB RAS, pr. Akademichesky 2/4, Tomsk 634021, Russia
[3] Institute of Materials Testing, Materials Science and Strength of Materials, University of Stuttgart, Germany

E-mail: bakulin@ispms.tsc.ru

Abstract. We present a comparative *ab-initio* study of atomic and electronic properties of Al/TiC(N) and Al/VC(N) interfaces performed using DFT with the projector-augmented-wave method. The most stable configuration of metal film on the ceramic substrate was determined. The work of separation of metallic films in dependence on cleavage plane was calculated. The analysis of the electronic properties confirms a stronger interaction in the case of the Al top position over metalloid which indicates the dominant role of the covalent contribution in the chemical bonding at these interfaces. We demonstrate that point defects at the interface (metal, carbon and nitrogen vacancies) change the adhesion at the metal-ceramic interfaces significantly and cause redistribution of the electron properties across the interface.

1. Introduction

As well known, metal-ceramic interfaces are very common in modern technologies. The influence of interfacial energies is critical in a large number of industrial applications. Metal-ceramic interfaces were in a focus of both experimental and theoretical studies in the last decade. Much attempts of researches were aimed at understanding metal-oxide interfaces [1,2], less is known about metal–ceramic interfaces involving transition metal carbides or nitrides [3-6]. To our knowledge, there have been a few studies in which the influence of point defects on the interfacial energy was studied. In particular, the influence of nitrogen vacancies on the properties of the Fe/VN interface was considered in the paper [7] in dependence on the number of VN layers. It was shown that the presence of nitrogen vacancies raises the interface energy. First principles methods can provide a deep insight into the microscopic nature of the metal-ceramic bonding and their usage increases the understanding of materials properties considerably.

In this paper, we present a comparative *ab-initio* study of adhesion at both Al/TiC(N) and Al/VC(N) interfaces in dependence on metalloid and point defects in the interfacial layers.

2. Computational details

Atomic and electronic structure calculations of the Al-ceramic interfaces were performed using the projector-augmented-wave method [8,9] implemented in the Vienna *ab-initio* simulation package (VASP) [10,11]. Generalized gradient approximation GGA-91 [12] was used for exchange-correlation

[4] To whom any correspondence should be addressed

functional. An energy cutoff of 400 eV for plane wave basis set, and an 8×8×1 Monkhorst-Pack *k*-points mesh for integration over the Brillouin zone were used in the calculations. The total energies were converged up to 10^{-4} eV. The metal-ceramic (001) interfaces were modeled using a supercell approach with periodic boundary conditions. A sandwich configuration of the (001) interface with a seven layer ceramic slab between two metal slabs of five layers each with a vacuum region not less than 8 Å between the two metal slabs was used. It is known that many transition metal carbides or nitrides are crystallized in the NaCl (rocksalt) structure. Note that each metal carbide or nitride (001) face is a non-polar surface with equal numbers of metal and metalloid atoms within the atomic layers parallel to the interface. Theoretical lattice parameters for bulk Al (4.044 Å), TiC (4.339 Å), TiN (4.256 Å), VC (4.158 Å) and VN (4.120 Å) are in good agreement with experimental data. The lattice mismatch of 1.9% (2.8%) is small in the case of VN (VC) but it increases to 5.2% (7.3%) for TiN (TiC). The surface energies of 0.85 J/m^2 for VN(001) and 1.17 J/m^2 for VC(001) are slightly less than those (0.95 J/m^2 and 1.28 J/m^2) obtained in [6] whereas the value 1.27 J/m^2 for the TiN(001) surface agrees well with 1. 25 J/m^2 from [6]. The surface energy of 1.61 J/m^2 in the case of TiC(001) surface was also found to be in good agreement with the result from [3]. The higher surface energies in the case of transition metal carbides compared with their nitrides indicate more reactive surfaces that can influence on the bonding at the interface.

The work of separation is a fundamental quantity, which controls the mechanical strength of the metal-ceramic interface. It is the work needed to separate interface into two free surfaces and can be defined as

$$W_{\text{sep}} = (E_{me} + E_{\text{carbide(nitride)}} - E_{me\text{–carbide(nitride)}})/2S\,, \tag{1}$$

where $E_{me\text{–carbide(nitride)}}$ is the total energy of supercell containing the multilayered slabs; E_{me} and $E_{\text{carbide(nitride)}}$ are the total energies of the same supercell containing a single slab of the metal or carbide (nitride), respectively; S is the interface area and the factor 2 takes into account the presence of the two equivalent interfaces within a unit cell. The equation (1) is appropriated for the defective slabs also.

3. Results and discussion

In order to identify the most stable interface geometry we considered two different stacking sequences with interfacial Al atom above the Ti/V-atoms (Ti/V-top site) and above the C/N atoms (C/N-top site). Our calculation of ideal work of adhesion (table 1) showed that the interfaces with Al atom in the C/N-top configuration exhibit the largest W_{sep} values for all ceramics. The preference for the metalloid adhesion site is in agreement with an earlier study [4]. In general, ours values are found to be slightly less than those obtained for Al/VC(N) interfaces in [4]. Note that the pseudopotential approach was used in the paper [4].

Table 1. Work of separation (in J/m^2) at ideal and defective Al/TiC(N) and Al/VC(N) interfaces.

Interface	Al/TiC Ti-top	Al/TiC C-top	Al/TiN Ti-top	Al/TiN N-top
Ideal	0.66	2.49	0.71	1.44
Ti-vacancies	4.26	3.14	1.50	2.21
Al-vacancies	0.48	2.02	0.62	1.30
N(C)-vacancies	0.81	2.49	0.85	1.76
Interface	Al/VC V-top	Al/VC C-top	Al/VN V-top	Al/VN N-top
Ideal	0.76, 0.96 [4]	1.99, 2.14 [4]	0.59, 0.91 [4]	1.31, 1.79 [4]
V-vacancies	3.38	2.60	0.80	1.98
Al-vacancies	0.83 (1.74)	1.72	0.63	1.38
N(C)-vacancies	0.95	2.23	1.53	1.60

A stronger adhesion for metalloid (C/N) site is consistent with an increase of interfacial distance in comparison with Ti/V top sites. For example the interfacial distances for Al/VC(N)(001) and Al/TiC(N)(001) interfaces are 0.55-0.74 Å (0.73-0.96 Å) and they are shorter in the case of C(N)-top sites in comparison with Ti(V) sites. The degree of rumpling of the ceramic interfacial layer is approximately the same as for carbide (nitride) surfaces. In each system C or N atoms relaxes towards the Al film while Ti or N atoms move slightly deeper into the bulk region. The C(N)-V distances in Al/VC(N) are 0.21 (0.28) Å which agree well with results of [4]. The distance C(N)-Ti is slightly shorter (0.15-0.17 Å) in the case of Al/TiC(N) interfaces. Local densities of states (DOS) are presented in figure 1 for both Al/TiC and Al/VC systems. As can be seen from figure 1 a large peak in the range of -12 – -10 eV in the Al local DOS is induced by the interaction between C and Al states. In both systems (Al/TiC and Al/VC) the bonding is due to hybridization of the Al *s,p* states with the C *s* states. This indicates the dominant role of covalent contribution in the chemical bonding at the interfaces. Similar trends were obtained for both Al/TiN and Al/VN systems as well.

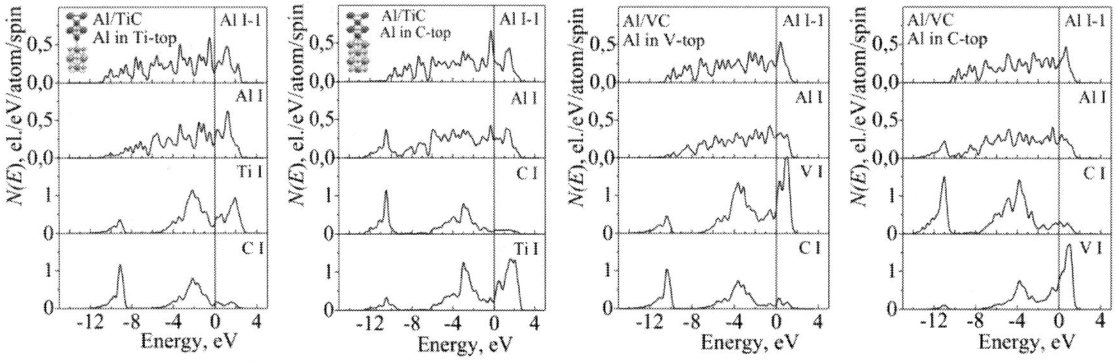

Figure 1. Local density of states of Al, Ti/V and C atoms at Al/TiC (left) and Al/VC (right) interfaces. Symbols I denotes the interfacial atoms and I-1 was used for atoms of next from interface layer.

The charge density difference ($\Delta\rho(\vec{r}) = \rho_{me}(\vec{r}) + \rho_{\text{carbide(nitrade)}}(\vec{r}) - \rho_{me\text{-carbide(nitride)}}(\vec{r})$) is shown in figure 2. It should be noted that electronic structures of considered metallic carbides and nitrides have common features which result in similar charge distributions at the interfaces. One can see from figure 2 that the interfacial charge redistribution has localized character and it is restricted by the atomic layers closest to interface. The redistribution of charge density is more pronounced for the Al location in the C-top site. There is a charge depletion region near the Al interfacial layer which indicates a reduction of lateral bonding in the Al film but the formation of new bonding across the interface (i.e. the accumulation of negative charge along C(N)-Al bonds) is more favorable). The larger values of W_{sep} in the C-top configuration in comparison with Ti-top site can be explained by the higher charge density along the Al-C bond.

In order to simulate vacancies in the interfacial layers the volume of the computational cell was increased twice in the planar direction perpendicular to the interface. As can be seen from table 1 in the case of Al vacancy the adhesion slightly decreases or increases depending on the system under question. In the case of the Al/TiN or Al/TiC interfaces W_{sep} decreases slightly compared with ideal interface but it behaves oppositely at the Al/VN interface. This behavior of W_{sep} correlates with the change of the interatomic and interfacial distances. It should be stressed that in the case of the Ti or V vacancies where the Al atom is located initially in the *Me*-top site the diffusion of Al atom towards the vacancy takes place alongside with the shift of the Al atoms in the film from Ti/V-top configuration into C-top. This results in significant increase of the adhesion at the interfaces. If the Al atom belongs to the substrate during cleavage of the interface, the value of W_{sep} (it is given in brackets in table 1) is practical the same as that for the case of the Al-vacancy in the Al film (C-top configuration). In the case of the metalloid vacancy the shift of Al atoms also takes place but it is significantly less (~0.3 Å) in comparison with the case of Ti(V) vacancies mentioned above.

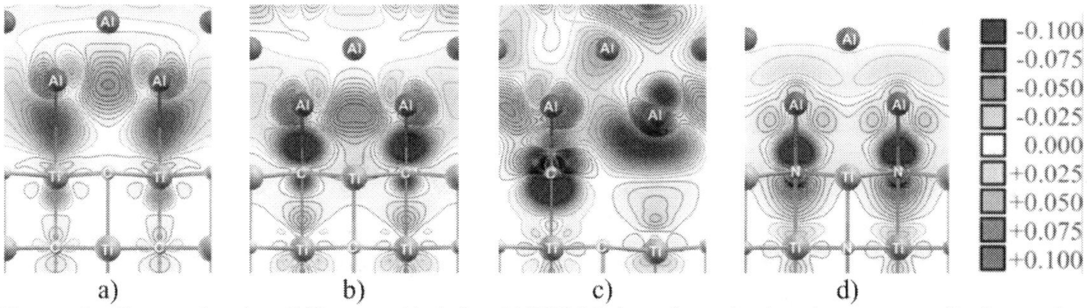

Figure 2. Charge density difference ($\Delta\rho$) for Al/TiC(N) interfaces in the plane perpendicular to the interface intersecting both Al and C/N interfacial atoms: a) Al in the Ti-top site, b) Al in the C-top site, c) Al in the C-top site with C-vacancy in the interfacial layer, d) Al in the N-top site.

It should be also stressed that the formation energy of the Al substitution impurity on the C-site is higher than that on Ti or V sites in the bulk and surface as well. Therefore, it is unfavorable for Al atoms to occupy the C site at the interface also. The trend mentioned above is valid at the Al/TiN interface with Ti vacancy in the nitride substrate as well but in this case the Al film does not change its Ti-top configuration. In general, the formation of both C and N vacancies inside ceramic substrates lead to increase of the adhesion at the considered interfaces due to the mechanism of diffusion mentioned above.

4. Conclusion

A systematic investigation of the adhesion, structure and bonding at the Al/TiC(N) and Al/VC(N) interfaces was performed using PAW-GGA method. It was found that for all considered systems the most stable configuration is C/N-top site where Al atoms are located above the metalloid sites. Obtained values of the work of separation are higher at the interfaces with transition metal carbides than those for metal-nitrides interfaces which can be attributed to their high surface energies. An analysis of the interfacial electronic structure reveals that the bonding at the interface is due to formation of the covalent Al *s,p* – C(N) *s* bonds. The influence of metal, carbon and nitrogen vacancies on the adhesion was studied. It was shown that in the case of Ti(V) vacancies in ceramic substrate the diffusion of aluminium inside substrate is observed alongside with the change of the metal film configuration from metal-top to metalloid-top which results in increase of adhesion. In general, the performed investigations allow us to reveal a relationship between electronic, geometrical factors and mechanical properties of considered metal-ceramic interfaces.

References

[1] Finnis M W 1996 *J. Phys: Condens. Matter.* **8** 5811
[2] Muñoz M C, Gallego S, Beltrán J I and Cerdá J 2006 *Surf. Sci. Rep.* **61** 303
[3] Dudiy S V, Hartford J and Lundqvist B I 2000 *Phys. Rev. Lett.* **85** 1898
[4] Siegel D J, Hector L G and Adams J B 2002 *Acta Materialia* **50** 619
[5] Siegel D J, Hector L G and Adams J B 2002 *Surf. Sci.* **498** 321
[6] Siegel D J, Hector L G and Adams J B 2003 *Phys. Rev.* B **67** 092105
[7] Hartford J 2000 *Phys. Rev.* B **61** 2221
[8] Blöchl P E 1994 *Phys. Rev.* B **50** 17953
[9] Kresse G and Joubert J 1999 *Phys. Rev.* B **59** 1758
[10] Kresse G and Hafner J 1993 *Phys. Rev.* B **47** 558
[11] Kresse G and Furthmüller J 1996 *Comput. Mater. Sci.* **6** 15
[12] Perdew J P and Wang Y 1992 *Phys. Rev. B* **45** 13244

IOP Conf. Series: Materials Science and Engineering **38** (2012) 012015 doi:10.1088/1757-899X/38/1/012015

The effect of Zn vacancies and Ga dopants on the electronic structure of ZnO: *Ab initio* simulations

Alexandre V Sorokin[1], **Yuri F Zhukovskii, Juris Purans and Eugene A Kotomin**

Institute for Solid State Physics, University of Latvia, 8 Kengaraga Str., Riga, LV-1063, Latvia

E-mail: as08504@lu.lv

Abstract. Zinc oxide modified by metal dopants can be used as a low-cost material for production of transparent conducting films. Its optical and electronic properties vary with the type and the concentration of dopants. In this study we have performed first-principle calculations on ZnO with Zn vacancies and that with Ga dopants in wurtzite type hexagonal morphology using density functional theory approach. Dependence of the electronic properties on the concentration of dopants has been studied using supercells of different sizes.

1. Introduction

ZnO continues to attract enhanced technological interest as a suitable material for manufacturing of transparent conducting films for electrodes and other microelectronic applications. The wide accessibility of ZnO single crystals [1] make ZnO a promising low-cost material for synthesis of nanostructures with a large variety of morphologies and for substrates. Under normal conditions ZnO exists in the wurzite-type hexagonal polymorph, being a wide-gap semiconductor with a band gap E_g = 3.4 eV [2, 3]). When ZnO is doped by Ga, Al or transition metal atoms, its electrical conductivity increases retaining high optical transparency. Doped ZnO can be considered as a good substitute for indium–tin oxide in manufacturing of transparent electrodes [1]. Fabrication of doped and undoped ZnO thin films is accompanied by formation of O and Zn vacancies and interstitials [4] and formation of nitrogen complexes [4], as well as hydrogen adsorption [5], which contributes to n-type conductivity.

In recent years energetics of formation of O and Zn vacancies and interstitials was calculated using hybrid QM/MM approach [3, 6], DFT-PAW approach with semi-local GGA [7] and LDA/GGA parametrization [8]. Acting as deep acceptors or deep donors, no intrinsic defects can cause conductivity [9]. In contrast, adsorbed hydrogen is readily incorporated as an H^+ interstitial, ensuring n-type conductivity [9]. Sokol et al [6] have shown that metal atom dopants usually act as deep donors, reluctantly taking part in the formation of conductivity.

In this study we have performed first-principles calculations on the atomic and electronic structure of pure and doped ZnO with different concentrations of defects in order to analyze the effect of Ga dopant and V_{Zn} on different properties.

2. Computational details & preliminary calculations

DFT calculations have been performed using basis sets of linear combination of atomic orbitals (LCAO) combined with hybrid DFT-HF exchange–correlation functional PBE0 as implemented

[1] Corresponding author

Table 1. Lattice constants a and c (Å), bulk modulus B (GPa), cohesive energy E_{coh} (eV) and band gap E_g (eV) for ZnO compared to other theoretical and experimental data.

	This study	DFT	Expt
a	3.2581	3.2498 [3]	3.2417 [11]
c	5.2218	5.2029 [3]	5.1876 [11]
B	160.4	160.2 [3]	135–183 [12]
E_{coh}	13.07	10.64 [13]	—
E_g	3.61	3.18 [14]	3.44 [2]

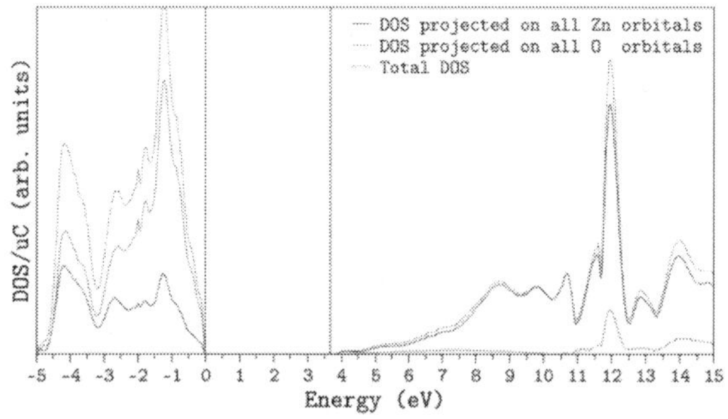

Figure 1. DOS for a pure ZnO, projected on Zn (blue) and O (red) orbitals, as well as total DOS (cyan). Black dashed lines mark band edges. All the energies are wrt TVB. Band gap $E_g = 3.68$ eV

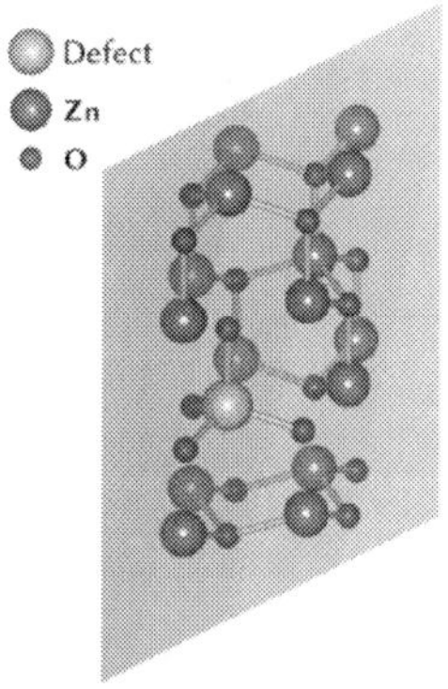

Figure 2. Characteristic vertical cross-section plane (V) (shaded) with the Miller indices (740).

into CRYSTAL09 code [10]. Total energy convergence criterion in all the calculations was 10^{-7} Ha. The effective charges on atoms have been calculated using Mulliken population analysis. The $\boldsymbol{k}$-space integration has been performed by sampling the Brillouin zone with Pack–Monkhorst mesh – $9 \times 9 \times 6$ for perfect ZnO bulk and $8 \times 8 \times 8$ for all defective structures. The geometry of all the structures was fully relaxed.

Calculations on the perfect ZnO bulk have been performed for reference, and various computed properties are compared with experiment and theory in table 1. The electronic structure was characterized by projected densities of electron states (PDOS; Figure 1).

Mulliken analysis demonstrates a considerable Zn–O bonds' covalency, its population being $0.14e$. Zinc and oxygen charges are $\pm 1.01e$, respectively. Oxygens are coordinated in almost perfect tetrahedrons, and Zn–O bonds have length $d_{\text{Zn–O}} \approx 1.99$ Å.

3. Main Results

A series of calculations has been performed using various supercells to alter concentrations of defects. To match experimental works (eg references [15, 16]), we used $3 \times 3 \times 3$ and $2 \times 2 \times 2$ supercells (1.85 % and 6.25 % of defects, respectively).

3.1. ZnO with V_{Zn}

The electronic and structural parameters for ZnO with zinc vacancies are compared in table 2 with pure bulk material. Projected densities of states (Figure 3(a)) have been calculated for a structure with 6.25 % of defects. It shows that this system is a p-type conductor. Similar plot for a more diluted structure not depicted here shows that the latter is an insulator. Electron density redistribution on the plane V (Figure 2) is shown in figure 4(a).

Table 2. Lattice parameters a, c (Å), Zn/Ga/V_{Zn}–O bond lengths d' in [001] direction and d'' in [$\bar{2}2\bar{1}$] direction (Å), band gaps E_g (eV) and charges on the defect (Zn for perfect ZnO) Q (e) of pure ZnO, V_{Zn}- and Ga-doped structures. Covalent radii for 4-coordinated ions are as follows: R_{Zn} = 74 pm, R_O = 124 pm, R_{Ga} = 61 pm.

	ZnO	ZnO:V_{Zn}	
		$2 \times 2 \times 2$	$3 \times 3 \times 3$
a	3.2581	3.2644	3.2636
c	5.2218	5.2211	5.2042
d'	1.99	2.31	1.99
d''	1.98	2.15	1.98
E_g	3.61	—	3.44
Q	+1.01	−0.05	−0.05

	ZnO	ZnO:Ga	
		$2 \times 2 \times 2$	$3 \times 3 \times 3$
a	3.2581	3.2641	3.2619
c	5.2218	5.2434	5.2302
d'	1.99	1.89	1.87
d''	1.98	1.89	1.88
E_g	3.61	—	—
Q	+1.01	+1.57	+1.63

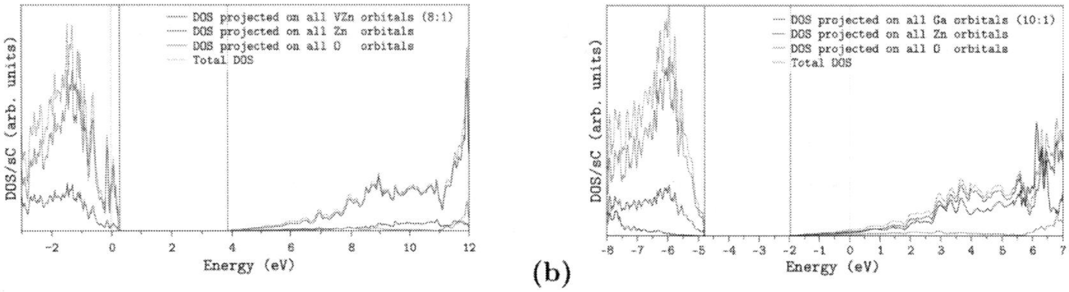

(a) (b)

Figure 3. DOS for $2 \times 2 \times 2$ supercells of (a) ZnO:V_{Zn} and (b) ZnO:Ga, projected on Zn (blue), O (red) and Ga/V_{Zn} (black) orbitals, as well as total DOS (cyan). Boundaries of the band gap are shown with black vertical lines. Fermi level (green dashed line) corresponds to zero energy.

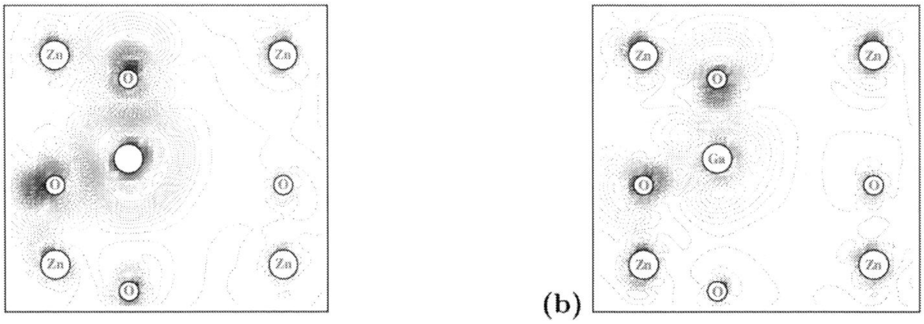

(a) (b)

Figure 4. Differential electron charge density in (a) $2 \times 2 \times 2$ ZnO:V_{Zn} and (b) $2 \times 2 \times 2$ ZnO:Ga, projected on plane V of figure 2. Increment for isolines is $0.001e$ within the interval of electronic charge $(-0.1 \div +0.1)e$.

V_{Zn}–O dangling bonds are substantially stretched for $2 \times 2 \times 2$ structure, increasing their length by 0.2–0.3 Å. Bond lengths in the $3 \times 3 \times 3$ structure coincide with those for the parent material. Effective charge on the vacancy is almost zero ($0.05e$) and does not depend on defect concentration. To estimate vacancy creation costs, its formation energy should be calculated: $E_V^F = E_D + E_{Zn} - E_P$, where E_D, E_P and E_{Zn} are the energies of defective system, perfect system and isolated atom, respectively. E_V^F = 15.8 eV for ZnO:V_{Zn} (6.25 %) and E_V^F = 10.6 eV for ZnO:V_{Zn} (1.85 %). The structures with lower defect concentrations are found to be insulating, while more concentrated ones are p-type conductors.

3.2. Ga-doped ZnO
Same properties were calculated for ZnO:Ga and are shown in table 2. For ZnO:Ga (6.25 %) a projection of differential charge density on V plane (figure 2) is shown in figure 4(b), and densities of states are shown in figure 3(b).

After structural relaxation in ZnO:Ga (6.25 %) Ga–O bond length decreased by *ca* 0.11 Å. The effectve electronic charge on Ga atom exceeds the charge on the substituted Zn atom by $\sim 0.6e$. The same trends are found for more diluted systems, but trends therein are generally less pronounced. Atomic charge of Ga behaves inversly and grows with the concentration reaching $+1.63e$ at 6.25 %. Densities of states for the ZnO:Ga (6.25 %) show that the system is an n-type conductor. Ga impurity levels are delocalized over conducting and valency bands of ZnO. The defect formation energies for ZnO:Ga (6.25 %) and ZnO:Ga (1.85 %) were found to be 4.52 eV and 2.98 eV, respectively. As expected, making less concentrated Ga solutions in ZnO is energeticaly more favourable.

4. Conclusions
Our calculations show that the possible formation of large concentrations of Zn vacancies is limited by high energy costs. Lower concentrations of vacancies are formed more readily. States of zinc vacancy are delocalized mainly over its conduction bands. Large concentration of Zn vacancies can make the material conducting, whereas small one does not give rise to conductivity. The lattice distortions associated with introduction of zinc vacancies in large concentrations are drastic and can make a whole structure unstable. With defect dilution, this problem disappears, and the geometrical properties of a defective material tend to those of the perfect one. Ga-doped ZnO is found to be an n-type conductor. Increase of Ga impurity concentration leads to an increase of the dopant effective charge wheras distances to the nearest neighbour ions decrease. Due to the fact, that the supercells studied are relatively small and the concentrations are relatively large, defects cannot be considered isolated and interact with one another.

Acknowledgments
The authors would like to thank D Gryaznov for fruitful discussions and ERAF 2010/0272/2DP/2.1.1.1.0/10/APIA/VIAA/088 project for financial support.

References
[1] Ü Özgür, Alivov Y I, Liu C, Teke A, Reshchikov M A, Doğan S, Avrutin V, Cho S J and Morkoç H 2005 *J. Appl. Phys* **98** 041301
[2] Reynolds D C, Look D C and Jogai B 1999 *Phys. Rev. B* **60** 2340–2344
[3] Catlow C R A, French S A, Sokol A A, Al-Sunadi A A and Woodley S M 2008 *J. Comput. Chem.* **29** 2234–2249
[4] Look D C, Farlow G C, Reunchan P, Limpijumnong S, Zhang S B and Nordlund K 2005 *Phys. Rev. Lett.* **95** 225502
[5] Lavrov E V, Börrnet F and Weber J 2005 *Phys. Rev. B* **72** 085212
[6] Sokol A A, French S A, Bromely S T, Catlow C R A, van Dam H J J and Sherwood P 2007 *Faraday Discuss.* **134** 267–282
[7] Chan J A, Lany S and Zunger A 2009 *Phys. Rev. Lett.* **103** 016404
[8] Wróbel J, Kurzydłowski K J, Hummer K, Kresse G and Piechota J 2009 *Phys. Rev. B* **80** 155124
[9] Van de Walle C G 2000 *Phys. Rev. Lett.* **85** 1012–1015
[10] Dovesi R, Saunders V R, Roetti R, Orlando R, Zicovich-Wilson C M, Pascale F, Civalleri B, Doll K, Harrison N M, Bush I J, D'Arco P and Llunell M 2009 *CRYSTAL09 User's Manual* University of Torino Torino URL `http://www.crystal.unito.it`
[11] Albertsson J, Abrahams S C and Å Kvick 1989 *Acta Crystallogr. B* **45** 34–40
[12] Liu H, Ding Y, Somayazulu M, Qian J, Shu J, Häusermann D and Mao H k 2005 *Phys. Rev. B* **71** 212103
[13] Cai J and Chen N 2007 *J. Phys.: Condens. Matter* **19** 266207
[14] Oba F, Togo A, Tanaka I, Paier J and Kresse G 2008 *Phys. Rev. B* **77** 245202
[15] Liu X C, Chen Z Z, Chen B Y, Shi E W and Liao D Q 2010 *J. Cryst. Growth* **312** 2871–2875
[16] Li Z Z, Chen Z Z, Huang W, Chang S H and Ma X M 2011 *Appl. Surf. Sci.* **257** 8486–8489

IOP Conf. Series: Materials Science and Engineering **38** (2012) 012016 doi:10.1088/1757-899X/38/1/012016

The role of nanopowder particle surfaces and grain boundary defects in the sintering of ZnO ceramics

F Muktepavela[1,3], R Zabels [1], V Sursajeva[2], L Grigorjeva[1], K Kundzins[1]

[1]Institute of Solid State Physics, University of Latvia, Riga LV1063, Latvia
[2]Institute of Solid State Physics, Russian Academy of Sciences, 142432 Chernogolovka, Moscow district, Russia

E-mail:famuk@latnet.lv

Abstract. This work focuses on the characteristics of sintered ZnO ceramics and explores the role of source powder morphology in the process of sintering. The source ZnO powders had grained (d= 100 nm) and tetrapod-like (d=50-100 nm, l=3-10 μm) morphologies, they were compacted and sintered at 1200° C. The results have shown that ceramics sintered from the grained powder exhibit relatively high (8%) porosity at grain boundaries and as cavities within grains, which facilitates brittleness. Photoluminescence spectra for these ceramics besides a narrow exitonic band contain a broad "green" luminescence band attributed to defect states. The second ceramics sintered from the tetrapod-like powder has lower porosity (<2 %) Photoluminescence spectrum at 12 K revealed a narrow exitonic band with satellite peaks (1LO_Ex states) and almost negligible "green" band which is an indication of high quality of this ceramics. Experiments on bicrystals have shown that in the stage of grain growth the motion of grain boundaries is directly correlated to triple-junction mobility. The obtained results have been discussed from the point of contact phenomenon and evolution processes of the grain boundary microstructure at different sintering stages.

1. Introduction.

ZnO, a direct wide band gap semiconductor, is in the focus of extensive research for its potential in fabrication of visible light emitters, UV laser diodes, sensors and scintillators, etc. Recently the attention of research community has increasingly turned to ZnO ceramics. Its properties can be comparable to a single crystal but at the same time the production is by far simpler. As varistors, ZnO ceramics are known for a long time but for optoelectronic devices this material is new. To implement ZnO ceramics in optoelectronic applications the knowledge about its structure and grain boundary properties is of high importance.

Typically ZnO ceramics are made from nano -and micropowders via different sintering methods including high temperature annealing without pressing, and force methods such as ultrasound, magnetic-impulse, microwave and hot pressing [1]. From listed methods a high temperature sintering is a relatively simple and economical method. However, it is well-known that individual crystalline nanoparticles in the

[3] To whom any correspondence should be addressed

nano-powder compacts tend to agglomerate and form strong inter-particle bonds [2, 3]. The agglomeration of nano-powders is due to high surface-to-volume ratio and is commonly attributed to the high amplitude of attractive Van der Waals forces acting between nanoparticles [3]. The properties of ceramics are strongly influenced by newly formed grain boundaries during the manufacturing process. Since most source materials are strongly aggregating micro- or nanograined powders, the high temperature sintering results in that previous (old) enclosed boundaries are saturated with difficult to remove impurities and coagulating pores, which can be titled as a hereditary influence. Thus, development of advanced power application technologies is primarily aimed to crush old boundary networks and obtain a good conjugation for qualitatively new boundaries. The choice of parameters and operational modes to develop these technologies is a complicated process. However, it is known that ZnO has an appealing quality: the variety of nano- and micro-morphological forms. A wide variety of 2D and 3D ZnO nanostructures such as nanowires, nanobelts, tetrapods, nanorings and nanocombs have been successfully and repeatedly grown. Such nanostructures in the direction of growth have large free surfaces with minimal energy [4], its surface-to-volume ratio is low and, one can assume they will not cling together. In contrast to typical granular powders for these oddly shaped powder particles other sintering processes may take place in all sintering stages.

In this work the role of free surface and grain boundary (GB) defects (pores, facets, triple junctions (TJ)) in the process of ZnO ceramics formation from nanopowders with grained and tetrapod-like morphology at high temperature sintering has been studied.

2. Experimental

ZnO powders were obtained by a simple method of Zn vapor condensation and oxidation in a tube furnace. Filings of Zn (99.99%) as initial material were used. This was done to accelerate the phase formation of zinc vapor in the whole volume and quickly move to the area of the counter-flow oxygen. Powders with grained (d=100-200 nm) and multipod-like (tetrapods) (d=50-100 nm, l=3-10 μm) morphology were obtained at 950 °C temperature by varying conditions of oxygen flow. Powders were compacted at identical conditions: pressed at P=0.5 GPa at 20 °C and then sintered at 1200 °C for 3h in a tube furnace. Obtained powders and sintered ceramics were investigated by XRD, SEM (CARL ZEISS EVO) with EDS, AFM (VEECO CPII), Photoluminescence (excitation: nitrogen laser (337nm, 10ns), photon counting head HAMAMATSU H8259) and optical microscope Nikon Eclipse L120.

3. Results and discussions.

On Fig.1 the structure of two obtained ZnO powders with different morphology as initial materials for sintering ceramics are presented. The powder N1 has tetrapod-like structure with dimensions d=50-100 nm, l=3-10 μm. As can be seen, structural elements of powders are not inclined towards an adhesive interaction. The powder consists of spatially separated needles. Powder N2, on the contrary, has a grained structure (d=100-200 nm) and is a conglomerate of interacted particles.

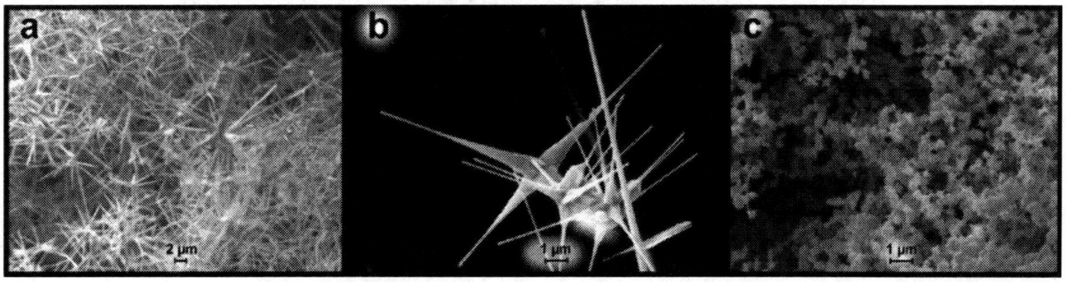

Figure 1. SEM images of ZnO powders structure: tetrapod-like - N1 (a,b) and grained powder –N2 (c).

Results of X-ray diffraction and EDS spectra are evidence that both powders N1 and N2 have identical ZnO phases, no other elements or Zn phase were observed (Fig.2). However, given the method of oxidation to produce these powders, the presence of non-stoichiometric zinc is difficult to avoid.
On the Fig.3 the structures of sintered ceramics from both powders N1 and N2 are presented.

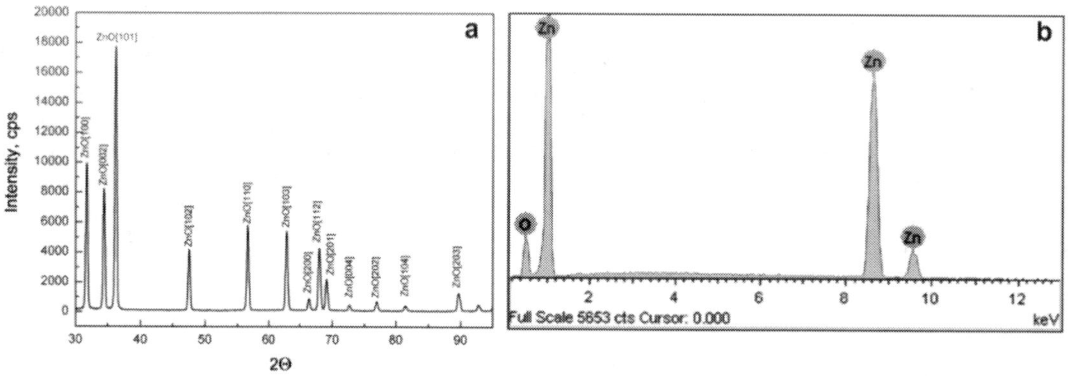

Figure 2. X-ray diffraction pattern (a) and EDS spectrum (b) of obtained ZnO powders.

The ceramic obtained from the tetrapod-like powder has small grained (d=2-4 μm) structure and contains small individual micropores in the vicinity of triple junctions (TJ), porosity was <2% (Fig.3,a).

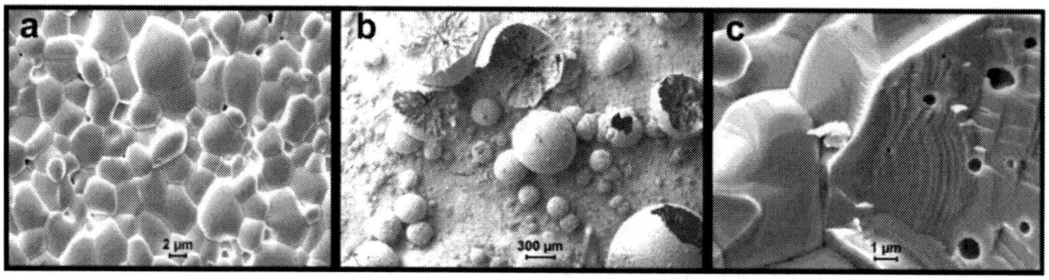

Figure 3. SEM images of ZnO ceramic structure: from tetrapod-like powders (a), and from grained powders (b,c).

Due to rapid recrystallisation $(d-d_0)^n \sim t$ (n=4) of nanograined materials, (where d and d_0 are the final and initial grain sizes) ceramic from powder N2 has quite large and inhomogeneous grains (d=6-10 μm). The surface of the ceramic is covered with large ZnO spheres which are a product of a rapid oxidation of an escaped excess Zn vapor (Fig.3,b). N2 ceramics have high (8%) porosity at GB as cavities inside grains. Previous grain boundaries are locations of accumulation of impurities, non-stoichiometric Zn and micropores, which altogether facilitates brittleness (Fig.3,c).
For N1 ceramics (from tetrapod-like powders) luminescence spectrum at 12 K revealed a narrow excitonic band with a probable satellite peak and an almost non-existent "green" luminescence band that agree with data for ZnO single crystal and is an evidence of high quality of this ceramics (Fig.4 a). For N2 ceramics a broad "green" band, which is attributed to defect states, was observed in PL spectrum (Fig.4 b). PL results are in agreement with SEM data.

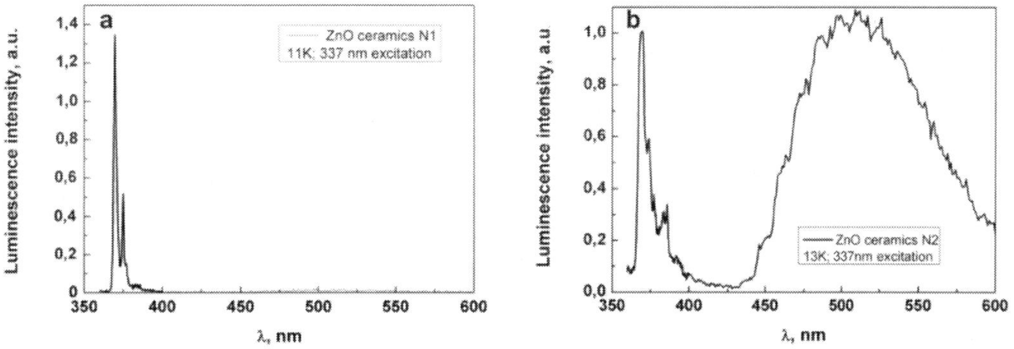

Figure 4. PL spectra at 12K for ceramic from tetrapod-like powders (a) and for ceramic from grained (b)

Elimination of defects in N2 ceramics (micropores, impurities within grains) by pushing them towards the surface can be accomplished only by volume diffusion at sintering temperatures 0.8-0.9 Tm.
Sintering of tetrapod-like powders occurs without influence of previous grains. As was shown earlier [5], at 800 ^{0}C the transformation of tetrapods into grains is promoted due to surface diffusion. In the solid phase adhesion contact involves clean surfaces of grains. The stage of recrystallization in N1 ceramic occurs at higher temperatures (950-1000 ^{0}C) that results in small grains.
Thus, the observed differences in the structure of the investigated ceramics are related to the influence of external and internal interfaces in the stages of adhesion and recrystallization processes during sintering.
To improve the structure of ceramics N2, it is necessary to control the defects and structural elements at the boundaries during the sintering. Model experiments have shown that in the stage of grain growth triple junctions have a substantial influence: at low temperatures they slow down, whereas at high temperatures ($\approx$ 0.8 Tm) – speed up the motion of grain boundaries [6].

4. Conclusions.

Ceramics obtained with high temperature sintering method from nanograined and tetrapod-like powders have different properties. The observed effect of morphology can be explained in terms of contact phenomena. The hereditary influence of impurities and micropores prevent the good contact formation, which leads to the accumulation of the non-stoichiometric Zn and determines the high porosity and brittleness of ceramics based on the nanograined powder. Ceramics based on tetrapod-like powder in which hereditary factor is absent have better quality of the structure that reflects on luminescent properties. For these ceramics the positive influence comes from the surface diffusion and structure transformation processes.
The obtained results show a perspective for developing a technology based on high temperature sintering of tetrapod-like ZnO powders.

5. References.

[1] Lu K. Sintering of nanoceramics. 2008 Intern.Mater.Rev. **53** 21
[2] Wang Z L 2008 ACS NANO **2** 10 1987
[3] Singhal A, Skandan G, Wang A, Glumac N, Kear BH and Hunt R D 1999 NanoStr. Mater. **4** 545
[4] Raichman Y, Kazakevich M, Rabkin E, Tsur Y 2006 Adv. Mater. **18** 2028.
[5] Muktepavela F, Bakradze G, Grigorjeva L, Zabels R, Tamanis E 2009 Thin Solid Films **518** 1263
[6] Sursajeva V G, Straumal B B, Gornakova A S, Shvindlerman L S, Gottstein G 2008 Acta mater. **56** 2728

Acknowledgments
This work has been supported by the Latvian Council of Science (project number 10.0032).

Shear banding mechanism of plastic deformation in LiF irradiated with swift heavy ions

J Maniks[1], R Zabels, I Manika
Institute of Solid State Physics, University of Latvia, Latvia

E-mail: manik@latnet.lv

Abstract. The effect of ion irradiation on the behavior of plastic deformation at micro- and nanoindentation on (001) face of LiF has been investigated. The irradiation was performed using heavy ions (U, Au, Ti and S) with energy in the range from 3 MeV to 2 GeV at fluences up to $5x10^{13}$ ions/cm^2. In non-irradiated LiF, the indentation produces dislocation gliding on the {110} planes along the <100> and <110> directions. At high fluence irradiation, the resource of the dislocation slip along the preferable directions becomes exhausted due to immobilization of dislocations by radiation defects and their aggregates. The present study demonstrates the change of the mechanism of plastic deformation from homogenous dislocation slip to localized shear banding in samples irradiated to high fluences. The factors facilitating of the localization of deformation have been analyzed.

1. Introduction

Modification of the structure, physical and mechanical properties of solids irradiated with swift heavy ions (SHI) is a topic that has attracted considerable interest over recent decades. Numerous papers have been devoted to radiation damage studies in lithium fluoride and related solids with strong ionic bonding [1]. Such materials maintain crystallinity even under conditions of severe irradiation. However, nanostructuring on surface and in bulk of LiF samples accompanied by strong (up to threefold) increase of hardness under irradiation with SHI was reported [2, 3]. The ion- induced change of the mechanism of plastic deformation is suggested [4].

In this work the behavior of plastic deformation at indentation on LiF crystals irradiated with SHI to high fluences has been investigated.

2. Experimental

Irradiations were performed on LiF crystals (Korth Kristalle, Germany) to high fluences (10^{11}-10^{13} ions/cm^2) using ^{238}U, ^{197}Au, ^{49}Ti and ^{32}S ions. Exposure to high energy (GeV) ion radiation was performed at the UNILAC linear accelerator within GSI, Darmstadt, Germany. A portion of LiF samples were irradiated with Au ions in the 3-15 MeV energy range at the Tandetron accelerator in Porto Alegre, Brasil. For the indentation measurements two methods were utilized. The instrumented indentation was performed using nanoindentation unit MTS Nano G200 with a Berkovich diamond tip. For higher loads a microhardness tester PMT-3 with a precision loading mechanism and Vickers diamond indenter was used.

[1] To whom any correspondence should be addressed

Structural defects around the indentation imprints were revealed by a chemical etching in saturated aqueous $FeCl_3$ solution. The surface was studied using optical and atomic force microscopy.

3. Results

3.1 Nanoindentation study

Figure 1 shows MeV ion-induced hardening in LiF and evolution of the deformation pattern around indents. Substantial changes in hardness and deformation mode are observed at high fluences ($>10^{13}$ ions/cm^2), when saturation and aggregation of radiation defects occur.

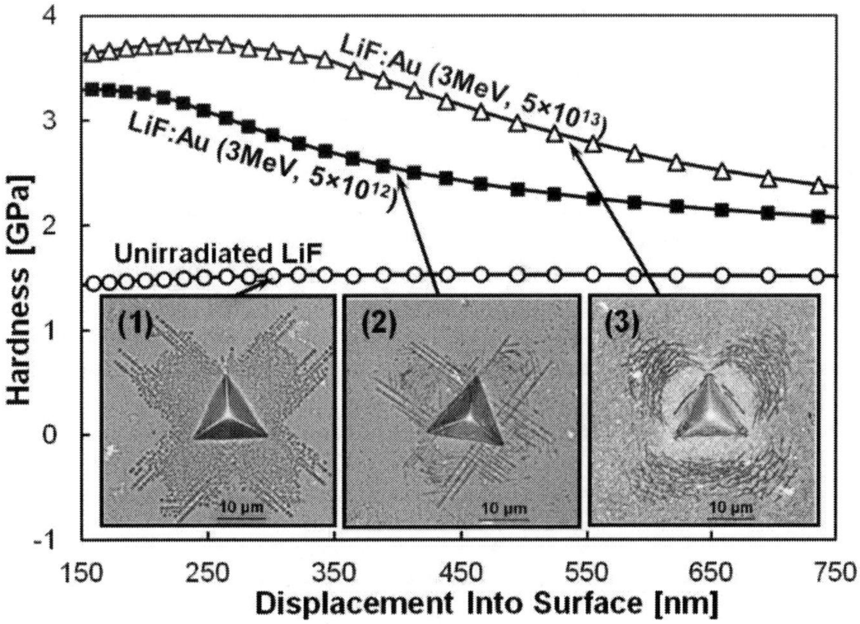

Figure 1. Hardness as a function of indenter penetration depth for LiF samples irradiated with 3 MeV ^{197}Au ions. The inset shows optical images of imprints and deformation zone after chemical etching.

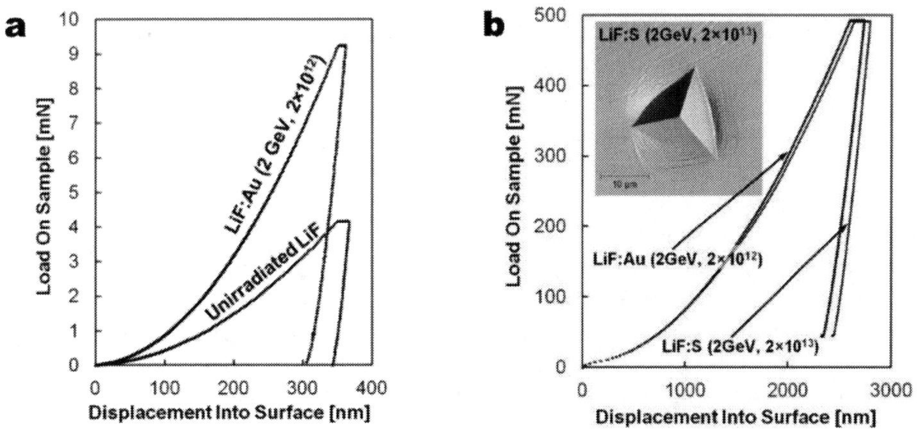

Figure 2. Load-Displacement data from instrumented indentation on LiF (a) – before and after irradiation with 2 GeV Au ions, (b) on LiF irradiated with Au and S ions. The inset shows the AFM image of deformation zone around the imprint for the sample irradiated with S ions.

Figure 2a shows loading-unloading curves for unirradiated and with Au ions irradiated LiF. The indentation has been performed at identical depth – 350 nm. When relative values of plasticity are compared, the irradiated material is only by 10% less plastic than the unirradiated crystal – 81.2% versus 92% plasticity. Fig.2 b contains a comparison of Load-Displacement data for LiF samples irradiated with 2 GeV Au ions at 2×10^{12} ions/cm^2 and S ions at 2×10^{13} ions/cm^2. In both cases the radiation dose was similar – 2.75 MGy. Light S ions cause comparable structural changes, which reflect both on hardness as well as plasticity. Even the high load of 500 mN didn't cause failure of these highly irradiated materials – no signs of brittleness could be observed on the loading curves. The inset of Fig.2 b shows Berkovich imprint on the sulphur irradiated LiF – shear bands around indent are clearly visible, similar as in the case of Au ions. Regarding hardness, GeV ions produce similar effects of hardening as MeV ions – only the modified layer is thicker – proportional to the ion penetration depth.

3.2 Structural study of the deformation zone

In an unirradiated LiF, the rows of dislocation etch pits can be seen extending along the <100> and <110> directions (Fig.1, inset 1) which are the directions of easy glide for dislocations [5]. The reduction of the dislocation arm length with increasing the fluence is observed. At high fluences the change of the deformation pattern from classic rosette formed by dislocation slip to a pattern of localized shear banding in zones of maximum stress against the centre of indenter faces has been observed (Fig.1, inset 3, inset of Fig.2b). The transition stage between them shows features characteristic to both – dislocation gliding as well as localized deformation (fig.1, inset 2, Fig3 b).

Deformation bands around an imprint on heavily irradiated LiF become very similar to those observed in amorphous metals (Fig.3 a). After such irradiation the material maintains a reasonable plasticity. No cracks are formed at light loads. Brittle cracking is observed only in heavily irradiated samples at high loads (Fig.3c).

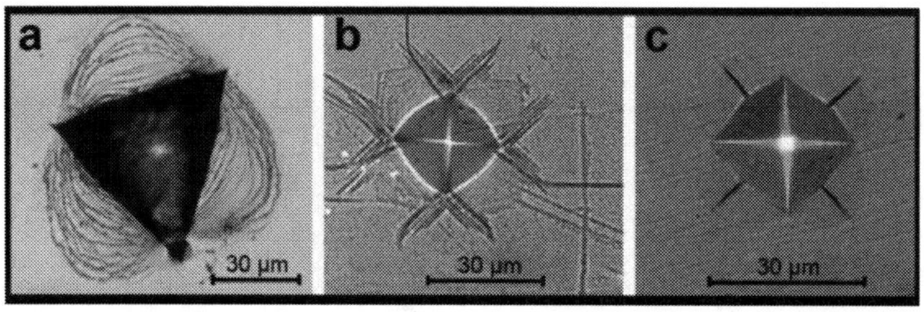

Figure 3. Deformation zones around imprints on (a) metallic glass [6], (b) LiF irradiated with 0.6 GeV Ti ions, fluence 5×10^{12} cm^{-2}, (c) LiF irradiated with 2 GeV U ions, fluence 5×10^{12} cm^{-2}.

4. Discussion

The swift-ion-induced damage processes in LiF are localized in tracks exhibiting comparatively large (up to few tens on nm) halo region of point defects. At room temperature the most significant type of defects are F centers, F_2 centers and complementary hole centers [1]. At high-fluences, the tracks overlap that leads to saturation and aggregation of point defects. The interaction of dislocations with defect aggregates as strong obstacles affects the deformation behavior and leads to increase of indentation hardness. Single defects as weak obstacles play a minor role [7]. The investigation of the structure of deformation zone in irradiated crystals (see inset of Fig.1 and Fig.2 a, b) shows a change of the mechanism of plastic deformation. The typical for LiF mechanism of dislocation glide is replaced by the mechanism of localized shear banding. In a transition stage the two mechanisms coexist.

The shear banding mechanism at definite conditions is observed for different materials, such as irradiated metals [8], nanocrystalline materials [9] and metallic glass [6]. Shear banding mechanism was reported also for LiF crystals at high temperature deformation [10]. In the latter case the deformation is ensured by thermo-activated climb of dislocations over obstacles. In our experiments performed at room temperature the contribution of thermo-activated processes to deformation is small as it follows from negligible indentation creep observed for irradiated crystals.

Formation of localized shear bands in irradiated metals is based on a specific mechanism named the channeling of deformation, which includes periodical formation of defect-free local zones followed by softening [4]. In such case, the instability of deformation is observed and cyclic bursts appear on the loading curves at indentation. Formation of defect free zones during deformation is never observed for ionic crystals. Confirming this, the loading curves on irradiated LiF show no signs of plastic instability.

Shear banding is well known as a main mechanism of plastic deformation in metallic glasses, were movement of assembles of atoms under high stresses becomes possible due to free volume in the structure of glassy materials. A noticeable contribution of similar non-dislocation mechanisms of deformation is suggested also for crystalline solids including LiF [11]. Appearance of the deformation zone in irradiated LiF and metallic glass is quite similar (see Fig.3 a). It should be taken into account that LiF crystals exhibit comparatively strong ion-induced effect of swelling that indicates formation of free volume. So, favorable conditions for the contribution of non-dislocation plasticity exist also for LiF crystals irradiated to high fluences. On the other hand, the nanostructuring of LiF under irradiation with SHI is observed [2, 3]. In nanostructured materials the deformation can be ensured by grain boundary sliding. A high concentration of ion-induced glissile prismatic dislocations facilitates such deformation mode.

The development of shear banding is known to occur without strain-hardening. Its initiation leads to saturation of hardness in irradiated crystals. Further increase of fluence shows small or no effect on flow stress or hardness. The present work and our earlier results show that the upper limit of hardness in ion-irradiated LiF reaches about 4 – 4.5 GPa.

5. Conclusion

In LiF crystals irradiated to high-fluences the change of the mechanism of plastic deformation from homogenous dislocation glide to localized shear banding is observed. Generally, the results indicate a complex nature of plastic deformation of ion-irradiated LiF. The non-dislocation mechanisms of deformation and grain boundary sliding are suggested to contribute the shear banding. The initiation of the shear banding mechanism limits further ion-induced increase of hardness. The irradiated crystals maintain reasonable plasticity. It follows from the obtained results that formation of localized shear bands at indentation is facilitated by ion-induced strengthening, which ensures high applied mechanical stress, swelling-induced formation of free volume and ion-induced nanostructuring.

Acknowledgements: This work has been supported by the Latvian government grant No 09.1548 and European Social Fund within the project «Support for Doctoral Studies at University of Latvia».

References

[1] Perez A, Balanzat E and Dural J 1990 *Phys Rev B* **41** 3943
[2] Maniks J, Manika I, Grants R, Zabels R, Schwartz K, Sorokin M and Papaleo RM 2011 *Appl Phys A* **104** 1121
[3] Maniks J, Manika I, Zabels R, Grants R, Tamanis E and Schwartz K 2012 *Nucl. Instr. Meth. B* **282** 81
[4] Manika I, Maniks J, Schwartz K, Toulemonde M and Trautmann C 2003 *Nucl. Instr. Meth. B* **209** 93
[5] Nadgorny EM and Stepanov AV 1963 *Soviet Phys Solid State* **5** 732
[6] Keryvin V 2007 *Acta Mater* **55** 2565
[7] Manika I, Maniks J and Schwartz K 2008 *J Phys D: Appl Phys* **41** 074008
[8] Byun T and Hashimoto N 2006 *Nucl Eng & Technol.* **38** 619

[9] Meyers MA, Mishra A and Benson DJ 2006 *Progr Mater Sci* **51** 427
[10] Skvorcova NP 1996 *Cryst Res Technol* **31** 373
[11] Golovin YuI, Tyurin AI and Farber BYa 2002 *Mater Sci* **37** 895

IOP Conf. Series: Materials Science and Engineering **38** (2012) 012018 doi:10.1088/1757-899X/38/1/012018

Dimensional effects of sample geometry and microstructure of MnZn and NiZn ferrites

J Jankovskis[1], N Ponomarenko[1,3], N Mironova-Ulmane[2], D Jakovlevs[1]

[1]Riga Technical University, Latvia, Riga, Kalku Str. 1, LV-1048

[2]Institute of Solid State Physics, Latvia, Riga, Kengaraga Str.8, LV-1063

Abstract. The effects of sizes of polycrystalline ferrite ring cores on magnetic characteristic – complex initial permeability (CIP) are studied. In the case of MnZn ferrite under uncontrolled microstructure (grain sizes) clear size effect is observed: dispersion of smaller core sizes is at higher frequencies and vice versa. In the case of controlled microstructure (NiZn ferrite) no size effects were observed.

1. Introduction

At the present time for soft magnetic materials there is demand not only for "high permeability and low loss" [1] but for their predictability and stability in different applications as well.

Permeability specified as frequency f function $\mu(f)$ of magnetic materials depends on variety of factors, several of which act still in an unknown form. In the case of polycrystalline ferrites (PF) the picture is additionally complex since they can be put into use over very broad range of frequency and thus different frequency effects may appear. Among the acting factors there are geometrical ones as well which may disclose themselves at macroscopic level as the physical size (core dimension) effects and at microscopic level – as grain size (more broadly - microstructure) ones. In the former case $\mu(f)$ dependence on dimensions usually is related to the dimensional resonance [1] which manifests itself in a very rapid dropping of $\mu(f)$ with f, when within cross section of PF sample there is formed electromagnetic standing wave. Thus the dimensional resonance should manifest itself mainly near the corresponding resonance frequency f_0 ; nevertheless in the original article [3] for MnZn ferrite the cross section of samples are influencing the static permeability $\mu(0)$ as well (the lower value for smaller cross section). Additional confusion comes from data of more modern MnZn ferrite, e.g. T37 [4]: there for smaller cross section rings $\mu(0)$ is higher (≈6500) and inversely for higher ones is smaller (≈6000). In the line of this is the fact that manufacturers are presenting several characteristics (especially the frequency dependence of complex initial permeability, CIP $\dot{\mu}(f) = \mu'(f) - j\mu''(f)$) only for definite core size allowing one to guess what corrections (if any) are needed for another.

All these facts as well the ones mentioned in particular studies [5, 6] show that dimensional effect is a reality the details of which are worthy of consideration. The aim of this study is to clear up the specific details of size-effect in PF firstly at macroscopic level in frequency range (by the use of appropriate CIP spectra experimental data) with the later on attempt to provide some grounding in theory coming from microscopic level approach to PF samples.

[3] To whom any correspondence should be addressed

2. Samples and experimental results

To clear up the aforementioned dimensional effect, the toroidal samples of different composition and sizes were studied. The study herein first of all comprises the measurement of $\mu(0)$ and CIP spectra up to sufficient high frequencies to include the most characteristic dispersion region of PF samples – the large amplitude one. Experimental samples – ferrite toroidal cores with single layer windings of an appropriate number of turns were used for determination of CIP spectra. The meters include: phase meter (samples treated as 2-poles); vector network analyser (ZVRE from Rhode and Schwarz; samples treated as 2-poles or as 4-poles or simply as a ferrite ring in special coaxial line). Correctness of spectra measured were verified by their agreement with Kramers-Kronig relations [2] using special computer programme.

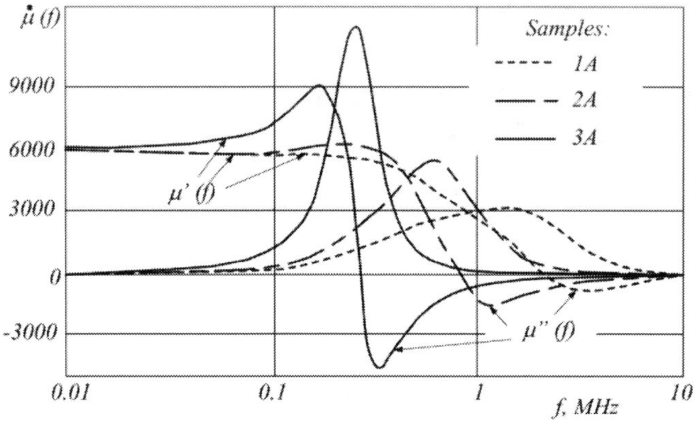

Figure 1. CIP spectra of MnZn ferrite T37

The first series of experiments (samples 1A…3A, Tab.1.) deals with standard ring cores [4] of MnZn ferrite T37. The geometrical characteristics (every geometry there is represented by three samples of same sort) as well as $\mu(0)$ for sets used are given in Tab.1 (where A stands for cross section, but $\mu(0)_{exp}$ – averaged over samples experimental value). Representatives within each geometry give similar, but non-identical CIP spectra; to grasp the main direction of changes in relation to geometry there are presented in Fig.1 the averaged over given geometry spectra. The analysis of these spectra involves difficulties since the samples may have different microstructure.

Table 1. Data of ring samples used in experiments

No	Dimensions (mm)	A (mm^2)	$\mu(0)$ Data [4]	$\mu(0)_{exp}$	No	Dimensions (mm)	A (mm^2)	$\mu(0)$ Data [7]	$\mu(0)_{exp}$
1A	R12.5x7.5x5	12.5	6500	5800	1B	R12x8x6	12	2000	2200
2A	R20x10x7	35	6500	5900	2B	R20x10x6	30	2000	2100
3A	R40x24x16	128	6000	6000	3B	R31.5x20.5x6	33	2000	2125
					4B	R40x23.5x6	51	2000	2140

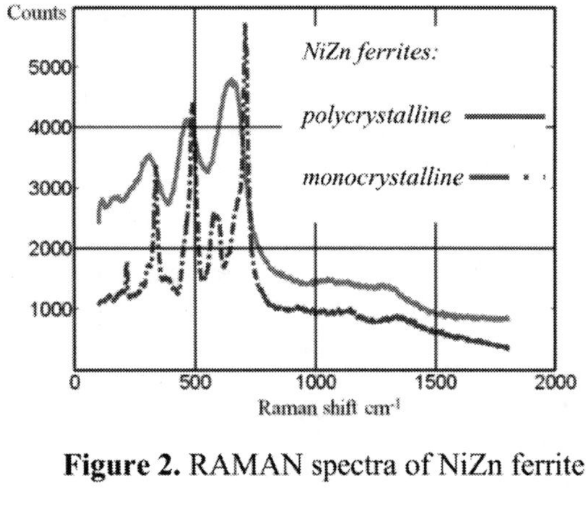

Figure 2. RAMAN spectra of NiZn ferrite

The second series of experiments (samples 1B…4B, Tab.1.) deals with non-standard ring cores with controlled microstructure. The rings of different dimensions but with the same microstructure have been made from single Ferroxcube 4S60 tile [7] by waterjet assisted mechanical cutting. Since the abrasives of cutting may contaminate the ferrite, from different spots of it the RAMAN spectra (Fig.2) and composition (Fig.3) were recorded (on "Nanofinder - S" using the 532nm excitation line; the elemental composition was assessed by energy dispersive X-ray spectrometry: ESD,

Oxford Instruments 7378). In reality the spectra and all elements analysis don't reveal any appreciable contamination. Microstructural characteristics of ferrite were obtained from micrographs (examples are shown in Fig.4). They reveal that density function *f(D)* of grain sizes *D* of PF actually is log-normal:

$$f(D) = \left(\sqrt{2\pi}\sigma D\right)^{-1} exp\left[-(lnD - lnD_{med})^2/2\sigma_{lnD}^2\right] \quad (1)$$

where D_{med} is the median, and σ_{lnD} is the standard deviation of *lnD*. From micrographs (using Saltikov's method [8]) the parameters of distribution (1) follow: $D_a = 7.9\mu m = D_{med} exp(\sigma_{lnD}^2/2)$ and $\sigma_{lnD} = 0.55$ (this value of σ_{lnD} shows that microstructure – aggregation of polycrystal grains is that of normal grain growth, when $\sigma_{lnD} \approx 0.5$ [9]).Micrographs show as well that there are clearly observable defects within numerous of grains. CIP spectra of these samples are shown in Fig.5.

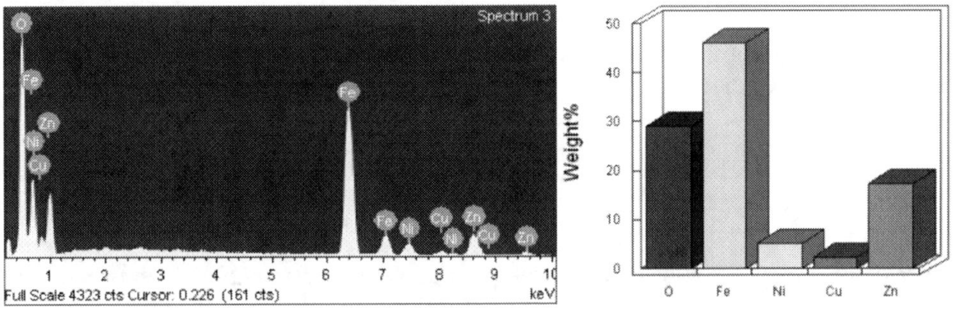

Figure 3. Elemental analysis of samples

3. Analysis of experimental data

Experimental data of previous subsection require more substantial examination and generalization.

Characteristically, the overall appearance of most of CIP spectra are similar (Fig1; 5): they are formed in fact as the combination both of relaxation attributes (dispersion and absorption within a broad range of frequency) and resonance ones (positive ups and negative downs on $\mu'(f)$ – the dispersion curves). It was shown in [10] that in such a case the most appropriate approximation comes from account of unhomogenous broadening of absorption caused in fact by the distribution of grain sizes (1). This accont results in, as it is called, symmetrical (by visual evaluation along *logf* axes from $\mu''(f)$ standpoint) spectrum:

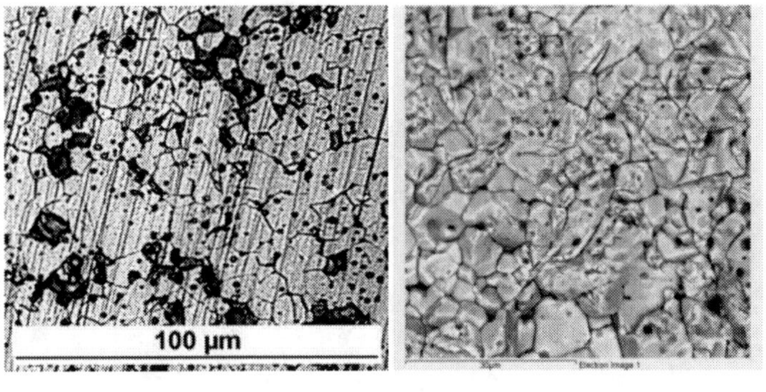

Figure 4. Micrographs of NiZn ferrite 4S60

$$\mu''(f) = \mu''_{max} exp[-(logf - logf_u)^2/2\sigma^2] \quad (2)$$

where f_u and $\mu''_{max} = \mu''(f_u)$ are parameters of maximum absorption, and $\sigma = 2\sigma_{lnD}/ln10$. For a given spectrum the parameters of (2) follow from the experimental spectrum: μ''_{max} and f_u - directly, but σ (there used for all frequencies$f$) – from the relation:

$$\sigma_{A(B)} = |log f_u/f_\sigma| / \sqrt{2ln\left(\mu''_{max}/\mu''(f_\sigma)\right)} \tag{3}$$

where f_σ is a frequency well away from f_u (often such as to fulfil $\mu''(f_\sigma) \approx 0.5\mu''_{max}$).

Nevertheless, the majority of spectra are asymmetrical one. Such spectra is possible to present by (2) as well if formally two values of σ are used: σ_A for $f \leq f_u$and σ_B for $f \geq f_u$. The values of these characteristics come from the experimental spectrum and (3) as well using it two times: firstly with $f_\sigma = f_A$ for $f_A < f_u$ and then with $f_\sigma = f_B$ for $f_B > f_u$.

In the case of standard cores of ferrite T37 the data clearly demonstrates the dependence of CIP spectra on the geometry (Fig.1): the dispersion shifts to lower frequencies as dimensions growth. Since $f_u \propto 1/D^2_{med}$ one can conclude that: small changes of grain sizes could give rise to considerable shifts of dispersion; formally the shifts in Fig.1 signalls about different microstructure in different geometry of standard cores. This means that smaller rings are with finer grains and vice versa. But the form of spectra is changing from practically relaxation type (sample 1A: $\sigma_A = 0.55$; $\sigma_B = 0.24$) to resonance one (sample 3A: $\sigma_A \approx \sigma_B \approx 0.1$). These rather radical changes are not possible to explain by dissimilarity of microstructure; obviously in the sample 3A conditions for dimensional resonance are realizing. In the case of non-standard cores with fixed microstructure no dimensional effects are observed (Fig.5) – different geometry spectra are practically identical. It is particularly notable that these spectra are pronounsly asymmetrical. Asymmetricy of the spectra and the values $\sigma_A = 0.417$ and $\sigma_B = 0.841$ show that grains of this PF contains considerable amount of intergrain defects (partitially seen in Fig.4 as well).

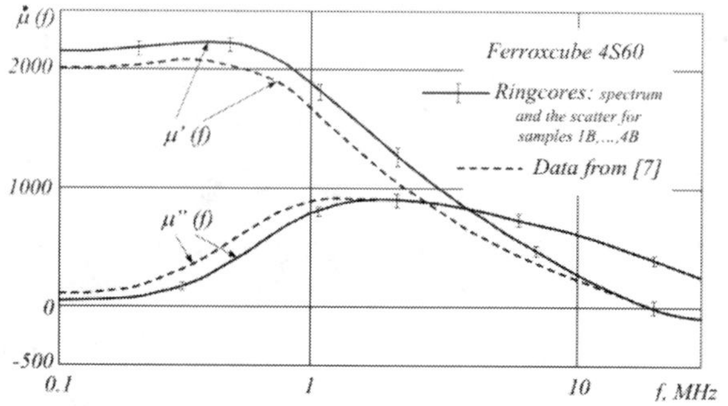

Figure 5. CIP spectra of NiZn ferrite

4. Conclusions

Thus it is possible to summarize that particular size-effects can appear in MnZn ferrites first of all because of changes in microstructure of samples, but more clearly size-effect demonstrates itself as electromagnetic effect in the samples of reasonably large dimensions. In NiZn ferrites no size-effect is observed under conditions of similar microstructure at least for medium sized cores. It is possible to say as well that CIP spectra can be used as an important adjunct to other methods in the non-destructive identification of microstructure of PF.

5. References

[1] Chikazumi S *Physics of Magnetism*. J. Wiley & Sons, 1964
[2] Lucarini V, Saarinen J J, Peiponen K-E and Vartiainen E M *Kramers-Kronig relations in optical materials*. Springer, 2005
[3] Brockman F G, Dowling P H and Steneck W G *Phys. Rev.* 1950 **77** 85
[4] Ferrites and accessories. Data book 2007, EPCOS AG, 2006
[5] Skutt G R, Lee F C *Characterization of dimensional effects in ferrite-core magnetic devices*. PESC'96, 27th ann. IEEE record, 1996, pp. 49-54
[6] Skutt G R *High frequency dimensional effects in ferrite-core magnetic devices*. PhD dissertation, Blacks burg, Virginia, 1996
[7] Ferroxcube Data book, 2008
[8] Saltikov S A *Stereometricheskaja Metallografija. Izd. Metallurgija*, Moskow, 1976 (in Russian)
[9] Kurtz S K, Carpay F M A 1980 *J. Appl. Phys.* **51** 5725
[10] Jankovskis J 2004 *J. Magn. Magn. Mat.* **272** e1847

Characterization of $LiFePO_4$/C Composite Thin Films Using Electrochemical Impedance Spectroscopy

G Bajars[1], G Kucinskis, J Smits, J Kleperis, A Lusis

Institute of Solid State Physics, University of Latvia, Riga, Latvia

E-mail: gunars.bajars@gmail.com

Abstract. The composite $LiFePO_4$/C thin films were prepared on steel substrate by radio frequency (RF) magnetron sputtering. Electrochemical properties of the obtained thin films were investigated by cyclic voltammetry charge-discharge measurements and electrochemical impedance spectroscopy (EIS). The films annealed at 550 °C exhibited a couple of redox peaks at 3.45 V vs. Li/Li^+ characteristic for the electrochemical lithium insertion/extraction in $LiFePO_4$. At low current rate such composite thin film showed a discharge capacity of over 110 mAh g^{-1}. The dependence of charge transfer resistance, double layer capacitance and lithium diffusion coefficients on applied electrode potential were calculated from EIS data. Determined values of lithium diffusion coefficient were in the range from $8.3 \cdot 10^{-13}$ cm^2 s^{-1} to $1.2 \cdot 10^{-13}$ cm^2 s^{-1} at 3.4 V and 3.7 V, respectively.

1. Introduction

Since Goodenough et. al. [1] first reported about reversible insertion and extraction of Li ions in lithium iron phosphate, $LiFePO_4$ has been considered as one of the most promising cathode material for lithium-ion batteries. In spite of such attractive features as low cost, environmentally benignity and thermal stability, $LiFePO_4$ requires further modifications to overcome limitations involving low electronic conductivity and slow lithium-ion diffusion [1-5]. Much effort has been devoted to improve the electrochemical properties of $LiFePO_4$ by reducing the particle size [6-9], by coating $LiFePO_4$ particles with carbon or performing synthesis of $LiFePO_4$/C composite [3, 8-11]. However, up to now, it has been difficult to describe the kinetic behavior of $LiFePO_4$ materials used for lithium-ion batteries because many factors including particle size, ionic and electronic conductivity, as well as phase transition kinetics affect their electrochemical performance [12].

The use of thin film electrodes enables a more detailed investigation of electrochemical properties of active material [13-18]. This is especially useful for poorly conductive materials because the thickness can be reduced to an extent which does not significantly affect the electrochemical behaviour [18]. In this study, $LiFePO_4$/C composite thin films prepared by RF magnetron sputtering were investigated by electrochemical impedance spectroscopy (EIS), which is a very powerful method for determining the rate of individual electrode kinetic steps.

2. Experimental

The $LiFePO_4$ (from Linyi Gelon New Battery Materials Co.) powder and carbon black (from Aldrich) in an amount of total carbon 5 wt% were used as precursor materials to prepare the sputtering target. The mixture of reagents was first ball-milled for 30 minutes. After this step,

[1] To whom any correspondence should be addressed

the sputtering target was prepared by cold pressing the powder on a copper pad (12 cm in diameter) lined with indium for 5 minutes. A stainless steel lid was used to apply the pressure.

Thin films of $LiFePO_4$ were prepared on a stainless steel substrate by RF magnetron sputtering. The sputtering time was varied, with the average sputtering rate of 11 nm min^{-1} and the applied power of 300W. The power was raised gradually in three steps (100W → 200W → 300W) to avoid crack formation on the target surface. The subsequent annealing of the prepared $LiFePO_4$ film was done at 550 °C for 1 h in argon atmosphere to increase the crystallinity of the thin films. The thicknesses of the obtained films were in the range of 400 – 500 nm.

Electrochemical measurements for the $LiFePO_4$/C thin films were performed in three-electrode cell using metallic lithium as the counter and reference electrodes. The electrolyte used was 1 M $LiPF_6$ ethylene carbonate/dimethyl carbonate (1:1 in volume) solution. The cell was assembled in an argon filled glove box. A VoltaLab 40 potentiostat/galvanostat (from Radiometer Analytical) was employed for the cyclic voltammetry and charge-discharge measurements of the annealed $LiFePO_4$ thin films on the stainless steel substrate. Electrochemical impedance spectroscopy (EIS) measurements were carried out at various electrode potentials using a computer controlled frequency response analyzer FRA with module Autolab PGSTA-30. The impedance spectra were measured in a frequency range from 0.002 Hz to 10 kHz by applying a 10 mV harmonic perturbation signal. Before each measurement the system was stabilized at constant voltage for several hours. Parameter fitting for equivalent circuit elements was performed by using the simulation program "RCNTRANS" and "EQUIVCRT" based on non-linear least squares method.

3. Results and discussion

The x-ray diffraction pattern of the as-sputtered 400 nm thick $LiFePO_4$/C thin film does not show any detectable peaks; annealed 400 nm thick $LiFePO_4$/C thin film, however, displays peaks corresponding to normal material similar to what has been mentioned previously [19], suggesting a possible increase of the $LiFePO_4$ grain size.

Both annealed and non-annealed thin films display an initial open circuit voltage (OCV) of 3.25 V, suggesting smaller lithium content in the as-sputtered $LiFePO_4$. Nevertheless, after several cycles vs. lithium anode the OCV has increased to about 3.4 V. The cyclic voltammogram of the annealed 400 nm thick $LiFePO_4$/C film recorded at 1 mV s^{-1} scan rate exhibits a typical couple of anodic and cathodic peaks with an equilibrium potential at 3.45 V vs. Li/Li^+ (Figure 1). These redox reaction peaks characterize the electrochemical lithium insertion and extraction in $LiFePO_4$ [1]. The ratio between anodic and cathodic peaks current is close to 1, implying good reversibility of lithium intercalation and de-intercalation from $LiFePO_4$/C thin film. The difference between the anodic and cathodic peaks at 1 mV s^{-1} scan rate is 0.2 V. This indicates that the polarization of the electrochemical system is not very high and characterizes a quasi-rapid system [18]. Absence of other redox peaks in this region indicates that the films annealed at 550 °C consist of single-phase $LiFePO_4$ [20].

Charge-discharge curves of the annealed 400 nm $LiFePO_4$/C (5 wt%) thin film are shown in figure 2. When charging and discharging were performed at 5 μA cm^{-2}, obtained discharge capacity is 110 mAh g^{-1}. With the decrease of current density higher values of discharge capacity are obtained. Measured discharge capacity at 1 μA cm^{-2} reaches 127 mAh g^{-1}. Although below the theoretical (170 mAh g^{-1}), this value is close to the capacity reported by other authors [14,15,20,21].

Figure 3 shows the Nyquist plots of the impedance spectra obtained from $LiFePO_4$/C thin film for five different electrode potentials representing both charged and discharged states. The spectra consist of semicircle at high frequency range corresponding to the charge transfer process of the lithium intercalation reaction and of straight line at the low frequency range representing the Warburg diffusion impedance (Z_W) associated with lithium ion diffusion in $LiFePO_4$/C thin film particles.

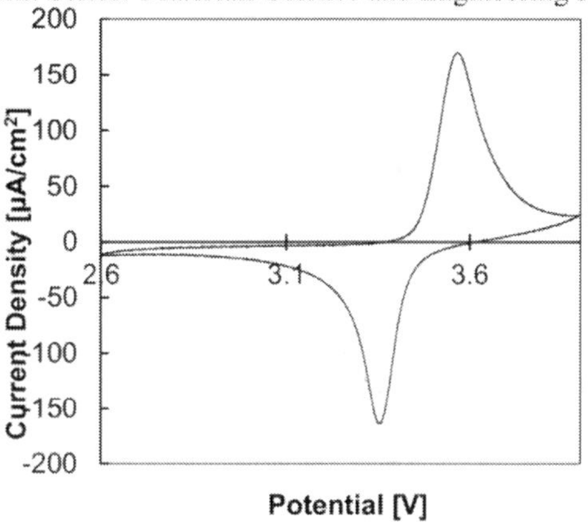

Figure 1. Cyclic voltammogram of annealed $LiFePO_4$/C (5 wt%) thin film, scan rate 1 mV s^{-1}

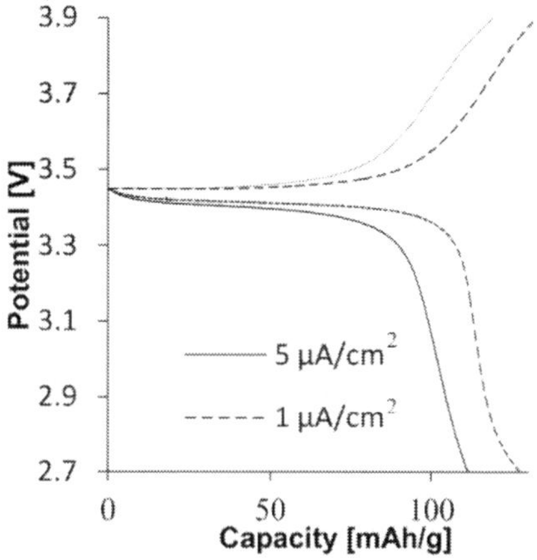

Figure 2. Charge and discharge curves of $LiFePO_4$/C thin film at 5 µA cm^{-2} and 1 µA cm^{-2}

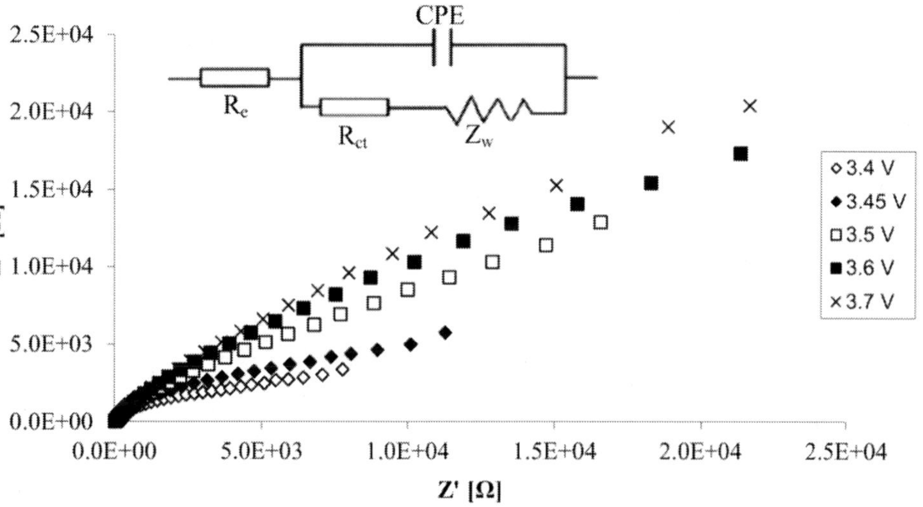

Figure 3. Nyquist plots for $LiFePO_4$/C at various electrode potentials vs. Li/Li^+ and the equivalent circuit used for fitting EIS data of $LiFePO_4$/C thin film

Table 1. Numerical values of the elements from equivalent circuits.

	3.7 V	3.6 V	3.5 V	3.45 V	3.4 V
R_e, Ω	20.3	19.2	19.1	18.3	20.1
R_{ct}, kΩ cm^{-2}	14.1	12.8	7.49	4.86	2.64
CPE, mF cm^{-2}	0.165	0.165	0.167	0.198	0.209
CPE n	0.726	0.720	0.719	0.699	0.707
W_2, mΩ $s^{-1/2}$ cm^{-2}	0.184	0.263	0.491	1.03	1.37

A simplified equivalent circuit model (figure 3) is applied to analyze the impedance spectra. A constant phase element (CPE) is introduced to represent the double layer capacitance as well as an influence of the passivation processes on electrode/electrolyte interface and other heterogeneities. The equivalent scheme parameters obtained by computer simulations are summarized in Table 1.

Evidently, charge transfer resistance (R_{ct}) increases with lithium ions extraction from the thin film material. The maximum R_{ct} value $1.4 \cdot 10^4$ Ω cm^{-2} is obtained for the most charged state at 3.7 V. Obviously the highest charge transfer resistance corresponds to the situation when $FePO_4$ phase is dominating in $LiFePO_4/C$ electrode. Contrary, the minimum value of charge transfer resistance $2.6 \cdot 10^3$ Ω cm^{-2} is acquired in a discharged state at 3.4 V which represents the higher concentration of Li^+ ions in lithium iron phosphate. Similar results were obtained in our previous research of $LiFePO_4/C$ thin films obtained by combined direct current and radio frequency magnetron sputtering [17].

As it was mentioned above CPE corresponds to the double layer capacitance C_{dl}. The variations of double layer capacitance with the electrode potential are not significant and retains about $2 \cdot 10^{-4}$ F cm^{-2} in the full range of applied voltages. In spite of negligible changes, obviously, the highest C_{dl} value $2.1 \cdot 10^{-4}$ F cm^{-2} is obtained for the discharged state at 3.4 V when the concentration of lithium ions in $LiFePO_4/C$ thin film is higher. Values of double layer capacitance decrease with the lithium ions extraction from the thin film material. There is an almost constant value of double layer capacitance $1.65 \cdot 10^{-4}$ F cm^{-2} in the charged state (3.6 – 3.7 V).

Data from EIS measurements were used to calculate Li^+ ion diffusion coefficient D_{Li} in the $LiFePO_4/C$ thin films. The equation for the calculation of D_{Li} is [22]:

$$D_{Li} = \frac{1}{2}\left[\left(\frac{V_m}{FS\sigma}\right)\cdot\left(\frac{dE}{dx}\right)\right]^2,$$

where V_m is the molar volume of $LiFePO_4$, F is the Faraday constant, S is the surface area of the electrode, dE/dx is the slope of the electrode potential vs. composition roughly estimated from figure 2, σ is the Warburg factor obtained by linear fitting of Z' or –Z'' vs. $\omega^{-1/2}$ (ω is an angular frequency) in the low frequency region [12,16,23]. Determined Li^+ ion diffusion coefficients are shown in Table 2.

Table 2. The D_{Li} dependence on the cell potential

Electrode potential vs. Li/Li^+, V	D_{Li}, cm^2 s^{-1}
3.7	$1.2 \cdot 10^{-13}$
3.6	$1.4 \cdot 10^{-13}$
3.5	$2.2 \cdot 10^{-13}$
3.4	$8.3 \cdot 10^{-13}$

The calculated Li^+ ion diffusion coefficients in $LiFePO_4/C$ thin film at voltages from 3.4 to 3.7 V are in the range from $8.3 \cdot 10^{-13}$ cm^2 s^{-1} to $1.2 \cdot 10^{-13}$ cm^2 s^{-1}. The obtained values of lithium ion diffusion coefficient in $LiFePO_4/C$ thin film are in accordance with data reported in similar studies [16,23] for the compositions corresponding to the states where one phase - $LiFePO_4$ or $FePO_4$ is dominating. In present work D_{Li} is not calculated at 3.45 V, where $LiFePO_4$ probably shows a two phase co-existence and dE/dx is uncertain. Therefore values of Li^+ ion diffusion coefficient in two phase region are not reliable since the variation of D_{Li} is mainly determined by dE/dx.

The dependence of Li^+ ion diffusion coefficient value on the electrode potential is very similar to that dependence of double layer capacitance. The highest $D_{Li} = 8.3 \cdot 10^{-13}$ cm^2 s^{-1} value is obtained for the discharged state at 3.4 V when the concentration of lithium ions in $LiFePO_4/C$ thin film is the highest. The values of Li^+ ion diffusion coefficient decrease with lithium ions extraction from the thin film material. There are very close values of $D_{Li} = 1.2 \cdot 10^{-13} – 1.4 \cdot 10^{-13}$ cm^2 s^{-1} obtained in charged state (3.6 – 3.7 V).

4. Conclusion

$LiFePO_4/C$ thin films were prepared by RF magnetron sputtering. Cyclic voltammogram of the crystallized $LiFePO_4/C$ thin film showed the typical red-ox reaction peaks characterizing the electrochemical lithium insertion/extraction reactions in $LiFePO_4$. The obtained discharge capacity of $LiFePO_4/C$ thin films was 110 and 127 mAh g^{-1} at the current densities 5 and 1 μA cm^{-2}, respectively. Three element equivalent circuit which contains charge transfer resistance R_{ct}, Warburg impedance Z_W and constant phase element CPE is applicable to analyse the EIS results of $LiFePO_4/C$ thin film. The charge transfer resistance increases in relation with an appearance of the $FePO_4$ phase. The minimum value of charge transfer resistance is acquired in discharged state which corresponds to the maximum concentration of Li^+ ions in lithium iron phosphate. Contrary to R_{ct}, values of double layer capacitance C_{dl} decrease with the lithium ions extraction from the thin film material. The obtained lithium ion diffusion coefficients are in the range from $1.2 \cdot 10^{-13}$ to $8.3 \cdot 10^{-13}$ cm^2 s^{-1} and show a similar dependence on the applied voltage as C_{dl}.

5. Acknowledgements

The authors acknowledge Taiwan – Latvia – Lithuania cooperation project "Materials and processing development for advanced Li ion batteries" and National Program in Energetic for financial support. The technical assistance and consultations of AS Sidrabe, Juris Katkevics and Arturs Viksna are acknowledged.

References

[1] Padhi A K, Nanjundaswamy K S, Goodenough J B 1997 *J.Electrochem.Soc.* **144** 1188
[2] Andersson A S, Kalska B, Häggstrom L, Thomas J O 2000 *Solid State Ionics* **130** 41
[3] Prosini P P, Zane D, Pasquali M 2001 *Electrochim.Acta* **46** 3517
[4] Chung S-Y, Bloking J T, Chiang Y M 2002 *Nat.Mater* **1** 123
[5] Takahashi M, Tobishima S, Takei K, Sakurai Y 2002 *Solid State Ionics* **148** 283
[6] Yang S F, Zavalij P Y, Whittingham M S 2001 *Electrochem.Commun.* **3** 505
[7] Prosini P P, Karewska M, Scaccia S, Wisniewski P, Pasquali M, 2003 *Electrochim.Acta* **48** 4205
[8] Yamada A, Chung S C, Hinokuma K 2001 *J.Electrochem.Soc.* **148** 224
[9] Delacourt C, Poizot P, Levasseur S, Masquelier C 2006 *Electrochem.Solid State Lett.* **9** A352
[10] Raver N, Chouinard Y, Magnan J F, Besner S, Gauthier M, Armand M 2001 *J.Power Sources* **97/98** 503
[11] Hu Y, Doeff M M, Kostecki R, Fiñones R 2004 *J.Electrochem.Soc.* **151** 1279
[12] Gao F, Tang X Y 2008 *Electrochimica Acta* **53** 5071
[13] Sauvage F, Baudrin E, Morcrette M, Tarascon J M 2004 *Electrochem.Solid State Lett.* **7** A47
[14] A. Eftekhari 2004 *J. Electrochem. Soc.* **151** 1816
[15] Chiu K-F, Chen P Y 2008 *Surface & Coatings Technology*, **203** 872
[16] Xie J, Imanishi N, Zhang T, Hirano A, Takeda Y, Yamamoto O 2009 *Electrochimica Acta* **54** 4631
[17] Bajars G, Kucinskis G, Smits J, Kleperis J 2011 *Solid State Ionics* **188** 156
[18] Sauvage F, Baudrin E, Gengembre L, Tarascon J M 2005 *Solid State Ionics* **176** 1869
[19] Smits J, Kucinskis G, Bajars G, Kleperis J 2011 *Latvian Journal of Physics and Technical Sciences* **48** 27
[20] Yada C, Iriyama Y, Abe T, Jeong T K, Inaba M, Ogumi Z 2005 *J.Power Sources* **146** 559
[21] Chiu K-F 2007 *J. Electrochem. Soc.* **154** 129
[22] Ho C, Raistrick I D, Huggins R A 1980 *J. Electrochem. Soc.* **127** 345.
[23] Tang K, Yu X, Sun J, Li H, Huang X 2011 *Electrochimica Acta* **56** 4869.

Study the effects of moisture content on the electrical properties of technical textiles by impedance spectroscopy

A Lusis[1], E Pentjuss, G Bajars, J Gabrusenoks , R Janeliukštis, J Zandersons

Institute of Solid State Physics, University of Latvia, Latvia

E-mail: lusis@latnet.lv

Abstract. Application of metal coatings for the functionalization of technical fibres and fabrics faced with influence of moisture on functional properties, e.g., the impedance of the metal coated K-glass fabrics have strong dependence of content absorbed water or moisture. The paper devoted to develop methodology for characterisation functional materials based on fabrics and model for interpretation of the electrical impedance spectra to obtained functional characteristics of technical textile fabrics. Model based on analyses of 3D plot of imaginary part of complex modulus spectra versus sample mass. Methodology helps to control content of adsorbed water in fabric and influence of moisture on the functional characteristics.
Keywords: glass fabrics, moisture, metal coatings, electrical impedance spectra.

1. Introduction

Recently many researchers recognized there is strong interest in new nanostructured porous materials based on fabrics or textiles for the different technical applications [1-2]. The fabrics have wide application as support material of functional species in ionic devices as well as materials for screening EMR, heat reflecting, protecting clothes, air purification etc. [3]. Nanostructured textiles where conducting substances coated on porous textile fibers, are serving as conductive three-dimensional frameworks for electrodeposition of functional species (metals, oxides). Such 3D porous networks not only permit large loading of active electrode materials but facilitate easy access of electrolytes to the electrodes. The transport of liquids (electrolytes) into textile depends on its permeability, water absorption and desorption characteristics. The actual for textiles are last topics which textiles are used both as functional construction material and as a support of functional species for ionic devices.

The metal coatings are widely used for functionalization of textiles (woven and nonwoven). Fabrics are porous medium which adsorbs some species (water, gasses) from environment. The influence of environment moisture on physical and chemical properties is topics for research functionality of technical textiles. In the moist environment the coatings is exposed to corrosion.

[1] To whom any correspondence should be addressed

The fabrics have to be characterized before functionalization as well as after functionalization for technical applications.

There are problems with the interpretation of impedance data. One of ways is to find relations between metal coated fabric constitution, pore and metal type, moisture content and water sorption/desorption kinetics. The fabric is two or three phase's hetero system based on textile yarn regular network and adsorbent or/and metal particles on fiber surface. The adsorbent and deposited metal particles can be localized inter fibers and inner in pores.

The water content has strong influence on complex impedance Z(Z', Z'') spectra and measured spectra of the system M/GF/M as prepared has metallic like behavior up 100 kHz and after that has inductance behavior which had been described in our paper [4]. The absorbed water reduces electrical conductivity and after drying impedance has capacitance behavior. The wet metal coated samples |Z| spectra have similar spectral behavior.

The wet K-glass fabric in equilibrium state in air atmosphere and room temperature has water content up to 20 wt% (TGA data). From the leached fiber Raman spectrum the concentration of Si-OH bonds is more than 20 mol% and is active centers on pores surface for H_2O absorption. The surface of no leached glass fiber is smooth and for leached fiber the pore sizes are larger than 30 nm and depth of the surface pore is at least 10 nm [4].

This paper is devoted to study the influence of moisture on the electrical properties of the fabrics based on glass fibers. The adjacent aim of our current investigations is related to methodology have to control water influence on electrical properties of metal coated and uncoated fabrics.

2. Experimental

There are used leaching, etching and coating technologies to modify properties of the fabrics based on glass fibers (GF). The fabrics of K-glass [composition: $(18\text{-}22)Na_2O(3\text{-}5)Al_2O_3$ $(73\text{-}79)SiO_2$] had been leached in water and dried at 220 °C. The leaching method is described in [5]. The leaching process is changing a surface morphology of glass fibers as well as internal nanostructure of glass fibers by forming pores [4]. The fabrics and yarns have been modified in the technological processes in vacuum: etching with argon plasma and deposition of metal particles (Al, Ni, Cu) in magnetron process.

The both sides of leached and non-leached K-glass fabrics had been coated with metal by DC magnetron sputtering. The density of deposited metal on the fabrics is 0,5-1,5 mg/cm^2.

The electrical impedance spectroscopy (EIS) has been used to study properties as well as the influence of moisture content in metal coated fabrics. The electrical properties have been studied by impedance analyzers HP 4194A (100 Hz–15 MHz) with Dielectric Test Fixture for solids HP 16451B. The Test Fixture for solids HP 16451B have been used in three versions - electrodes with or not isolation film (PET 33 μm): (2i) – on both electrodes, (1i) –

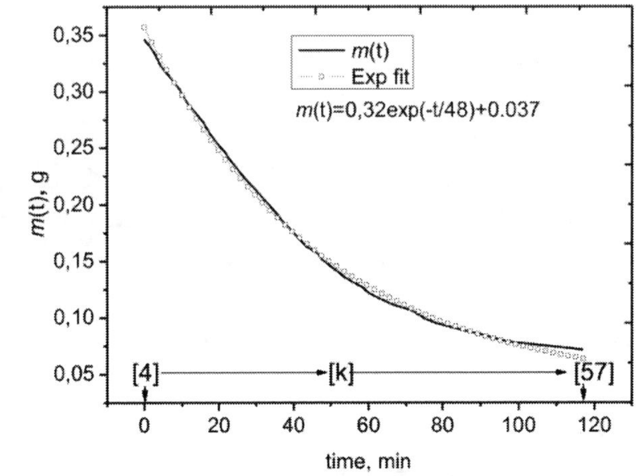

Figure 1 The drying kinetics of sample **n-n**. [k] - measurement sequence number for sample **n-n** in series of measurements [4]…[k]…[57].

on one electrode and (0i) - no. Fixture electrode area is 11,3 cm^2. EIS had been carried out for uncoated sample **n-n** and coated with metal M samples **M/GF/M.** Before impedance measurements glass fabric samples had been kept in the desiccators at 100 % RH and room temperature for 24 h. The starting density of absorbed water for sample **n-n** was 0.031 g/cm^2. The drying kinetics $m(t)=m[k]-m[57]$ of sample **n-n** is exponent [Fig.1]. EIS measurements were done with a period (0i, 1i, 2i) in the measurement sequence [4] - [57]. The impedance spectra have been measured during samples drying. For the analyses data of the electrical impedance spectra components have been used ZView2 program (Scribner Associates Inc.).

3. Results and Discussion

The fabric samples can be characterized as mechanical mixture of ionically and electronically conducting phases. The water in such hetero system serves as electrolyte which involved in electrochemical reactions with sample fixture electrodes as well as with metal particles. The impedance such cases have components related to electrode reactions, ion transfer over double layer and bulk (yarn framework) as well as relaxation of dipoles, ions and electrons (polarization of ions and H_2O). The sample fixture electrodes can be turned as blocking electrodes by thin insulator film (PET 33 μm; fabric thickness 450-500 μm).

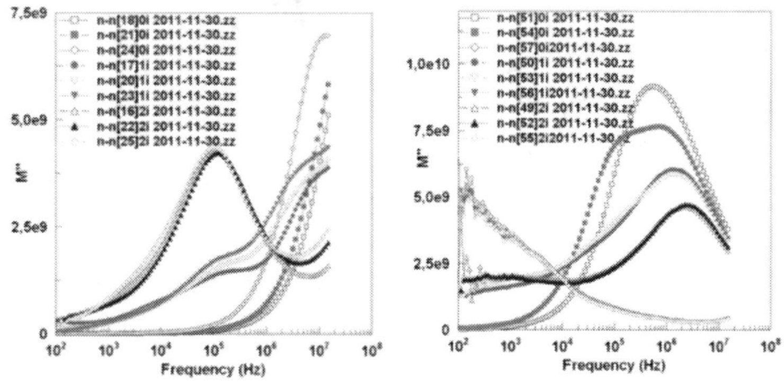

Figure 2. M″(ω, m) for measurements n-n[0i, 1i, 2i]

In this paper has been used complex modulus method for analyses impedance spectra [6]: $M(M', M'')=j\omega C_0 Z(Z', Z'')$, where C_0 – geometric capacitance of fixture electrodes, $M'=\omega C_0 Z''$ and $M''= \omega C_0 Z'$. From the last formula followed $M''= (Cp/\varepsilon_0)\ \omega\tau/[1+ (\omega\tau)^2]$, where ε_0 – vacuum dielectric constant, τ=CpRp – Maxwell relaxation time of charge carriers in bulk of sample, Cp, Rp - parallel circuit elements.

For wet sample ImM = $M''(\omega, m)$ spectra is function of ω and water content m. In Fig.2 is presented $M''(\omega, m)$ data of measurement series **n-n**[4-57] for no leached fabric sample and with three electrode arrangements (2i, 1i and 0i). The sequences of curves are

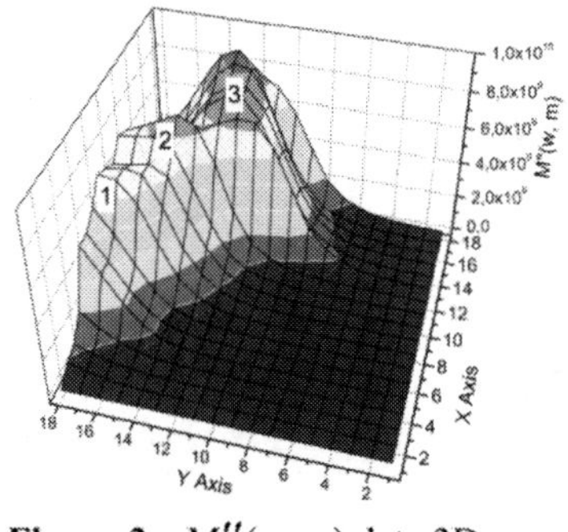

Figure 3a. M″(ω, m) data 3D presentation for (0i) regions:
1-τ1=3E-8 sec
2-τ2=2E-7 sec
3-τ3=2E-6 sec

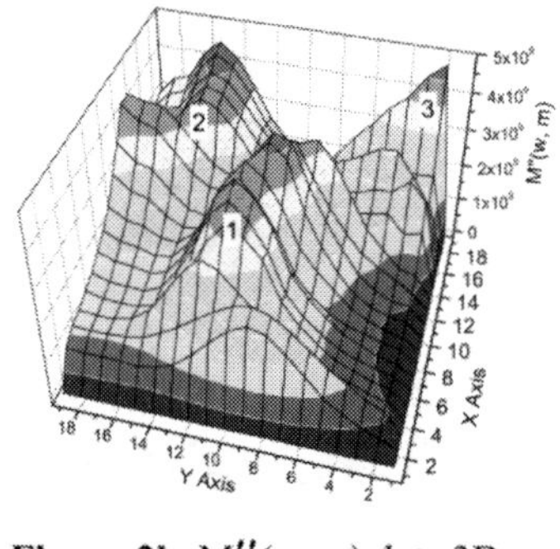

Figure 3b. M″(ω, m) data 3D presentation for (2i) regions:
1-τ1=2E-5 sec
2-τ2=7E-7 sec
3-τ3=1E-4 sec

X Axis: matrix data for frequency 100 Hz – 15 MHz
Y Axis: matrix data for water content 0.35 – 0,08 g

reflecting complicate character of water desorption kinetics and role of blocking electrodes. The insulator films application allows to separate peaks of M″ spectra (Fig.2). On the 3D presentation of spectra of M″(ω, m) can be distinguished three regions with different τ (Fig.3). In 3D graph xy plane represented matrix of 19 rows for sample mass values for 19 groups of (0i, 1i, 2i) and 19 colons for fixed frequencies in range 100 Hz -15 MHz.

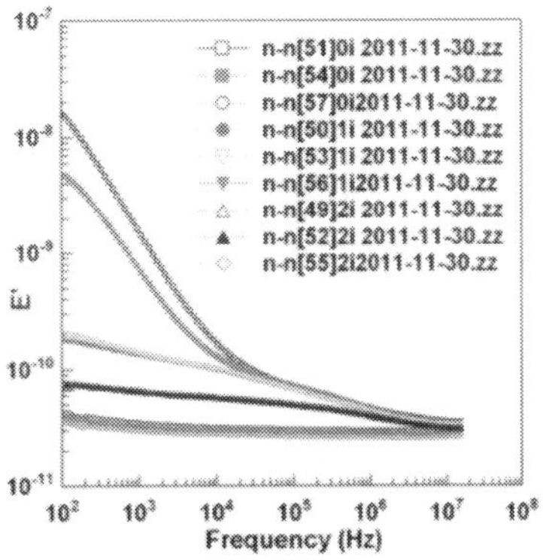

Figure 4a. Real dielectric function E′(ω, *m*)

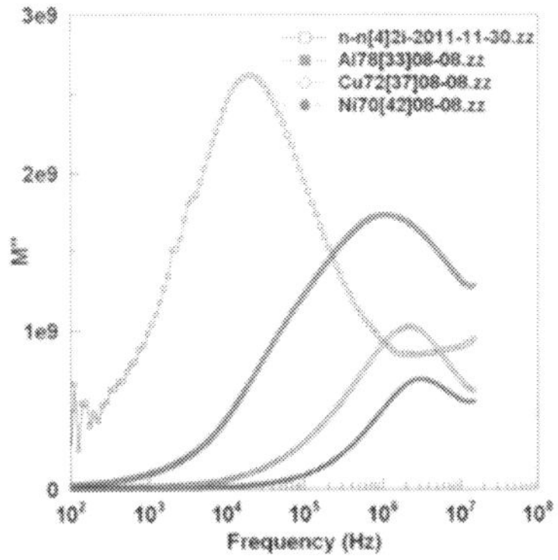

Figure 4b. M″(ω, *m*) spectra of glass fabrics coated with metal particles

The isolation for dry state for all cases (0i, 1i and 2i) have very small influence on the real part complex dielectric function E′(ω, *m*) (Fig. 4a).

The water have influence too on the M″(ω, *m*) spectra of glass fabrics coated with metal particles (Fig.4b). Water masked metal particle contribution in spectra (0i) measured without blocking electrochemical reactions at fixture electrodes. In spectra M″(ω, *m*) for (2i) case dominate peak at 20 kHz, indicative of water polarization contribution, and which correlated with high value of real part of the complex dielectric function E′(ω, *m*) in Fig. 4a.

4. Conclusions

Application of metal coatings for the functionalization of technical fibers and fabrics faced with influence of moisture on functional properties, e.g., the impedance of the metal coated K-glass fabrics have strong dependence of content of absorbed water or moisture.

Impedance spectroscopy is powerful tool for moisture control and to study the interaction of environment moisture with metal particles on the fibers and the influence on physical and chemical properties (corrosion of coatings) and functionality of technical textiles.

The fixture electrode isolation method can helped to identify influence of moisture on the electrical properties of the fabrics based on glass fibers. The 3D plot of imaginary part complex modulus M″(ω, m) can be good methodology to control water influence on electrical properties of metal coated and non coated fabrics.

Acknowledgments

This work was financially supported by National Research Program “IMIS” and “Cooperation project of. Latvian Council of Science 10.0040 “Investigation of Latvian renewable raw materials – flax and hemp products for development of innovative technologies and new functional materials”.

References

[1] Hu L, Pasta M, Mantia F L, Cui L F, Jeong S and Deshazer H D 2010 *Nano Lett.* **10** 708–714
[2] Hecht D S, Hu L, Gruner G 2007 *Current Applied Physics* **7** 60–63
[3] http://www.swicofil.com/textile_metallization.html
[4] Lusis A, Pentjuss A, Balodis J, Janeliukštis R and Zandersons J 2011 *IOP Conf.Series:*

Materials Science and Engineering **25** 012011
[5] Veveris G, Eglitis V, Lusis A and Pentjuss E 2006 *Latvian Journal of Physics and Technical Sciences* **4** 54
[6] Hodge I M, Ingram M D and West A R 1976 *J.Electroanal. Chem.* **74** 125-143

Electrochemical impedance and moisture content of glass fabric

E Pentjuss[1], A Lusis, G Bajars, J Gabrusenoks, L Jekabsone

Institute of Solid State Physics, University of Latvia

E-mail: pentjuss@cfi.lu.lv

Abstract. The glass fiber fabrics have application at the wet conditions. Impedance measurements of sodium alumosilicate glass fabric in dependence on its moisture content are presented. The impact of pores of glass fibres and fabric components to electrochemical impedance of fabric are investigated and discussed.

Keywords: glass fibres fabric at wet conditions, pores, charge double layer, electrolyte

1. Introduction.

The glass fibres fabrics have wide application including wet circumstances. It is accepted [1,2] that interaction of alkali silicate glasses with mineral acids and water proceeds by ionic exchange by diffusion of Na^+ ions to the surface and H^+ or H_3O^+ into the bulk of material, which fill the vacancies of Na^+ ions [1]. There is decreased concentration of Na^+ ions or they fully absent in the leached layer [1]. As a result the Na^+ ions are replaced by H^+/OH^- ions in 3Si-ONa groups [2]. Some vacancies remain unfilled [1] and, consequently, leached layers are porous. In addition, pores may contain moisture. There are presented results of impedance measurements of sodium alumosilicate glass fabric in dependence on its moisture content to determine impact of pores in fibres and fabric components to impedance of fabric.

2. Measurements.

The partial leaching was fulfilled in a bath of distilled water during 1 h at 20 ^{0}C. As maximally leached samples there are used the samples leached by acid in factory and additionally by water during 24 h in laboratory. The moisture of fabric above its equilibrium content was created by soaking it with water using capillarity of fabric and the moisture changed using free water evaporation to environment. The moisture content below equilibrium was created by sample gradual heating up to 40 ^{0}C- 50 ^{0}C. Moisture content was determined by sample weighing before and after impedance measurement. It is assumed that heating of sample during 2 h at 150 ^{0}C took away all water from samples. The impedance measurement frequency range was 100 Hz-15 MHz. To prevent the charge transfer through the interface of metallic (Ni) electrode and wet fabric and electrochemical reactions there in some cases two isolating polymer films are placed between them.

---

[1] To whom any correspondence should be addressed

3. Results and discussions

3.1. Equilibrium moisture content of glass fabric

Fig. 1 shows the equilibrium weight levels for untreated, partially leached and heated glass fabric sample A and untreated and heated sample B. The leaching procedure lowered permanently the sample A equilibrium weight. At the same environment conditions the equilibrium weight decrease $\Delta_L(A)$ was almost equal to the mass value of leached matter in a bath after water evaporation. It indicates to direct leaching of matter in a leaching water. In this case the water (crystalline and free) leaved the fabric together with substances its containing that causes the decrease of equilibrium weight of sample and definite its irreversibility. The XRD showed that main leached substances after water evaporation at room conditions was a mixture of $Na_2CO_3 \cdot H_2O$ and $Na_3H(CO_3)_2 \cdot 2H_2O$, that are soluble in water and forms porous shell (the existence of shell was observed in [3]) on elementary glass fibres. These substances leave the fabric and change it looks from matte to bright. The pores after leaching procedure have to be full of water. The heating procedure (2h at 150 ^{0}C) decreases equilibrium weight of fabric comparably to leaching procedure ($\Delta_H(B) \geq \Delta_L(A)$).

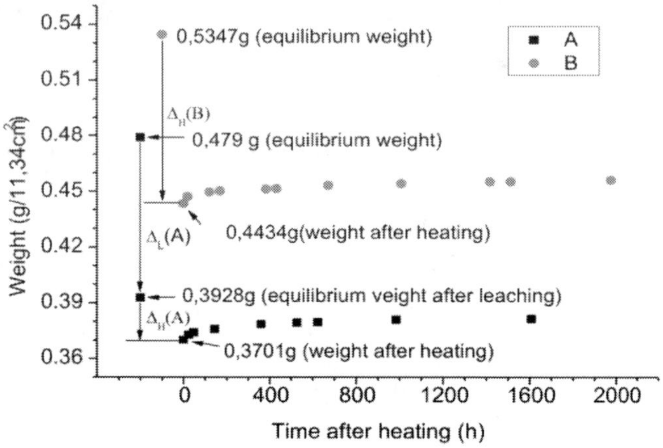

Figure 1. Equilibrium weight of leached (A) and unleached (B) glass fabric samples versus time after heating (150 ^{0}C) procedure. $\Delta H(A)$=0,0227g. $\Delta L(A)$=0,0862g that was approximately equal to leached matter mass in a bath after water evaporation (0,086g) and $\Delta H(B)$=0,0913g on 11.34cm^2 (area of electrode).

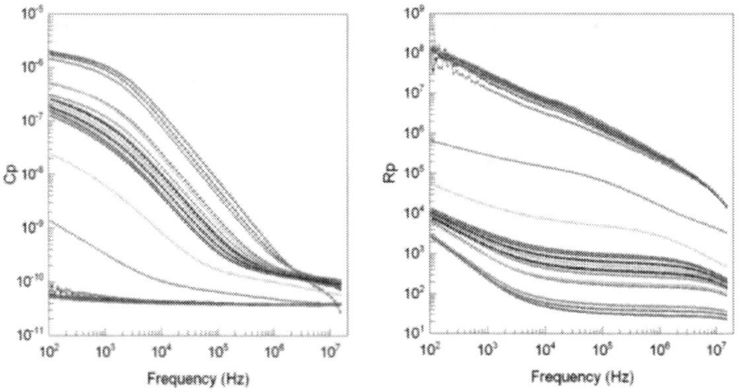

Figure 2. Capacitance Cp and resistance Rp of fabric sample at the different moisture content versus frequency. The moisture content changes from 0,1397 g/11,34cm^2 (upper curve of Cp and lower curve of Rp) to zero (lower curve of Cp and upper curve of Rp).

If fabric was leached before heating (sample A), the heating impact was small, i.e., $\Delta_H(A) \ll \Delta_H(B)$. It indicates that leachable (but not leached) matter of fabric irreversible lost the weight by heating also. It may be associated with dehydration at temperature as high as 150 ^{0}C (dehydration temperature of $Na_2CO_3 \cdot H_2O$ is 100 ^{0}C). The second reason of weight loss may be described by splitting reaction $2Na_3H(CO_3)_2 \cdot 2H_2O = 2(Na_2CO_3 \cdot 2NaHCO_3 \cdot 2H_2O) \rightarrow 3Na_2CO_3 + CO_2\uparrow + 5H_2O\uparrow$ at (140-160) ^{0}C [4]. The third reason of weight loss must be the free water evaporation from all part of fabric including pores. In addition to weight loss during the heating there was observed the ability loss of fabric to incorporate water. This phenomenon have to be associated with crystalline water loss that is renewed during the months (an increase of equilibrium weights of samples A and B after heating in Fig.1) and should be associated with corresponding volume changes of pores in the shell of sodium carbonates. In connection with the second reaction mentioned above, there have to be the equilibrium weight loss also. In a case of maximal leaching the all shell of carbonates should be dissolved in water and there are no more substances for dehydration during heating up to (110-150)^{0}C and for hydration after heating also. It explains the absence of equilibrium weight loss during heating and equilibrium weight increase after heating that exist in Fig.1 in a case of unleached and partially leached fabric.

3.2. Impedance characteristics of leached fabric

Moisture in Fig.2 increases capacitance Cp and lowers resistance Rp for parallel Rp and Cp connection. The high relative permittivity ε_r (above 100000) at low frequencies (Fig.3), experimentally observed the known Randles-circuit type characteristics in Z'-Z'' plane and charge transfer resistance directly indicate to charge double layers at the both contacts, ions diffusion and charge transfer through the metal–fabric interfaces. Both metallic contacts are seriously connected by water electrolyte containing Na^+ and other ions. When water content is small, the fabric converts to dielectric with small capacitance Cp (the lowest curves Cp in Fig.2) and ε_r <2 (Fig.3), because the great part of fabric volume is an air. In her turn the resistance Rp (Fig.2) has highest values and are mainly associated with dielectrics loss.

3.3. Pores in leached glass fibres and fabric components impact to impedance

Water evaporation from unheated samples decreases ε_r (Fig.3) and an active (parallel) conductivity Yp (Fig.4a). The three regions are seen in relations in Fig.3 and Fig.4 in dependence on moisture content and two decrease regions between them. The last sharp and high decrease of ε_r (Fig.3) and Yp (Fig.4a) cannot be associated with electrochemical reaction at the interface of metal (Ni) contact or charge transfer through it, because the similar decreases at the same moisture content exist at the isolated polymer films between contacts and fabric. The almost stable conductivity in region of moisture from the sharp conductivity fall to zero content indicates that as last are empted from water the pores. Water of pores cannot take part in current conduction of fabric because they are enclosed in electrically isolating glass fibres. In this region of water content there is no continuous path (media) for current conduction between elementary fibres and electrodes. The rest conductivity at the very low moisture content in Fig. 4 can be explained by dielectric loss.

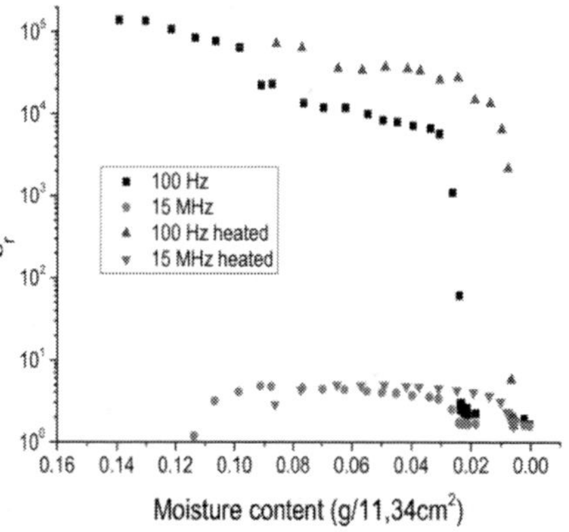

Figure 3. Relative permittivity ε_r of unheated and heated samples at frequencies of 100 Hz and 15 MHz versus moisture content

Fig.4 shows that heated fabric conductivity decreasing point is shifted by value Δ to less moisture content that is associated with hydrated water and substance mass loss during heating. If the fabric do not encounter substances that leave the fabric by heating (except water), then volume of pores can be determined. Using moisture values before raising of conductivity and considering that pores are fulfilled by capillary water, then in a case of heated fabric (Fig.4b) the pore volume is less than $1{,}22\cdot10^{-2}$ cm^3/g of fabric equilibrium weight before leaching and heating and may be are located in a shell of carbonates. In a case of maximally leached fabric the pores volume are about $2{,}6\cdot10^{-2}$ cm^3/g of a glass fibres.

There are 400 elementary glass fibres in the threads of fabric. The sharp decrease of conductivity in Fig. 4 should be associated with moisture loss from the surface of elementary fibres that breaks the current paths between electrodes. The two parts of slow conduction decrease in Fig.4 should be associated with water evaporation from free space between elementary fibres and evaporation from free surfaces of metallic contacts and rest threads at the highest moisture content of fabric. The shape of conductivity in Fig.4 is similar to the shape of relative permittivity in Fig.3 and indicates that charge double layer formation is restricted by ions transport through the volume of threads of elementary glass fibres.

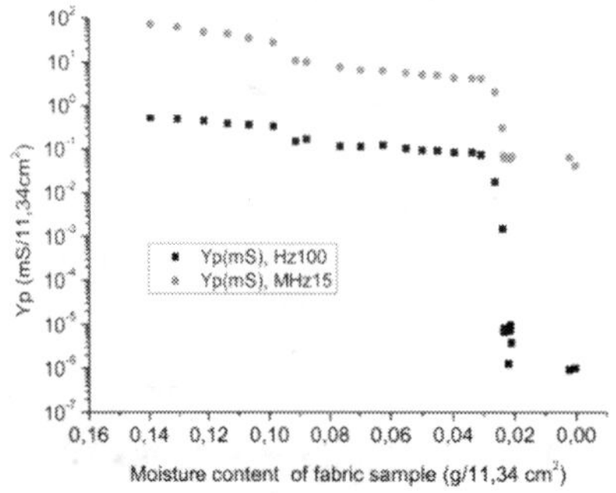

Figure 4a. Active conductivity Y_p of a partly leached and unheated sample A.

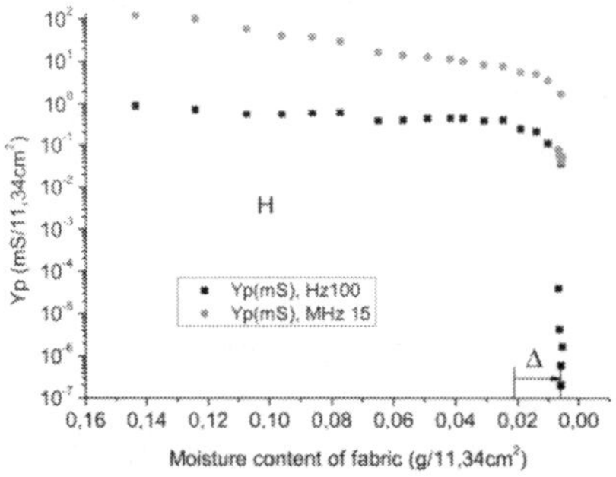

Figure 4b. Active conductivity Y_p of a partly leached and heated sample A.

4. Conclusions

1. Water in Na Al Si glass fabric is located in leachable substances and in pores of elementary glass fibres.
2. Conductivity of glass fabric has sharp decrease when surface of elementary glass fibres lost the water (electrolyte) and there are lost the continuous paths for ions conduction.
3. The heating of fabric leads to shift of conductivity decline to less water content that is associated with hydrated water and substance loss. In a case of unleached and partly leached samples the hydration restores during the months.
4. The high relative permittivity value of metal-fabric-metal cells (above 100000) at low frequencies and Randles-circuit type characteristics directly indicates to charge double layers at the both metallic contacts, to ions diffusion and charge transfer through the metal–fabric interfaces, when the continuous electrolyte paths exist.
5. There is proposed the way to determine volume of pores based on fabric water content and electrical conductivity measurements.

References

[1] Veal B W, Lam D J and Karim D P 1980 *Nuclear Technology* **51** 136
[2] Tsong I S T, Houser C A, White W B, Power G L and Tong S S C 1980 *Non-Cryst.Solids* **38-39** 649
[3] Veveris G, Eglitis V, Lusis A and Pentjuss E 2006 *Latvian Journal of Physics and Technical Sciences* **4** 54
[4] Gärtner R S and Witkamp G J 2007 *Hydrometallurgy* **88** 75

IOP Conf. Series: Materials Science and Engineering **38** (2012) 012022 doi:10.1088/1757-899X/38/1/012022

Cotton Fabric Surface Modification by Sol-Gel Deposition of ZnO Thin Films

S Vihodceva[1], S Kukle

Riga Technical University Institute of Textile Materials Technology and Design, Riga, Latvia

E-mail: Svetlana.Vihodceva@rtu.lv

Abstract. One of the main tasks of research is to impact the additional value on natural textiles by adding to them ultraviolet (UV) absorption and antimicrobial protection properties with ZnO nano-level coatings. ZnO shows high absorption in the UV region of the light spectrum, in comparison with organic absorbers conventionally used in the textile industry shows no significant degradation, is stable and classified as non-toxic material. Nanosols were prepared by using the sol-gel process. In this work comparison of samples coated by nanosols with zinc acetate (Zn $(CH_3COO)_2 \cdot 2H_2O$) and zinc sulphate ($ZnSO_4$) was made. Scanning electron microscopy (SEM) was used to examine the nature of the surface modification with ZnO coating by the sol-gel technique as also after exploitation of samples; energy dispersive X-ray spectroscopy was used for the analysis of elemental composition of coated fabric samples.

1. Introduction

The UV radiation is composed of three types: UV-A (320-400nm), UV-B (290-320nm), and UV-C (100-290nm). The UV-C radiation is absorbed by the ozone layer, however, the UV-A and UV-B reach the earth surface and cause serious health problems such as skin cancer, sunburn, and photo-aging. [1, 3-8] The UV-B radiation penetrates the upper skin layers (epidermis) while the UV-A radiation also reaches the deeper areas also (dermis). [1, 3, 5, 7] Therefore, special attention has been focused recently on the UV transmission of textile because of the growing demand in the marketplace for light-weight apparel that offers protection from UVR, while fostering comfort. [9]

To prevent harmful UV rays transmission, textiles need to treat with UV absorbers, such as ZnO, what shows strong absorption in the UV region of the light spectrum, in comparison with the organic absorbers conventionally used in the textile industry, it shows no significant degradation, is stable and classified as non-toxic material. [2, 9, 10] ZnO is harmless, that is why it is used in cosmetics such as suncreams. For the above-mentioned reasons, ZnO seems to be ideal for the preparation of UV-absorbing, nanosol based coatings. [2]

2. Materials and Methods

Raw woven plain wave 100% cotton fabric with the surface density 147.36 g/m^2 from yarns of linear density 33.6 Tex has been used in the experiment. The thickness of the textile fabric is ~ 0.36 mm; the

[1] To whom any correspondence should be addressed

measurement was taken by the textile thickness tester ("TH-25", "Zapadpribor", Ukraine). Cotton is a cellulosic fibre with a high ratio of hydroxyl groups that make it hydrophilic and that are available for polar interaction or potential surface reactions.

Tetraethoxysilane (TEOS, $C8H20O_4Si$), ethyl alcohol (C2H5OH), hydrofluoric acid (HF), deionized water and in the first case $Zn(CH_3COO)_2{\cdot}2H_2O$ and in the second case $ZnSO_4$ have been used for nanosols preparation.

2.1. Nanosol preparation and fabric treatment

Nanosols were prepared by a controlled hydrolysis, by adding ethyl alcohol slowly into TEOS with continuous stirring, after adding deionized water and hydrofluoric acid, stirred for 30 minutes, after mixed in the first case with zinc acetate and in the second with zinc sulphate with continuous stirring 10 minutes. The process was performed at room temperature. The obtained nanosols were clear and homogeneous. To evaluate the coating quality different fabric samples were prepared with TEOS concentrations 2% and various zinc acetate and zinc sulphate concentrations 2.5%, 5%, 7.5% and 10%.

Fabric samples were dipped into the prepared nanosols, soaked for 10 minutes at room temperature, then samples were dried at 50 °C for 10 minutes and after cured in an oven at 100 °C and 120 °C for 5 minutes.

3. Results and Discussion

3.1. SEM micrograph analysis

In this research morphological changes as a result of nanosol treatment of natural textile have been investigated using SEM (Tescan, Mira//LMU Schottky).

Figure 1. Cotton textile samples after nanosol treatment: zinc acetate concentration 2,5%

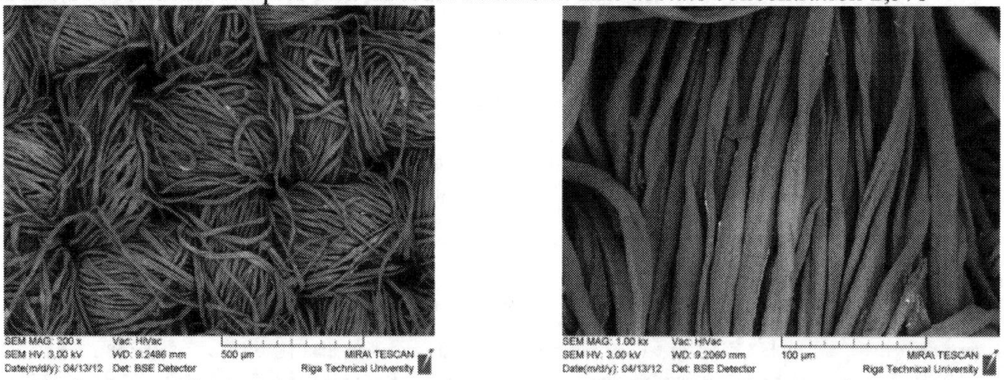

Figure 2. Cotton textile samples after nanosol treatment: zinc sulphate concentration 2,5%

In figure 1 and figure 2 there are micrographs of cotton textile samples after nanosol treatment. Figure 1 represent SEM micrographs of cotton samples after nanosol treatment with zinc acetate concentration 2.5%, samples have sufficiently even coating with a few microdefects. Figure 2 represent micrographs of cotton samples after nanosol treatment with zinc sulphate concentration 2.5%, samples have even coating, without visible defects. The same results are received for samples treated by nanosol with zinc acetate and zinc sulphate concentration 5%.

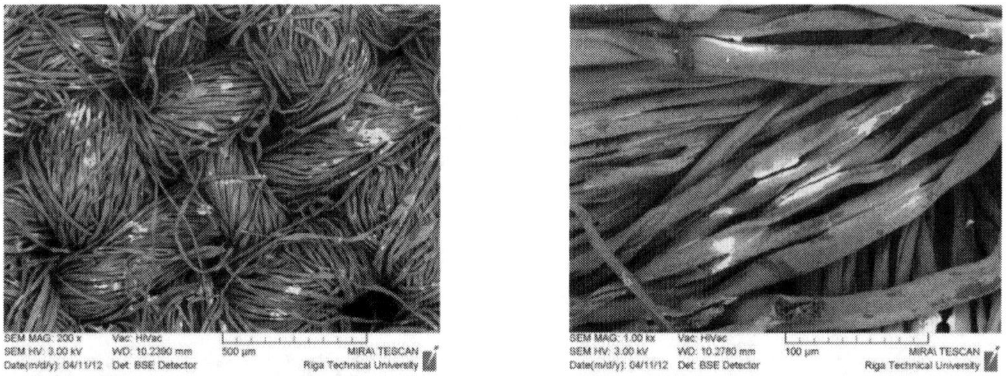

Figure 3. Cotton textile samples after nanosol treatment: zinc acetate concentration 10 %

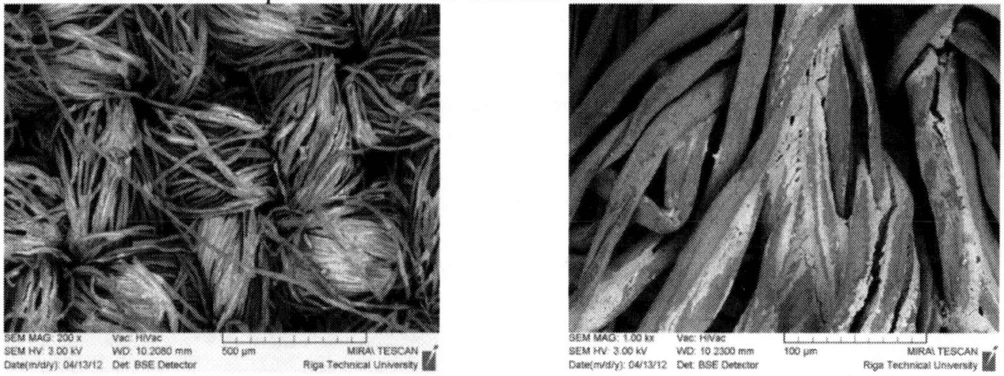

Figure 4. Cotton textile samples after nanosol treatment: zinc sulphate concentration 10 %

Figure 3 and figure 4 represent SEM micrographs of cotton samples after nanosol treatment with zinc acetate and zinc sulphate concentration 10%, both samples have coating with microdefects, and with a larger concentration of zinc acetate and zinc sulphate on the sample surface were formed cracks and bridges between fibres, what can be explained with rather high concentration of zinc acetate and zinc sulphate or insufficient TEOS concentration, or too long dipping time. The same results are received for samples treated by nanosol with zinc acetate and zinc sulphate concentration 7.5%.

3.2. Energy Dispersive X-ray (EDX) Analysis

For the analysis of elemental compositions of samples after treatment was used EDX analysis.

Table 1. EDX data – elemental weight in %

Samples	C	O	F	Si	Zn
Zinc acetate 2.5%	25.58	13.82	10.97	1.73	47.90
Zinc acetate 5%	18.24	14.09	14.43	3.23	50.01
Zinc acetate 7.5%	23.00	16.09	11.01	-	49.99
Zinc acetate 10%	28.10	13.56	10.04	-	48.30

Zinc sulphate 2.5%	62.94	21.22	4.96	3.36	7.52
Zinc sulphate 5%	59.18	20.80	3.17	2.94	13.91
Zinc sulphate 7.5%	54.28	21.34	1.27	-	23.11
Zinc sulphate 10%	17.16	20.76	2.82	-	59.26

From EDX analysis, of samples treated with nanosol prepared with zinc acetate and zinc sulphate are visible that the main elements of the surface species of treated cotton textile samples are C, Zn, O, F and Si presented at samples coated by nanosols with zinc acetate and zinc sulphate concentrations 2,5% and 5% (table 1).

3.3. Thermal aftertreatment

The nanosol process has to be adapted for the treatment of textiles because of textile low heat resistance; a thermal aftertreatment at higher temperatures should be avoided to reduce degradation and deformation of the textile materials, and has to be adapted with regard to the type of textile coated and to the applied nanosol. Experiment results show that a moderate thermal aftertreatment at 120°C after the sol applied to the cotton textile is appropriate for samples coated by nanosol prepared with zinc acetate, but is to large for samples coated by nanosol prepared with zinc sulphate, at higher temperature samples were changed it color to black, that was reduced, diminished aftertreatment temperature till 100°C.

4. Conclusion

It is concluded that the dipping time 10 min is sufficient to get sufficiently even coating on the fabric surface with zinc acetate or zinc sulphate concentrations 2.5% and 5% on condition that TEOS concentration is 2%. With larger concentrations of zinc acetate and zinc sulphate 7.5% and 10% on the sample surface were formed cracks and bridges between fibres, what can be explained with rather high concentration of zinc acetate and zinc sulphate or insufficient TEOS concentration, or too long dipping time.

The nanosol coating technique used in this research to implement ZnO coating on the fabric surface is a simple process that can be easily transferred to the textile industry, nanosol can be also applied by conventional coatings techniques used in the textile industry – the application can be implemented by simple dipping, padding or spraying process.

Acknowledgment

This work has been supported by the European Social Fund within the project «Support for the implementation of doctoral studies at Riga Technical University».

References

[1] Mahlting B, Textor T 2008 *Nanosols and Textiles* (Singapore: World Scientific Publishing)
[2] Wei Q 2009 *Surface modification of textiles* (UK: Woodhead Publishing)
[3] Farouk A, Textor T, Schollmever E et al 2010 Sol-gel-derived inorganic-organic hybrid polymers filled with ZnO nanoparticles as an ultraviolet protection finish for textiles *J. AUTEX Rsrch.* **10** 58
[4] Abidi N, Hequet E, Tarimala S 2007 Cotton Fabric Surface Modification for Improved UV Radiation Protection Using Sol-Gel Process, *J. Appl. Polym. Sci.* **104** 111
[5] Scott R A 2005 *Textiles for protection* (UK , Woodhead Publishing)
[6] Saravan D 2007 UV protection textile materials *J. AUTEX Rsrch.* **7** 53
[7] Putnina A, Kukle S, Turigina D 2010 UV Radiation Effect on UV-Absorbing Protection, *Sci. J. Mater. Sci. and Cloth. Tech.* **5** 157
[8] Grancarić A M, Tarbuk A, Dumitrescu I, Bišćan J 2006 UV Protection of Pretreated Cotton – Influence of FWA's Fluorescence *AATCC Rev.* **6** 44
[9] Kathirvelu S, D'Souza L, Dhurai B 2009 UV protection finishing of textiles using ZnO nanoparticles *J. of Fibre & Textil. Rsrch.* **34** 267

[10] Wong Y W H, Yuen C W M, Leung M Y S , Ku & Lam H L I 2006 Selected applications of nanotechnology in textiles *J. AUTEX Rsrch.* **6**

IOP Conf. Series: Materials Science and Engineering **38** (2012) 012023 doi:10.1088/1757-899X/38/1/012023

Preparation and Characterization of CNT/TiO2 Based Transparent Fiber Electrodes

M Paalo[1], T Tätte[1], M Hussainov[1], K Hanschmidt[1], M Lobjakas[1], A Lõhmus[1], U Mäeorg[2] and I Kink[1]

[1]Institute of Physics, University of Tartu, Estonia,

[2]Institute of Chemistry, University of Tartu, Estonia

E-mail: madis.paalo@fi.tartu.ee

Abstract. In the present work, it is shown that carbon nanotube-doped transition metal oxides are potential candidates for use as ceramic transparent electrode materials. Electrodes in shape of fibers are obtained via inexpensive and low temperature sol-gel method. Fiber electrodes are characterised by SEM-FIB and rheological analizes. Due to extraordinary electrical and optical properties of CNT-s and good chemical and physical stability of metal oxide ceramics, resulting composites could be an interesting subject for industry.

1. Introduction

During the past decade, significant efforts have been made to transfer the extraordinary and unique properties of carbon nanotubes (CNT) to bulk materials in order to achieve novel functionality. Remarkable mechanical and chemical properties of CNTs allow using them as low-density engineering material with significantly increased resistance to cracking [1]. Electrical conductivity of CNT materials enables their use as an alternative to transparent electrode materials. The field is of very high importance as the most widely used transparent and electrically conductive electrode material, indium tin oxide (ITO), is becoming progressively more expensive. The applications of transparent electrodes include numerous technological solutions as energy storage and conversion devices [2], organic light emitting diodes (OLED-s) [3] and others.

CNT doping is widely acknowledged as a means of improving tensile strength and electrical properties of organic polymers. Compared to undoped matrix, tensile strengths of organic polymers containing CNT-s are usually by some orders of magnitude higher while carbon nanotube loadings vary from 0,1 - 50 wt % [4]. Electrical conductivity of a polymer can be increased by orders of magnitudes by small amounts (0,4 wt%) of CNT additives as shown by Kim et al. [5]. Despite the aforementioned improvements, the low thermal stability and inferior UV resistance of the polymer matrix still remain as serious drawbacks for the materials in many important applications. Contrary to organic polymers, thermal stability and high hardness of metal oxide ceramics make them attractive candidates to be alternatively used as a host medium in CNT composites. Most common methods for obtaining CNT composites with matrices like SiO_2, Al_2O_3, MgO, SiC, TiN, $BaTiO_3$ etc. [6] are based on in situ growth of CNT-s by chemical vapor deposition (CVD), colloidal or electrophoretical processing and sol-gel method.

Rheological properties of liquid state of CNT composites used as an initial state in sol-gel processing have a fundamental role in both processing and the comprehension of their final microstructure and properties. As rheology of CNTs dispersed liquids depends on many factors, it is important to know what kind of CNTs have been used and are they chemically functionalised or not. It is also very important to know CNTs aspect ratio, concentration of CNTs in the host medium, nature of the suspending media (Newtonian or Non-Newtonian), degree of dispersion, method of dispersion (mechanical mixing or sonication) etc. The extensional rheology (often used to evaluate rheological behaviour of materials) is especially important in terms of understanding the processing behaviour of CNT suspensions in fibre spinning, inkjet printing, and curtain coating.

2. Experimental and discussion

In the current paper we describe our studies of the preparation and characterization of ceramic composites obtained by using gelation of CNT doped titanium alkoxide nano sols. The studies are motivated by the simplicity, low cost and low processing temperatures of sol-gel technology. Compared to aforementioned methods, sol-gel process has the advantage of the possibility of creating intimate dispersion of CNT-s in matrices. Low process temperatures were applied to avoid carbon nanotubes oxidation in air. It is known that CNTs oxidise in temperature range from 400 to 500 °C in the presence of oxygen. In our previous studies we have shown that it's possible to obtain self-standing fibers and films from high viscosity transition metal oxides by using sol-gel method [7].

Multi-walled carbon nanotubes (MWCNT) (product of CheapTubes Inc.; diameter of 10-20nm and length 10-30 μm) were ultrasonically dispersed into propanole [8] and were then added to liquid titanium(IV)propoxide ($Ti(OPr)_4$). To evoke polymerization, water was added to the alkoxide as a water-propanol 10% solution. To obtain homogeneous CNT dispersion in viscous alkoxide material, sonication was used before and during the removal of solvent and alcohol.

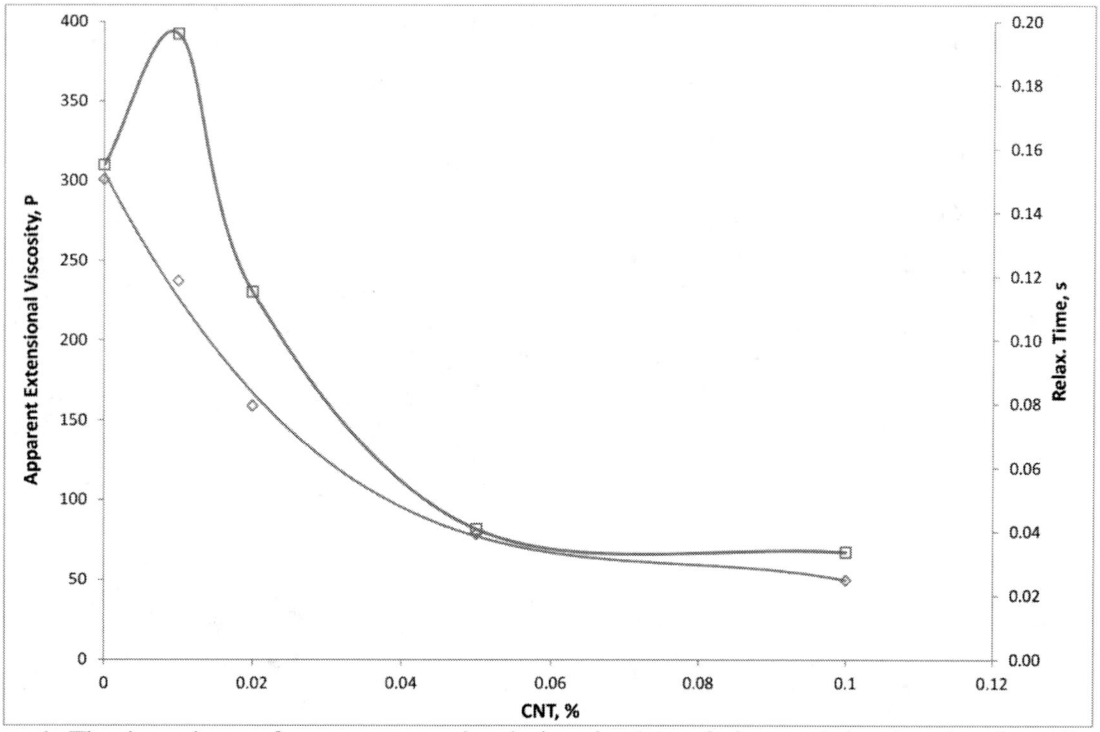

Figure 1. The dependence of apparent extensional viscosity (□) and characteristic relaxation time (◊) on the concentration of CNT.

The effect of addition of CNTs to the rheological behaviour of alkoxide-based precursors was studied experimentally. As alkoxides are viscous and very highly reactive liquids species, specially constructed device was used to carry out measurements directly inside the synthesis vessel. The results of the study showed that initially pure precursors are weakly strain-hardening viscoelastic liquids, which confirms our previous study [9]. It can be concluded that an increase in the concentration of CNT, composites behave more Newtonian: the apparent extensional viscosity and characteristic relaxation time dramatically decrease with increasing of the CNT concentration (Figure1). Increased extensional viscosity of samples, loaded with small concentration of CNT-s, is probably related to CNT-s natural tendency to self-form ropes and bundles. As interaction with surrounding matrix is not enough to break these bundles and ropes under applied external forces, these remain to stay against flowing of the fluid. When more CNT-s are added, bundles and ropes start to interact with each other in addition to matrix, which under external forces cause bundles to break into smaller units which orients to flowing direction and lower the viscosity. Still, this effect isn't studied enough to make full conclusions and should be simulated for final clarification.

After the rheology measurements fiber-shape electrodes were prepared by pulling jets from polymerized CNT-alkoxide mass. Liquid jets react with water in humid environment and rapidly become completely solid while low concentrations organic additives and water remain in the material. The diameter of fibers depends on the visco-elasticity of material, applied pulling speed, humidity of the surrounding air and nanotube aggregates remained in the matrix after sonication. Prepared fibers were left to age for 24 hours in lab atmosphere with relative humidity of 20-50%. To remove remains of organics from the material, thermal annealing up to 350 °C was carried out. As a result, 10-50 mm in length, transparent or partially transparent and partially crystallized composite fiber material was obtained. Fiber diameters ranged from 5 to 70 microns. Similarly, the composites can be prepared from oxides of tin, hafnium and zirconium etc. in final form as films, structured films, fibers, etc. [10]

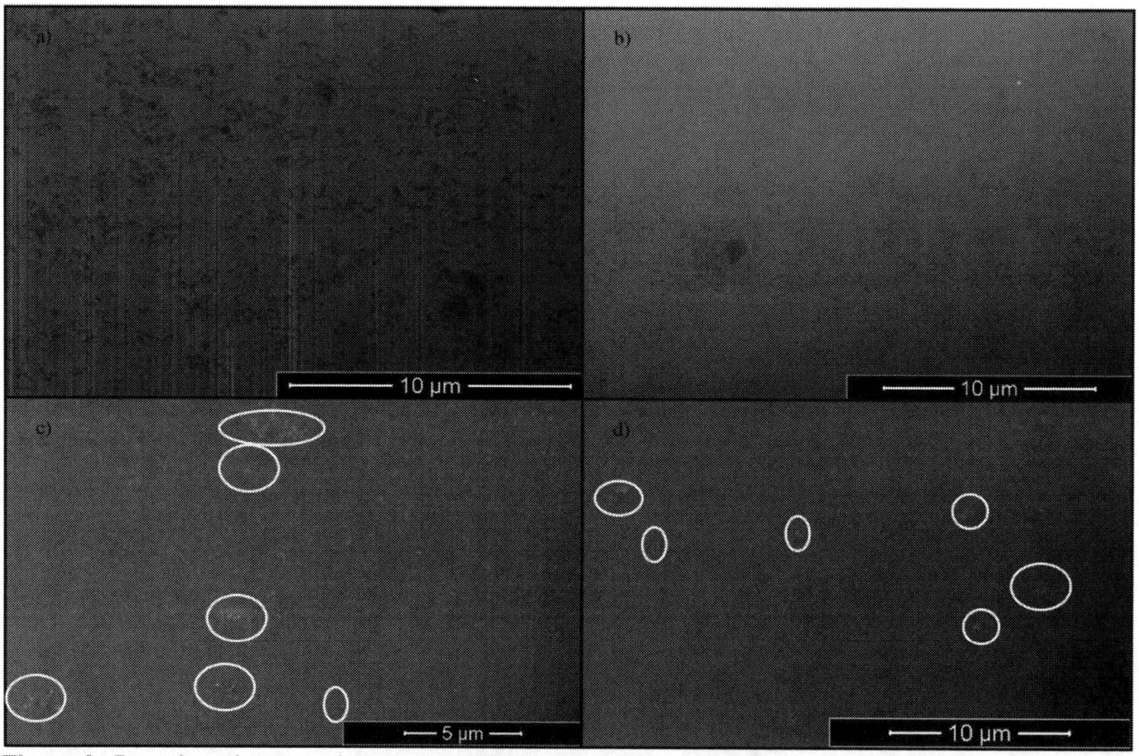

Figure 2. Scanning electron microscope (SEM) images of 1 wt% (a), 0,1 wt% (b), 0,05 wt% (c) and 0,01 wt% MWCNT doped composite fibers. The fibers were treated with focused ion beam (FIB) to reveal nanotubes and nanotube aggregates inside the material.

SEM and FIB analysis revealed that although the fibres contained some MWCNT aggregates, the overall dispersion of CNTs inside the material was good (Figure 2.) The electrical properties of the materials were measured by 4-point method using In-Ga alloy contacts. Metallic contacts ensure a firm and easily achieved electrical and physical contact between probes and fiber material. Only the fibers consisting 1% MWCNTs achieved detectable electrical conductance up to 15 S/m (electrical conductance of undoped titanium dioxide is approximately 10^{-16} S/m). This result was predictable also from the SEM-FIB analysis where it could be seen that MWCNT network was formed trough the TiO_2 matrix (Figure 2a). Higher conductivity of the sample also enabled higher imaging contrast. Although some CNT aggregates inside fiber electrodes could be observed in the optical micrographs, in general, obtained materials remained visually transparent.

3. Conclusions

Current study proves that metal oxide/CNT composite fibers have potential in transparent electrode application. SEM and FIB analysis revealed that the overall dispersion of CNTs inside the material was good. Our results prove that CNT-doping significantly improves the conductivity of the fiber electrodes (up to 15 S/m).

Acknowledgments

This work was supported by the Estonian Nanotechnology Competence Centre, Estonian Science Foundation (Grants ETF 7612, 9292, 8377, 8420) and by the European Union through the European Regional Development Fund (Centre of Excellence "Mesosystems: Theory and Applications", TK114).

References

[1] Raki L, Beaudoin J, Alizadeh R, Makar J and Sato T 2010 *Materials* **3** 918
[2] Zhang W D, Xu B and Jiang L C 2010 *J. Mater. Chem.* **20** 6383
[3] Bansal M, Srivastava R, Lal C, Kamalasanan M N and Tanwar L S 2009 *Nanoscale* **1** 317
[4] Spitalsky Z, Tasis D, Papagelis K and Galiotis C 2010 *Prog. Polym. Sci.* **35** 357
[5] Kim H M, Kim K, Lee C Y, Joo J, Cho S J, Yoon H S, Pejakovic D A, Yoo J W and Epstein A J 2004 *Appl. Phys. Lett.* **84** 589
[6] Cho J, Boccaccini A R and Shaffer M S P 2009 *J. Mater. Sci.* **44** 1934
[7] Tätte T, Paalo M, Kisand V, Reedo V, Kartushinsky A, Saal K, Mäeorg U, Lõhmus A and Kink I 2007 *Nanotechnology* **18** 531
[8] Ham H T, Choi Y S and Chung I J 2005 *J. Colloid. Interf. Sci.* **286** 216
[9] Hussainov M, Tätte T, Paalo M, Part M, Talviste R, Kiisk V, Mändar H, Põhako K, Pehk T, Reivelt K, Natali M, Gurauskis J, Lõhmus A and Mäeorg U 2011 *Sci. Technol. Adv. Mater.* **12** 034412/1
[10] Kisand V, Shulga J, Tätte T, Visk U, Natali M, Mistura G, Paalo M, Lobjakas M and Kink I 2007 *Mat. Sci. Eng.* B **137** 162

Pseudoisochromatic test plate colour representation dependence on printing technology

K Luse[1,3], S Fomins[2] and M Ozolinsh[1,2]
[1]University of Latvia, Optometry and Vision Science Dept., Ķengaraga street 8, Riga LV-1063, Latvia
[2] Institute of Solid State Physics, Ķengaraga street 8, Riga LV-1063, University of Latvia

E-mail: kaiva.luse@gmail.com

Abstract. The aim of the study is to determine best printing technology for creation of colour vision deficiency tests. Valid tests for protanopia and deuteranopia were created from perceived colour matching experiments from printed colour samples by colour deficient individuals. Calibrated *EpsonStylus Pro 7800* printer for ink prints and *Noritsu HD 3701* digital printer for photographic prints were used. Multispectral imagery (by tunable liquid crystal filters system *CRI Nuance Vis 07*) data analysis show that in case of ink prints, the measured pixel colour coordinate dispersion (in the *CIExy* colour diagram) of similar colour arrays is smaller than in case of photographic printing. The print quality in terms of colour coordinate dispersion for printing methods used is much higher than in case of commercially available colour vision deficiency tests.

1. Introduction

Colour vision deficit affects approximately 8% of male and 0.4% of female population [1]. Colour deficiency tests are widely used to diagnose the type and severity of colour vision deficiencies. Among the most frequently used are pseudoisochromatic test plates (in text term “test plates” will be used for simplicity). The colours of test plates are chosen to create different perception for individuals with normal (trichromatic) colour vision and affected colour vision [2]. Among the most popular and most easy to create designs for colour vision test plates are so called “vanishing” designs – the test object is camouflaged for the specific type of colour vision deficit individuals and visible for others [1,2].

Colour vision anomalies are divided in three categories monochromacy, dichromacy, and anomalous trichromacy [3]. Protan and deutan perception was studied and tests created. In case of protan, deutan and tritan defects, colour confusion line defines sets of colours that are indistinguishable for each type of deficiency. Colour confusion lines in *CIExy* colour diagram converge at a point which is called the copunctal point. For the three different types of colour vision deficiency there are three different copunctal points [4].

Currently there are no regulations regarding the choice of printing technology or paper type to be used for colour vision test plates. Previous studies have shown that the printing technology impacts the

[3] To whom any correspondence should be addressed

colour display accuracy of the test plates, resulting in a wide distribution of colour coordinates along the desired points on the confusion lines in *CIExy* colour space diagrams [5].

2. Methods and experiment setup

2.1. Printing technology

For clinical purposes sets of printed colour test plates are used. The aim of the study is to search for most suitable printing technology with smallest dispersion of colours. Commercially available colour vision tests are made using offset print. Previous research shows that colours from the same test plates but different editions have significant chromatic variations that result in clinical misevaluation [6].

2.1.1. Ink printing

In case of ink printing, the final image is produced by applying series of ink dots for resulting appearance of colour. Ink printing is the most common technique of printing. Colour samples and test plates were created using a calibrated *EpsonStylus Pro 7800* printer. Printer uses nine *UltraCrome K3 Epson Ink* cartridges (4 types of black, light Cyan, light Magenta, Cyan, Magenta, Yellow) and *Premium Semimatte photopaper 260* (resolution 2880 dpi).

2.1.2. Photographic printing

A process of producing a final image on paper, using chemically sensitized paper. In Photographic printing process, the tiny dots produced by chemical reaction are observable under 100 x magnification. Colour samples and test plates were created using a calibrated *Noritsu HD 3701* digital printer (640 dpi resolution). Printing was done on silver-halide (AgX) color-negative *Kodak Professional ENDURA* paper (F-glossy paper type).

2.2. Colour matching experiments for creation of colour vision plates

3 deutan (2 deuteranope, 1 deuteranomalous) and 2 protanope male subjects (age 24±4 years) participated in the experiment. 270 chromatic and 106 grey different photographic colour print samples were created using an algorithm of display RGB channel value adjustment (sample size 3x3 cm). Samples were created to appear on or closely by those specific protan and deutan colour confusion lines crossing the achromatic point in *CIExy* colour diagram. Samples were of various saturation and lightness. Colour matching tasks were conducted in *Qualitest CT-100W1* light booth under D65 illumination (T=6500K). Subjects were given two tasks: first sort samples in piles of grey and chromatic and second, sort grey samples in piles of similar lightness. 44 pairs (22 for each type of deficiency) of chromatic – achromatic samples for creation of colour vision tests were obtained. 1 colour vision test plate (hidden symbol number "8") for protan and 1 for deutan was created and printed using photographic and inkjet printing technology (see figure 1). Tests were verified and proven valid for protanope and deuteranope testing in D65 and daylight conditions. Prior to printing, description of colour sets was mathematically equal. Results section analyses print induced deviations in colour.

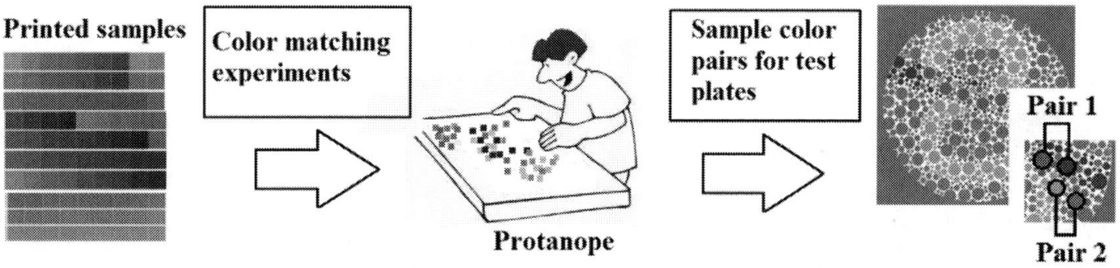

Figure 1. Schematic representation of creation of colour vision plate. Example for protanope subject is given where he sorts samples and pseudoisochromatic colour samples pairs are acquired.

3. Results and discussion of plate multispectral analysis

For multispectral imaging *CRI Nuance Vis 07* multispectral camera with *Nikon AF Micro-Nikon 60mm f2.8D* objective was used. 1290 x 920 pixel spatial images in range of 420 to 720 nm (step 10 nm) were taken (light booth, D65 light source, colour T=6500K). Images were transformed using cone sensitivities into cone excitation images. Selected image areas of 100x100 pixels were analysed and *CIExy* diagram [x,y,Y] values acquired for each pixel. For each print technology the chromaticity coordinates x,y of 9 areas are analysed in this paper (4 light chromatic, 2 light achromatic, 1 dark chromatic, 1 dark achromatic and 1 background (white paper)).

3.1. Sample relations to confusion lines

Figure 2 shows alignment of dark chromatic (1), light chromatic (2) and light achromatic (3) areas in a protan plate. The measured pixels of the image of the plate are in a close proximity to a single deutan colour confusion line.

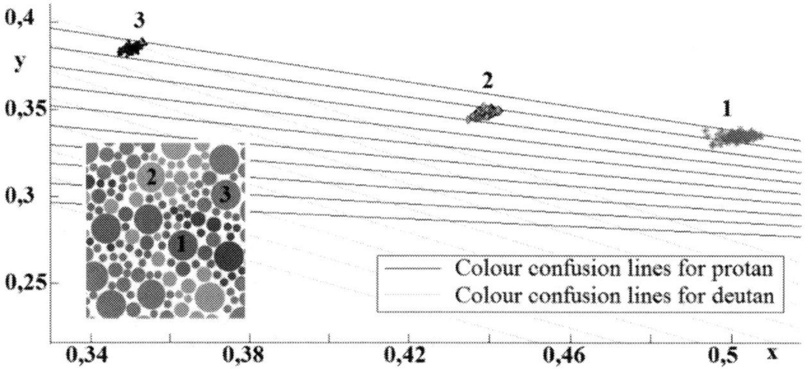

Figure 2. Sample relations to colour confusion lines.

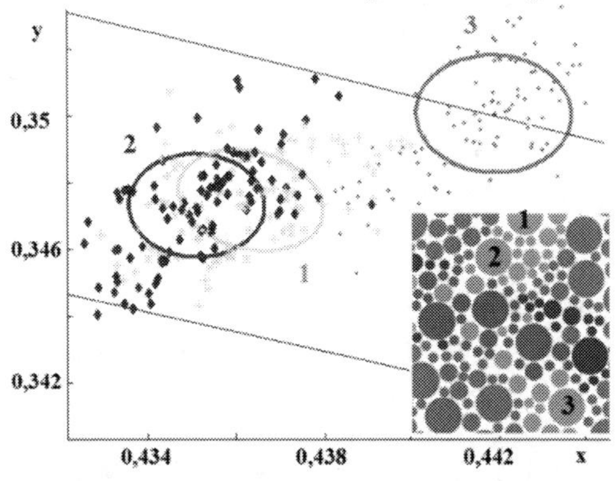

Figure 3. Dispersion, photographic print.

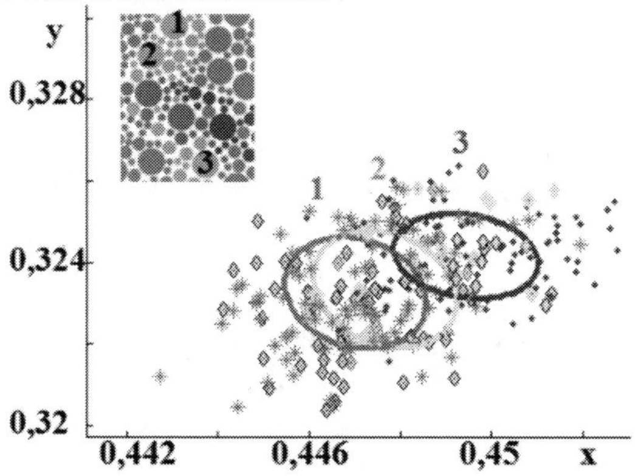

Figure 4. Dispersion, ink print.

3.2. Colour coordinate dispersion analysis

Printing quality was analysed in terms of image pixel colour coordinate mean value dispersion in *CIExy* diagram for printing method used. For mean value comparison statistics tool *independent sample t-test* was used with confidence level 95%. For dispersion description Gaussian curve standard deviation (SD) of data was calculated.

3.2.1. The overlap of pixel data sets

The pixel values of 4 similar colour but different position areas on the plate were analysed for overlap. In case of ink print x coordinate, 2 out of 4 data sets overlapped (0 in case of photographic print). In case of ink print y coordinate, 3 of 4 data sets overlapped (1 in case of photographic print). The probability of data sets to overlap is higher in case of ink print. See figure 3 for three data sets in case of photographic print and figure 4 for ink prints. Fourth data set in not included in the pictures to avoid overcrowding.

The SD's of 32 data sets were compared (8 coloured samples for each print type (2) for each chromaticity coordinate [x,y]). Statistically the mean photographic print colour dispersion (x=0.0017; y=0.0018) is larger compared to ink print SD's (x=0.0016; y=0,0014) along both x and y axis. Colour pixel dispersion in case of commercially available tests is much larger than for both printing methods tested [5]. We have shown that the area size containing each set of data points in each case of print technology differs along the x and y direction in the *CIExy* colour diagram.

3.2.2. The distance of ellipse centres representing pixel data sets

To analyse weather colours in different places of the plate vary for both printing techniques, the distance between pixel sets in *CIExy* diagram was described using ellipses (radius of the ellipse can be defined as SD of data in x and y directions). 4 similar chromatic and 2 achromatic areas in case of both printing technologies were analysed for ellipse shift differences. Meaningful differences were obtained for both directions (x and y) in both chroma saturations in favour of ink prints (figure 3 and 4).

4. Conclusions

1) The RGB value adjustment method for the print samples and the results of the colour matching experiments allow to create plates containing colour pairs on the lines of confusion for the colour deficient individuals. Plates were valid for testing colour deficit individuals in D65 and daylight conditions.
2) In case of inkjet prints (for the created plates) the mean sizes of ellipses representing pixel dispersion of each individual colour array of a plate are smaller than in case of photographic prints. The mean sizes of ellipses representing pixel dispersion of each individual colour array for printing methods used is much smaller than in case of commercially available colour vision deficiency tests.
3) In case of inkjet prints (for the created plates) the mean distances between the centres of ellipses representing pixel dispersion of two or more equal colour arrays in different spatial positions of a plate are smaller than in case of photographic prints.

5. Acknowledgments

Support for this work was by the ESF within the projects "Support for Doctoral Studies at Univ. of Latvia–2" and No. 2010/0259/2DP/2.1.1.1.0/10/APIA/VIAA/137.

6. References

[1] Birch J, 1993 *Diagnosis of Defective Colour Vision* (London) Butterworth Heinemann, 184 p., 42, 55.
[2] Dain S J, 2004 *Clin Exp Optom* **87** 276
[3] Gegenfurtner K R, Sharpe L T 2000 *Color Vision From Genes to Perception*, Cambridge University Press, 492p., 24-30.
[4] Smith V C, Pokorny J 1975 *Vision Research* **15** 161
[5] Fomins S, Ozolinsh M 2011 *Materials Science (Medžiagotyra)* **17** 104
[6] Lee D Y, Honson M 2003 *Color research and Application* **28** 267

International Conference on Functional Materials and Nanotechnologies (FM&NT2012) IOP Publishing
IOP Conf. Series: Materials Science and Engineering **38** (2012) 012025 doi:10.1088/1757-899X/38/1/012025

The expression of cytokines and β –defensin 2, - 3, -4 in rabbit bone tissue after hydroxyapatite (HAp), α–Tricalcium phosphate (α-TCP) and polymethylmethacrylate (PMMA) implantation

J Vamze[1,3], M Pilmane[1], A Skagers[2]

[1]Riga Stradins University, Institute of Anatomy and Anthropology, Kronvalda 9, LV-1010, Riga, Latvia

[2]Riga Stradins University, Department of Oral and Maxillofacial Surgery, Dzirciema 20, LV-1007, Riga, Latvia

E-mail: j.vamze@inbox.lv

Abstract Bone loss induced by inflammation is one of the complications after biomaterial implantation. There is no much data on expression of cytokines and defensins into the bone tissue around the implants in literature. The aim of this work was to investigate the distribution and appearance of interleukin (IL)-1, IL-6, IL-8, IL-10 and β – defensin (BD)-2, BD-3, BD-4 after the implantation of different biomaterials. Bone developing zones, signs of bone-implant contact and low expression of pro-inflammatory cytokine IL-1, IL-6 and anti-inflammatory cytokine IL-10 in experimental tissue with pure HAp and unburned HAp implants indicate a potential advantage of this material in terms of its biocompatibility over the other materials used in our study.

1. Introduction

Bone loss induced by inflammation is one of the complications after biomaterial implantation into the hard tissue. Pro-inflammatory cytokines interleukin-1 (IL-1) and interleukin-6 (IL-6) exhibit a pro-osteoclastogenic effect in response to implant-derived wear particles. IL-6 recruits and activates osteoclasts and may therefore leads to bone resorption and the loss of implant fixation [1]. Interleukin (IL-8) has been described as a pro-inflammatory mediator in gingivitis [2]. Interleukin-10 (IL-10) is an anti-inflammatory cytokine with pleiotropic effect on immunoregulation and inflammation [3]. The role of β – defensin (BD)-1, BD-2, BD-3 in immune defence against oral infection in cases of biomaterial implantation is not completely defined. There are varying data regarding the expression of cytokines and defensins into the bone tissue around the different bioceramic implants is little discussed.

---

[3] To whom any correspondence should be addressed

The aim of this experiment was to research changes in distribution and appearance of pro-inflammatory cytokines and anti-inflammatory proteins in the lower jaw and tibial bone of experimental rabbits 3 months after the implantation of different biomaterials.

2. Material and methods

The Animal Ethics Committee of Latvian Food and Veterinary Administration approved the use of 4 Californian male rabbits for this morphofunctional study. Commercial biomaterial and biomaterials produced in Riga Technical University Biomaterial Innovation and Development centre were used in the experiment.

The rabbits received general anaesthesia with Sol. Ketamini (15 mg/kg) and Sol. Dormicum (2 mg/kg), local anaesthesia with Sol. Midazolami (0.5 mg/kg) and Sol. Articaini (5 mg/kg). After raising mucoperiosteal flap, the channel was drilled using 2.0 mm diameter trephine and intraosseal implantation of the following biomaterials were performed: pure hydroxyapatite (HAp) tablet, HAp tablet covered by polycaprolactone (PCL), unburned HAp granules, HAp granules burned under 1150^{0}C, commercial polymethylmethacrylate (PMMA) bone cement (Biomet Bone Cement R), PMMA cement, α-Tricalcium phosphate (α-TCP) cement I with pH6 and α-TCP cement II with pH7. The control site was similarly prepared and closed but no biomaterials were inserted.

Three months afterwards, the euthanasia of rabbits was performed by air embolisation and blocks of bone and soft tissue were harvested from experimental and control sites using diamond disc. Tissue were embedded into the paraffin, cut into 5 μm thick slices and prepared for detection of the interleukins: interleukin-1 (IL-1, working dilution 1:1000, abcam, UK), interleukin-6 (IL-6, working dilution 1:50, Santa Cruz Biotechnology, Inc), interleukin-8 (IL-8, working dilution 1:50, Santa Cruz Biotechnology, Inc), interleukin-10 (IL-10, working dilution 1:400, abcam, UK) using biotin-streptavidin immunohistochemistry (IHC) [4]. Innate immunity in rabbits' bones was detected by β – defensin 2 (BD-2, working dilution 1:100, RD systems, UK), β defensin 3 (BD-3, working dilution 1:100, Life Span Biosciences), β defensin 4 (BD-4, working dilution, 1:150, Santa Cruz Biotechnology, Inc) using IHC. Routine staining for haematoxylin and eosin (H/E) was performed for each case [5]. Structure quantification was done applying the semi – quantitative counting method [6].

3. Results

Routine histological examination revealed bone developing zones and signs of bone-implant contact in the experimental tissue with such biomaterials as pure HAp, HAp tablet covered by PCL, commercial PMMA bone cement and unburned HAp granules (Figure 1). In bone tissue samples with α-TCP cements biomaterial, the regions of ossification were less pronounced but in tissue with uncommercial PMMA cement, the ossification zones were not detected at all. Expression of defensins and interleukins in the experimental and control tissue was variable. Marked expression of BD-2 and BD-3 was observed in the tissue with pure HAp (Figure 2).

BD-4 (β-defensin-4) expression was more pronounced in the tissue with commercial PMMA bone cement (Figure 3) and with HAp covered by PCL. Large number of IL-1 positive cells were noticed in tissue with both types of PMMA, α-TCP cement I and HAp tablets covered by PCL, whereas a low expression of IL-1 was detected in control tissue. Little difference in number was found between IL-6 and IL-1 positive bone cells. IL-6 positive osteocytes in tissue with uncommercial PMMA cement were detected in less amount but in tissue with burned HAp granules (Figure 4), IL-6 positive cells were found more than IL-1 positive osteocytes.

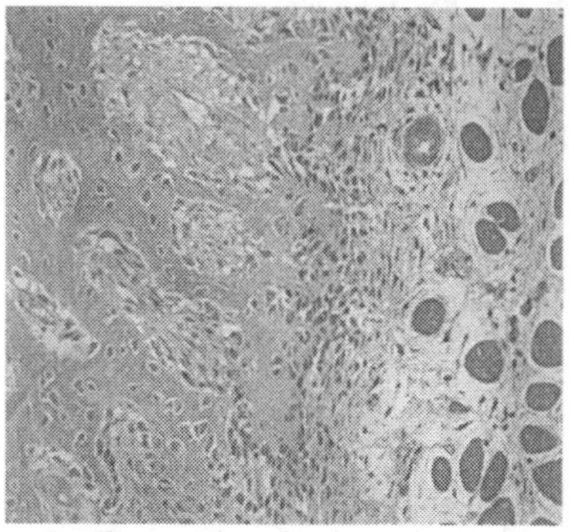

Figure 1. Bone developing zones in tissue using unburned HAp granules; H/E, x250

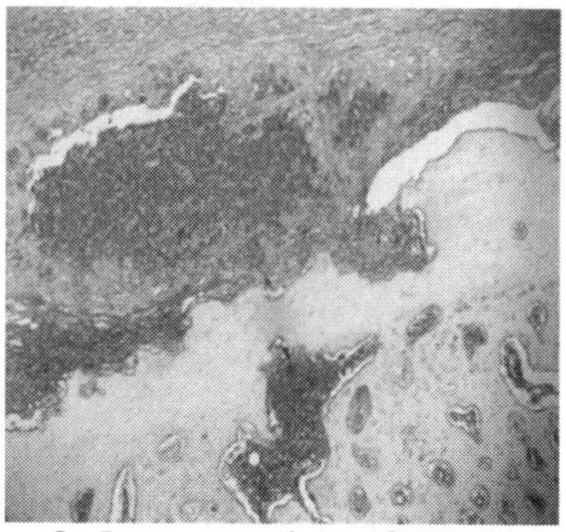

Figure 2. Large number of BD-3 positive osteocytes in bone tissue with pure HAp biomaterial; IHC, x200

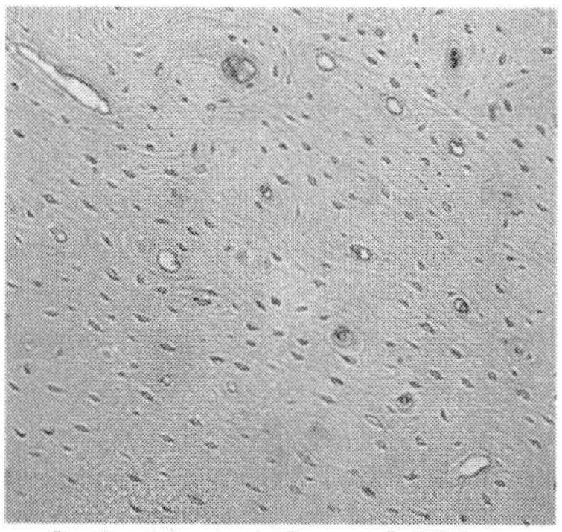

Figure 3. Large number of BD-4 positive osteocytes in bone tissue with commercial PMMA cement; IHC, x250

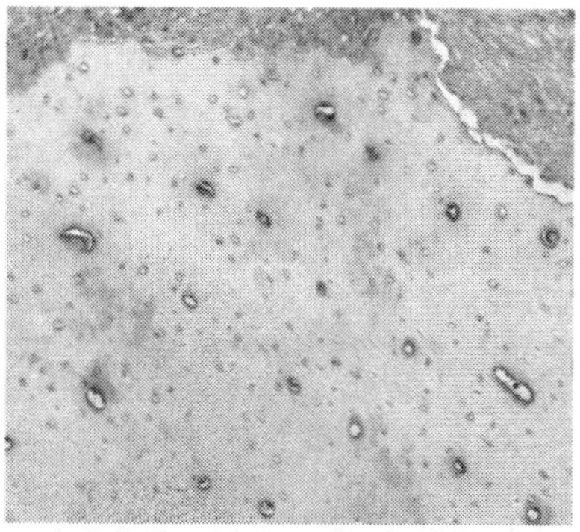

Figure 4. Moderate number of IL-6 positive osteocytes in bone tissue with HAp granules; IHC, x250

However, control tissue showed a significant number of IL-6 positive cells. Low expression of IL-8 was noticed in all experimental (Figure 5) and control tissue, except in the case of commercial PMMA bone implant where IL-8 positive osteocytes were detected in moderate number. Large number of IL-10 positive osteocytes were detected in experimental tissue with HAp tablets covered by PCL (Figure 6), commercial PMMA cement and α-TCP bone cement I.

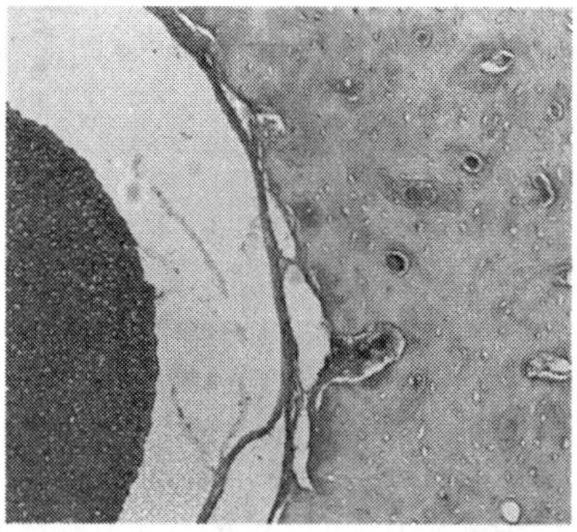

Figure 5. Some IL-8 positive osteocytes in bone tissue with burned HAp granules; IHC x200

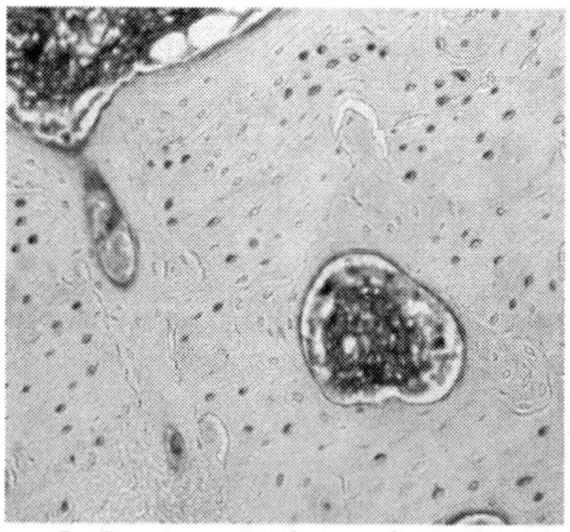

Figure 6. Large number of IL-10 positive osteocytes in bone tissue with HAp covered by PCL; IHC, x250

4. Discussion

Our experimental study showed that biomaterials such as commercial PMMA cement, HAp covered by PCL and α-TCP cement I influence the increase in the expression of IL-10 and IL-6. Weak expression of IL-10 we revealed in experimental tissue with pure and unburned HAp material. As IL-10 is true anti-inflammatory cytokine, its absence suggests the absence of inflammation in the bone. However, any variation of IL-10 level should be evaluated as individual response from host tissue. The research of Ninomiya (2001) demonstrates that HAp and HA/TCP (hydroxyapatite/tricalcium phosphate) particles are capable of stimulating the expression and secretion of cytokines IL-6 and proteases that increase bone resorption [7].

The augmentation in IL-8 expression was revealed only in bone tissue with commercial PMMA, whereas in the experiment with other biomaterials it was very weak. The expression of pro-inflammatory cytokine IL-1 in our experimental material was similarly distinct in both commercial and uncommercial PMMA bone cements, α-TCP cement (pH6) and HAp tablet covered by PCL. IL-1 is essential protein for the bone resorption but IL-8 is suggested to be less potential for this function [8].

β-defensins 1-3 were first described in the osteoblast cell cultures [9] with dominance of defensin 1 and 2. With regard to the reaction of BD-2, BD-3 and BD-4 in our study, the activity was varying. We suggest a role of defensins in traumatic injury could be important in case of biomaterial implantation in the bone.

5. Conclusion

Bone developing zones, signs of bone-implant contact and low expression of pro-inflammatory cytokine IL-1, IL-6 and anti-inflammatory cytokine IL-10 in experimental tissue with pure HAp and unburned HAp implants indicate a potential advantage of this material in terms of its biocompatibility over the other materials used in our study.

References

[1] Holt G, Murnaghan C, Reilly J, Meek R M 2007 *Clin Orthop Relat Res.* **460** 240
[2] Haake S K, Huang G T J 2002 Molecular Biology of the host-Microbe Interaction in Periodontal Diseases (Selected Topics) In: Newman, Takei, Carranza, eds. *Clinical Periodontology,* Philadelphia: W.B.Saunders Co 162
[3] Eskdale J, Kube D, Tesch H, Gallagher G 1997 *Immunogenetics* **46** 120
[4] Hsu S M, Raine L, Fanger H 1981 *J Histochem Cytochem* **29** 577
[5] Kiernan J A 2008 *Histological and Histochemical Methods: Theory and Practice* In: Bloxham, UK: Scion 141–174
[6] Pilmane M, Luts A, Sundler F 1995 *Thorax* **50** 551
[7] Ninomiya J T, Struve J A, Stelloh C T, Toth J M, Crosby K E 2001 *J Orthop Res.* **19** 621
[8] Schmidt C, Steinbach G, Decking R, Claes L E, Ignatius 2003 *Biomaterials* **24** 4191
[9] Bissell J, Joly S, Johnson G K, Organ C C, Dawson D, McCray P B, Guthmiller J M 2004 *J Oral Pathol Med* **33** 278

The impact of light polarization on the direct relief forming processes in As_2S_3 thin films

U Gertners[1], **J Teteris**

Institute of Solid State Physics, 8 Kengaraga Str., Riga, LV-1063, Latvia

E-mail: gertners@gmail.com

Abstract. In this report direct a photo-induced formation of surface relief gratings (SRG) in thin layers of arsenic sulfide (As_2S_3) are shown. This anisotropic light-induced mass transfer phenomenon has been discussed with the special attention focused on the polarization and intensity of the corresponding light. The experimental setup for the SRG recording is straight-forward consisting of ~10μm optical slit through which an unfocused beam of light is projected on the surface of sample. The verified sets of experimental parameters (light intensity, polarization and slit opening) were chosen. Additionally the evolution of surface relief in dependence from the recording time and polarization has been investigated in detail. The processes of SRG formation and mass transfer which are based on the photo-induced plasticity in As_2S_3 have been discussed.

1. Introduction

Holographic recording is a perspective method for obtaining high quality diffraction gratings amongst other applications. The possibility of direct surface patterning without the necessity of development and etching is even more appealing thus the effects of surface forming processes in holographic recording are of vital importance.

Non-contact mass transfer is known for a few decades now – it can be found in optical particle traps and optical tweezers (first reported in 1970 by A.Ashkin [1]). The idea behind this is the movement of individual particles with light intensity gradient. One might think that somewhat similar processes apply to holographic recordings (first reported in 1975 by Chomat *et al* [2]) but further investigations are required. The look in this direction has been thrown and there are works showing relief forming in dependence of focused light, polarization and so forth [*e.g.*, 3, 4]. The results on influence of light polarization though are quite inconclusive (as shown in the [5]) and require further investigation.

From earlier works it is known that an effective materials for obtaining direct surface formation are vitreous chalcogenides – As_2S_3 in particular is attractive because of sample preparation simplicity and its optical properties. This well known group of materials has been studied for many years and various results have been obtained but less attention has been paid to the significance of light polarization. Thus the aim of this work was to investigate the influence of light polarization on mass transfer processes in A_2S_3 thin films.

[1] To whom any correspondence should be addressed

2. Experimental

Amorphous As_2S_3 films on glass with thickness of 0.5-4.0 μm were obtained by thermal evaporation. A single beam recording setup (shown in Fig.1.) was utilized to obtain surface modulation in the As_2S_3 samples, which were illuminated by homogenized 532 nm laser light through 10 μm wide optical slit. An optional assisting incoherent 532 nm laser source was used for softening purposes of As_2S_3 film thus improving the recording efficiency. Polarization of recording and assisting beams was varied by halfwave and quarterwave plates. The obtained surface relief was mapped by atomic force microscopy (AFM) using *Veeco CPII*.

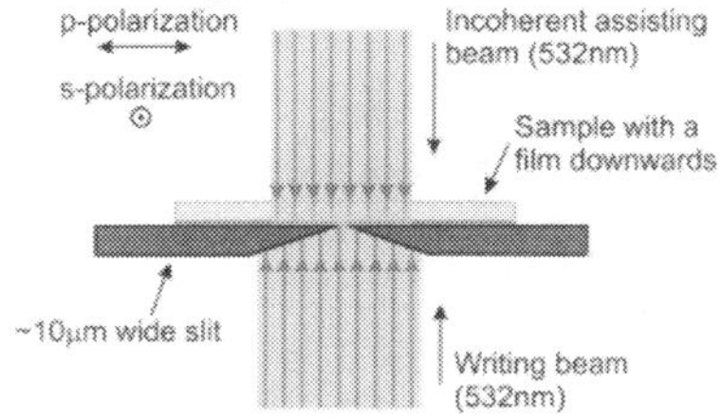

Figure 1. Experimental setup (top view) for formation of surface-relief structures by illumination through an optical slit

3. Results

When recording by using a holographic setup it is difficult to understand the mass transfer processes because of the periodicity of the surface relief – it is hard to define the point of origin and analyze the mass transfer direction. To avoid this limitation a simplified single beam recording setup was introduced. The main object of which was a narrow optical slit. Through this slit a sample was exposed by a uniformly distributed and unfocused monochromatic light with a particular polarization. The resulting intensity distribution then is well defined and the formed surface relief thus has to be symmetric which allows introducing well defined coordinate system.

The first obtained recordings independently from the polarization were insignificant (few nm in height) and could be attributed to photo-induced expansion effect [6]. As previous experiments with direct holographic recordings showed that relief forming efficiency can be significantly improved by additional incoherent assisting beam [7], the same approach was utilized for further experiments. It has been determined that the polarization of the optional assisting beam causes isotropic or anisotropic softening which in conjunction with polarization of the recording beam might show some peculiarities. In Fig.2. the summary of obtained results has been presented as table. Certain cells are carrying a value *n/a* – this means that with such polarization configuration the obtained relief is insignificant (few nm) and basically represents smooth surface. The remaining cells which do carry values different from zero are represented by AFM images and profile height distributions.

As can be seen from combinations that produced an observable surface relief modification, the mass transfer can manifest in different ways. If the polarization of recording beam coincides with the direction of optical slit (defined as s-polarization), then independently from the polarization of the assisting beam (in the 1st row: CP- circular, p- perpendicular to the slit or 45 degree) a mass transfer away from the recording region can be observed – formation of a groove. The efficiency of recording or the depth of groove changes from the best case, when the polarization of assisting beam is perpendicular to that of the recording beam, then follows the 45 degree and finally CP polarization. The same applies when the polarization of recording beam is perpendicular to the slit (the 2nd row in Fig.2.), only in this case mass transfer is directed towards recording region – formation of a ridge.

The polarization of recording beam is not the only factor, which determines the direction of mass transfer. Independently from the polarization of recording beam, consistent formation of a ridge, when the assisting illumination is s-polarized (1st column in Fig.2.) or formation of a groove, when assisting illumination has p-polarization (2nd column in Fig.2.), can be observed.

Therefore all recording combinations are symmetrically aligned in respect to the main diagonal of the table. This observation means that the polarization of both – recording as well as assisting beam

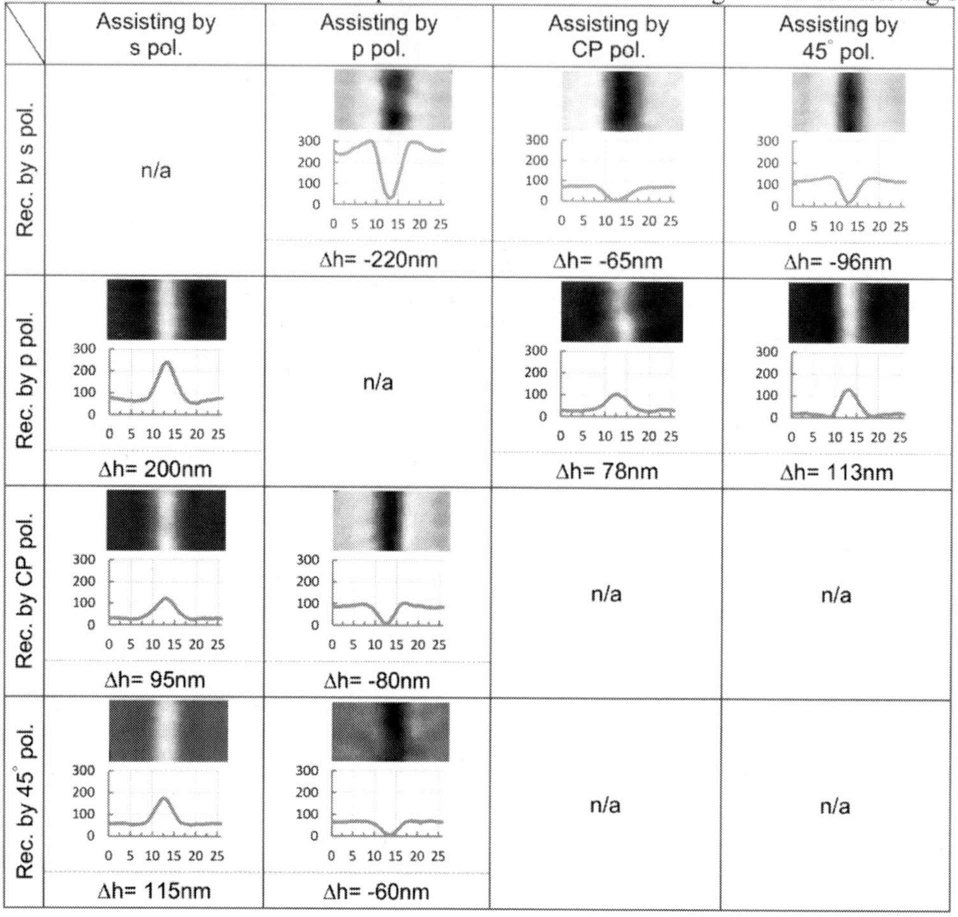

Figure 2. The correlation between polarization and light-induced mass transfer processes in 2.1 μm thin layer of As_2S_3 sample for different polarization combinations of the writing and assisting beams (1000 mW/cm^2 and 1500 mW/cm^2 for 1 h respectively) in optical slit experiments. The inset contains the 2D 26×13μm AFM topography pictures and its profiles in the 25μm×300nm frame

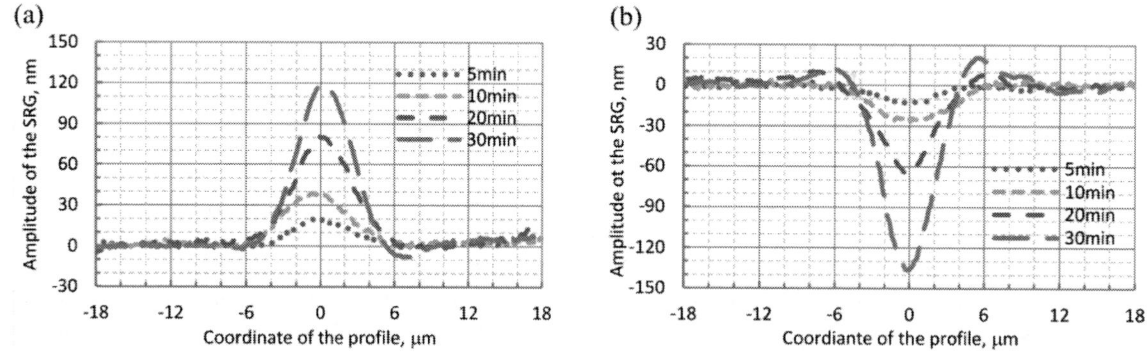

Figure 3. The time evolution of SRG profiles extracted from the AFM measurements. Recording performed on the As_2S_3 sample with a: (a) p/s and (b) s/p setup. The energy densities for the writing and assisting beam were 4.24W/cm^2 and 0.37W/cm^2 respectively, Corresponding exposure lengths: 5, 10, 20 and 30 minutes

play a significant role in the process of mass transfer. As can be seen the most effective mass transfer occurs when s-p polarization combinations between recording and assisting illumination beams are implemented. Thus the time dependent evolution of direct SRG formation for those cases is shown in Figure 3. As mentioned above, in the case of different polarization of the writing beams the direction of mass transfer is opposite. For the s polarization (s/p setup: polarization of the writing beam is parallel to the slit) the mass is transported away from the illuminated area thus forming a groove, but for the p polarization (p/s setup) this process is inverted – the mass is transported into the illuminated area, forming a ridge. Note that the assisting beams were polarized orthogonally to the writing. The symbol *a/b* denotes the possible combinations of polarization, *a* – for the recording beam and *b* – for the assisting beam. It is clearly seen that for the both recording setups (p/s and s/p) the mass transfer starts from the middle of 10 μm optical slit used in these experiments. For the short time exposure (in this case up to 10 min) A- or V-shaped profiles appear. If the recording time exceeds 10 minutes an active mass squeezing takes place and the profiles are changing to W- or M- shaped.

The behaviour of mass transfer and thus the resulting recording could be related to interaction between the polar photo-induced defects and the polarized electric field of recording and assisting illumination. Depending on the configuration of the resulting electric field, mass transfer can be directed both ways – towards or away from the electric field intensity gradient.

4. Conclusion

It has been shown that the formation of surface relief grating in amorphous As_2S_3 films closely depends on the superposition of electric field of recording and assisting beams. The recording efficiency in an arbitrary recording setup using linearly polarized light can be significantly improved by introducing additional incoherent orthogonally polarized assisting illumination.

A direct recording technique of SRG is a comparatively new solution for lithography and, as shown in this article, provides new experimental techniques for better understanding of the interaction between the light and matter. The obtained gratings are very stable at room temperature, so this method can replace some of the chemical etching techniques and find a practical application in the applied physics.

5. Acknowledgments

This work has been supported by the European Social Fund within the project «Support for Doctoral Studies at University of Latvia».

References

[1] Ashkin A 1970 *Phys. Rev. Lett.* **24(4)** 156–159
[2] Chomat M, Lezal D, Gregora I, Srb I 1976 *J. Non-Cryst. Solids* **20(3)** 427–437
[3] Tanaka K, Terakado N, Saitoh A 2008 *J. of Optoelectron. Adv. Mater* **10(1)** 124–130
[4] Kryshenik V M, Trunov M L, Ivanitsky V P 2007 *J. of Optoelectron. Adv. Mater* **9(7)** 1949–1964
[5] Ishitobi H, Tanabe M, Sekkat Z, Kawata S 2007 *Opt. Express* **15(2)** 652
[6] Hamanaka H, Tanaka K, Matsuda A, Iizima S 1976 *Solid State Commun.* **19** 499–501
[7] Gertners U, Teteris J 2010 *Opt. Mat.* **32(8)** 807–810

IOP Conf. Series: Materials Science and Engineering **38** (2012) 012027 doi:10.1088/1757-899X/38/1/012027

UV optical record and electron beam lithography in polymer films.

A Gerbreders[1,3], O Shimane[2], V Kolobjonoks[2], J Teteris[1]
[1]Institute of Solid State Physics, University of Latvia, Latvia
[2]G.Liberts' Innovative Microscopy Centre, Daugavpils University, Latvia

E-mail: andrejmah@gmail.com

Abstract. Possibility of UV optical record and electron beam lithography in different type of organic polymer films was studied. Mechanisms of molecular structure changes: photoisomerization, destruction, cross-linking and oxidation have been discussed. The results of UV illumination of polyurethanes, polyacrilates, and some block-copolymers are described. The element analysis of polyisoprene block copolymer was performed before and after UV illumination, and the changes in transmission spectra of the polymer film were measured. The resolution of electron beam lithography on polymer films was studied. In the polyisoprene block copolymer film the oxidation polymerization was ascertain at UV-illumination.

1. Introduction

Formation of determined micro and nano surface relief structures in the different materials has great role in producing of modern electronic and optical devices. Today double-step lithography is using for it: a material is illuminated by light of suitable wavelength and then etching has been used. However, etching operation is related with some problems. Firstly, it is an additional operation, which makes more difficult a process of devices preparation. Secondly, the active chemical reagents are used in process, which needs in further utilization.

Often for modern lithography and for information record UV-light is used. The main reason for it is possibility to obtain higher resolution. The great group of materials, which has an absorbtion edge in UV-area, is an organic polymer. Study of UV-light influence on polymers continues for many years. Primary, the goal of such research was a study of polymer destruction under natural UV-radiation, later it was a research of photo-polymerization [1,2] and UV-ablations [3,4]. In result, besides ordinary lithography, which has very limited use for high-molecular systems, today the two main methods are used for primary (no replication) microrelief formation in polymer films by UV-illumination:

- Photopolymerization of monomer mixture by illumination through a mask or modulator. In illuminated area a monomer has been polymerized, but from non illuminated area it is dissolved out by special solvent [5].
- Illumination of polymer films by powerful UV-light through a mask results a decomposition of macromolecules in illuminated area to low-molecular volatile products [6] or re-polymerize to another crosslink polymer [7].

Recently many papers were published about direct surface relief formation in different polymeric media by holographic recording [8,9]. According to [10,11], the mechanism of this process is based on

[3] To whom any correspondence should be addressed

matter transfer due to electric field gradient influence on cis-trans isomerization process. For relief formation the polarization holography method is widely used [11]. However the answer on the possibility of this mechanism of relief formation on the polymer surface by using polarization holography method with UV-light is vague.

In this work we have studied the features of direct relief formation in three different polymer types: polymethylmetacrylate, polyurethane and block-copolymer styrene and isoprene by pulse excimer laser radiation with wavelength of 248 nm.

2. Experimental

Polymethylmetacrylate (PMMA) (Aldrich, CAS# 9011-14-7, M_w=350 000, T_g=105°C) was dissolved in n-butylacetate (Aldrich, CAS#126-86-4) and 1-methoxy-2-propanol (ROTH, CAS# 107-98-2) (1:1 wt) mixture. Polystyrene-block-polyisoprene-block-polysterene (PS/PISP) (Aldrich, CAS# 25038-32-8, styrene 17 wt%, M_n=1,900) was dissolved in toluene (Aldrich, CAS# 108-88-3). As polyurethane polymer was used the product G-TDI, it synthesis and solution preparation was described in [12]. Fabrication of films from the polymer solutions were performed by company BYK Gardner GmbH applicator. A glass and quartz plates were used as substrates. The parameters of films fabrication are presented in tab. 1. The thickness of dry films was determined by the Veeco Dektak 150 surface profiler.

Table 1. Parameters of film fabrication.

Polymer	Concentration of polymer solution, wt. %	Thickness of wet film, μm	Drying parameters	Thickness of dry film, μm
PMMA	7	30	130 °C, 15 min	1,5
G-TDI	70	90	100 °C, 120 min	10
G-TDI	50	60	100 °C, 120 min	2,5
PS/PISP	10	60	120 °C, 15 min	5
PS/PISP (for EB lithography)	10	30	120 °C, 15 min	2,5

For all films the spectra of optical transmission were measured by the Ocean Optic HR4000CG spectrometer. The measurements were performed with the films on quartz substrate.

Illumination of the films was realized through 50, 100 and 250 μm mask by Illumination of films through 50, 100 and 250 μm mask was realized by excimer laser 248 nm 5 ns pulse with the repetition rate of 11 Hz. S- and p- polarizations of laser beam were used, polarization was changed by λ/2 plate. All films were illuminated in the range of exposure of 10 – 2200 J/cm^2.

For microanalysis, the film PS/PISP with a thickness of 5 μm was illuminated through 3x6 mm mask with exposure of 10-100 J/cm^2.

Microanalysis of the films was performed with module of X-ray spectroscopy "INCA x-stream and mics" (Oxford Instruments Analytical), which was assembled on bases of "TESCAN-VEGA VGU" scanning electron microscope (SEM). An electron beam was scanned over a defined area of the sample. The beam can be localized on a certain point in the image and used to measure energy dispersive X-ray (EDX) spectrum. The element-specific spectral lines are then identified to obtain the local elemental composition. The sample was placed in chamber with 11.088 mPa pressure and exposed by an electron beam focused to 914 nm diameter. The experiments were performed using point scan mode at 20 keV accelerating voltage and 47,1 μA emission current.

The sample of PS/PISP for electron beam (EB) lithography before experiment was covered with 20 nm gold layer by vacuum evaporation method. The surface relief formation experiments were performed using a scanning electron microscope (SEM) with installed EB lithography system. EB

parameters were: diameter - 91 nm, exposure 600 - 1000 μA*s*cm^2. Experiments were carried out at room temperature, in vacuum 10^{-3} Pa.
After experiments, the surface relief of the samples was analyzed by the Veeco Dektak 150 surface profiler and Atomic Force Microscope (AFM).

3. Result and discussion

3.1. Relief formation under UV illumination.

The figure 1 shows the changes in optical transmission spectra after 248 nm beam illumination due to the changes in molecular structure of polymer films. More probable explanation was presented in [13-15] where the processes of oxidation destruction leading to the changes of chemical compound and mass of the macromolecules were described.
The studies showed that the surface relief of illuminated films in the first phase has approximately the same profile for all studied materials. The surface relief formation for all films during illumination is shown in the figure 2. No dependence from the films thickness and mask configuration (diameter of illuminated area was in the range of 100 – 250 μm) was observed. An increase of exposure dose causes the changes in a height and a character of central area of illumination. The probable explanation for the formation of "hill" or "cavity" could be the type of destruction products: non volatile or volatile destruction products, respectively.
The formation of cavities with a depth of 300-700 nm was observed on the border between illuminated and non-illuminated film areas for all studied polymer compositions. The possible reason of its appearance according to [16] is the electric field gradient forces of light.

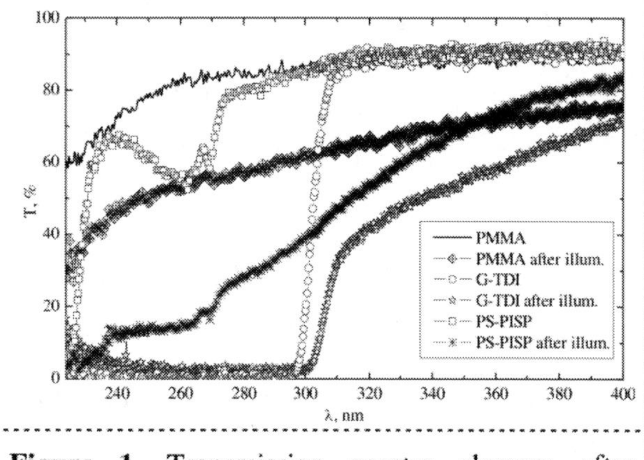

Figure 1. Transmission spectra changes after illumination. E (PMMA) = 188 J/cm^2; E (G-TDI) = 11 J/cm^2 ; E(PS/PISP) = 24 J/cm^2

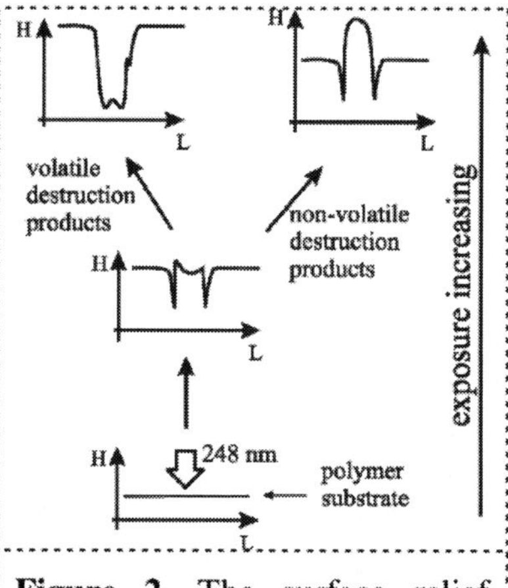

Figure 2. The surface relief profiles of the polymer films.

3.2. UV light influence on PS/PISP films.

The copolymer PS/PISP was selected due to a presence of C=C double bonds in the macromolecule. Primary we expect that the cis-trans isomerisation and photoinduced birefringence in UV region of spectra area is possible. Such possibility according to [17] has been indirectly confirmed. For understanding the mechanism of photoinduced changes, we performed the elemental analysis of illuminated and non-illuminated film. An oxygen appearance in chemical composition of polymer films was detected after illumination. The 20% (from atom quantity) of carbon was replaced by oxygen in macromolecules. The difference in chemical dissolving kinetic of illuminated and non-illuminated film was detected. This process takes place at low exposure doses (from 10 J/cm^2) and can be explained by the formation of oxygen bridges (–O–) between polymer macromolecules.

3.3. Direct EB lithography in PS/PISP films.
Unexpected results were obtained at EB illumination of PS/PISP. Even in glass substrate without conducting layer, direct relief formation on the film surface was detected. After 20 nm gold layer deposition, surface relief with depth 200 nm and 1 μm resolution was obtained (figure 3). In illuminated area the "cavity" appears. We assume that the destruction process of polymer with low molecular products formation takes place.

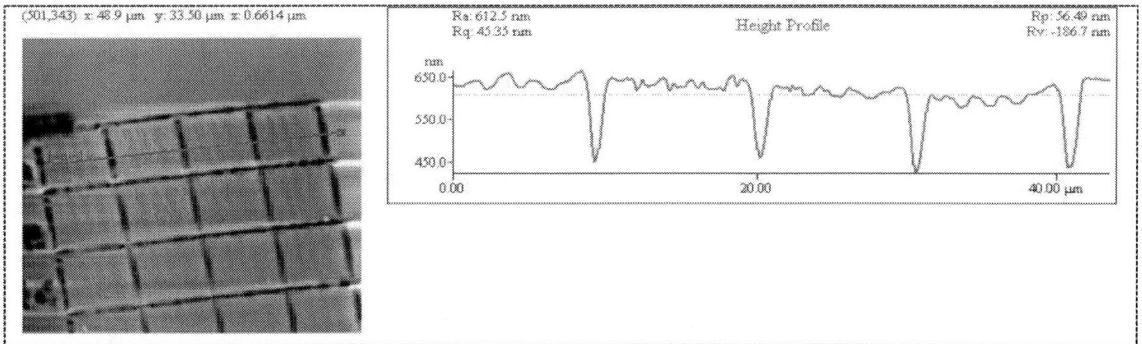

Figure 3. AFM image and relief of EB illuminated PS/PISP film

4. Conclusion
The direct surface relief formation on polymethylmetacrylate, polystyrene-block-polyisoprene-block-polysterene and polyurethane polymer thin films was observed under 248 nm eximer laser light and electron beam exposure was observed. In the PS/PISP film the oxidation polymerization was ascertain at UV-illumination with exposure from 10 J/cm^2. The direct surface relief formation with a depth up to 700 nm by EB lithography in the films was established.

Acknowledgment
This work has been supported by the European Regional Development Fund within the project Nr. 2010/0275/2DP/2.1.1.1.0/10/APIA/VIAA/124.

5. References
[1] Lawrence J, O'Neill F and Sheridan J 2001 *Optik* **112** 449
[2] Sun H, Kawata S 2004 *Advances in Polymer Science* **170** 169
[3] Bityurin N 2005 *Annu. Rep. Prog. Chem., Sect. C.* **101** 216
[4] Bounos G 2006 *J. Appl. Phys.* **100** 114323
[5] Yang C-R 2004 *J. Micromech. Microeng.* **14** 1126
6] Rubahn K 2004 *Appl. Phys. A* **79** 1715
[7] Shaw J 1997 *IBM J.Res.Develop.* Vol. 41 No. 1/2 81
[8] Ozols A 2008 *Thin Solid Films* **516** 8887
[9] Gerbreders A 2008 *Proc. of SPIE* **7142** 714212
[10] Yang K 2006 *Phys. Rev.* B **73**, 165204
[11] Nikolava L, Ramanujam P 2009 "Polarization holography." Cambridge University Press, 239 p.
[12] Gerbreders A 2011 *Proc. SPIE* **8306**, 83061A
[13] Kaczmarek H 2000 *European Pol. J.* **36** 767
[14] Gupta A 1980 *Macromolecules* ***13*** **(6)** 1696
[15] Newman R, Forciniti D 2001 *Ind. Eng. Chem. Res.* **40** 3346
[16] Teteris J, Gertners U and Reinfelde M 2011 *Phys. Stat. Sol.* **C8** 2780
[17] Golub M 1957 *J. of Pol. Sc.* **XXV** 373

IOP Conf. Series: Materials Science and Engineering **38** (2012) 012028 doi:10.1088/1757-899X/38/1/012028

Photoinduced phenomena in azo-dyed gelatine films

J Aleksejeva[1], **J Teteris**
Institute of Solid State Physics, University of Latvia

E-mail: aleksejeva.jelena@gmail.com

Abstract. In this work a photoinduced properties of azo-dye doped gelatine films were studied. Gelatine is the oldest polymeric matrix, which is used in holography and photography. Preparation of these films is simple and cheap, but ingredients aren't toxic. Films were synthesized using gelatine and azo-dye –Disperse Red13. Non- aqueous solvents - dimethyl sulfoxide and acetic acid, were used. Films were coated on glass plates using applicator with layer thickness 30, 60, 90 μm and then were heated in vacuum chamber. Azo-dyed gelatine films were illuminated by 532nm laser radiation. Photoinduced birefringence and photoinduced dichroism dependence on films ingredients ratio, laser light intensity and sample treatment temperature were investigated. Holographic recording on these films was performed in dependence on recording beam intensity and polarization. Volume polarization gratings and direct surface relief formation were studied by measuring diffraction efficiency with 660nm diode laser radiation. The obtained surface relief was investigated by AFM.

Introduction

Azo-dyed polymeric matrix in holographic recording was studied due to its possible application in diffractive optical elements fabrication. In this work gelatine as polymeric matrix was used. Gelatine is successfully used in silver-halide and bichromated gelatine films, where good resolution and high diffraction efficiencies can be obtained. There are few papers regarding to azo-dyed pre-etched gelatine films, where volume polarization gratings were recorded [3]. In this work we synthesised azo-dyed gelatine films, investigated their photoinduced processes – photoinduced birefringence and dichroism upon laser radiation. We performed holographic recording on the prepared films with different recording beam polarization state and intensities.

Experimental

The synthesis of azo-dyed gelatine: Unflavoured gelatine purchased at grocery store with *pH*>7 was used. 2.5g gelatine was dissolved in 20ml dimethylsulfoxide (DMSO) (films type I) and acetic acid (AA) (films type II) then solution was mixed by magnetic stirrer until all the gelatine dissolved [2]. In obtained solution azo-dye - Disperse Red13 (DR13) and citric acid (CA) in type I was added. Then solution was mixed again for several hours. During the stir, solution was treated at increased temperature. Prepared solution was applied on glass plate by applicator with variable thickness – 30, 60, 90 μm. Samples were placed into vacuum chamber, where were dried at 55°C for two days. After drying the films lost ~95% of their thickness.

[1] To whom any correspondence should be addressed

Experimental set-up for photoinduced dichroism investigation is shown in Fig. 1a. An incoherent semiconductor laser with wavelength λ=532nm was used. The laser beam was split in two beams by glass plate. One beam had very weak intensity and was used as a probe beam while another one with high intensity (I=0.53 W/cm^2) was used as a pump beam. A polarization state of both beams were improved with polarization cubes and set linearly orthogonal. Polarization state of each beam was varied with half-wave plates. Both beams were symmetrical to the sample normal and formed a small angle (~3°). Transmittance changes of pump and probe beams were measured with photodiodes *D*. Photoinduced dichroism were evaluated using formula: $D = A_{\parallel} - A_{\perp}$ (2), were $A_{\parallel}$ - absorbance of pump beam; $A_{\perp}$ - absorbance of probe beam.

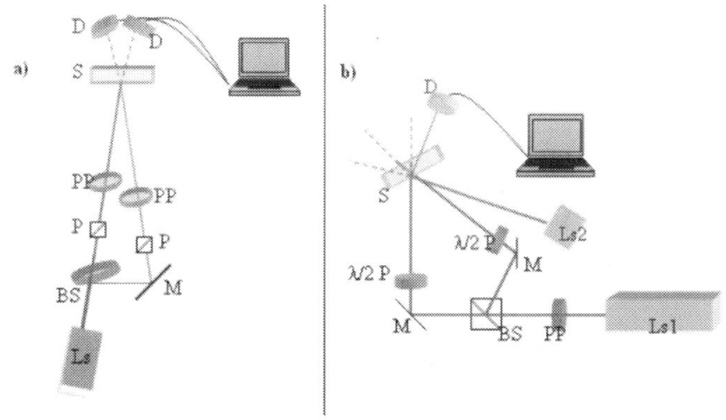

Figure 1 - Experimental set-up for a) dichroism measurements, b)holographic recording

Experimental set-up for holographic recording is shown in Fig. 1b. Solid-state diode-pumped laser Verdi-6 (Ls1) was used for holographic recording. Laser beam was splitted into two equal intensity beams by beam splitter (BS). Using mirrors (M), beams were directed toward the sample. Angles between beams and normal were $\theta = 15.4°$ thus period of the

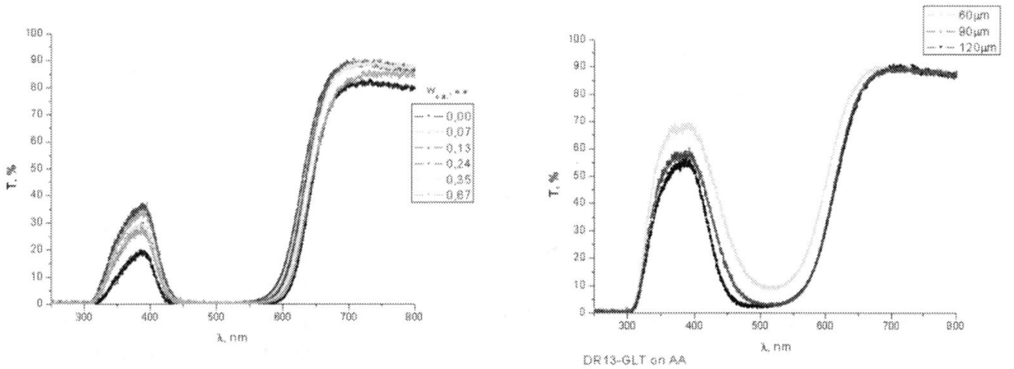

Figure 2 - Transmittance spectrum for DR13-gelatine film a) type I with CA, b) type II

grating was Λ=1μm according to Bragg law: $\Lambda = \dfrac{\lambda}{2\sin\theta}$ (3), where Λ – period of the grating, λ – wavelength, θ – angle between the recording beam and normal. Polarization state of recording beams was varied with half-wave or quarter-wave plates (PP). Diffraction efficiency of recorded grating was measured in transmittance mode by diode laser (Ls2) with 660nm wavelength. The polarization state of probe beam was *s* – vertically linearly polarized with respect to grating lines. The ±1 diffraction maxima were measured with photodiode (D).

Results and discussion

Transmittance spectra of prepared samples are shown in Fig. 2a and 2b. In samples type I and II DR13 characteristic absorbance bond between 450-600 nm is seen. The type I sample had higher viscosity than type II, thus type II samples have higher transmittance in the absorption edge region.

Adding CA leads to *pH* decreasing and viscosity increasing [1,2,6], thus transmittance increases in samples with higher CA rate.

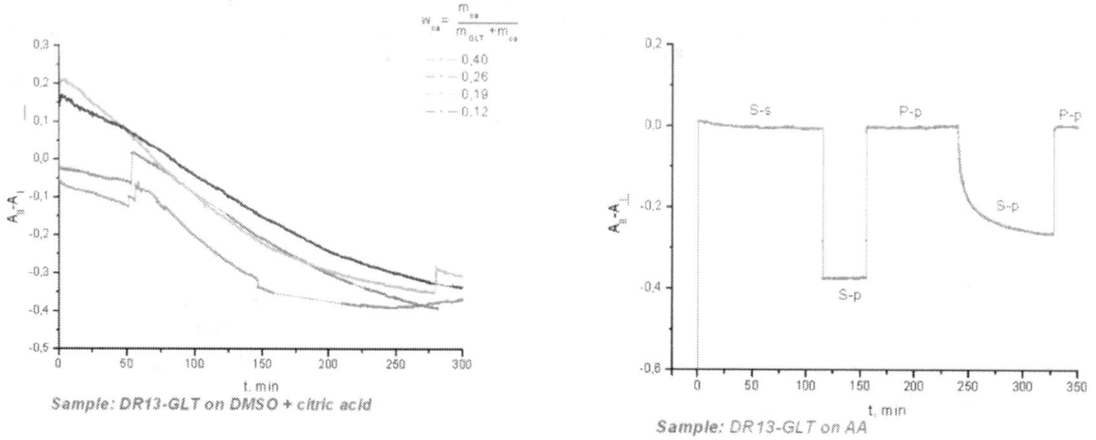

Figure 3. Photoinduced dichroism dependence on time in DR13-gelatine a) type I with CA films; b) type II films

Under the laser radiation and presence of free charges in the volume gelatine molecule degradation was observed as photobleaching of the sample. Irradiated area became yellowish and didn't regain its original colour. Increasing of CA in the sample provided more free charges in the sample, thus photobleaching speed decreases with increment of the CA mass ration in the sample. Although CA mass ratio higher than w=0.3 gave unstable sample, because after few days CA crystallized making sample uneven. Samples with lower mass ratio stayed stable for months.

Photoinduced dichroism was observed in samples – type I + CA and type II.

In the Fig. 3a photoinduced dichroism curves for the samples type I + CA is shown. Increment of CA mass ratio, initial dichroism decreases and became negative with further CA mass ratio increment. With low CA mass ratio (0.12) in the sample curve of the photoinduced dichroism is even and no "jumps" was observed. With higher CA mass ratio (0.19) after large exposition dichroism value decreases. Under light radiation length of gelatine molecule decreases and inert strength in the sample appears. Changes of dichroism value could be explained with release of the strength in the volume. At higher CA mass ratio (0.26) dichroism value decrement is observed at much lower exposition, due to higher free charges ratio. If CA mass ratio is higher than w=0,3 (0.40) than several dichroism value decrement and increment can be observed.

No initial dichroism was observed in the sample type II (Fig. 3b). As long both pump and probe beams had the same polarization state (*S-s* or *P-p*) no dichroism was observed and transmittance of both beams was equal. If polarization state of probe beam was changed to the orthogonal one (*S-p*), dichroism appeared and reached D=-0.37 value. If polarization state of the pump beam is changed, first fast increment of dichroism value is observed due to photoisomerisation of DR13 molecules, and then slower increment due to bond breaks orthogonal to the pump beam radiation polarization state in gelatine molecules. Such kinetics for dichroism changes can be repeated several times with no changes.

Holographic recording have been performed on prepared samples. Changing of polarization states of recording beam showed that grating recorded with polarization state, which gives polarization modulation on the surface (*-45/+45* and *R-L/L-R*), weak volume grating and no surface relief grating was observed. The best results were obtained using polarization state, which gives intensity modulation on the surface (*p-p*, *s-s*, *R-R/L-L*). Surface relief and volume grating was obtained. Thus for further experiments *s-s* polarization state was chosen.

Holographic gratings were recorded with intensity I=1.10 W/cm^2 in each recording beam. On the samples with $w_{c.a}$=0.19, 0.26 and 0.33 CA mass ratio similar volume gratings were recorded. Obtained diffraction efficiency was 1.4-1.6% on transmittance mode for 660nm wavelength reading laser. Volume grating in the sample with very low CA mass ratio (0.07) formed slower due to slower processes and lower free charges rate in the sample.

A weak surface relief grating formation (Fig. 5) was observed only if recording beams polarization state gave intensity modulation, thus relief formation can be explained by gelatine swelling [6].

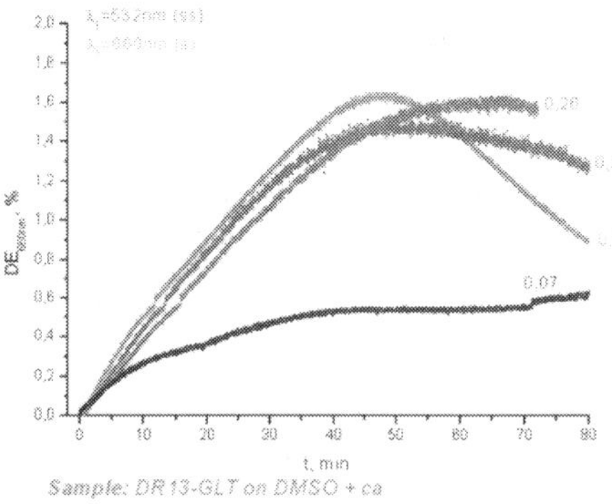

Figure 4 The first-order transmittance mode diffraction efficiency in DR13-gelatine type I.

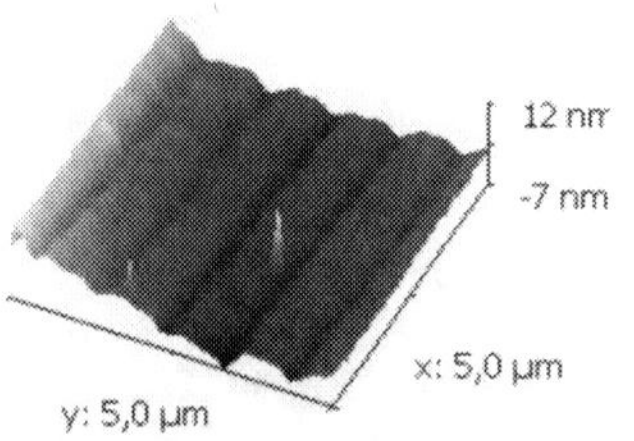

Figure 5 – The surface relief in DR13-gelatine on DMSO with CA film.

Conclusion

In azo-dye gelatine films photoinduced birefringence induced at 532nm wavelength reached Δn=0.006. There was observed a photoinduced dichroism in those films and difference between absorbance of perpendicularly and parallel linearly polarized beam reached D=-0.37. CA adding to the sample gave films with faster response to the laser radiation. Diffraction efficiency of volume grating reached 1.6% at the first maxima. Weak surface relief modulation was observed.

Acknowledgement

This work has been supported by the European Regional Development Fund within the project Nr. 2010/0275/2DP/2.1.1.1.0/10/APIA/VIAA/124

References:

[1] Solano C 1989 *Applied Physics* **28** 3524
[2] Choktaweesap N, Rayanarakul K A, Aht-Ong D, Meechaisue C and Supaphol P 2007 *Polymer Journal* **29** 622
[3] Ivanov M, Eiju T 2001 *Optical review* **8** 315
[4] Vygovskii Yu N, Konop S P, Malov A N, Malov S N 1998 *Laser physics* **8** 901
[5] Aoki H, Taguchi T, Saito H, Kobayashi H, Kataoka K, Tanaka J 2004 *Materials Science and Engineering C* **24** 787
[6] Oujja M, Rebollar E, Abrusci C, Amo A Del, Catalina F, Castillejo M 2007 *Journal of Physics: Conference Series* **59** 571

IOP Conf. Series: Materials Science and Engineering **38** (2012) 012029 doi:10.1088/1757-899X/38/1/012029

Iron complexes embedding influence on PMAA hydrogel

L Khenkin[1], T Baluyan, A Novakova, I Rebrin, E Makhaeva

M.V.Lomonosov MSU, physics department, Moscow, Russia

E-mail: levkhenkin@gmail.com

Abstract. The interaction of weakly cross-linked gel of poly(methacrylic acid) with iron complexes results in Fe^{2+} absorption, additional cross-links and therefore gel collapses. Lattice dynamics of frozen investigated samples were examined by means of Mossbauer spectroscopy in order to find out the mechanism of metal-polymer bonding. It was established that iron complexes form strong bonds with polymer functional groups contracting gel network.

1. Introduction

Poly(methacrylic acid) (PMAA) groups can show specific interactions, such as hydrogen bonding, acid–base interactions and ionic interactions of opposite charges, which can help to prepare novel self-organized materials. Numerous studies have been also devoted to the interaction of PMAA with metal ions, since knowledge of association phenomena of metal ions with charged macromolecules is of importance for the understanding of their physicochemical behavior in environmental and biological systems. PMAA could be successfully used as a component of characteristic 'intelligent' organic–inorganic hybrid materials.

The interaction of weakly cross-linked PMAA gel with iron complexes results in their absorption and gel collapse [1, 2]. This collapse occurs due to iron ions/polymer functional groups bonds that form additional cross-links contracting gel network. The aim of the present work is study of molecular lattice dynamics and metal-polymer bonds strength in PMAA gel/iron complexes systems by means of Mossbauer spectroscopy.

2. Experimental section

PMAA gels were prepared through radical polymerization in solution with N,N'-methyl-bis-acrylamide as a cross-linker. The procedure of PMAA gels treatment was the same as that described previously [3]. The degree of the polymer gel swelling (Φ) in complex solutions is defined by the ratio of the mass of polymer gel after immersion in iron complex solution to the initial gel mass.

In the present study PMAA gel was incubated in two different media: aqueous solution of iron(II) chloride and aqueous solution of ferroin (Fe^{2+}–phenanthroline complex).

Temperature-dependence of their Mossbauer spectra can serve to elucidate the bonding and dynamical behavior of the iron ions in such systems [4]. Mossbauer lattice temperatures θ_M and effective vibrating mass M_{eff} were estimated from temperature dependences of recoilless fraction and second order Doppler shift δ_{SOD}. Mossbauer spectroscopy data for all incubated PMAA gels and for

[1] To whom any correspondence should be addressed

used frozen solutions were obtained in the temperature interval from 118K up to the Mossbauer effect disappearance temperature.

Mossbauer spectra were obtained in a standard transmission geometry using a $^{57}Co(Rh)$ gamma source. The gamma radiation was detected with a NaI(Tl) scintillation detector. The spectrometer was calibrated by a standard α-Fe absorber. The spectra were analyzed using the UNIVEM MS program (2006), which ensures fitting by a combination of Lorentzians with adjustable Mossbauer parameters: isomer shift (δ), quadrupole splitting (Δ), peak width, and peak area under the curve (A).

3. Results and discussion

The swelling behavior of PMAA hydrogels was studied in [2]. The degree of contraction depends on the amount of monomers in the parent PMAA gel and the amount of absorbed Fe ions. The samples under investigation after week incubation in iron complex solution were collapsed. Degree of their swelling was Φ=0,22 for incubation in ferroin aqueous solution and Φ=0,12 for incubation in iron chloride aqueous solution.

Mossbauer spectra at T=118K and images of investigated samples are shown in figure 1. In iron salt solution all Fe^{2+} ions form $[Fe(H_2O)_6]^{2+}$ coordination entities of octahedral geometry. So the spectra of all samples comprise one component: doublet with big quadrupole splitting for high-spin Fe(II) ions in chloride salt solution and doublet with small quadrupole splitting for low-spin Fe(II) ions in ferroin solution (see table 1 for values).

Mossbauer spectral parameters (isomer shift and quadrupole splitting) for both iron complexes appeared to be the same in investigated aqueous solutions and in gel incubated in these solutions. The invariance of the spectral parameters reflects the iron ions environments uniformity. Both ferroin and aqua-complex remain holistic and stable after embedding to gel matrix.

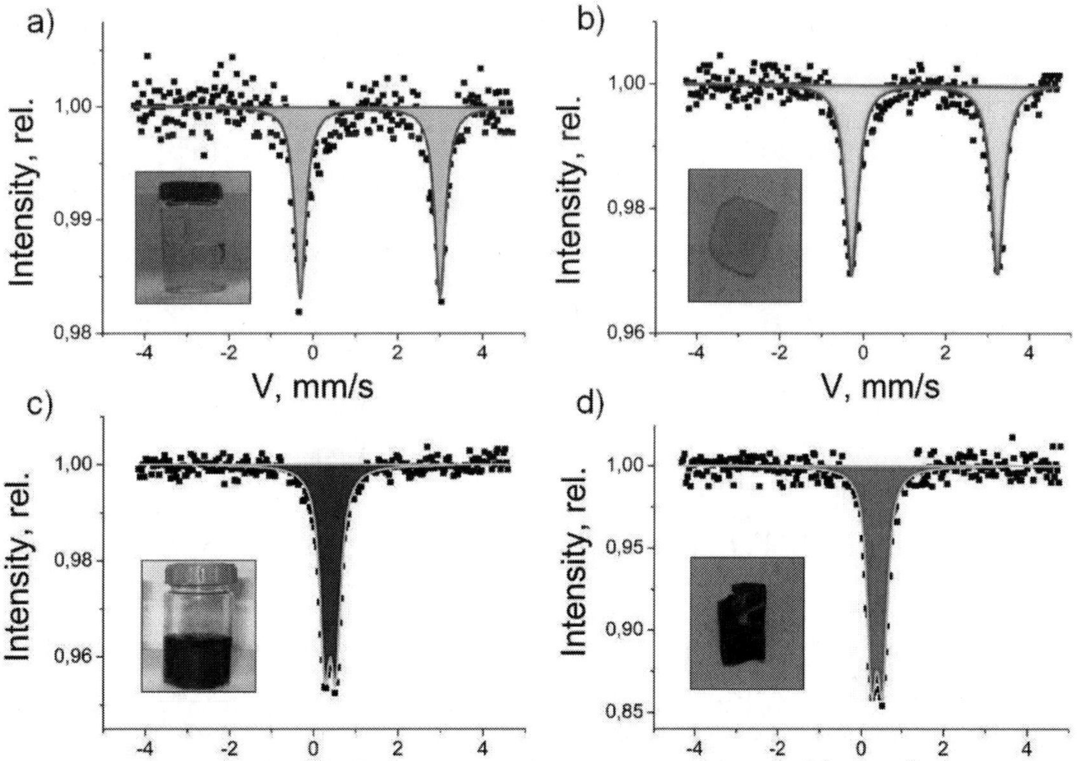

Figure 1. Mossbauer spectra at T=118K and photo of: a) $FeCl_2$ aqueous solution b) PMAA gel in $FeCl_2$ aqueous solution c) ferroin aqueous solution d) PMAA gel in ferroin aqueous solution

Temperature dependences of all the samples under investigation spectrum intensity (area under curve) characterizing recoilless fraction are shown in figure 2. All dependence curves have contrary flexure point at temperatures near 190K. This feature is induced by diffusion motion of iron complexes that will be discussed later.

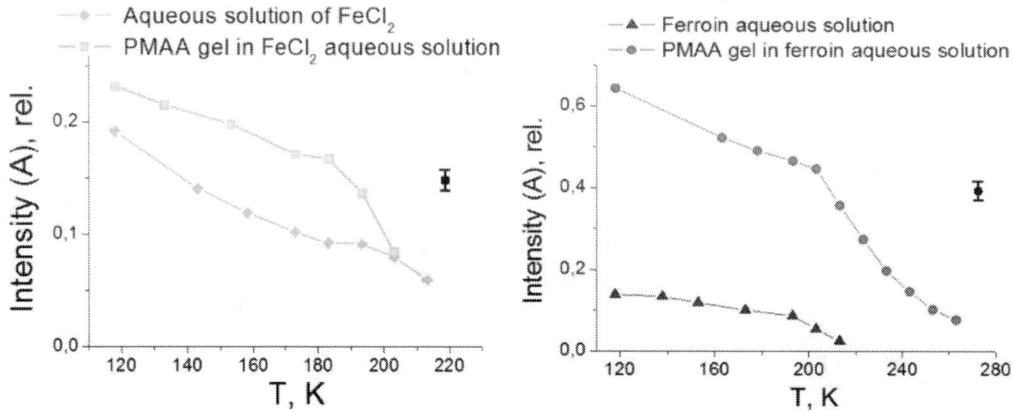

Figure 2. Temperature dependence A (area under curve) for a) $FeCl_2$ aqueous solution, PMAA gel in $FeCl_2$ b) ferroin aqueous solution, PMAA gel in aqueous solution

For molecular crystals the normal modes of vibration are usually separated into two groups: the weaker intermolecular modes related to vibrations of the molecular center of mass and the stronger intramolecular modes of much higher frequencies within the molecule.

If the molecule is an ideal rigid body, the mass value in the expressions for f and δ_{SOD} would just be the molecular weight. In reality, the molecule is not entirely rigid due to some degree of internal vibration; the mass value may be replaced by an effective vibrating mass M_{eff}. Its upper limit is the molecular weight and lower limit is the mass of the Mossbauer iron isotope (57 Da); the latter corresponds to the situation of Fe atoms as monatomic vibrating entities in a solid.

For the lower temperature area theory gives the following equations [4, 5]:

$$\ln f = -\frac{3E_\gamma^2}{M_{eff}c^2k_B\theta_M^2}T = \ln A(T) + const\,, \qquad (1)$$

$$\delta_{SOD} = -\frac{3}{2}\frac{k_BT}{M_{eff}c} + const\,, \qquad (2)$$

where A is the area under the resonance curve ($\ln A \propto \ln f = \exp(-k^2\langle u^2\rangle)$ for an optically "thin" absorber, $\langle u^2\rangle$ is the mean square amplitude of the metal atom vibration), k_B - Boltzmann's constant, c - velocity of light, E_γ - the energy of the emitted γ-ray.

Both the δ and ln A temperature dependencies for this compound are well fitted by linear regressions (ln A for the temperature diapason from 118K to curve contrary flexure point). Curve slopes values of the fitting lines are given in table 1. After substitution these values to (1) and (2) we obtained effective vibrating masses and lattice temperatures for all our samples (table 1).

The deviation of the M_{eff} values from 57 Da reflects the covalency of the bonding force between the Fe atom and its neighbours [6]. A comparison of M_{eff} values between complexes in frozen solution and in the PMAA gel matrix leads to conclusion that strong chemical bonds appear between the iron complexes and the functional groups of polymer (COO^-). Moreover, temperature Θ_M should consider as a quantitative measure of such bonds. Thus, stable metal - polymer complexes formed.

Table 1. Mossbauer parameters and derived quantities of investigated samples. Measurement errors are indicated in brackets.

Sample	δ, mm/s at 118K	Δ, mm/s at 118K	dδ / dT, 10^{-4} mm/sec•K	d lnA/ dT, 10^{-3} K^{-1}	M_{eff}, Da	Θ_M, K
$FeCl_2$ aqueous solution	1,41(2)	3,39(3)	7,5 (5)	10,1 (5)	111 (7)	83 (7)
PMAA Gel in $FeCl_2$ aqueous solution	1,38 (2)	3,33(3)	6,7 (4)	4,6 (3)	124 (7)	116 (9)
Feroin aqoueous solution	0,40 (2)	0,26(3)	3,7 (2)	8,1 (4)	224 (9)	65 (6)
PMAA Gel in ferroin aqueous solution	0,40 (2)	0,27(3)	3,4 (2)	4,1 (3)	244 (9)	87 (8)

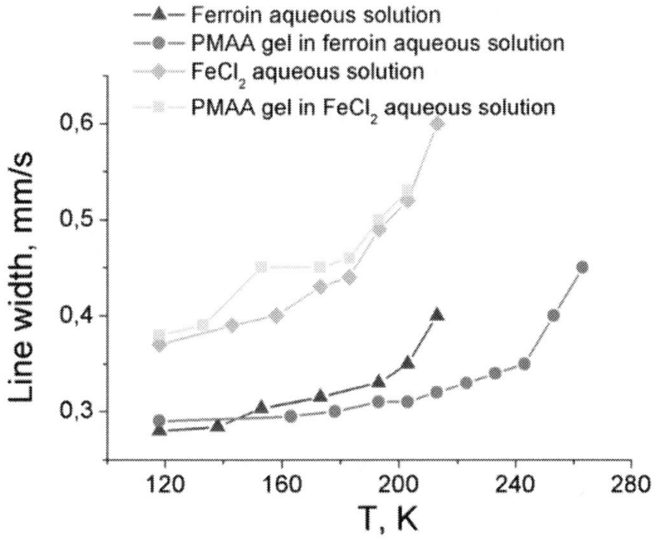

Figure 3. Line width temperature dependence for all investigated samples

Temperature dependences of Mossbauer spectra line width (figure 3) give the data about increasing of diffusion motion with increasing temperature. Spectra of samples in the chloride solution had significantly broader lines than spectra for ferroin solution because of higher molecular mobility of aqua molecules from nearest iron surrounding. On warming towards the melting point the relaxation time for diffuse jumps of the resonance atoms becomes of the same order of magnitude as excited state lifetime [7]. So at elevated temperatures diffusive motion increases rapidly. Line broadening in frozen PMAA gel incubated in ferroin solution started at the temperature on 20K higher.

4. Conclusion

All obtained data allow us to conclude that investigated iron complexes (ferroin and aquacomplex) embed in PMAA gel network without destruction and form strong bonds with polymer functional groups. These bonds are stronger than iron complexes bonds in frozen aqueous solutions crystal. And these bonds have higher degree of covalence. Such stability of iron complexes inside the polymer gel matrix could be widely used in practical applications.

References

[1] Grigor'ev T E, Nguen K H, Skryabina I V, Makhaeva E E and Khokhlov A R 2008 *Polym. Sci.* A **50** 68
[2] Starodoubtsev S G, Khokhlov A R, Sokolov E L and Chu B 1996 *Macromolecules* **28** 3930

[3] Khenkin L V, Shishakov A I, Novakova A A, Kozhunova E Yu and Makhaeva E E 2011 *Inorganic Materials* **47** 1271
[4] Chen Y L and Yang D P 2007 *Mossbauer Effect in Lattice Dynamics* (Weinheim: Wiley-VCH)
[5] Herber R H and Nowik I 2008 *J. Nucl. Radiochem. Sci.* **9** 33
[6] Long G J, Hautot D, Grandjean F, Morelli D T and Meisner G P 2000 *Phys. Rev.* B **62** 6829
[7] Singwi K S and Sjolander A1960 *Phys Rev* **120** 1092

IOP Conf. Series: Materials Science and Engineering **38** (2012) 012030 doi:10.1088/1757-899X/38/1/012030

Mechanical, structural and magnetic properties of polypropylene/iron ferrite magnetic nanocomposites

I Reinholds[1], V Kalkis[1,4], J Zicans[2], R Merijs Meri[2], A Grigalovica[2] and M Maiorov[3]

[1] *Faculty of Chemistry of University of Latvia, 48 Valdemara St., Riga LV-1013*

[2] *Institute of Polymer Materials, Riga Technical University, 14 Azenes Str. 14, Riga LV-1048*

[3] *Institute of Physics of University of Latvia, Miera Str. 32, Salaspils, LV-2169*

E-mail: Valdis.Kalkis@lu.lv

Abstract. In this research, composites of polypropylene (PP) and a relatively small concentrations of iron ferrite (Fe_3O_4) nanopowder equal to 2, 5 and 10 wt.% have been made. An methacrylate monomer 2,2'-bis-[4-(2-methylprop-2-enoyloxy)phenyl]-propane (BAD) was added at a small amount (3 wt.%) to part of the composites during the thermoplastic mixing. The BAD has been used as a compatibilizing agent on the phase interaction between the filler and the matrix of polypropylene. It has been found that composites have adjustable mechanical, structural and magnetic properties dependent on the content of magnetic filler. Increase of concentration of ferrite particles affects a notable increase of elastic modulus and reduces the deformability in comparison to pure polypropylene, whereas the modification with BAD has mainly a plasticizing effect, affecting a small decrease on stiffness but a notable increase of strain at break. An important role of Fe_3O_4 adhesion in polymer matrix and interphase effects on the changes of crystallinity and magnetic properties has been found.

1. Introduction

In last years, there has increased interest attributed to magnetic composites based on elastomeric or thermoplastic polymers filled with several micro and nano homogeneity particles, as lightweight design elements with desired mechanical, optical properties, high thermal stability and customizable electrical and magnetic properties suitable in engineering for application in electronic devices as radio frequency absorbing systems and other ones [1-3].

In our research polymer nanocomposites of polypropylene (PP) with a relatively small content (up to 2 vol.%) of iron ferrite Fe_3O_4 nanofiller have been made. It is known that photo, radiation or thermally initiated grafting of acrylate monomers can improve the properties of polyolefin based polymers [2-3]. For such a purpose 3 wt.% of methacrylate monomer 2,2'-bis-[4-(2-methylprop-2-enoyloxy)phenyl]-propane (BAD) was added to one part of PP/Fe_3O_4 blends during the thermoplastic mixing to improve the compatibility between the filler particles and the polymer matrix. The

[4] To whom any correspondence should be addressed

mechanical properties of composites have been estimated, showing an increase of stifness as reinforcement effect of nano filler [2], an increment of plasticity for composites modifed with BAD has also been found.

The method of DSC has been used to investigate the effect of filler dispersion on the crystallinity and the calorimetric properties. The magnetic properties have also been researched

2. Materials and investigation methods

2.1. Materials

Polypropylene homopolymer (PP) of trademark Moplen HP456J (Lyondell Basell) with a melt flow index 3.4 g/10 min and Fe_3O_4 nanopowder: spherical particles with a diameter <50 nm, ρ - 4.8 g/cm^3 (Sigma Aldrich) were used as the component materials. The blends of HDPE and PP with concentrations of magnetic filler Fe_3O_4 equal to 0, 2, 5 and 10 wt. % (that is 0, 0.4, 1 and 2 Vol.%) were obtained by thermoplastic mixing and pressing at 190°C. 3 wt.% of 2,2'-bis-[4-(2-methylprop-2-enoyloxy)phenyl]-propane (bisphenol-A-dimethacrylate, BAD) monomer was added during the thermoplastic mixing to one part of the blends polyolefin matrices to increase the compatibility between the filler and the polypropylene. Polymer plates with

2.2. Methods

The tensile tests of composite specimens were performed at 23±2°C on a Tinius Olsen H1K-S UTM at a deformation speed of 1 mm/min for determination of elastic modulus E and at a deformation speed of 50 mm/min for stress-strain characteristics in accordance to ISO 527-2 standard.

The method of DSC (Mettler Toledo DSC1/200W, temperature range 25-300°C under N_2 atmosphere at the heating rate 10°C/min) was used to determine the effect of filler concentration on degree of crystallinity calculated from the second heating scans.

Magnetization curves $M(H)$ of Fe_3O_4 nanopowder and magnetic composites were obtained at room temperature with a maximum applied field of 10^4 G on highly sensitive vibration sample magnetometer designed at the Institute of Physics, University of Latvia.

3. Results and discussion

3.1. Tensile tests

The typical stress-strain $\sigma(\varepsilon)$ curves of pure PP and PP blends filled with different concentrations of Fe_3O_4 in the presence and without the BAD modifier are compared in Figure 1. The unmodified PP has a typical crystalline polymer character of tension with a clearly seen maximum load at 38 MPa and a broad area of yielding with later increase of strength up to 40±0.5 at the break point at the deformation of break equal up to 704±27%.

The unmodified blends with the concentration of Fe_3O_4 greater than 2 wt.% tend to break up to failure immediately after the maximum load due to the formation of the stiff interphase between the particles an the PP matrix, while some deformability remains at the content of Fe_3O_4 equal to 2 wt.%. It is also evident that increase of the stiffness affects a noticeable decrease of elongation at break up to 63 times for composite with the content of filler 10 wt.% comparing to that of pure PP.

In comparison, a noticeable effect of BAD additive is seen on the character of elastic behavior of blends modified with BAD – the curves bend to lower angle that indicates increment of mechanical stiffness what coincides with proportional increment of elastic modulus with increase of the filler concentration in PP matrix. The value of elastic modulus decreases by 14% for entire PP modified with BAD while at the content of Fe_3O_4 10 wt.% the decrement it decreases only by 9%.

The values of E increase by 8% and 14% at the content of Fe_3O_4 5% and 10% for composites modified with BAD, whereas for unmodified PP the increment of E reaches 3% and 7% at the same concentrations of the filler that is less than 2 times in comparison to that of BAD modified ones. That indicates an important role of methacrylate modifier on the phase morphology and improvement of mechanical properties of PP/Fe_3O_4 blends. Moreover, the composites modified with BAD show an increased deformability while the yielding behavior decreases that may be due to the presence of bisphenol group of BAD, which tend to have a plasticization effect.

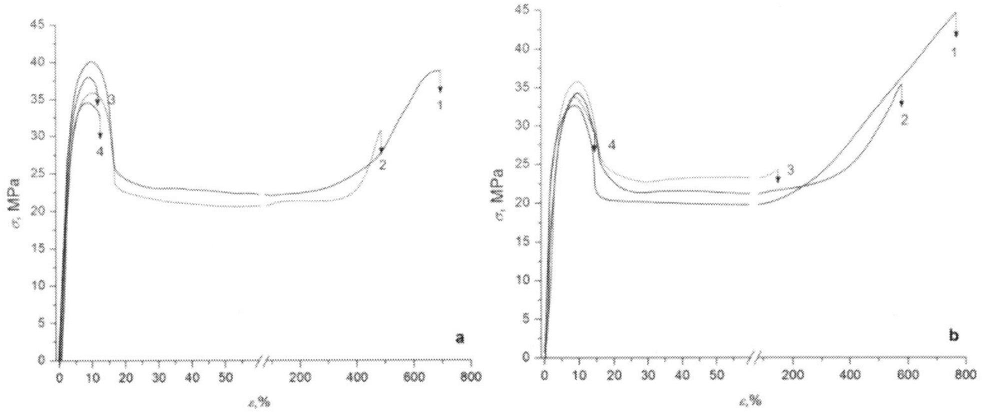

Figure 1. The Stress-strain curves of composite blends: PP/Fe_3O_4 (a) and PP/Fe_3O_4/BAD (b) at different wt.% concentrations of filler: 0% (1), 2% (2), 5% (3) and 10% (4).

3.2. Degree of crystallinity

The thermal history during the first heating cycle and after the cooling the second heating scans were done which were used for further analysis The ratio of melting heat ΔH_{m2} and the melting heat of completely crystallized polypropylene ($\Delta H_{100\%PP}$ = 207 kJ/g) were used to calculate the relative degree of crystallinity (χ_c, %) of PP in composites, which was recalculated according to the mass content of polymer matrix (Figure 2). The relative crystallinity of PP and its composites remain almost unchanged and is about 33 % for pure PP, while the values of χ_c noticeably increase for the composites at the presence of BAD modifier.

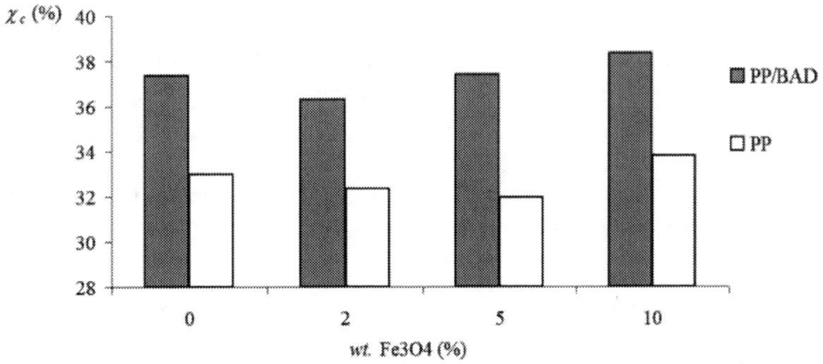

Figure 2. Relative crystallinity of PP samples in dependence of filler concentration for unmodified and modified with BAD composites.

3.3. Magnetic properties

The magnetization curve of Fe_3O_4 nanopowder showed a total magnetization saturation M_s equal to 72 emu/g_{tot}, a coercive field equal to 89 G and a ratio of remanence and saturation equal to 0.15.

Figure 3 compare the magnetization curves of PP/Fe_3O_4 composites at different filler contents. They indicate a small hysteresis and a slow approach to saturation. The small ratio of M_r/M_s indicates a formation of multi-domain aggregates due to the nucleation process [1-3].

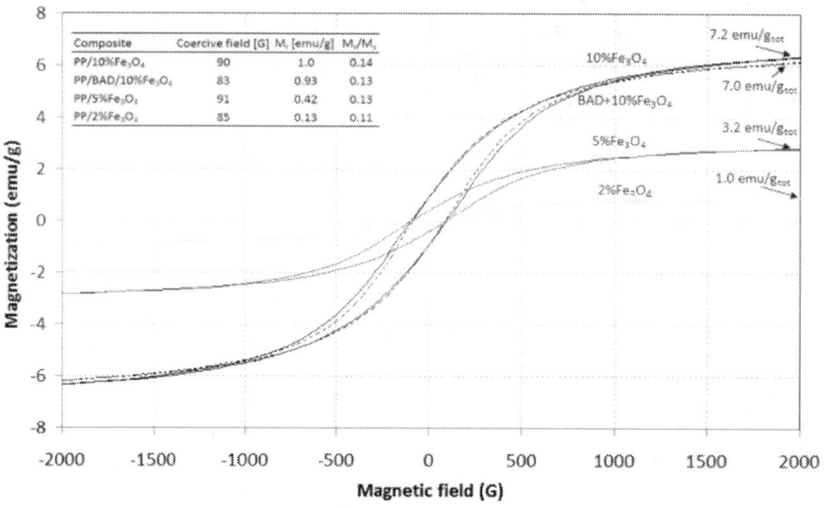

Figure 3. Magnetization curves of PP/Fe_3O_4 composites with 2, 5 an 10 wt.% filler content and a curve of composite modified with BAD at a content of Fe_3O_4 equal to 10 wt.%.

4. Conclusions

The mechanical properties have shown a typical reinforcement effect of Fe_3O_4 nanofiller with increased stiffness and a significant decrease of deformation at break with increase of the filler content.

The modification of polymer matrix with a compatibilizing agent consisting of methacrylate and bisphenol moieties (BAD) can effectively improve the deformability of polymer composites, whereas a small reduction of elastic modulus is observed. The changes of stress-strain character at maximum load indicate a notable effect of BAD on the phase morphology due to the interaction with polymer matrix.

The magnetic properties indicate a ferromagnetic behavior of both the polyolefin nanocomposites with Fe_3O_4 that is linearly dependent on the content of magnetic filler phase. The ratio of remanence and saturation M_r/M_s equal up to 0.15 indicate a close interaction between the magnetic particles affecting formation of agglomerates. The use of organic coating agents would be desired to gain a better compatibility.

Acknowledgements: This work has been supported by the European Social Fund within the project "Support for Doctoral Studies at University of Latvia".

References

[1] Carosio F, Fina A, Coisson M 2010 *Polym. Bull.* **65** 681
[2] Gomes de Barros G, Galembeck F 1986 *Polym Bull.* **16** 499
[3] Dallas P, Georgakilas V, Niarchos D, Komninou P, Kehagias T and Petridis D 2006 *Nanotechnology* **17** 2046
[4] Kihal A, Fillion G, Bouzabata B and Barbara B 2012 *Phys. Status. Solidi. B* **249** 604

Synthesis highly active platinum tri-metallic electrocatalysts using "one-step" organometallic chemical vapour deposition technique for methanol oxidation process

Qi-Ling Naidoo[1,2], Sivapregasen Naidoo[3], Leslie Petrik[1], Alexander Nechaev[1], Patrick Ndungu[4], Guntars Vaivars[5, 6]

[1]Environmental and Nanosciences Research Group, Department of Chemistry, University of the Western Cape, Bellville, South Africa

[2]HySA/Catalysis Competence Centre, Chemical Engineering Department, University of Cape Town, Rondebosch, South Africa

[3]Environmental and Process Systems Engineering, Department of Chemical Engineering, University of Cape Town, South Africa

[4]School of Chemistry, University of KwaZulu-Natal, Westville Campus, Durban, South Africa

[5]Institute of Solid State Physics, University of Latvia, 8 Kengaraga iela, LV-1063, Riga, LATVIA

E-mail: Guntars.Vaivars@cfi.lu.lv

Abstract. A simple solvent free method for the synthesis of tri-metallic platinum electrocatalysts on carbon nanotubes is presented. By investigating the platinum alloy electrocatalysts, it was showed that the additional metals of platinum alloys could reduce the metal particle sizes and produce larger chemical-active surface area, as well as the higher methanol oxidation activity of the catalysts. The organometallic chemical vapour deposition method was successfully applied to produce multiple samples of PtRuFe, PtRuCu and PtRuV. The electrocatalysts were characterized by ICP, XRD, HRTEM and the catalytic activity was determined by cyclic voltammetry (CV).

1. Introduction

Fuel cells are new clean energy systems for power to be generated at the point of use utilizing a variety of fuels [1]. Benefits arising from the use of fuel cells include efficiency and reliability, as well as economy, unique operating characteristics, planning flexibility and future development potential [2]. In the context of fuel cell engineering, platinum is the most popular electrocatalyst for fuel cells. When this platinum electrocatalyst is in contact with one of the electrodes in a fuel cell, it increases the rate of oxygen reduction to water (or hydroxide or hydrogen peroxide) [3]. To enhance catalytic performance [4-5], stability [6-7], durability [8-9] and reduce operational costs [10] of platinum

[6] To whom any correspondence should be addressed

electrocatalysts, advanced electro-catalyst design relying on the bi-functional mechanism, and using hybrid support carbon materials has been proposed. It has been found that the methanol electro-oxidation activity of PtRu/C is the highest among the binary Pt-based alloy, which is attributed to the ability of the Ru in the alloys to form active oxygen species (-OH) at low electrode potential (~0.2V) that can remove poisonous carbon monoxide on the Pt sites [11]. Despite the PtRu catalyst exhibiting good methanol oxidation activity, the state-of –the –art PtRu catalyst needs further improvement for higher catalytic activity and replacement of expensive nobel metals, Pt and Ru. The most successful way to achieve both these goals is to add a third metal to the PtRu catalyst. Ternary catalysts PtRuFe, PtRuNi, PtRuCo, PtRuW and PtRuIr [12-13] were found to have high activities in methanol oxidation reaction. Demirci used theoretical calculations and agreed that incorporation of a third metal might be the best way to improve the methanol oxidation activity [14]. The mixing of Pt with the third metal leads to weakly adsorbed CO on Pt sites and OH strongly adsorbed only on the third metal sites. Such tri-metallic catalysts favour the formation of OH_{ads} species which assist in the oxidation of CO_{ads} [15].

In this paper we present a simple OMCVD method that can be used to synthesize highly dispersed tri-metallic electrocatalysts supported on carbon nanotubes. The procedure is a modified dry-mix method that produces particles with a very narrow size distribution, and results in stable nanoclusters with excellent electrochemical activities.

2. Experimental Methods

Commercial MCNTs were pre-treated and catalysts were prepared by a modified dry – mix method using organometallic chemical vapour deposition as reported previously [16].

Samples were characterized using XRD, ICP and transmission electron microscopy (TEM). The electrochemical (EC) activity and active surface area of the catalysts were evaluated by cyclic voltammetry (CV), which was using 0.5 M H_2SO_4 as the initial electrolyte and then modified by the addition of 0.5 M methanol. The measured potential range was at a sweep rate of 20 mV/s. All CV cycles were done in a de-aerated solution (bubbled nitrogen gas through the solution for 30 minutes), and under a blanket of nitrogen. Electrocatalyst inks were prepared by mixing 10 mg catalyst powder in 10.3 ml ultrapure water with 3.24 ml isopropanol and addition 54 µl of 5% Nafion® solution. The ink was allowed to sonicat for 30 minutes. 10µl of the ink was deposited on a 5 mm glassy carbon electrode and dried at 50 °C [17]. This results in a catalyst loading of 15 µg/cm^2.

3. Physico-Chemical Characteristics

3.1 Physico-Chemical Characteristics

Table 1 presents a summary of the various prepared catalysts, and a comparison of the theoretical loading versus the values obtained using ICP. A good agreement between the theoretical and measured values is observed. These results demonstrate that the CVD method can be used to prepare the electrocatalysts with a well-defined atomic ratio. XRD results for catalysts are presented in Figure 1 and they are compared with commercial PtRu/C catalyst.

Table 1. Summary of metal loading and electrochemical activities for catalysts

Catalyst	Theoretical metal content (wt. %)	Metal content from ICP (wt. %)	Metal loading (mg/cm^2)	Qr (µC)	Se (m^2/g)	Se (m^2/g)	Particle size by TEM (nm)	Direct potential (V)	I_B/I_A *100%
JM Pt/C	-	-	0.0042	169	190	190	2.01	0.78	75.0
Pt:Ru:Fe/CNT	20:10:10	20:9:10	0.0033	139	200	200	2.31	0.72	74.6
Pt:Ru:Cu/CNT	20:10:10	19:10:10	0.0033	154	222	222	1.75	0.74	71.4
Pt:Ru:V/CNT	20:10:10	19:9:10	0.0032	160	238	238	1.66	0.71	70.6

For the samples of tri-metallic PtRuFe/CNT, PtRuCu/CNT and PtRuV/CNT catalysts, the third metal or metal oxide diffraction peaks are not observed, indicating that alloying was complete [18]. Nevertheless, all peaks were shifted to higher 2θ values than that of the commercial Pt/C catalyst which is ascribed to the incorporation of the third metal. This is because the third metal atom is smaller than Pt, thus the interaction of the metal would cause shrinkage of the lattice parameter, which results in the peak shift to the higher 2θ values [19]. In addition to the alloy peaks, there are separate Ru peaks on the in-house samples at 2θ values of 43.5° and 78.5°, which are ascribed to Ru (101) and Ru (103) of hcp. The broadening of the peaks on the in-house samples is due to the smaller particle size of the metal deposits [20]. The calculated mean sizes according to the diffraction peak of Pt (111) were calculated for PtRuFe, PtRuCu and PtRuV as 1.88 nm, 2.12 nm, 1.42 nm and 1.30 nm, respectively. It should be noted that as the crystallite size is related to the area of coherent diffraction, the particle size obtained from XRD can be smaller than the true particle size in general [21].

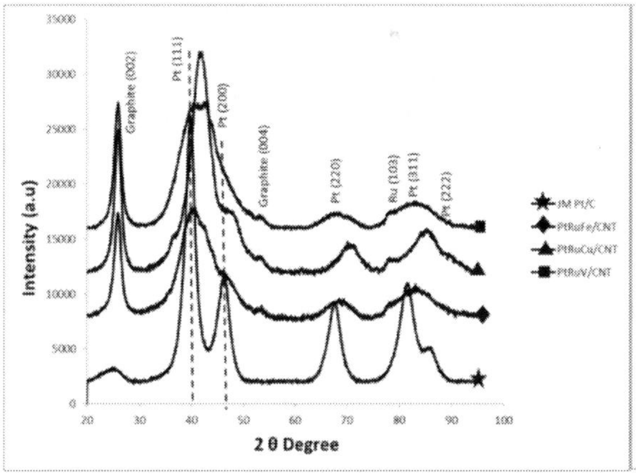

Figure 1. XRD patterns for electrocatalysts.

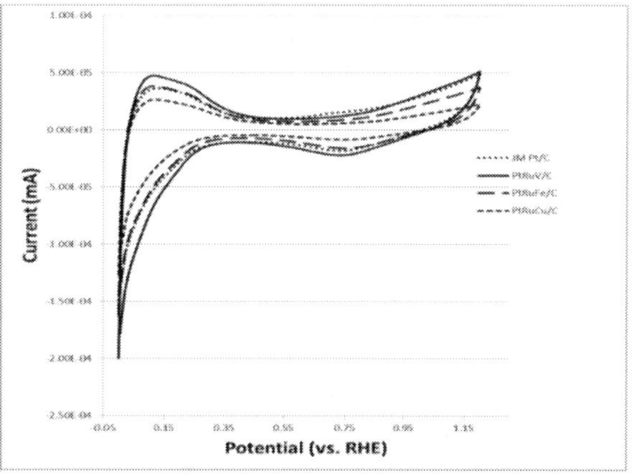

Figure 3. Cyclic voltammograms of PGM/CNT and JM Pt/C electrocatalysts.

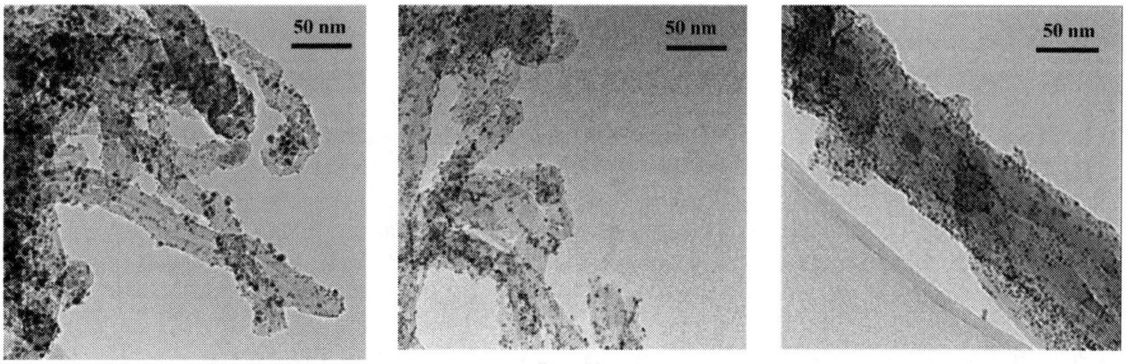

Figure 2. HRTEM graphs of (a) PtRuCu/CNT, (b) PtRuFe/CNT and (c) PtRuV/CNT.

The TEM images are given in Figure 2 (a)-(c) of catalysts. It confirms that original CNTs have an average outer diameter of 20 – 30 nm and a wall thickness of ~ 10 nm decorated with small black metallic particles and all catalysts were dispersed uniformly upon CNT supports and no agglomeration was observed. The catalysts showed well-separated nanoparticles with narrow-size distribution. The average particle sizes are 2.31, 1.66 and 1.58 nm for PtRuCu/CNT, PtRuFe/CNT and PtRuV/CNT, respectively. These results are in agreement with particle sizes obtained from application of the Scherrer equation based upon the X-ray diffraction pattern of the catalysts.

3.2 Electrochemical activity of catalysts

The EC active surface area graphs for the JM Pt/C and home-made tri-metallic Pt catalysts are presented in Figure 3. The electrochemical-active surface areas were calculated from the charge of the hydrogen desorption region (Q_r). There is no change in the voltammograms during this initial cycling procedure. The base voltammetry of the catalysts are essentially identical. All catalysts showed the typical Pt-Hydride desorption peak in potential range from -0.21 to 0.10 V and the Pt-Oxide reduction peak in potential range from 0.30 to 0.80 V. It is also can be noted that the onset of oxide formation shifts to more positive potentials with the incorporation of other metals. From these results it can be concluded that all catalysts were active for hydrogen and oxygen reactions.

Table 1 summarize the respective charges (Q_r) and electrochemical active surface areas determined from the hydride desorption peaks of the catalysts. It shows that PtRuV/CNT presented the highest activity and surface area compared to the other catalysts. The high activity may be attributed to the smallest particle size and highest surface area of the catalyst. Catalyst PtRuCu/CNT displayed the smallest active surface area when compared with the other home-made catalysts as it had the largest particle size of the prepared catalysts. It can be seen from Table 1 that the smaller particle size leads to the bigger active surface area of catalysts even when the catalysts contained a similar metal loading. It also can be concluded that for the home-made catalysts, the additional metals incorporated with Pt can reduce the particle size and increase the active surface area of the catalysts. This conclusion also accorded with the TEM results. The interaction of the second or third metal can cause an enhanced positive electrochemical activity of the catalysts.

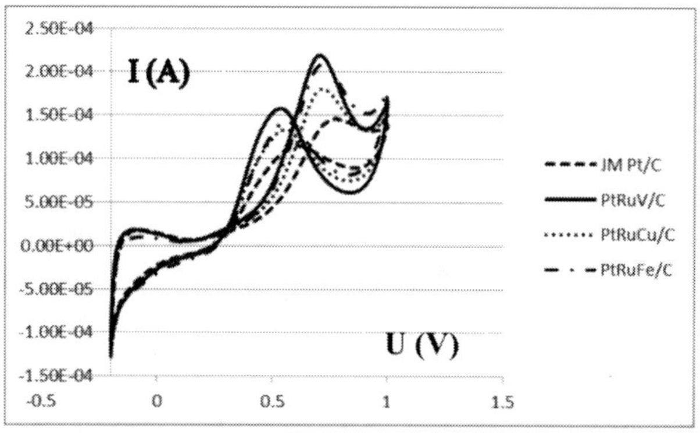

Figure 4. Cyclic voltammograms of JM PtRu/C catalyst and home-made PGM/CNT electrocatalysts in 0.5 M H2SO4 + 0.5 M MeOH.

The methanol oxidation activities of the prepared PGM/CNT catalysts were compared to the JM Pt/C catalyst and shown in Figure 4. The highest direct current of the PtRuV/CNT catalyst was observed in Table 1. The order of the current amplitude corresponded to that of the chemical active surface area of the catalysts. It can be seen that the direct oxidation peak (A) potentials for methanol oxidation of catalysts are slightly reduced with increasing current. The lower the peak potential, the better the electro-catalytic activity, thus these results indicate that the PtRuV/CNT is the most active catalyst of all the catalysts that were prepared in this study and all the home-made plurimetallic catalysts performed better than the commercial catalyst standard. Furthermore, it is well known that the direct scan peak current (I_A) displays the methanol oxidation and the reverse scan peak current (I_B) relates to the methanol re-oxidation. The higher ratio of I_B/I_A suggests the lower poisoning effect and the better tolerance of the electrode towards strongly adsorbed intermediates [22]. Table 1 also lists the oxidation potentials (V) for the direct conversion of methanol to CO_2, as well as the corresponding current intensities when the scan is switched towards cathodic potentials. The feature points to the highest poisoning tolerance of PtRuCu/CNT and the lowest tolerance of PtRuV/CNT home-made catalyst compared to the commercial catalyst. This fact could be associated with smaller particles presented on the carbon support surface, which would favour the surface poisoning by CO [23]. It indicates that the oxidation of methanol to CO_2 is performed in a more selective way on the catalyst with bigger particle size than that on the catalysts with smaller particles. This is because the second and third metal added to Pt enhanced CO removal from the surface, since the addition metals promote a weakening of the Pt-CO bond producing easily the CO oxidation reaction to CO_2 via a bi-functional mechanism.

4. Conclusions

A relatively simple solvent free technique has been successfully applied to produce tri-metallic PGM/CNT catalysts. Results showed that the OMCVD can produce highly active PGM catalysts via a very simple synthesis procedure and produced results consistent with uniform particle size, high dispersion and high electrochemical activity and the added additional metals of platinum alloys could reduce the particle sizes and resulted in a larger chemical-active surface area and higher methanol oxidation activity of the catalysts.

5. References

[1] Johnson Matthey Annual Report & Accounts 2000 - Fuel Cells.files , [Online], Available: http://www.matthey.com/AR00/06fuel-cells.html
[2] Holland B J, Zhu J G and Jamet L, [Online], Available: http://services.eng.uts.edu.au/~joe/subjects/eet/AUPEC01_111.pdf
[3] Catalysis 2008, [Online], Available: http://en.wikipedia.org/wiki/Catalyst
[4] Wang C Y, Du H Y, Tsai Y T, Chen C P, Huang C J and Chen L C 2007 *J. Power Sources* **171** 55
[5] Pereza H, Morin A, Akroura L, Cremona C, Baret B and Haccoun J 2009 *Electro. Acta* 2009
[6] Pannopard P, Khongpracha P, Probst M and Limtraku J 2009 *J. Molecular Graphics and Modelling* **28** 62.
[7] Sasaki K, Wang J X, Balasubramanian M, McBreen J, Uribe F, Adzic R R 2004 *Electro Acta* **49** 3873
[8] Sebastia´n D, Suelves I, La´zaro M J and Moliner r 2009 *J. Power Sources* **192** 51
[9] Shao Y, Yin G and Gao Y 2007 *J. Power Sources* **171** 558
[10] Jafria R I, Sujathaa N, Rajalakshmib N and Ramaprabhu S 2009 *Int. J. Hydrogen Energy* **34** 6371
[11] Ocampl A L, Miranda-Hernández M, Morgado J, Montoya J A and Sebastian P J 2006 *J. Power Sources* **160** 915
[12] Strasser P, Fan Q, Devenney M, Weinberg W H, Liu P and Norskov J K 2003 *J. Phys. Chem.* B **107** 11013.
[13] Lu G, Cooper J S and McGinn P J 2006 *J. Power Sources* **161** 106
[14] Jeon M K, Cooper J S and McGinn P J 2008 *J. Power Sources* **185** 913
[15] Marquez-Návarro Y, Galicia L, Lara V H and Del Ángel G 2008 *J. Mexico Chem. Soc.* **52** (1) 67
[16] Naidoo Q L, Naidoo S, Petrik L, Nechaev A and Ndungu P 2012 *Int. J. Hydrogen Energy* XXX I – II
[17] Jafria R I, Sujathaa N, Rajalakshmib N and Ramaprabhu S 2009 *Int. J. Hydrogen Energy* **34** 6371
[18] Brewer L 1967 *Phase Stability in Metals and Alloys* (McGraw-Hill, New York) p 39
[19] Jeon M K, Cooper J S and McGinn P J 2008 *J. Power Sources* **185** 913
[20] Ocampl A L, Miranda-Hernández M, Morgado J, Montoya J A and Sebastian P J 2006 *J. Power Sources* **160** 915
[21] Pozio A, De Francessco M, Cemmi A, Cardellini F and Giorgi L 2002 *J. Power. Sources* **105** 13
[22] He D M, Yang L X, Kuang S Y and Cai Q Y 2007 *Elecrochem. Commun* **9** 2467
[23] Santiago E I, Batista S M and Assaf E M 2004 *J. Electrochem. Soc.* **151** A944

Formation of nanosized strontium substituted hydroxyapatites

K A Gross[1,4], A Jeršova[2], D Grossin[3], C Rey[3], A Vīksna[2]

[1] Institute of Biomaterials & Biomechanics, Riga Technical University, Latvia
[2] Department of Analytical Chemistry, University of Latvia, Latvia
[3] Université de Toulouse, CIRIMAT, France

E-mail: kgross@rtu.lv

Abstract. Incorporation of specific elements into calcium phosphates offers the combination of a bioactive material and a therapeutic effect. This is important for improving the integration of implants as well as treating medical conditions. Strontium is a suitable candidate and displays the ability to stimulate bone growth and reducing bone resorption. This study investigated the formation of strontium carbonated hydroxyapatite nanoparticles from an amorphous phase. Crystallization of carbonated hydroxyapatite occurred at 585 °C, but samples with an intended 25% and 75% replacement of calcium with strontium crystallized at 624 °C. Heat treatment at the crystallization temperature revealed that strontium free apatite does not crystallize in 5 minutes, but an increasing strontium concentration leads to a higher rate of crystallization. X-ray diffraction patterns suggest that it may be difficult to include strontium, but higher strontium concentrations are possibly included with ease in the lattice. This work has produced a nanosized apatite accompanied by an amorphous phase after a short heat-treatment time. This offers a range of features that collectively show great promise for significantly enhancing the release of strontium for improved bone therapeutic effects.

1. Introduction

Hydroxyapatite (HAp) is a widely used biomaterial for bone regeneration and provides a surface for bone attachment to implants. The crystal lattice of hydroxyapatite [$Ca_{10}(PO_4)_6(OH)_2$] can accommodate many divalent cation substitutions such as strontium (Sr), magnesium (Mg), barium (Ba), zinc (Zn), etc. These influence the processability of hydroxyapatite and also impact the biological properties. The osteoconductive properties of Sr make it a very interesting candidate for inclusion in the lattice for the preparation of nanosized apatite. Research will be directed towards the production of a nanosized Sr HAp.

Strontium increases bone formation and reduces bone resorption, and therefore dose control is important. Strontium may be administered orally as strontium renelate or included in the structure of apatite where a complete substitution for calcium is possible [1]. In

[4] To whom any correspondence should be addressed

hydroxyapatite, it is thought to favour the osteogenic differentiation of mesenchymal stem cells [2]. Studies in rats have shown that Sr improves the microarchitecture of bone, but is not as effective as bisphosphonate at improving the bone bonding to polymethylmethacrylate implants [3]. Another study with hydroxyapatite coated implants in osteoporotic rats has shown that Sr does however improve bone bonding to the implant [4]. A combination of Sr and hydroxyapatite could thus provide a synergistic effect in promoting bone growth and providing a good surface for attachment. The release of Sr may need to be increased and this can be provided by the more soluble carbonated apatite [5].

Previous work has shown the use of Sr enriched hydroxyapatites as powder [6], bone cements [7] and coatings made by plasma spraying [8] or laser ablation [9] for stimulating cell response. This work will prepare nanosized Sr substituted carbonated hydroxyapatite that can further increase the solubility, and determine whether it is possible to also include an accompanying amorphous phase. This is the first attempt to change the chemistry, decrease the crystal size and use an amorphous phase of the same composition as the basis for increasing the solubility.

The objective of this work is to a) produce strontium apatites with a different degree of calcium replacement, b) determine the influence of Sr on the crystallization of an amorphous precursor, and c) determine conditions for producing nanosized Sr carbonated hydroxyapatites. As an explorative study, the focus will be placed on 0% Sr, 25% Sr and 75% Sr replacement of calcium in apatite.

2. Materials and Methods

Strontium substituted carbonated hydroxyapatites were crystallized from an amorphous calcium phosphate precursor. The amorphous calcium phosphate was prepared from reagent grade $Ca(NO_3)_2 4H_2O$ and $(NH_4)_2HPO_4$ and $(NH_4)_2CO_3$. Strontium nitrate was substituted for calcium nitrate to adjust the concentration for a 25% and a 75% replacement of calcium ions. Reactants were dissolved in deionized water at room temperature and the pH of the calcium containing solution was adjusted to 9.4. Both solutions were mixed for 10 minutes. And then washed with deionized water before filtration. Water was removed by freeze drying for 2 days.

The crystallization temperature was determined within a differential thermal analyzer (DTA), an EXSTAR TGA/DTA 6300. Heating was conducted in air from 30 °C to 900 °C at a heating rate of 10 °C min^{-1} with the samples in zirconia crucibles. A cylindrical furnace was then heated to 5 °C higher than the crystallization temperature and a powder placed on a platinum foil for 5 mins, 15 mins, 30 mins and 60 minutes.

The phase of the heat treated powder was determined in an X-ray diffractometer (D8 Advance, Bruker) operated with Cu K□$_\alpha$ radiation (λ=1.54056 Å) generated at 40 kV and 40 mA and the diffracted intensity detected with a scintillation counter. Each sample was scanned from 10 to 60° 2θ at a 0.02° step.

The chemical composition of Sr apatites was determined using wavelength dispersive X-ray spectrometry (S8 TIGER, Bruker) using a Rh lamp at 4 kW. Samples were analyzed in a powder form under a 5 μm X-ray polypropylene film, using an 8 mm collimator in a helium atmosphere. Results are expressed in oxide formula units.

3. Results and Discussion

Three compositions for investigation included a 0% Sr, 25% Sr and 75% Sr where the percentage indicates the degree of intended calcium replacement with strontium, as set by the reactant concentrations. This allows comparison of a strontium depleted, a low substitution and high substitution of strontium for the calcium.

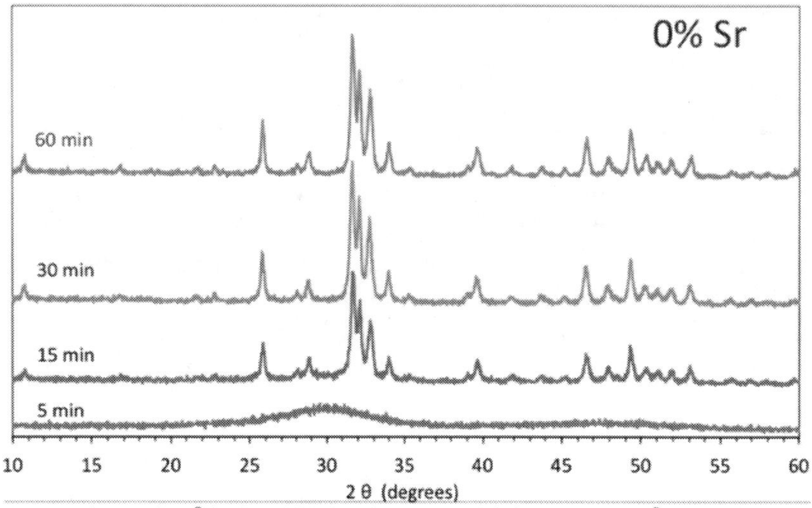

The exothermic peak on the DTA for 0% Sr extended over a range of 60 ^{o}C, indicating slow ordering of calcium and phosphate for forming a crystalline state. A similar trend was observed for both Sr concentration samples. Crystallization occurred at 585 ^{o}C for 0% Sr, but was higher at 624 ^{o}C for both 25% Sr and 75% Sr. The large size of the Sr ion must reduce the mobility necessary to form a denser assembly. Heat treatment was chosen at 5 ^{o}C higher than the crystallization peak. A comparison of all three compositions shows that 0% Sr does not crystallize after 5 minutes, but 25% Sr and 75% Sr reveal a combination of the starting amorphous phase and an apatite, Fig 1. The amount of apatite is higher for 75% Sr. No other calcium phosphate phases are present. Sr delays the onset of crystallization with a

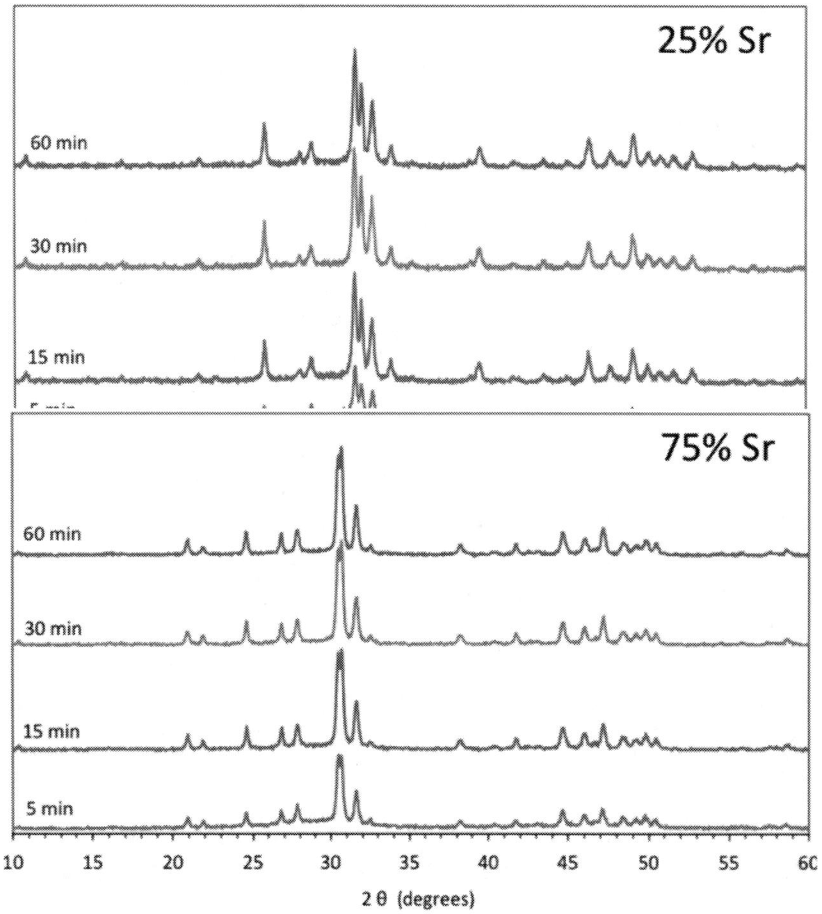

contribution from carbonate, but favours faster crystal growth.

Figure 1. X-ray diffraction patterns of 0% Sr, 25% Sr and 75% Sr heated for different periods. A dashed line is included at $2\theta = 31.7^{\circ}$ to show the shift to lower angles for 75% Sr.

The characteristics of the crystallized carbonated hydroxyapatite is set by the heat treatment at 15 minutes. No substantial change is noted at longer heating times, Fig 1. The diffraction peaks for the 25% Sr seems very similar to 0% Sr, but a clear shift to lower angles is apparent for 75% Sr. An absence of a concurrent display of peaks from high and low strontium concentration apatites suggests a uniform solid solution for all compositions. The X-ray diffraction patterns validate the X-ray fluorescence results that show an absence of Sr for 0% Sr, a small concentration for 25% Sr and a large concentration for 75% Sr, Fig 2. It appears that less than the added strontium may have entered the lattice for the 25% Sr sample and more than 75% Sr has entered the lattice for the higher Sr containing carbonated hydroxyapatites. Further work is needed to determine the extent of incorporation in the lattice. This is more readily ascertained from a peak shift in samples heated treated at a higher temperature, to produce narrower X-ray diffraction peaks.

Further work needs to produce a more complete range of Sr containing samples. This could include a 50% Sr and a 100% Sr composition. These samples will facilitate a better understanding whether strontium is washed away during the rinsing stage and whether it is possible to produce a pure Sr carbonated apatite using the new approach.

Further characterization could employ Fourier transform infra-red spectroscopy to analyze for the carbonate and hydroxyl groups. DTA/TGA could reveal the thermal stability and reveal the suitability of the powder for sintering or other thermal processing routes.

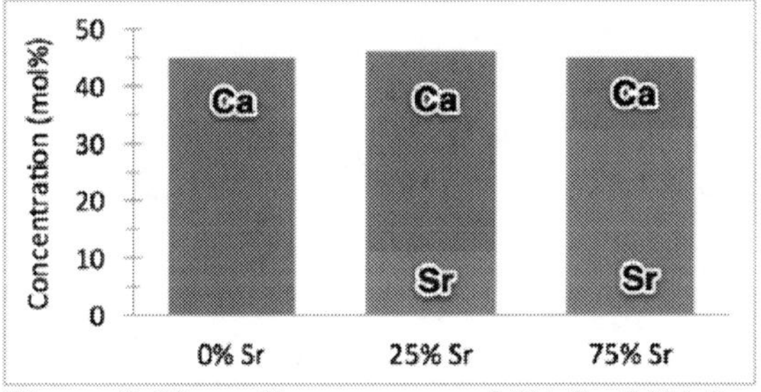

Figure 2. X-ray fluorescence of 0% Sr, 25% Sr and 75% Sr.

The inclusion of carbonate in the apatite lattice, the production of a small crystal size and the presence of an amorphous phase, as seen by heating for 5 minutes provides many factors to extend the apatite solubility and hence the strontium delivery to the surrounding tissue. Further work is necessary to measure the surface area and solubility of the nanopowders and the nanosized apatite together with the amorphous phase.

Conclusions

Crystallization of an amorphous phase has successfully led to the introduction of Sr inside the apatite lattice. While carbonated hydroxyapatite crystallizes at 585 °C, strontium containing compositions require further heating to 624 °C. Nanosized crystals can be obtained by heating between 5 and 10 minutes just above the crystallization temperature.

Acknowledgement. The authors acknowledge the Taiwan-Latvia-Lithuania project "Nano-scaled functional materials for biotechnological and optical applications".

References

[1] Schoenberg H P 1963 *Biochem. Biophys. Acta* **75** 96
[2] Yang F 2011 *Stem Cells* **29** 981

[3] Linderback P 2012 *Bone* **50** 350
[4] Li Y F 2010. *J Orthop Res* **28** 578
[5] Landi E 2008 *Acta Biomater* **4** 656
[6] Xue W C 2006 *J. Biomed. Mater. Res.* **79A** 804
[7] Ni G X 2006 *Biomaterials* **27** 4348
[8] Xue W C 2007 *Surf Coatings Technol.* **201** 4685
[9] Capuccini C 2008 *Acta Biomater.* **4** 1885

Role of paramagnetic polyconjugated clusters in lignin antioxidant activity (*in vitro*)

T Dizhbite[1], J Ponomarenko[1], A Andersone[1,4], G Dobele[1], M Lauberts[1], J Krasilnikova[1,2], N Mironova-Ulmane[3] and G Telysheva[1]

[1]Latvian State Institute of Wood Chemistry, Dzerbenes 27, Riga LV-1006, Latvia
[2]Riga Stradins University, Latvia
[3]Institute of Solid State Physic, University of Latvia, Riga, Latvia

E-mail: ligno@edi.lv

Abstract. Using physico-chemical methods (EPR, SEC, Py-GC/MS and UV/VIS spectroscopy) and wet chemical analysis, the characteristics of 6 hardwood lignins in terms of functionality, molecular weight and composition of lignin substructures were determined and considered together with the results of DPPH$^•$, ABTS$^{•+}$ and $O_2^{•-}$ antioxidant assays with the aim to understand the relationships governing antioxidant properties of lignin. The strong positive linear correlation between lignin antioxidant capacity in the three assays used and the extent of conjugation of paramagnetic polyconjugated clusters in lignin macromolecules was found. The biological activity of the most active alkaline lignins was assessed by *in vitro* experiment with human blood.

1. Introduction

The application of natural antioxidants instead of the synthetic ones is a subject widely investigated with emphasis on reduction of the ecological problems. Phenylpropanoid polymer lignin *in situ* (Figure 1) serves to protect plants against chemical, biological and mechanical stresses. Antioxidant activity of technical lignins obtained from plant biomass processing in multi-tonnage scale are well documented [2-4]. Polymeric antioxidants due to their higher compatibility with polymeric systems and slower degradation rate often can be successfully applied in those fields in which the employment of a single molecule with antioxidant activity is inefficient. For example, lignins, comparing to

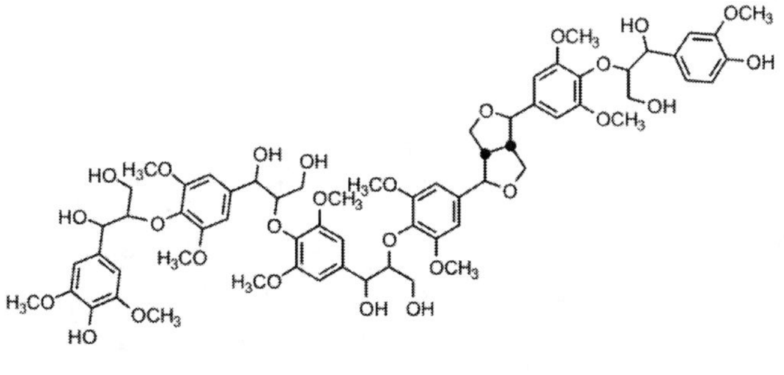

Figure 1. Potential structure of a repeated unit in macromolecule of hardwood lignin *in situ* [1]

[4] To whom any correspondence should be addressed

flavanoids, have longer circulation time in human and animal organisms and are much less toxic.

However, due to molecular complexity of lignins, it is difficult to assign their antioxidant activity to specific structural elements and further search of the relationship "structure-activity" is needed to determine directions for goal-oriented tuning of lignin structure, in order to use these naturally originated products successfully. The chemical structure of lignins (functionality, molar mass, cross linking density, etc.) depends on a number of factors including the botanical origin, the environmental conditions of growth and also the conditions of isolation from plant tissue due to all the isolation techniques significantly influence lignin intermonomeric linkages. Technical lignins, in particular those obtained from processing of lignocelluloses under conditions promoting double bonds formation (e.g. alkaline delignification, acid hydrolysis, thermal and hydrothermal treatment), are characterized with formation in their macromolecules of sub-structures with spatially extended π-bonding systems, so called polyconjugated systems (PCS). The polymers with high-developed PCS display high antioxidant, radioprotective and photoprotective properties [5]. The specific feature of PCS is the presence of paramagnetic centres of high-stability responsible for scavenging (deactivation) of labile free radicals [3,6].

The aim of the present work was to assess antioxidant and biological activity *in vitro* for a set of hardwood lignins and to investigate the relevance of lignin structural features, in particular development of polyconjugated clusters, molecular weight, functionality (phenoxy- and methoxy groups) and lignin antioxidant activity.

2. Experimental

Lignins isolated from hardwood species (alder, aspen and ash-tree) by alkaline delignification (Alk-L) and fast pyrolysis (Py-L) were the objects of the study. EPR spectroscopy and analytical pyrolysis (Py-GC/MS) were used as the main analytical tools for obtaining direct information on the extent of conjugation ("delocalization length") for lignin PCS and macromolecule condensity, respectively.

Ash tree (*Fraxinus excelsior*), alder (*Alnus incana*), aspen (*Populus tremula*) wood sawdust (fraction 0.1-1.0 mm, moisture 8.5-9%) were used as raw-materials for lignins obtaining by the alkaline delignification and fast pyrolysis treatment [7].

The polyconjugated paramagnetic clusters in lignin samples were characterized in terms of extent of polyconjugation estimated using the value of the conjugation length of π-polyconjugation systems calculated from the EPR spectra as the number of CH fragments (n) included in the region of unpaired electron delocalization [3]. The EPR spectra were recorded at 293K using an EMX-6/1 spectrometer (BRUKER) working at X-band frequency with 100 kHz modulation.

The condensity of lignin macromolecules as well as the content of carbohydrates admixtures were characterized by Py-GC/MS with Shimadzu GC/MS-QP 2010, column RTX-1701, 60 m x 0.25 mm x 0.25 μm. The key functionalities of lignin macromolecule, in particular phenolic hydroxyl and methoxy-groups were determined, respectively, by conductometric titration and the Viebock-Schwappach method [8]. All analytical data are shown on oven dry samples. The molecular weight of lignins was determined by the size exclusion chromatography (SEC) method (Agilent 1100 system, UV-DAD/RI, 60°C, column AGI_PLgel 5μm MIXED-D 300x7.5 mm, eluent DMSO).

Antioxidant activity was evaluated using three radical scavenging assays: deactivation of $ABTS^{\bullet+}$ cation-radicals and $DPPH^{\bullet}$ radicals, inhibition of generation of superoxide anion-radicals $O_2^{\bullet-}$ in the biosystem hypoxanthine-xanthine oxidase. Radical scavenging activity was expressed by IC_{50}, i.e. the concentration of lignin required for 50% inhibition of the radical species. The values of IC_{50} were calculated from the dependence of the free radical inhibition on lignin concentration determined by UV/VIS spectroscopy [3,9]. Trolox (water soluble analogue of E vitamin) was used as the reference antioxidant.

Biological activity of lignins was tested in the experiment *in vitro* with blood samples taken from voluntaries. The blood markers relevant to oxidative stress, namely the indices of blood lipids peroxidation (MDA) and proteins oxidation (PC) as well as glucose level, were analyzed using standardized biochemical methods.

All the analyses were carried out five times and the data were compared by linear regression analysis using the SPSS Statistics 17,0. Differences were considered significant for $p < 0.05$. The graphs were plotted using mean values.

3. Results and Discussion
All investigated lignin samples revealed high dose-dependant radical scavenging activity in three tests applied (Figure 2). Pyrolytic lignins had the lowest IC_{50} values, i.e. the maximum antioxidant activity; their antioxidant activity was close and even higher than that for Trolox.

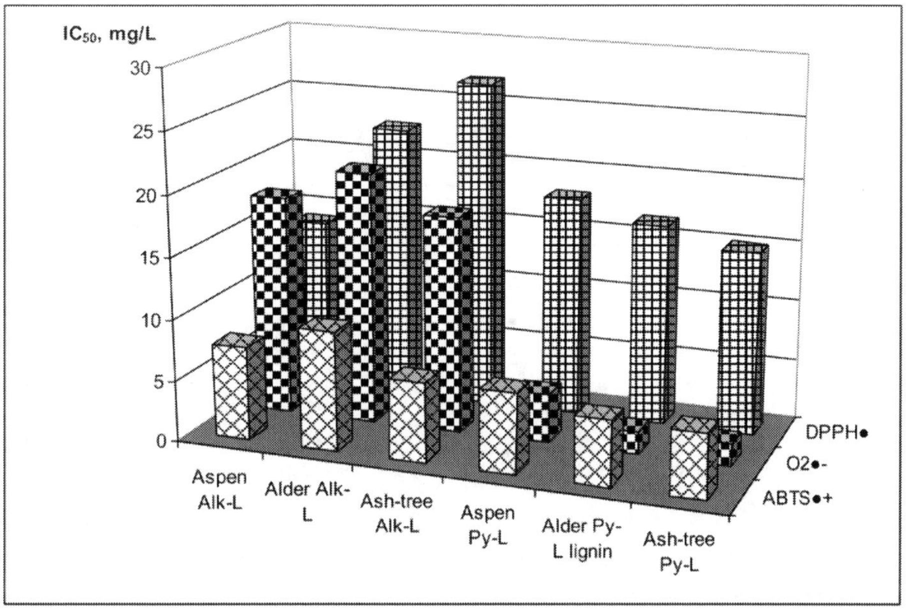

Figure 2. Half inhibition concentrations (IC_{50}) of lignins. For Trolox, IC_{50} values of 4.2, 4.8 and 17.7 mg/L in $ABTS^{•+}$, $DPPH^{•}$ and $O_2^{•-}$ assays, respectively, were found.

For all lignins investigated, the results of EPR spectroscopy show an intensive narrow single line spectrum with a g-factor of about 2.0030 and a line width of 0.3 – 0.5 mT that is characteristic for an unpaired electron localized in a conjugated polymer matrix. The comparison of the values of extent of conjugation calculated from the EPR spectra for alkaline and pyrolytic lignins (Table 1) gave evidence of the intensive development of polyconjugated clusters in lignin macromolecules as the result of fast pyrolysis. Formation of phenylpropanoid aroxyl radicals is an essential step in the realization of antioxidant activity of lignin. The stability of the aroxyl radicals strongly depends on unpaired electron delocalization. The development of polyconjugated clusters in lignin macromolecules leads to the extension of electron delocalization thus decreasing dangerous prooxidant potential of polyphenols.

The data obtained using Py-GC/MS and chemical analysis showed distinctions between the lignins in terms of condensity, carbohydrates admixtures content, OH_{phen} and OCH_3 groups contents (Table 1).

For searching of correlation between antioxidant activity of lignin and its structure parameters, the term of radical scavenging index (RSI), which was defined by as the inverse of IC_{50}, was used as a dependent variable. It was found (Table 2) that a growth of extent of conjugation of paramagnetic polyconjugated clusters in lignin resulted in linear increasing antioxidant capacity (the Pearson's correlation coefficient of 0.9932). OH_{phen} and OCH_3 groups contents are lower influential features. The heterogeneity of technical lignin had negative impact on the lignin radical scavenging capacity.

The results of the tests *in vitro* clearly demonstrated dose-dependent reducing action of the ash tree and alder lignins on the indices of blood lipids peroxidation (MDA) and proteins oxidation (PC). The decreasing action of the lignin on glucose level in blood (by 32%) was also found in tests with blood taken from voluntaries – patients with metabolic syndrom (initial glucose level of 7.4 ± 0.2 mkM/L). These results indicate the multifunctional activity of the investigated lignins.

Table 1. The characteristics of hardwood lignins structure and chemical composition

Sample	Index				
	Poly-conjugation extent, **n**	$-OH_{phen}$ content, %	$-OCH_3$ content, %	Condensity	Carbohydrates content, %
Aspen Alk-L	16	5.0	18.5	2.9	23.0
Alder Alk-L	14	4.7	17.1	4.0	28.5
Ash-tree Alk-L	17	5.6	19.5	3.2	10.0
Aspen Py-L	26	6.1	9.8	13.3	20.2
Alder Py-L	30	7.1	10.9	11.0	23.2
Ash-tree Py-L	30	7,3	12.2	11.7	16.9

Table 2. Correlation between structure parameters and radical scavenging indices of hardwood lignins

Independent variables	Pearson's linear correlation coefficient, r		
	Dependent variables		
	RSI in $DPPH^{\bullet}$ test	RSI in $ABTS^{\bullet+}$ test	RSI in $O_2^{\bullet-}$ test
$-OH_{phen}$ content	$+0.77^a$	$+0.92^b$	$+0,93^b$
Molecular weight, M_w	ns	ns	ns
$-OCH_3$ content	ns	ns	ns
Condensity	$+0.78^a$	ns	$+0.95^b$
Carbohydrates content	-0.97^b	ns	ns
Extent of conjugation	$+0,88^b$	$+0.88^b$	$+0.99^b$

$^a P < 0.05$; $^b P < 0.01$; «ns» indicates non-significant correlation ($P > 0.05$)

4. Conclusions

Lignins isolated from hardwood species by alkaline delignification and pyrolysis treatment demonstrated high antioxidant activity close to that of widely used reference material Trolox, which implies their application foreground as polymeric antioxidants of natural origin, capable of competing with low-molecular commercialized antioxidants. The extent of conjugation of paramagnetic clusters in lignin was found to be the major parameter predetermining antioxidant activity. Phenolic and methoxyl functional groups contents influence is of lesser importance. These regularities will bring the opportunity to determine directions for targeted modification of different types of lignins for the production of polymeric antioxidants utilizable in various technical and non-technical systems.

Acknowledgments

The financial supports from the Latvian budget, Research grant 1547 and Latvian National Program VPP-5,2,2.4 are gratefully acknowledged.

References

[1] Vanholme R, Demedts B, Morreel K, Ralph J and Boerjan W 2010 *Plant Physiol.* **153** 895
[2] Pouteau C, Dole P, Cathala B, Averous L and Boquillon N 2003 *Polym Degrad. Stab.* **81** 9
[3] Dizhbite T, Telysheva G, Jurkjane V and Viesturs U 2004 *Bioresour. Technol.* **95** 309
[4] Ugartondo V, Mitjans M and Vinardell M P 2008 *Bioresour. Technol.* **99** 6683
[5] Pini E, Bertacche V, Molinari F, Romano D and Gandolfi R 2008 *Tetrahedron* **64** 8638
[6] Byshneva L N and Senchuk V V 2001 *Appl. Biochem. Microbiol.* **37** 93
[7] Scholze B and Meier D 2001 *J. Anal. App. Pyr.* **60** 41
[8] Zaķis G F 1994 *Functional Analysis of Lignins and their derivatives* (Atlanta: Tappi Press)
[9] Baltrusaityte V, Venskutonis P and Cheksteryte V 2007 *Food Chem.* **101** 502

IOP Conf. Series: Materials Science and Engineering **38** (2012) 012034 doi:10.1088/1757-899X/38/1/012034

Nonlinear optical properties of low molecular organic glasses formed by triphenyl modified chromophores.

A Tokmakovs[1], **M Rutkis**[1,3], **K Traskovskis**[2], **E Zarins**[2], **L Laipniece**[2], **V Kokars**[2], **V Kampars**[2]

[1]Institute of Solid State Physics, University of Latvia, Latvia

[2]Institute of Applied Chemistry, Riga Technical University, Latvia

E-mail: martins.rutkis@cfi.lu.lv

Abstract. The series of organic molecular glasses have been studied as possible candidates for nonlinear optical (NLO) applications. Amorphous phase formation of investigated materials is ensured by the presence of bulky triphenyl substituents in molecular structure of NLO chromophores. Linear optical properties as well as NLO coefficients and thermal stability of NLO activity for the 13 molecular materials in glassy thin solid films have been determined. For the benzylidene-1,3-indandione chromophore containing compound the highest d_{33} value equal to 280 pm/V was measured under the 1064 nm excitation. Among the investigated compounds uppermost achieved thermal sustainability of NLO response was 108 °C. The relationship between number of triphenyl substituents and increased thermal sustainability of nonlinear response was observed.

1. Introduction

Organic nonlinear optical (NLO) materials are a subject of interest in fields as data processing and transmission due to wide modification possibilities and increased performance characteristics compared to inorganic materials [1]. The progress in this field is very much dependent on the development and characterization of new functional materials exhibiting high NLO efficiency and good long term stability. Organic glasses, class of materials where amorphous phase formation is achieved in low molecular weight molecules [2-4], from practical point of view are good candidates for such applications. These compounds are easily obtained, purified and characterized. However, some of important for practical use properties like thermal stability of NLO efficiency are worse than that in the polymer materials. Enhancement of thermal stability is a key issue for the employment of such molecular glasses in practical applications. In the given study we present the values of the key characteristics, NLO coefficients and thermal stability, for a series of low molecular weight glasses where amorphous phase formation is achieved by introduction of triphenyl substituents [5]. The analyzed structures contain wide variety of chromophores, as well as shape and design of amorphous phase formation enhancers are different. Screening of these results could reveal the connections between the structural properties and NLO characteristics useful for further improvements of materials.

[3] To whom any correspondence should be addressed

2. Experimental

2.1. Materials and thin film sample preparation

The synthesis and characterization of the investigated compounds (figure 1) are given elsewhere: for **A1**, **A2**, **B2**, **B3**, **B5**, **B7** see [5], for **A3**, **B1**, **B4**, **B6** [6], **C1** [7], **C2** [8]. Azochromophore **B8** was synthesized from 4'-[N-(2-hydroxyethyl)-N-methyl]amino-2-(2-hydroxyethoxy)-4-nitroazobenzene and 3,5-bis(2-(trityloxy)ethoxy)benzoic acid in N,N'-dicyclohexylcarbodiimide and 4-(dimethylamino)pyridine mediated esterification [8]. To obtain thin film samples an appropriate amount of glass forming compound was dissolved in analytical grade chloroform or chlorbenzene at a typical concentration of 100 mg/ml. The films from solutions were spin-coated with a Laurell WS-400B-6NPP/LITE spin-coater on ITO covered glass slides.

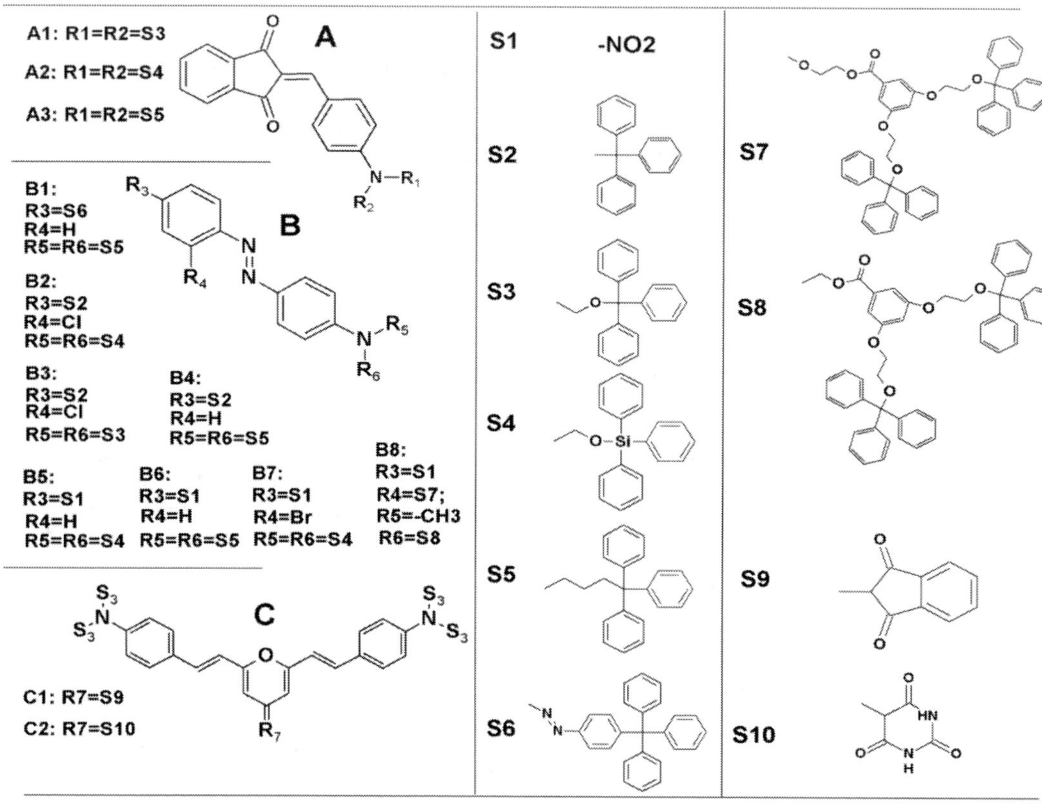

Figure 1. Structures of investigated glass forming compounds.

2.2. Corona poling procedure

To produce NLO active media thermo assisted electrical field poling procedure was applied to thin film samples via a custom built corona triode setup. The corona discharge was generated by a 9 kV voltage drop over a 15 mm gap between a tungsten wire needle (diameter 25μm) and a control grid. At a distance of 10 mm below the grid the ITO covered slide with spin coated glassy film was placed on a temperature controlled heater. All samples were poled through a mask with diameter of 0.8 cm by constant 2.7 kV grid to ITO layer potential. The corona discharge started in advance of heating the sample to poling temperature. At the point when the sample charging current reached an approximate steady state the poling was considered to be finished and the sample was cooled to room temperature under applied corona triode discharge. First of all samples where poled close to glass transition temperature ($\mathbf{T_{poling} \approx T_g}$). For these samples the temperature corresponding to the half vanishing of NLO activity ($\mathbf{T_{SHI50}}$) was evaluated from the SHI with temperature scans at 10 °C/min (see examples at figure 2). Finally fresh samples for further NLO coefficient measurements were poled at $\mathbf{T_{SHI50}}$.

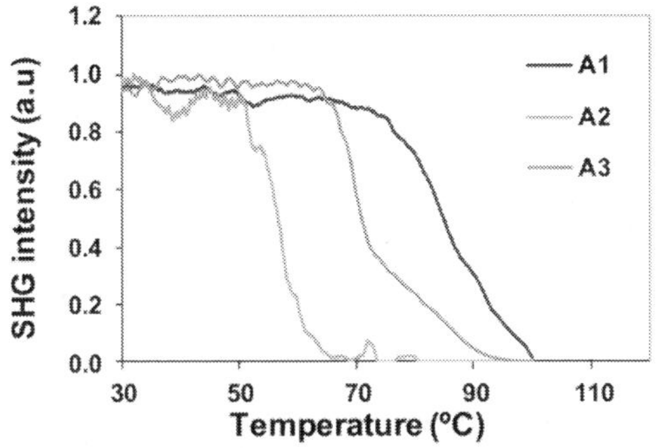

Figure 2. SH intensity with temperature scans for **A1**, **A2** and **A3** compounds.

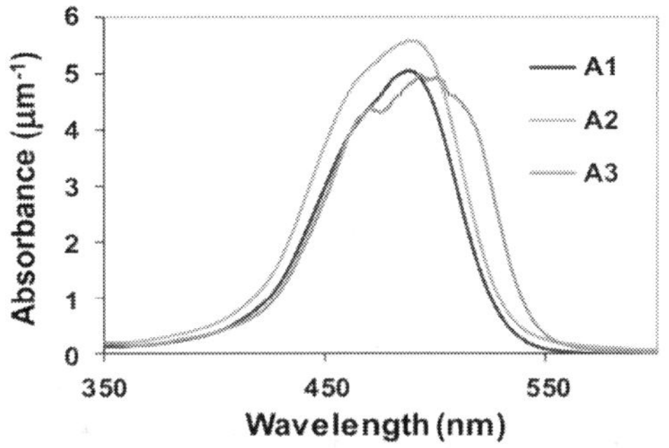

Figure 3. Absorption spectra of the thin films of **A1**, **A2** and **A3** compounds.

2.3. Measurements of linear and nonlinear optical properties of thin films

The absorption and reflectance spectra of the thin films were obtained with an *Ocean Optics HR4000CG-UV-NIR* based spectroscopic system. The thickness of investigated thin films was measured using a *Dektac 150* profilometer and refractive indexes were determined by a prism coupler *Metricon 2010*. In some cases thickness and refractive indices were evaluated by procedure described elsewhere [10] using interference fringe separation in the sample reflection spectrum. The experimental set-up for second harmonic generation is described elsewhere [11]. To avoid electric field-induced second harmonic generation (**EFISHG**) signal from charges trapped on the film surface the non-linear coefficients were usually measured 2 days after poling by Maker fringe technique. For corona poled films the $C_{\infty v}$ symmetry was assumed and the material could be characterized by three nonzero **NLO** coefficients – d_{33}, d_{31} and d_{15}. As it is usually done for the poled polymer films, we assume that $d_{31} = d_{15}$ [11] according to Kleinman symmetry. The **NLO** coefficients were acquired by a least squares fit of the experimental curves to the theoretical approximation. The theoretical value of second harmonic intensity (**SHI**) was calculated using the Herman – Hayden [12] approach, taking into account absorption of the film. The fitting was carried out in two steps: the value of d_{31} was evaluated from experimental ***s-p*** polarized **SHI**, then the d_{33} was calculated from the ***p-p*** **SHI**. An x–cut quartz crystal was used as reference (d_{11} = 0.3 pm/V) to calibrate the instrument response function [13]. The results of the measurements are given in table 1 where one could find also CT maxima wavelength (λ_{max}) and absorption coefficient (α_{max}) as well as refractive indexes at 532 and 1064 nm. Static field **NLO** coefficients $d_{33}(0)$ was obtained by extrapolation of experimental values to the zero frequency according to two level model.

Table 1. Linear and nonlinear optical properties of thin films.

Structure	λ_{max} (nm)	$\alpha_{max}\times10^{-4}$ cm^{-1}	RI (532nm)	RI (1064nm)	T_{SHI50} (°C)	Orientation T(°C)	$d_{33}(532)$ (pm/V)	$d_{33}(0)$ (pm/V)
A1	489	5.56	1.90	1.60	84	85	22.7	2.8
A2	490	5.03	1.94	1.59	60	56	41.9	6.2
A3	504	4.69	2.24	1.61	71	73	280.0	22.3
B1	507	3.08	2.55	1.63	83	82	24.2	1.7
B2	440	2.09	1.79	1.62	58	56	5.8	1.5
B3	440	2.71	1.84	1.64	84	84	7.6	2.0
B4	435	3.58	1.71	1.59	73	74	8.6	2.4
B5	488	3.41	1.93	1.61	52	50	62.9	7.9
B6	500	5.40	2.15	1.61	72	72	88.3	8.0
B7	510	4.85	2.34	1.69	53	48	214.0	13.3
B8	495	1.93	2.04	1.60	90	92	37.6	5.7
C1	533	4.93	2.20	1.60	108	110	106.0	0.3
C2	512	4.18	2.30	1.60	102	105	28.2	1.8

3. Discussion

As it can be seen from table 1, $\mathbf{T_{SHI50}}$ values of the compounds vary from 52 to 108 °C. In figure 2 we have presented examples of SHI with temperature scans for three molecular glass systems based on same chromophore. The lowest value is met when a triphenylsilyl fragment (**S4**) is present in an **A2** molecule. As our previous research revealed [5] it can be attributed to the fact that the given functional fragment has increased conformational freedom compared to other modifying groups and thus lowers amorphous phase thermal stability. More conformational rigid glassy phase forming bulk substituent **S3** raises thermal stability by almost 25 °C. As one can see from figure 2, rise of system flexibility caused by replacement of oxygen by carbon in chain linking chromophore with triphenylmethyl group reduces $\mathbf{T_{SHI50}}$ by 13 °C. The increased flexibility allows different chromophore packing of the **A3** in comparison with **A1** and **A2** and therefore red shift of the absorbance band takes place (figure 3). The highest thermal stability of NLO activity among the all investigated systems is observed in the cases of compounds **B8**, **C1** and **C2**. From structural point of view this can be explained by increased number of triphenylmethyl groups present given molecules. Obtained NLO efficiency values d_{33} for investigated molecular glasses vary significantly (see table 1). In most of the cases it is due to different NLO properties of chromophores used in particular organic glass system. For some of the chromophores (**C1**) charge transfer (CT) absorbance band coincide with second harmonic wavelength, for some others (**B1**, **B7**, **C2**) CT wavelength values are close to it. In all these cases resonance enhancement of NLO efficiency takes place. At same time certain low d_{33} values (**B2**, **B3**, **B4**) can be explained by small molecular hyperpolarizabilities and off resonance conditions for employed chromophores. However, if one compares the results for benzylidene-1,3-indandione fragment containing compounds **A1**, **A2** and **A3** the acquired values still differ remarkably. The difference by an order of magnitude for d_{33} is observed among compounds **A1** and **A3,** although in the both cases the same chromophore is used. One of explanations could be the different level of acentric order achieved in the corona poled samples. In the case of compound **A3** chromophore core and triphenyl moiety are connected by flexible C-C bridge what can attribute to easier reorientation in the poling process.

4. Summary

The nonlinear optical properties of the series of organic molecular glasses have been studied. Correlation between chemical structure of investigated compounds and obtained NLO properties was observed. It is apparent that increased molecular weight and number of triphenyl substituents in the molecule increases the thermal sustainability of NLO response. Increased flexibility of chromophore to triphenyl substituent linking chain reduces thermal stability. On the other hand such flexibility increases chromophore core mobility during corona poling what results in the improved NLO efficiency. Therefore careful optimization of structure should be done to avoid trade-off between thermal stability and NLO efficiency.

References

[1] Dalton L R, Sullivan P A and Bale D H 2010 *Chem. Rev.* **110** 25
[2] Choi C S, Moon I K and Kim N 2009 *Appl. Phys. Lett.* **94** 053302
[3] Kim T D *et al* 2007 *J. Am. Chem. Soc.* **129** 488
[4] G. Seniutinas *et al* 2012 *Dyes Pigments* **95** 33
[5] Traskovskis K *et al* 2012 *J. Mater. Chem.* DOI: 10.1039/C2JM30861D
[6] Traskovskis K *et al* 2012 Proc. SPIE **8434** 8434-59
[7] Zarins E, Jubels J and Kokars V 2011 *Adv. Mater. Res.* **222** 271
[8] Vembris A *et al* 2012 *Opt. Mater.* DOI: 10.1016/j.optmat.2012.02.051
[9] L.Laipniece, V.Kampars 2011 *Abstracts of the 52nd International Scientific Conference of Riga Technical University. Section: Material Science and Applied Chemistry* (Riga, Latvia) p 20
[10] Nitiss E and Rutkis M 2012 *Proc. SPIE*, **8430**, 8430-49
[11] Rutkis MA *et al* 2006 *Proc. SPIE* **6192** 61922Q
[12] Herman W N, Hayden L M 1995 *J. Opt. Soc. Am. B* **12** 416
[13] Jerphagnon J and Kurtz S K 1970 *Phys. Rev. B* **1** 1739

IOP Conf. Series: Materials Science and Engineering **38** (2012) 012035 doi:10.1088/1757-899X/38/1/012035

Hyper-Rayleigh scattering and two-photon luminescence of phenylamine–indandione chromophores

I Mihailovs[1, 2, 4], J Kreicberga[2], V Kampars[2], S Miasojedovas[3], S Juršėnas[3], L Skuja[1] and M Rutkis[1]
[1] Institute of Solid State Physics, University of Latvia, Riga, Latvia
[2] Institute of Applied Chemistry, Riga Technical University, Latvia
[3] Institute of Applied Research, Vilnius University, Vilnius, Lithuania

E-mail: igors.mihailovs_2@rtu.lv

Abstract. Based on results of quantum chemical (QC) screening, we put under research several phenylamine indane-1,3-dione derivatives including octupolar NLO chromophore A3BI. According to these calculations, molecular first hyperpolarizability (β_{HRS}) should be enhanced upon replacement of methyl substituents with phenyl groups. Surprisingly, β_{HRS} values obtained in our measurements by hyper-Rayleigh scattering (HRS) are much higher than one could expect from QC. Bearing in mind that two-photon luminescence (TPL) is usual cause for overestimation of β_{HRS} in HRS measurements, we have accomplished investigations of TPL properties of these compounds. All investigated triphenylamine derivatives exhibit strong TPL. In case of DANS and DMABI captured simultaneously, TPL spectra and HRS signal, reveal the least as clearly distinct on TPL background, thus TPL-free β_{HRS} values can be defined. Concurrently, sharp HRS signal of DPABI is sinking in TPL and background noise along with increase in concentration. Due to fairly intensive TPL for other investigated triphenylamino compounds, the HRS signal was rather indistinguishable and therefore correct β_{HRS} values could not be extracted.

1. Introduction

Last two decades were marked by outstanding interest in organic non-linear optical (NLO) materials, and especially noticeable are second-order non-linear effects. In particular, during the recent years several researchers paid their interest to multibranched and particularly octupolar NLO chromophores, as they are claimed to have advantages over dipolar counterparts in their applications in electro-optics [1][2].

Some of phenylamine–indandione derivates as promising in non-linear optics were already under our research for many years [3]. Others have come to our notice owing to results of quantum chemical (QC) calculations, performed during a kind of molecular screening for possibly NLO-effective structures. Multibranched derivatives of triphenylamine are known for cooperative effect between the branches that leads to enhanced electron delocalization in excited state [2]. In its turn, 1,3-indandione-2-ylidene group is known to be fairly good electron acceptor part of "push-pull" chromophores [3][4]. Considering all mentioned, we supposed several NLO-active indane-1,3-dione-based structures to be

[4] To whom any correspondence should be addressed

worth another study (see scheme 1).

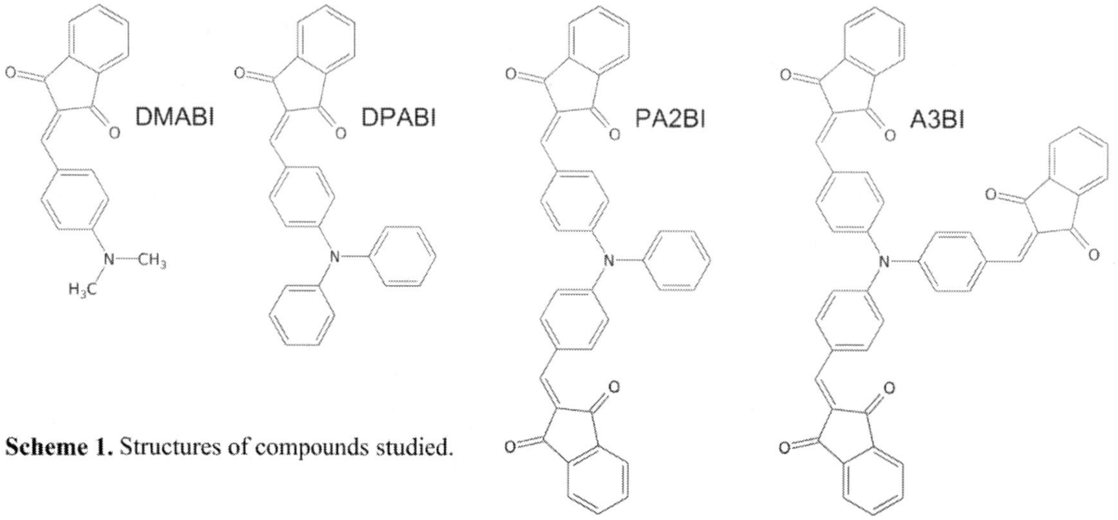

Scheme 1. Structures of compounds studied.

2. Experimental

QC calculations were undertaken in Gaussian 09W [5] employing RHF *ab initio* method with 6-31** basis set (using keyword "Polar CPHF=RdFreq"). Orientationally averaged second-order polarizability (β_{HRS}) values were calculated in chloroform environment.

To obtain β_{HRS} values of a compound experimentally, we usually use hyper-Rayleigh scattering (HRS) technique in $CHCl_3$. Typical solute concentration is of order $10^{-5}\div10^{-4}$ M. As excitation source EKSPLA NL640 is used, emitting NIR laser beam at 1064 nm with pulse duration 10 ns and repetition rate 1 kHz. The light is collected in perpendicular to the exciting beam, and its intensity is being detected by photomultipliers. More detailed description of experimental setup could be found in one of our previous papers [6]. It is important to mention that our methodology should allow us to eliminate from HRS signal unwanted two-photon luminescence (TPL) by two procedures used simultaneously. First of them is based on temporal discrimination of these phenomena based on the fact that HRS signal is coincident in time with excitation pulse, but TPL have certain decay time. Efficiency of that procedure drops in case when excited state lifetime becomes comparable with excitation pulse width. Another possibility is discrimination based on TPL and HRS spectral bandwidth characteristics. FWHM of TPL emission usually is a number of times wider than HRS. Therefore TPL and HRS extent in measured second-harmonic intensity depends on spectral bandwidth of detection system. In our measurements we are using two parallel detection systems with different spectral bandwidth (3 and 10 nm). To not fall in details, let us consider that our numerical deconvolution of HRS and TPL signals is based on conviction that β_{HRS} obtained by internal reference method using data acquired from both channels should be the same. HRS light intensities obtained after this deconvolution were used to obtain β_{HRS} values by external reference method (in our case well-known NLO-phore DANS). This procedure works well when TPL and noise levels are low enough – otherwise HRS signal could scarcely be recovered from highly intensive TPL background. To obtain β_{HRS}, we consider concentration dependence of a quadratic coefficient $Q = I_{2\omega}/I_{\omega}^2 = G\cdot N_{solute}\cdot\langle\beta_{HRS}^2\rangle_{solute}+G\cdot N_{solvent}\cdot\langle\beta_{HRS}^2\rangle_{solvent} = A+B\cdot C_{solute}$, and so can easily obtain ratio $B_{solute}/B_{DANS} = \langle\beta_{HRS}^2\rangle_{solute}/\langle\beta_{HRS}^2\rangle_{DANS}$ [7]. Here G is instrument constant, N and $C = N/N_A$ – number density and molar concentration, accordingly, and N_A – Avogadro constant.

One-photon luminescence (OPL) spectra were obtained on PerkinElmer LS 45 luminescence spectrometer. Concentrations of title compounds were adjusted so as to gain their absorption close to equal at excitation wavelength (chosen close to the absorption maximum). This way one could roughly compare quantum efficiency of the compound studied.

For two-photon luminescence spectra, first harmonic of YAG:Nd^{3+} EKSPLA PL-2140 (1064 nm, pulse duration 25 ps, repetition rate 10 Hz, maximum pump energy 500 μJ) laser was used as the excitation source. The size of the excitation spot was approximately 1 mm^2 for backscattering measurements. The emission from the solution (same concentration for all compounds was 10^{-4} M) was collected and dispersed by a monochromator with focal length of 40 cm.

Simultaneous hyper-Rayleigh scattering and two-photon luminescence spectra were obtained using the same experimental setup as in our conventional experiment. Exceptions were detection system, for which ANDOR Shamrock SR-303i-B imaging spectrograph was used. Due to lower sensitivity of detection system we are forced to use higher than previously solute concentrations (10^{-4}÷10^{-3} M). In this technique, HRS signal is being discriminated as an area of narrow 532 nm peak obtained by deconvolution of captured spectrum into Gaussian components. The processing of the results to obtain β_{HRS} value was the same as in our conventional HRS experiment.

3. Results and discussion

3.1. Quantum chemical calculations

For better ability to compare the calculated β_{HRS} values, we present them as relative to widely known NLO chromophore – 4-dimethylamino-4'-nitrostilbene (DANS). Expressed in this way, relative nonlinear efficiency of title compounds is 0.87; 1.04; 1.07 and 1.10 for DMABI, DPABI, PA2BI and A3BI, accordingly. As it can be seen, multibranched chromophores do show slightly higher β_{HRS} values. At same time for two-branched and three-branched chromophores it is clear that β_{HRS} per acceptor unit decreases each time when adding another branch. Similar behavior has already been reported [8] and it could be explained considering decreasing nitrogen atom donor ability (as the lone pair electrons upon excitation should become "shared out" between increasing number of acceptors) and growing steric hindrance between the branches. More interesting phenomenon, according to our QC calculations, is that a molecule with unsubstituted triphenylamine-based donor (DPABI) displays larger β_{HRS} than its dimethylphenyl counterpart DMABI. Quite usual in fact is the contrary situation where phenyl substituents delocalize the lone pair electrons, hence their shift along the main conjugation path during the excitation is encumbered. This should affect β negatively, and is usually overcome by substituting phenyl rings with electron donor groups [9]. In our case, according to our QC calculations, such negative impact of phenyl substituents on β is absent. It is also known from previous studies by *ZINDO/S* [3] that difference between ground and excited state dipole moments ($\Delta\mu_{ge}$) for DPABI is somewhat greater than for DMABI – 11.4 and 8.7 D, accordingly. The shape of the frontier orbitals plotted by AM1 (not shown here), supported greater charge redistribution upon excitation in DPABI than in DMABI. These effects in DPABI as well as already mentioned drop of NLO efficiency per branch upon further addition of acceptor units in case of other title compounds seemed for us well worth further experimental studying.

3.2. Hyper-Rayleigh scattering measurements

The results of HRS measurements done by our usual procedure were quite surprising. Reporting results in the same form as we did in QC calculations section, the relative to DANS experimental values for DMABI, DPABI, PA2BI and A3BI, accordingly, are 1.17; 2.58; 5.44 and 7.89. Thus, apparent (possibly overestimated) first hyperpolarizabilities of investigated compounds are much greater than it could be expected from QC calculations. Particularly, β_{HRS} of DMABI seems to be 1.17

times greater than that of DANS while calculated ratio is in fact reciprocal. Considering the point that triphenylamine derivates often display intense luminescence, including TPL [10], we concerned ourselves with possibility that β_{HRS} values obtained using our procedure are actually corrupted by effectively large TPL intensity. Therefore additional investigation of luminescence properties seems to be useful for us to clarify disagreement in the results of QC calculations and conventional HRS measurements.

3.3. Linear absorption, one- and two-photon luminescence studies

Attenuation coefficient spectra of the title compounds in $CHCl_3$ solution are shown on figure 1(a). One can easily notice that, according to the two-state model [11], resonance conditions in our HRS experiment (at second harmonic wavelength – 532 nm) are well met for PA2BI and A3BI, the remaining two compounds DMABI and DPABI are further from these. It seems at the same time – see figure 1(b) – that luminescence intensity at 532 nm is rather low, so probably opposing the assumption that the TPL could be a reason for overestimating β_{HRS} value in our HRS experiment. However, we still doubted if the luminescent light has not been reabsorbed in the measured media, especially at the short-wavelength side of luminescence band where it overlaps with the absorption band. Relative quantum efficiency (see experimental section) was found to be approximately 1:5:22:25 for DMABI:DPABI:A3BI:PA2BI. It can be seen, too, that DPABI luminescence possesses largest Stokes shift of 145 nm. In accordance with Lippert–Mataga equation [12], this may point to a large difference between the ground and excited state dipole moments in agreement with QC calculations (see above). Larger change of dipole upon an excitation, as supposed by the same two-level model, can lead to considerably increased β values of DPABI in comparison with DMABI.

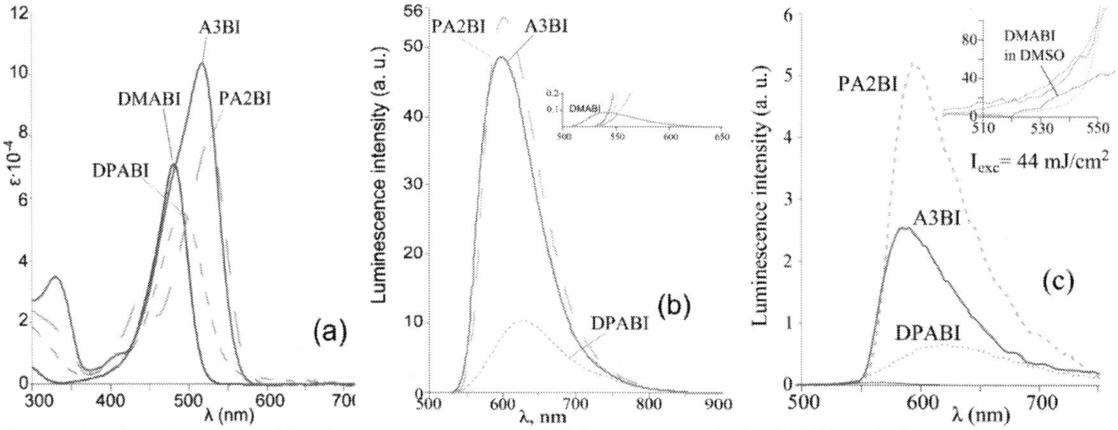

Figure 1. Absorbance and luminescence spectra of title compounds in $CHCl_3$ solution: (a) Attenuation coefficient ε; (b) One photon luminescence, representing quantum efficiency of compounds (see experimental section); (c) Two photon luminescence, at the same concentration for all solutes (see experimental section).

Further, we have performed experiments to acquire TPL spectra of compounds investigated – see figure 1(c). Same as in our previous HRS experiments, excitation wavelength 1064 nm and $CHCl_3$ as solvent were used. Unfortunately, under these experimental conditions DMABI produced no detectable signal at all when dissolved in chloroform, so the referred data are for DMABI solution in DMSO. Here relative intensities are 1:10:50:100 for the same compound sequence, thus the TPL intensity rises steeper than OPL intensity with the investigated structural changes. Two-fold increase in the TPL intensity for PA2BI in comparison with A3BI, which is absent for OPL, could be explained by two-photon absorption cross-section increase. Nevertheless luminescence intensity at wavelengths shorter than ~540 nm seems to be insignificant, concerns of TPL presence in light intensity measured at HRS wavelength (532 nm) still remain. To clear the uncertainty, it was decided to undertake an attempt to acquire spectra of HRS signal and TPL band simultaneously and based on deconvolution of such spectra in Gaussian components evaluate intensities of both emissions at 532 nm.

3.4. Simultaneous spectral study of hyper-Rayleigh scattering and two-photon luminescence

The spectra were acquired by measuring signal dependence both on excitation light intensity and solute concentration. Some of observed dependencies are shown on the figures 2–4.

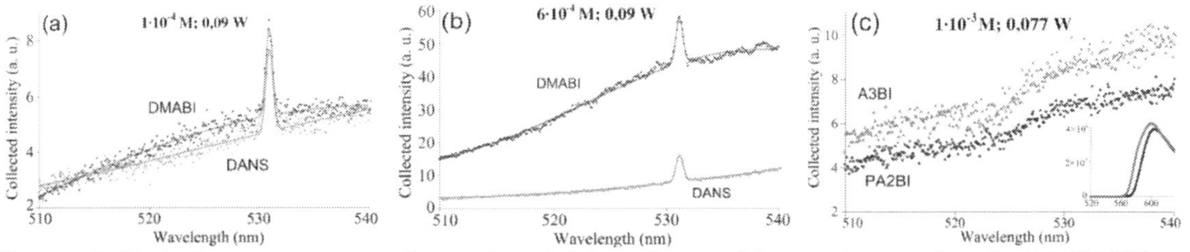

Figure 2. Emission spectra examples at the vicinity of second harmonic wavelength: (a) DANS and DMABI at $1x10^{-4}$ M/l; (b) DANS and DMABI at $6x10^{-4}$ M/l; (c) PA2BI and A3BI spectra (full recorded spectral range – on the inset).

As seen on figure 2(a)–(b), for DMABI and DANS HRS signal is easily distinguishable from the TPL and background noise. In case of higher concentrations DMABI signal appears stronger because DANS has twice larger ε_{532}. Unfortunately, there is no clear evidence of the HRS signal for PA2BI and A3BI at any investigated concentration. Typical examples of PA2BI and A3BI spectra are presented on figure 2(c). This could be readily attributed to the huge intensity of TPL, which is depicted in full range on the inset.

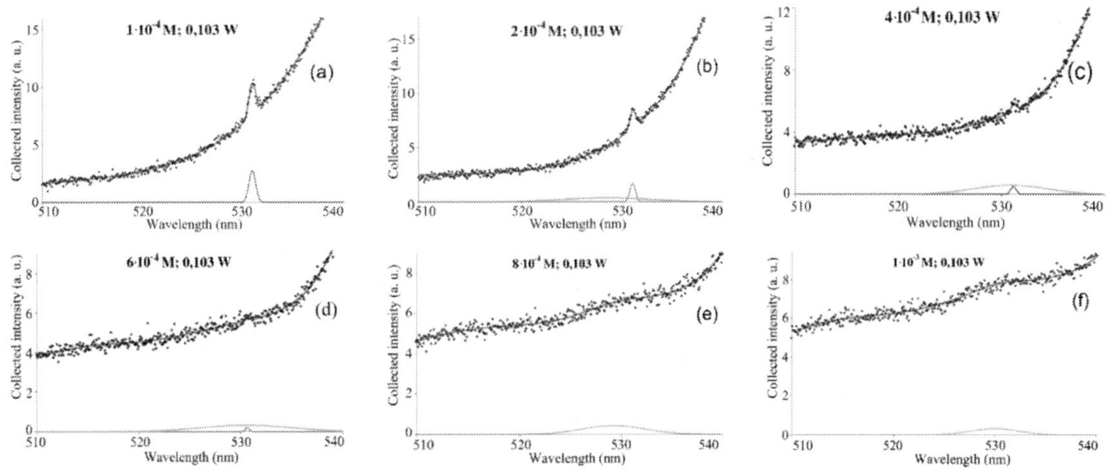

Figure 3. (a) – (f) Emission spectra at the vicinity of second harmonic wavelength of DPABI, depending on concentration. Deconvoluted Gaussian components of HRS peak and unexpected flat signals are shown.

Interesting is situation with DPABI, where the HRS signal is increasingly sinking in the tail of TPL band generated noise (see figure 3) along with increase in the solute concentration. This is not a surprise, as in case of DPABI attenuation coefficient $\varepsilon_{532} = 17600$ is rather high. Due to that at DPABI concentration ~10^{-4} mol/L just ~45% of scattered second harmonic light escapes the cell. Increase of DPABI concentration to ~10^{-3} mol/L almost completely (0.03 %) reduces detectable scattered SH light. Thus, HRS peak at higher concentrations obviously could not be visible. Broad band TPL is absorbed as well, what is indicated by descent of the band blue wing (see figure 3). Moreover, approximately at the wavelength of HRS signal another flat peak could be noticed. One of possible explanations for this could be stimulated luminescence caused by HRS light at 532 nm. Similar peculiar peak could also be hardly noticed in spectra of PA2BI and A3BI, as on figure 2(c).

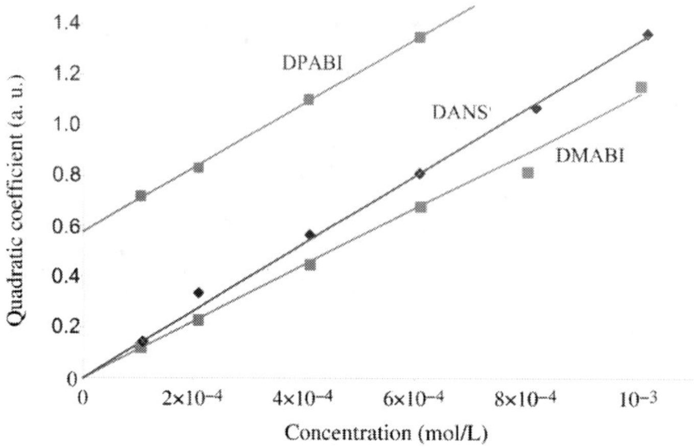

Figure 4. Concentration dependence of absorbance corrected quadratic coefficient Q for DANS, DMABI and DPABI;

As pictured on figure 4, for DMABI and DANS there is clearly linear concentration dependence of quadratic coefficient Q on solute concentration. This is the case, however, only when correction for absorbed HRS light is applied, as otherwise dependence curves would not be linear. Within experimental error obtained β_{HRS} ratio of DMABI and DANS (0.92) is close to the value predicted from QC calculations (0.87). In the case of DPABI puzzling is abnormally large intercept of absorption-corrected Q as a function of concentration, what makes us suspicious of employing these data in further analysis. Despite of this, as one also can observe on figure 4, Q correlates concentration rather well, what yields β_{HRS} value 0.98 relatively to DAN. This value is surprisingly close to QC predicted value 1.04, but one should consider this result only as provisional.

To finalize, our measurements according to this scheme yielded reliable β_{HRS} value for DMABI, what is in good agreement with quantum chemical calculations. Unfortunately, in case of other triphenilamino compounds measurements HRS signal was obscured by large absorbance at 532 nm and we are not able to obtain trustworthy β_{HRS} values for these chromophores. To evaluate β_{HRS} of these compounds we assume two possible solutions for that problem in future. One could be reduction of solute concentration range by one order, what will demand higher sensitivity of spectrometric detection system. Another possibility is to avoid coincidence of HRS and absorbance band by shifting excitation wavelength to IR. In this case unwilling TPL should be reduced proportionally to detune from resonance.

4. Conclusions

The molecular hyperpolarizability β_{HRS} measurements of the title compounds by HRS were found to be vulnerable on TPL influence. In case of intense TPL it is apparent that time- and spectral bandwidth-discriminated acquisition with further numerical deconvolution cannot satisfactory separate HRS light signal. Therefore such measurements lead to overestimated values. On the contrary, energy-resolved direct and simultaneous TPL and HRS signal spectra recording provide rather objective insight into the actual features of the compounds measured, as it helps to obtain the actual picture of the emission in the vicinity of second harmonic wavelength. Therefore, one could anticipate quite meaningful results using the technique we applied. However, it is worth to be mentioned, that the measurements needed by this approach are actually very time-consuming. In case when every single spectrum is accumulated for 5 minutes, it would take several hours to obtain set of spectra dependent on excitation light intensity and solute concentration, which is sufficient for β_{HRS} determination, for one single compound.

Acknowledgements
Authors gratefully appreciate support from Latvian Council of Science project 10.0032 „Development of research and technology potential for elaboration of new and nanostructured materials and related applications" subproject 10.0032.14 "Supramolecular nanostructuring of photonic materials – theoretical modeling and experimental investigations".

References
[1] Zyss J and Ledoux I 1994 *Chem. Rev.* **94(1)** 77–105
[2] Fakis M, Giannetas V and Mikroyannidis J 2010 *Dyes Pigm.* **87** 44–8
[3] Rutkis M, Vembris A, Zauls V, Tokmakovs A, Fonavs E, Jurgis A and Kampars V 2006 *Proc. SPIE - Int. Soc. Opt. Eng.* **6192** 61922Q
[4] Jursenas S, Kurilcik N, Karpicz R, Gulbinas V, Valkunas L, Rutkis M and Muzikante I 2008 *Thin Solid Films* **516** 8909–16
[5] M. J. Frisch *et al.*, Gaussian 09, Revision A.1, Gaussian, Inc., Wallingford CT, 2009.
[6] Traskovskis K, Mihailovs I, Tokmakovs A, Jurgis A, Kokars V and Rutkis M 2012 *J. Mater. Chem.* **22** 11268
[7] Verbiest T, Clays K and Rodriguez V 2009 *Second-Order Nonlinear Optical Characterization Techniques: An Introduction* (Boca Raton, FL: CRC Press)
[8] Sudharsanam R, Chandrasekaran S and Das P K 2003 *J. Mol. Struct.* **645** 51–9
[9] Cho M J, Choi D H, Sullivan P A, Akelaitis A J P and Dalton L R 2008 *Prog. Polym. Sci.* **33** 1013–58
[10] Chen L, Cui Y, Mei X, Qian G and Wang M 2007 *Dyes Pigm.* **72** 293–298
Lee K-S, Kim R H, Yang D-Y and Park S H 2008 *Prog. Polym. Sci.* **33** 631–681
[11] Oudar J L and Chemla D S 1975 *Opt. Commun.* **13(2)** 164–8
[12] Lippert E 1955 *Z. Naturforsch.* **109** 541
[13] Mataga N, Kaify Y and Koizumi M 1955 *Bull. Chem. Soc. Jpn.* **28** 690

IOP Conf. Series: Materials Science and Engineering **38** (2012) 012036 doi:10.1088/1757-899X/38/1/012036

Multiscale Approach to the Structure and Spectra of Nile Red Adsorbed on Polystyrene Nanoparticles

Alexandra Freidzon, Vyacheslav Tikhomirov, Alexey Odinokov, Alexander Bagaturyants

[1] Photochemistry Center, Russian Academy of Sciences, Moscow, Russia

e-mail: freidzon.sanya@gmail.com

Abstract. The structure and absorption spectra of the Nile Red (NR) dye adsorbed on the surface of a polystyrene (PS) nanoparticle is studied by DFT/TDDFT based on the results of previous molecular dynamics (MD) simulations. From NR/PS surface structures obtained from MD simulations, relatively small clusters were constructed composed of the adsorbed NR molecule and the nearest PS fragments. These clusters were treated either explicitly by DFT/TDDFT or using the effective fragment potentials (EFP) model. The absorption and emission spectra of the NR/PS system calculated using the EFP model are very close to those calculated with the explicit environment, so that the EFP model can be recommended for use in multiple large-scale calculations. The calculated positions of the absorption and fluorescence bands are in good agreement with experiment.

1. Introduction

Nile Red (NR, Scheme 1) is a lipophilic dye soluble in organic solvents and alcohols and poorly soluble in water. NR is a good solvatofluorochromic indicator, because its absorption and fluorescence spectra are highly sensitive to the environment.[1-3] Polystyrene (PS) is used as a matrix for the hydrophobic NR in optical chemical sensors. [4-6]

Scheme 1

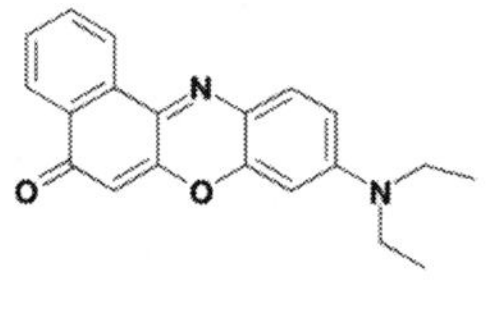

The NR solvatochromism and dual fluorescence observed experimentally is commonly explained by the charge-transfer character of its excited state, which, probably, involves a twist of the diethylamino group.[1, 7] In our recent study, it has been shown that the NR S_1 excited-state potential energy surface has two minima: one corresponding to a locally excited (LE) state with a planar structure and another one corresponding to a twisted intramolecular charge transfer (TICT) state. [8]

Four heteroatoms of the NR molecule can serve as potential hydrogen bond acceptors in its interaction with protic molecules. With aprotic molecules, NR can form intermolecular complexes through electrostatic and dispersion interactions.

PS is an irregular carbon-chain nonconjugated polymer, and its nanoparticles are used as hydrophobic matrices for indicator dyes. Previously, we have studied the adsorption of the NR molecule on a PS film or globule by molecular dynamics (MD).[9] No preferential orientation of the dye in the polymer surface has been found. This suggests that the absorption and emission spectra of NR on the polymer surface should exhibit an environment-induced shift and a noticeable inhomogeneous broadening.

The goal of this study was to demonstrate the possibility of simulating the absorption and emission spectra of NR adsorbed on the PS surface using the effective fragment potential (EFP) model of environment and the TDDFT method for the transition energies.[10, 11]

2. Computational Procedure

We modelled the fragments (links) of PS chains nearest to the NR molecule in the obtained surface structures by isolated toluene molecules. We suggested that the response of the local environment of the chromophore to excitation is fully electronic, while the nuclear displacements of the polymer links are negligible in the solid polymer matrix.

The details of the MD simulation of NR adsorbed on the PS surface were described in Ref.[9]

2.1. Constructing a cluster from a snapshot of the MD trajectory.

A snapshot of an equilibrium part of an MD trajectory was taken (Figure). The chromophore (shown in balls-and-sticks) and the neighbouring C_6H_5CH units (shown by thick lines) were selected. The unit was included in the cluster if the nearest contact between the chromophore and the monomer unit did not exceed 4 Å. The selected C_6H_5CH units were completed to $C_6H_5CH_3$ (toluene) molecules with a standard $C_{aliphatic}$–H bond length of 1.09 Å.

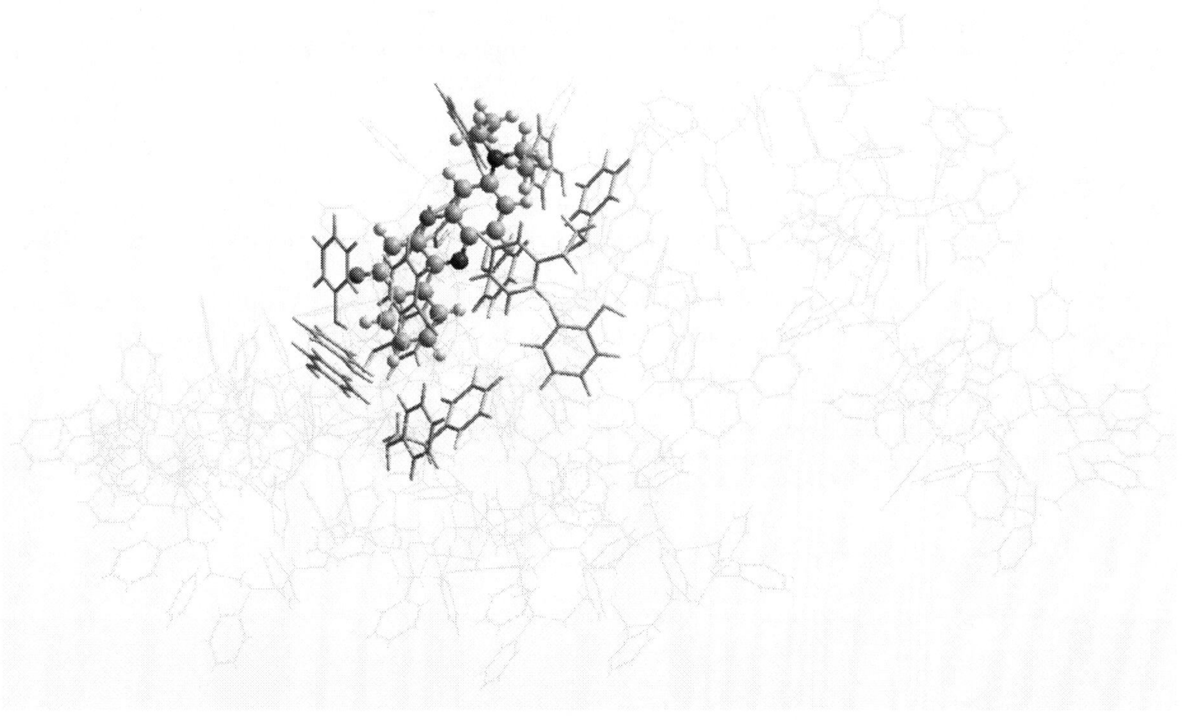

Figure 1. Constructing a cluster from a snapshot of the MD trajectory. The chromophore is shown by balls-and-sticks, thick lines denote PS links adjacent to the chromophore, and wire lines show the rest of the polymeric nanoparticle.

IOP Conf. Series: Materials Science and Engineering **38** (2012) 012036 doi:10.1088/1757-899X/38/1/012036

Next, the EFPs corresponding to toluene molecules were placed instead of the $C_6H_5CH_3$ units into their positions. A special care was taken to ensure that the distances between the effective fragments were no less than 1 Å.

2.1.1. *Generation of an effective fragment potential.* The EFP of toluene was generated using the MAKEFP procedure implemented in the GAMESS-US program package.[12, 13] The geometry of the effective fragments was optimized by B3LYP/PC1. Only monopoles, multipoles, and local static polarisable points generated by the MAKEFP procedure were used in the further TDDFT calculations, while the other data were discarded.

2.2. Quantum-chemical calculations.

All the ground-state structures were optimized by PBE0/6-31G(d,p), and the excited-state structures were optimized by TD-PBE0/6-31G(d,p) using GAMESS-US. We considered both NR and its complex with methanol coordinated to the carbonyl oxygen NR…MeOH on the surface.

Several clusters were treated completely by quantum mechanics (denoted by QM PS). The environment (toluene molecules) was frozen, and the dye structure was optimized by PBE0/6-31G(d,p). The difference in the dye structure optimized in the frozen environment and *in vacuo* was negligible. Later on, this allowed us to replace the dye structure obtained by molecular dynamics with that optimized *in vacuo* by PBE0/6-31G(d,p) in the clusters with effective fragments.

The absorption spectrum of the cluster treated by quantum mechanics was calculated by TD-PBE0/6-31G(d,p) and compared with the absorption spectrum of the same cluster with toluene molecules replaced by the effective fragments (denoted by EFP PS). In addition, the absorption spectrum of the chromophore was calculated by TD-PBE0/6-31G(d,p) in the polarisable dielectric continuum environment (D-PCM) using the Firefly program.[14]

2.3. Calculations of absorption and emission spectra.

The absorption spectrum was constructed by the following procedure: the spectra were calculated for 40 clusters constructed as described above. Each spectrum was presented as a Gaussian peak with full width at half maximum (FWHM) equal to 0.29 eV (homogeneous broadening)[15] and area corresponding to its oscillator strength. Finally, the spectra were summed and normalized to the average oscillator strength.

Our previous calculations[8] revealed two excited-state structures of NR: a planar locally excited (LE) structure and a structure with twisted amino group called twisted intramolecular charge transfer (TICT) state. These (fully optimized by TD-PBE0/6-31G(d,p)) structures were also used for calculating the emission spectra in the clusters with effective fragments. For the emission spectra, only average positions of the bands and standard deviation of the band maxima were calculated.

3. Results and Discussion

The absorption spectra of the clusters with NR and NR–methanol complex (see Figure 2) treated fully by quantum mechanics are compared with the absorption spectra of the same clusters with toluene molecules replaced by the effective fragments. The difference is rather small

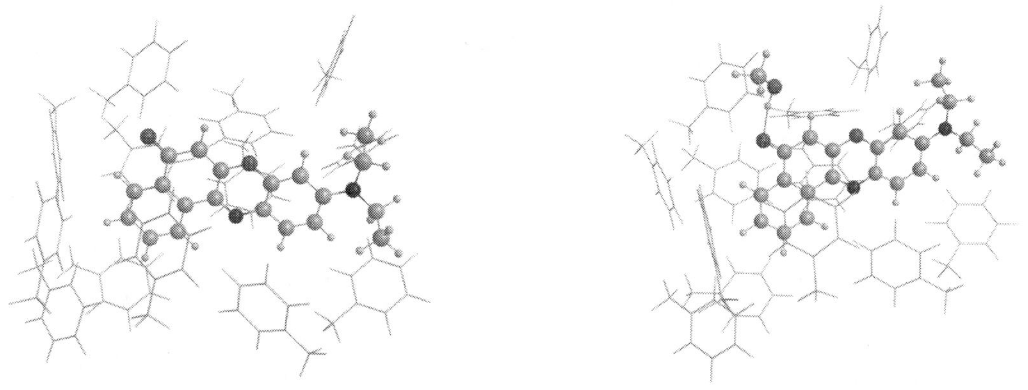

(

a b

Figure 2. Example clusters: Nile Red (a) and Nile Red–methanol complex (b) surrounded by EFPs mimicking the polystyrene units.

Table 1). The effect of the polarisable dielectric continuum environment is also similar to either that of QM PS or that of EFP PS. Therefore, the local effects of electrostatics included through the point charges, multipoles, and polarisabilities are substantially the same as those included explicitly or those caused by the dielectric continuum. The calculated absorption spectrum agrees well with the experimental spectrum of NR on PS nanoparticles.[5, 6]

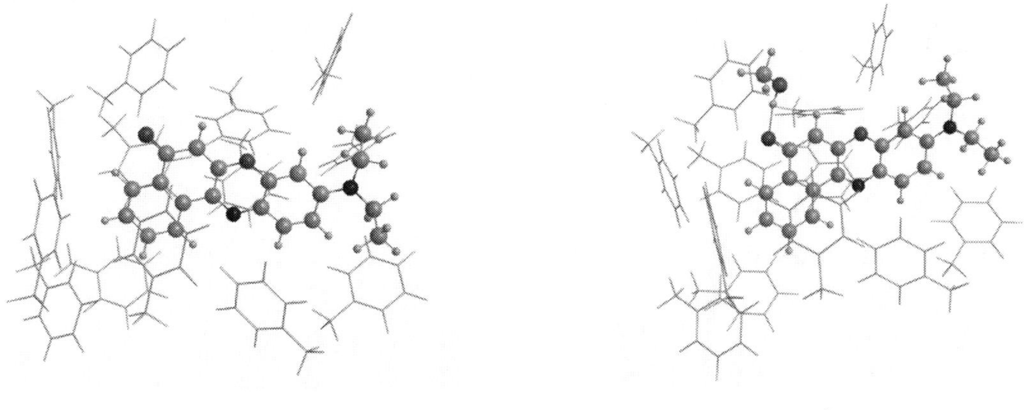

a b

Figure 2. Example clusters: Nile Red (a) and Nile Red–methanol complex (b) surrounded by EFPs mimicking the polystyrene units.

Table 1. Calculated absorption spectra of NR and NR...MeOH *in vacuo* and in PS environment

	free		QM		EFP		D-PCM		Exp.[5, 6]	
	eV	nm	eV	nm	eV	nm	eV	nm	eV	nm
NR	2.76	449	2.63–2.66	467–472	2.69	462	2.67	465	2.59	480
NR...MeOH	2.71	458	2.60–2.62	473–477	2.69	462	2.64	470		

The absorption spectrum of NR was constructed as described in the previous section. The resulting spectrum is shown in Figure 3. The small shift in the absorption band caused by the coordination of methanol to the dye is completely concealed by the environment-induced shift and inhomogeneous broadening.

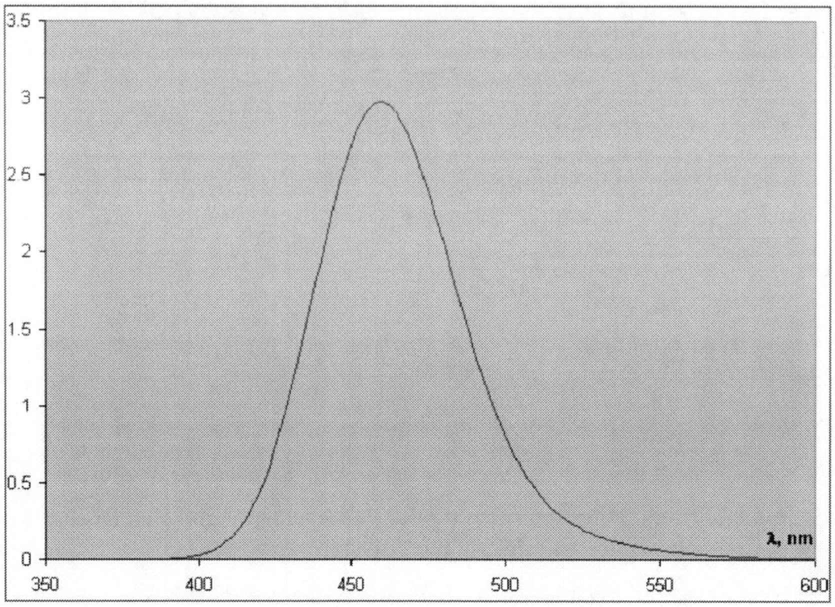

Figure 3. Calculated absorption spectrum of Nile Red in polystyrene including homogeneous and inhomogeneous broadening.

For the emission spectra, no calculated homogeneous broadening data is available to date. Therefore, only the average transition energy and standard deviation of the sample were calculated (Table 3). One can see that the LE fluorescence exhibits less scatter than the TICT fluorescence. Large dipole moment of the twisted structure in the S_1 excited state[8] induces large dipole moments in the environment, which, in their turn, affect the transition energy in the TICT state. This qualitatively agrees with the fact that the observed TICT emission bands are weak and broad. One can also see that the TICT state is more sensitive to complexation with MeOH than the LE state: even inhomogeneously broadened, the TICT fluorescence band exhibits a more pronounced shift upon complexation.

Table 3. Calculated emission spectra of NR and NR…MeOH *in vacuo* and in PS environment

	free			EFP[a]		
	eV	nm	(Δλ)	eV (FWHM[b])	nm	(Δλ)
NR LE	2.60	477		2.49 (0.12)	499	
NR TICT	1.92	646		1.88 (0.35)	660	
NR…MeOH LE	2.57	483	6	2.47 (0.13)	503	4
NR…MeOH TICT	1.77	701	55	1.77 (0.45)	700	40

[a] EFP PS data for NR are average of 30 clusters; for NR…MeOH, of 10 clusters.

[b] FWHM due to inhomogeneous broadening: $FWHM = \sigma\sqrt{\ln 4}$; σ is the standard deviation of the sample.

The agreement with the experimental emission data, however, is only semi-quantitative: the experimental emission spectra of NR in NR/PS nanoparticles strongly depend on the particle size. These spectra were recorded in polar solvents (water or alcohols). As a result, the emission spectrum was governed either by the NR molecules adsorbed on the surface (in small nanoparticles) or by those in the bulk (in large particles). The surface NR molecules can interact with the polar solvent, while the bulk ones interact only with nonpolar PS. The experimental emission spectrum of bulk NR qualitatively agrees with the calculated LE emission spectrum, while the emission spectrum of NR on the surface qualitatively agrees with the calculated TICT emission spectrum.

4. Conclusions

The indicator form of the dye is the TICT structure. Its emission spectrum is sensitive to the interaction with analytes because of its high dipole moment. The nearest environment of the chromophore can be described by EFPs to a reasonable accuracy. The distribution of transition energies and, thus, the shape and inhomogeneous broadening of absorption and emission bands can be obtained by sampling of the calculated MD trajectories.

5. Acknowledgments

This work was supported by the Russian Foundation for Basic Research, project no. 12-03-01103, and Russian Ministry of Education and Science, state contract no. 16.523.11.3004. The calculations were performed using the facilities of the Joint Supercomputer Center of the Russian Academy of Sciences.

References

[1] Dutta A K, Kamada K, Ohta K 1996 *J. Photochem. Photobiol. A: Chemistry* **93** 57–64.
[2] Dutta A K, Kamada K, Ohta K 1996 *Chem. Phys. Lett.* **258** 369–375.
[3] Cuiying L, Jianxi Zh, Li S 2009 *Acta Chimica Sinica* **67**(5) 381–386.
[4] Kim H-H, Song N W, Park T S, Yoon M 2006 *Chem. Phys. Lett.* **432** 200–204.
[5] Behnke T, Wurth C, Hoffmann K, Hubner M, Panne U, Resch-Genger U 2011 *J. Fluoresc.* **21** 937–944.
[6] Behnke T, Würth C, Laux E-M, Hoffmann K, Resch-Genger U, 2012 *Dyes and Pigments* **94** 247–257.
[7] Sarkar N, Das K, Nath D N, Bhattacharyya K 1994 *Langmuir* **10** 326–329.
[8] Freidzon A Ya, Safonov A A, Bagaturyants A A, Alfimov M V 2012 *Int. J. Quantum Chem.*, in press.
[9] Tikhomirov V A, Odinokov A V, Bagatur'yants A A and Alfimov M V 2010 *Theor. Exp. Chem.* **46**(6) 342–349.
[10] Jensen J H, Day P N, Gordon M S, Basch H, Cohen D, Garmer D R, Krauss M, Stevens W J *Effective fragment method for modeling intermolecular hydrogen bonding effects on quantum mechanical calculations* in *Modeling the Hydrogen Bond* ed D A Smith 1994 *ACS Symposium Series* **569** 139-151.
[11] Gordon M S, Slipchenko L, Li H, Jensen J H 2007 *Annual Rep. Comput. Chem.* **3** 177–193.
[12] Schmidt M W, Baldridge K K, Boatz J A, Elbert S T, Gordon M S, Jensen J H, Koseki S, Matsunaga N, Nguyen K A, Su S J, Windus T L, Dupuis M, Montgomery J A 1993 *J. Comput. Chem.* **14** 1347–1363.
[13] Gordon M S, Schmidt, M.W. in *Theory and Applications of Computational Chemistry, the First Forty Years*; Eds C E Dykstra, G Frenking, K S Kim, G E Scuseria 2005 (Amsterdam: Elsevier) p 1167.
[14] Granovsky A A *Firefly* version 8.0.0, http://classic.chem.msu.su/gran/gamess/index.html.
[15] The procedure for calculating the homogeneous broadening will be described elsewhere.

IOP Conf. Series: Materials Science and Engineering **38** (2012) 012037 doi:10.1088/1757-899X/38/1/012037

Near-band luminescence of CdZnTe detector crystals

L Grigorjeva[1,3], D Millers[1], V Ivanov[2], L Aleksejeva[2]

[1]Institute of Solid State Physics, University of Latvia, Riga, Latvia
[2]ZRF RITEC SIA, Riga, Latvia

E-mail: lgrig@latnet.lv

Abstract. The near-band luminescence due to both neutral donor and neutral acceptor bound excitons (D^0 Ex), (A^0Ex) and its phonon replicas as well as luminescence of donor-acceptor pair (DAP) and deep energy level due to undefined defect state (A band) were studied in CZT detector crystals. The near-band luminescence correlation with detector characteristics was analyzed and the results could be applied for nondestructive detector material selection. It is obtained that some parameters such as a presence and intensity of A and DAP luminescence bands, the decay kinetics in A^0Ex-1LO and DAP luminescence bands, are most important characteristics for detector material characterization.

1.Introduction

The cadmium zinc telluride, $Cd_{1-x}Zn_xTe$ (CZT) is a promising material for the room temperature X- and γ-ray detectors as well as for electro-optic modulators, solar cells and other devices. The high quality CZT crystals are requested for the radiation detectors manufacturing. Since the ternary compound CZT crystal growth process is very complicate the stoichiomety and defect distribution are different even along the same ingot. The CZT is direct band gap semiconductor with distinct exciton transitions at low temperature. The luminescence properties of CZT were described in [1-4] and the origin of principal luminescence bands is known. The low temperature near-band luminescence is due to exciton bound to neutral donor (D^0 Ex) and neutral acceptor (A^0Ex) and its phonon replicas (the LO phonon energy is 22 meV [1]), DAP luminescence and surface defect luminescence (A band) were observed also. The DAP luminescence is situated close to A^0Ex-1LO luminescence band and these two bands overlap [1].

In this paper we present the low temperature photoluminescence studies of CZT detector crystals. It is an attempt to clarify the main luminescence features that could be used for detectors performance prediction. For the first time the spectral-kinetic luminescence experiments were used for detector material testing.

2.Samples and experiments.

The set of CZT detector crystals from different producers and different ingots were used for luminescence studies. The sample surfaces were polished to optical quality. From the same samples

[3] To whom any correspondence should be addressed

the γ-ray detectors were prepared and the spectrometric characteristics of CZT quasi-hemispherical detectors were tested. The quasi-hemispherical detectors are the most simples ones in production and use, since do not require a special electronic circuit for their application [5]. For the detector quality characterization we use the 662 keV line of ^{137}Cs emission. The line FWHM characterizes the detector energy resolution.

The luminescence spectra and decay kinetics were measured under pulsed YAG:Nd laser excitation (266 nm, 1 ns, 0.3 μJ), spectral energy resolution was better than 0.25 meV over the measuring range. Luminescence measurements were carried out with a photon counting head (HAMAMATSU H8259-2) and photon counting board (module P 7888-1E) with minimal time bins 250 ps. The samples were cooled down to 12 K in a closed-cycle helium cryostat.

3. Results and discussion

Usually for the radiation detection the $Cd_{1-x}Zn_xTe$ crystals with x=0.1 was used. It is important to control the fundamental absorption edge (Eg) since the Eg depends on Zn concentration. The value of Eg was estimate from straight line on absorption spectrum in coordinates $(K^* h\nu)^2 = f(h\nu)$, where K is absorption coefficient; for this composition Eg was estimated 1.49-1.50 eV at 300K. Deviation of this parameter from this range means that $x \neq 0.1$.

The number of CZT samples was tested and the luminescence spectra and luminescence decay kinetics of three CZT typical detector crystals are shown in Figures 1–3. The peak position of D^0Ex luminescence at 12 K is in range 1.641-1.652 eV. Along with D^0Ex luminescence the A^0Ex luminescence was detected in studied CZT crystals. Usually, acceptors – V_{cd}; donors – Te_{Cd}, however another type of acceptors or donors may exist.

The decay of these two bands are fast and decay time was less than 2 ns (2 ns is time resolution of used equipment). The presence of mentioned excitonic bands and peak positions are evidence of high crystal quality.

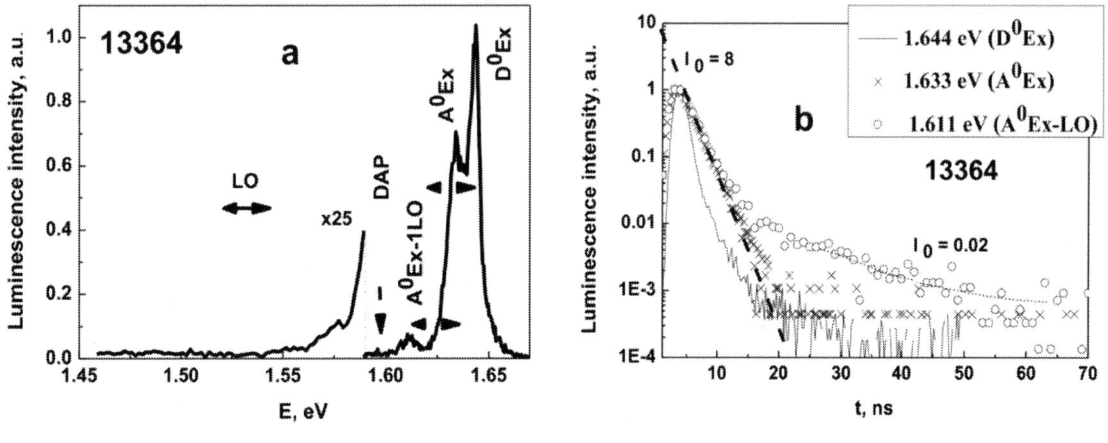

Figure 1. The luminescence spectrum (a) and luminescence decay kinetics (b) of different luminescence bands of sample 13364 at 12K.

In some samples (for example the samples in Fig.2 and 3) the DAP luminescence peaked at 1.595-1.599 eV was detected. The DAP luminescence has slow non-exponential decay kinetics. The decay kinetic straighten in coordinates lg(I)~lg(t), where I is luminescence intensity, t is time. The line slope was close to one and this is evident for DAP luminescence. The decay with slope close to one was shown in Fig.2b and 3b with strain lines. Since the A^0Ex-1LO and DAP luminescence peaks overlap and conclusion about the intensity of DAP luminescence will be drawn out from luminescence decay measurements. The example of kinetic analysis is shown in Fig.1b. The A^0Ex-1LO luminescence is fast with decay time 2-3 ns. The tail in decay kinetics at 1.611 eV is due to DAP luminescence. We estimate the contribution of slow component as less than 0.25 % of total intensity. The same estimation of DAP luminescence contribution in A^0Ex-1LO band (at 1.61 eV) from decay kinetics gives ~2.5%.

The decay kinetics for DAP luminescence was shown in Fig.2b and 3b. The existence of intensive DAP luminescence is one of the sure sign that material is not acceptable for good detector manufacturing. The decay kinetics of A luminescence band is also non-exponential and from decay kinetics it is concluded that tunneling recombination of donor-acceptor pairs occurs. According the [1,2] the defect states involved are some undefined surface defects. The presence of A band indicates that sample has some states for electron and hole trapping and recombination and these carriers are not involved in charge collection in radiation detectors.

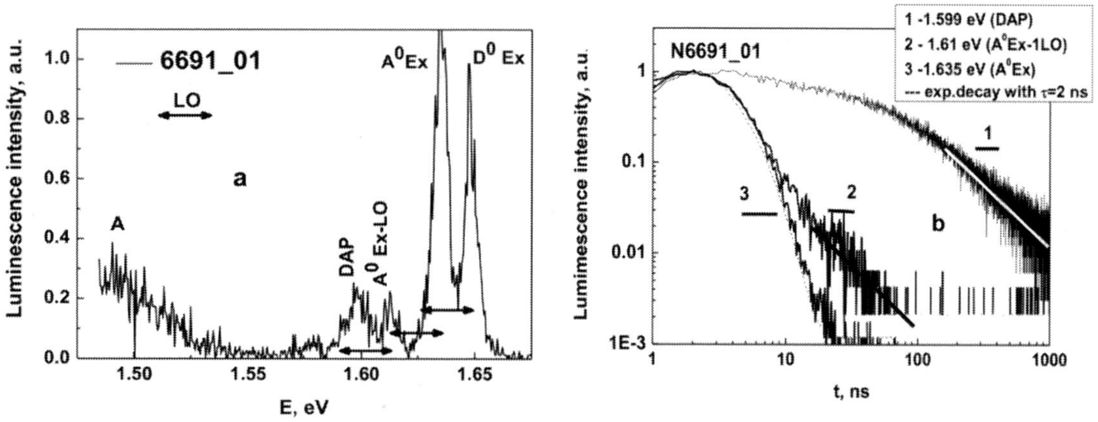

Figure 2. The luminescence spectrum (a) and luminescence decay kinetics (b) of different luminescence bands of sample 6691_01 at 12K.

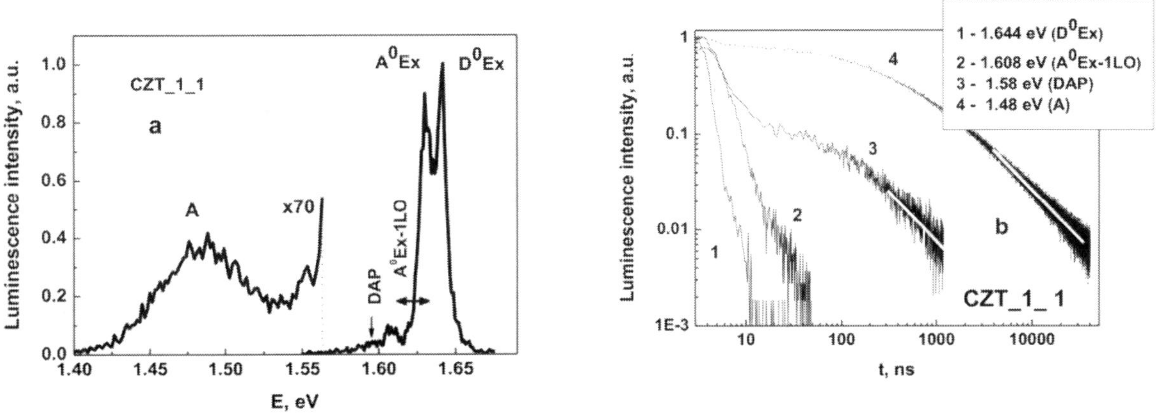

Figure 3. The luminescence spectrum (a) and luminescence decay kinetics (b) of different luminescence bands of sample CZT_1_1 at 12K.

We present here results for three CZT detector crystals showing different spectral resolution of 662 keV gamma rays. The parameters drawn out from results depicted in Fig.1-3 were collected in the Table 1. The analysis of obtained parameters shows that the Eg value could be the same or close for good and bad detectors. The peak positions of D^0Ex and A^0Ex luminescence bands also were very close for all studied detector crystals.

Table 1. The luminescence band position, band gap (Eg) and gamma detector energy resolution for three detector crystals.

Sample number	D0Ex (eV)	DAP (eV)	A	Eg (eV)	FWHM for 662 keV
13364	1.6437	1.598	-	1.4882	12.9
6691_01	1.647	1.6	+	1.4927	18.9
CZT_1_1	1.641	1.599	+	1.478	35.3

4.Conclutions

The low temperature near-band luminescence with time resolution in CdZnTe detector crystals was studied and analyzed. The study was important for the nondestructive detector material selection. The set of parameters such as the decay kinetics and intensity of A^0Ex-1LO and DAP luminescence bands were obtained. The presence of slow luminescence component in 1.59 eV - 1.64 eV spectral range means that DAP luminescence exist. Such material is unacceptable for detector manufacturing. The A luminescence bands are also important characteristic for material selection for gamma and X- rays detector performance however the further additional surface quality studies are needed.

Acknowledgements. The work was supported by LZP grant 09.1126.

References

[1] Hjelt K, Juvonen M, Tuomi T, Nenonen S, Eisslere E E and Bavdaz M 1997 *Phys.stat.sol.(a)* **162**, 747
[2] Schlesinger T E, Toney J, Yoon H, Lee E Y, Brunett B A, Frabks L and James R B 2001 *Material Sci. and Eng.* **32** 103
[3] Ju P, Jie W, Wang T 2011 *Nucl.Inst.&Methods in Phys.Res* ***A643*** 53
[4] Toney J E, Brunett B A, Schlesinger T E, Van Scyoc, James R.B, Schieber M, Goorsky, Yoon H, Eissler E, Jihnson C 1996 *Nucl.Inst.&Methods in Phys.Res.* **A380** 132
[5] Ivanov V, Alekseeva L, Dorogov P and Luchanskii A 2009 *IEEE Nucl. Sci.* **5402238** 1696

Synthesis of cubic and hexagonal $NaYF_4:Er^{3+}$

A Sarakovskis[1] , M Voss, G Doke, D Jankovica, and J Grube

[1]Institute of Solid State Physics University of Latvia, 8 Kengaraga str., Riga, LV-1063, Latvia

Anatolijs.Sarakovskis@cfi.lu.lv

Abstract. Up-conversion luminescence process, which is related to absorption of several light photons (usually infrared) followed by emission of light in the visible or even ultraviolet spectral regions, has attracted interest of scientists due to its potential practical use in various applications including biolabels, temperature sensors, light sources etc. Although observable in d- and f-ions doped materials, the highest efficiency of up-conversion luminescence is usually attained in lanthanides doped hosts. Among huge variety of materials suitable as up-conversion hosts the most prominent is considered to be $NaYF_4$, both due to its low phonon energy and multisite nature of the crystalline lattice. In the present report erbium doped $NaYF_4$ nanomaterial has been synthesized by a novel molten salt method. Depending on the synthesis temperature the structure of the material varied from solely cubic to exclusively hexagonal. It was found that the intensity of the up-conversion luminescence increased drastically as the structure of $NaYF_4:Er^{3+}$ changed from cubic to hexagonal.

1. Introduction

For many years rare-earth (RE) doped materials have been broadly used in a variety of applications. Due to their specific structure of energy levels many fields of science and technology like solid-state lasers, colour displays, optical telecommunication utilize the unique properties of these materials [1-3]. For some decades much attention has been paid for studies of up-conversion (UC) processes in RE doped materials, which involve the absorption of several photons and subsequent emission of one single photon [4]. UC effect could be used in different areas like solid state lasers [5], temperature sensors [6], white light simulation [7], improvement of the efficiency of solar cells [8] and others.

The efficiency of the conversion process strongly depends on the rate of non-radiative transitions within RE ion and therefore on the phonon energy of the medium: the higher the phonon energy, the lower the efficiency of the UC luminescence. Most fluoride crystals (LaF_3 [9], PbF_2 [10], CaF_2 [11], SrF_2, BaF_2 [12]) are attractive hosts for RE ions either due to their low phonon energy or multisite structure. Particular interest has been paid to RE doped $NaYF_4$, which has been widely studied [13-17]. In [13] $NaYF_4:Er^{3+}$, Yb^{3+}, Tm^{3+} is claimed to be one of the most efficient UC phosphors.

A huge variety of the synthesis methods ranging from rather simple solid state reaction [16] to some sophisticated methods like co-thermolysis of sodium and RE trifluoroacetates in some organic

[1] To whom any correspondence should be addressed.

compounds [18] has been reported to obtain hexagonal $NaYF_4$, however, the amount of the synthesized material is usually small (a few grams).

In this work we present a new cheap method for the synthesis of $NaYF_4:RE^{3+}$ which could be potentially used for mass production of this highly efficient phosphor.

2. Experimental

$NaYF_4:Er^{3+}$ (Er^{3+} 2.0 mol%) was prepared from Y_2O_3, Er_2O_3 (purity 4N), NaF (purity 4N) and hydrochloric acid (37%). For the synthesis of the material the rare-earth oxides were dissolved in hydrochloric acid yielding YCl_3 and $ErCl_3$. After evaporation of the liquid the chlorides were added to a mixture of sodium fluoride, $NaNO_3$ and NaCl. The mixture was heated at different temperatures (100°C - 300°C). Afterwards the mixture was dissolved in warm deionized water and filtered through a membrane (0.22mm). Later the precipitate was dried for 24h at about 70°C. Due to purely scientific interest in the samples the amount of the synthesized material was usually several grams but if required it could be easily increased to tens of grams.

The structure of the samples was detected by x-ray diffraction (XRD) measurements (XPert Pro MPD, Cu anode, operating at 40kV, 30mA). The UC luminescence was excited by a temperature controlled laser diode (λ_{em}=980nm, variable radiant power up to 1000 mW) from Thorlabs. Andor SR-303i-B spectrometer coupled to a CCD camera (Andor DU-401-BV) was used for the studies of the stationary luminescence. All the measurements were conducted at room temperature.

3. Results

XRD patterns of the samples obtained at different temperature are shown in Figure 1. One can see that at lower annealing temperatures the patterns are governed by the peaks attributable to a cubic $NaYF_4$ crystalline phase (JCPDS file number PDF 77-2042 in the figure indicated by arrows). As the annealing temperature grows the peaks attributable to hexagonal $NaYF_4$ appear (JCPDS file number PDF 28-1192 in the figure indicated by asterisks) and their intensity increases.

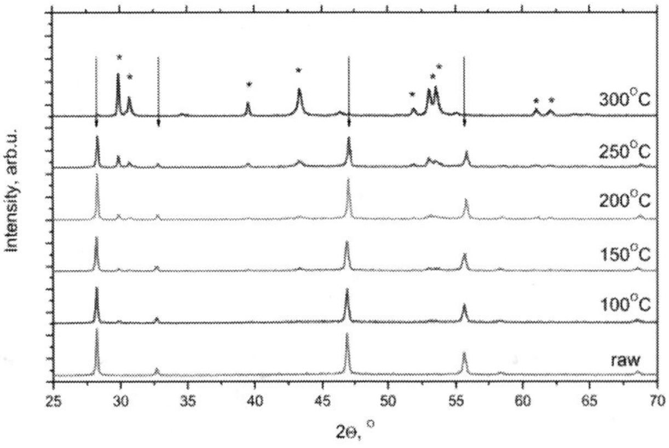

Figure 1. XRD pattern for $NaYF_4:Er^{3+}$ samples prepared at different synthesis temperatures. The peaks attributable to cubic and hexagonal structures are indicated by arrows and asterisks, respectively;

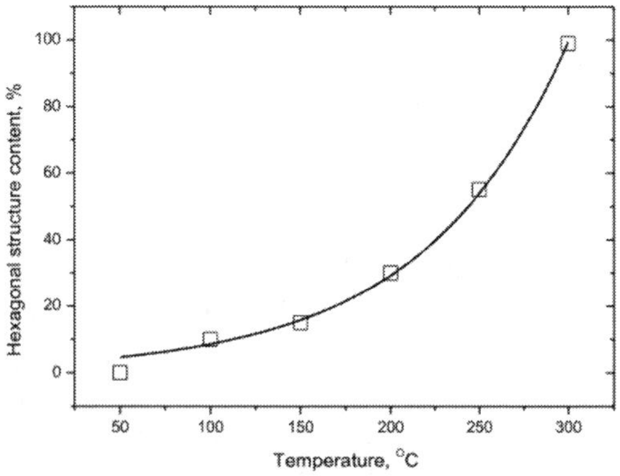

Figure 2. Content of hexagonal phase versus synthesis temperature

The average size of crystallites was determined by Sherrer relation. It was found that the grain size is below 100 nm almost irrespective of the synthesis temperature.

Figure 2 shows the relative content of the hexagonal $NaYF_4$ crystalline phase in the material calculated from the XRD patterns for the samples synthesized at different temperatures. Indeed, one can see that the heat treatment of the samples promotes the hexagonal crystalline phase and after the heat treatment at 300°C the structure of the samples is purely hexagonal.

UC luminescence spectrum under excitation at 980 nm for $NaYF_4:Er^{3+}$ synthesized at 300°C is shown in Figure 3. One can see traditional Er^{3+} luminescence bands at 410 nm ($^2H_{9/2} \rightarrow {}^4I_{15/2}$), 480 nm ($^4F_{7/2} \rightarrow {}^4I_{15/2}$), 520 nm ($^2H_{11/2} \rightarrow {}^4I_{15/2}$), 540 nm ($^4S_{3/2} \rightarrow {}^4I_{15/2}$) and 660 nm ($^4F_{9/2} \rightarrow {}^4I_{15/2}$). Different mechanisms of the up-conversion luminescence as well as energy level structure can be found in [19].

The inset of the Figure 3 shows the dependence of the intensity of the UC luminescence at 540 nm on the annealing temperature of the samples.

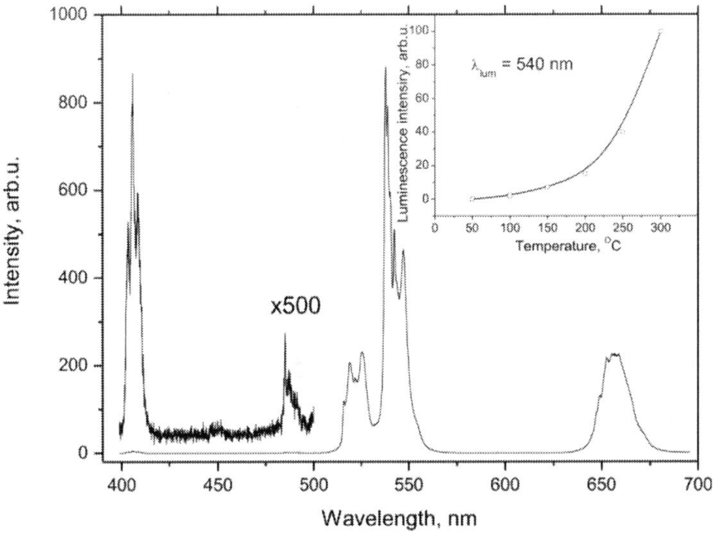

Figure 3. Up-conversion luminescence spectrum measured for $NaYF_4:Er^{3+}$ under excitation at 980 nm. In the inset: dependence of the green luminescence intensity on the synthesis temperature

One can see that the intensity of the green luminescence gradually increases as the synthesis temperature of the samples grows. The maximum intensity of the UC luminescence was observed for the sample synthesized at 300°C. This could be explained by the fact that the efficiency of the UC process in the hexagonal $NaYF_4:Er^{3+}$ is usually higher than that measured for the cubic structure [16]. Indeed the XRD measurements show that the relative content of the hexagonal $NaYF_4:Er^{3+}$ increases as the synthesis temperature grows, and so does the UC luminescence intensity.

It is still not fully clarified why the efficiency of the UC luminescence in erbium doped $NaYF_4$ is considerably higher in hexagonal crystalline phase rather than for the cubic structure. Among probable explanations multisite structure and low phonon energy of the hexagonal $NaYF_4$ are mentioned [20] however additional experiments are required to study this problem in detail.

4. Conclusions

$NaYF_4:Er^{3+}$ was synthesized by so-called molten salts technique. It was shown that depending on the synthesis temperature the structure of the material varied from purely cubic to hexagonal. The efficiency of the UC luminescence reached the maximum for the synthesis temperature of 300°C corresponding to purely hexagonal phase of $NaYF_4:Er^{3+}$.

Acknowledgments

The financial support of State Research Program IMIS (project 1) and ESF project 2009/0202/1DP/1.1.1.2.0/09/APIA/VIAA/141 is highly appreciated.

References

[1] Mendez-Ramos J, Tikhomirov V K, Rodriguez V D and Furniss D 2007 *Journal of Alloys and Compounds* **440** 328
[2] Schweizer T, Hewak D W, Payne D N, Jensen T and Huber G 1996 *Electronics Letters* **32** 666
[3] Payne D N and Reekie L 1988 *Proceedings Fourteenth European Conference on Optical Communication*, Vol. 1, (Brighton: Institution of Electrical Engineers, London) p 49

[4] Auzel F 2004 *Chemical Review* **2004** 139
[5] Tong F, Risk W P, Macfarlane R M and Lenth W 1989 *Electronics Letters* **25** 1389
[6] Dong B, Yang T and Lei M K 2007 *Sensors and Actuators* B **123** 667
[7] Silva J E C, de Sa G F and Santa-Cruz P A 2001 *Journal of Alloys and Compounds* **323-324** 336
[8] Shalav A, Richards B S and Green M A 2007 *Solar Energy Materials and Solar Cells* **91** 829
[9] Sarakovskis A, Grube J, Doke G, Springis M 2010 *Optical Materials* **32** 832
[10] Mortier M and Patriarche G 2006 *Optical Materials* **28** 1401
[11] Chen D, Wang Y, Ma E, Yu Y and Liu F 2007 *Optical Materials* **29** 1693
[12] Orlovskii Yu V, Basiev T T, Pukhov K K, Vorobiev I N, Papashvili A G, Pelle F, and Osiko V V 2001 *Journal of Luminescence* **94-95** 791
[13] Suyver J F, Grimm J, van Veen M K, Biner D, Kramer K W and Gudel H U 2006 *Journal of Luminescence* **117** 1
[14] Liu F, Wang Y, Chen D and Yu Y 2007 *Materials Science and Engineering:* B **136** 106
[15] Wei Y, Lu F, Zhang X and Chen D 2008 *Journal of Alloys and Compounds* **455** 376
[16] Kramer K W, Biner D, Frei G, Gudel H U, Hehlen M P and Luthi S R 2004 *Chemistry of Materials* **16** 1244
[17] Aebischer A, Hostettler M, Hauser J, Kramer K, Weber T, Gudel H U, Burgi H-B 2006 *Angewandte Chemie International Edition* **45** 2802
[18] Mai H-X, Zhang Y-W, Si R, Yan Z-G, Sun L-D, You L-P and Yan C-H 2006 *Journal of American Chemical Society* **128** 6426
[19] Sarakovskis A, Grube J, Doke G, Springis M 2010 *Journal of Luminescence* **130** 805-811
[20] Renero-Lecuna C, Martin-Rodriguez R, Valiente R, Gonzalez J, Rodríguez F, Kramer K W and Gudel H U 2011 *Chemistry of Materials* **23** 3442

Stability of color centers in $LiYF_4$ crystals at low temperatures

V M Lisitsyn[1], Y V Bikhert[2,5], V I Korepanov[1], L A Lisitsyna[3], A K Dauletbekova[2], A T Akylbekov[2], V M Reiterov[4]

[1] National research Tomsk polytechnic university, Tomsk, Russia
[2] Gumilyov Eurasian national university, Astana, Kazakhstan
[3] Tomsk state university of architecture and building
[4] INCROM Ltd, Saint-Petersburg, Russia

E-mail: bihertevgenij@rambler.ru

Abstract. This paper presents the results of examining the stability of induced radiation defects in crystals at 15K. The following regularities have been established. Defects induced at low temperatures in pure and Nd^{3+} doped $LiYF_4$ crystals are unstable: upon termination of radiation, even at 15K induced color centers are visibly destructed, exposure to the halogen lamp light increases the destruction speed significantly. Heating crystals up to 80K results in complete destruction of color centers induced at 15K.

1. Introduction

$LiYF_4$ crystals doped with rare-earth ions are promising materials for creating active optical elements on their base. Rare-earth ions enter crystals by substituting Y^{3+} ions, thus they can be introduced in big concentrations. One of the main features of optical materials is their radiation resistance. Changes in crystal clarity during exposure results in disastrous change of limits and efficiency of generation of optical media forced radiation. Thus the main focus of this paper is establishing the nature of processes causing the radiation changes of clarity in optically active materials and finding a way to create radiation resistant materials.

It is understood that exposure of $LiYF_4$ crystals provides additional optical absorption in the wide spectral range. Figure 1 shows electron irradiation at room temperature in undoped $LiYF_4$ crystals induces absorption bands with maximum limits of 4.5; 3.7; 2.8 and 2 eV. Transition absorption measured with high time resolution, is significantly different in central spectral range. Radiation at 80K results in induction of bands in regions 4.1; 3.7 and 2.7 or 2.1 eV depending on the orientation of crystal in relation to polarization plane of the probing light. Figure 2 shows that γ - radiation of $LiYF_4$ crystals at room temperature results in induction of absorption bands in region 260; 330; 440; 640 nm (4.7; 3.7; 2.8; 1.95 eV, respectively). Band 3.7 eV is believed to be due to F-centers induced in the crystal. Doping with ytterbium and neodymium reduces the efficiency of induction of these centers in the crystal. Figure 3 calculates the positions of bands corresponding to absorption F (3,9 eV) and V_k (3.02; 2.99; 2.88 eV) – centers close to experimentally measured band positions and supposedly accountable for these bands. It is known that in the course of time, after the exposure, the type of spectra change, due to transformation of defects into more stable forms [4].

[5] To whom any correspondence should be addressed

This paper focuses on research of kinetics of spectrum change of induced optical absorption in nominally "pure" and Nd doped $LiYF_4$ crystals when exposed to electron flows at low temperatures, 15K.

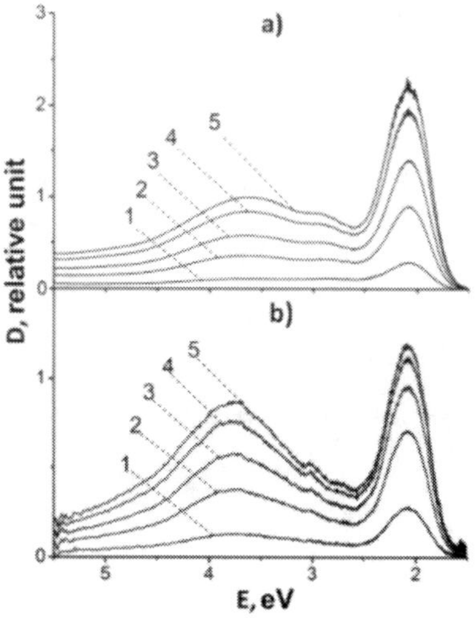

Figure 1. Spectra of induced and measured at 15K extra absorption in $LiYF_4$ crystals, pure (a) and containing 2 at% Nd (b), exposed to a series of electron impulses up to doses: 1- $5\cdot10^2$, 2- $2\cdot10^3$, 3-$4\cdot10^3$, 4-$8\cdot10^3$, 5-$1{,}2\cdot10^4$ Gy.

2. Experimental results and discussion

Pure and Nd^{3+} doped $LiYF_4$ crystals, made in INKROM CJSC, have been examined. Samples have been cut out from single crystals, alongside and perpendicular to optical axis c, then they have been polished. The absorption spectra of examined crystals had narrow absorption lines in region 1.3 – 2.3 eV, which is normal for $LiYF_4$ with Nd^{3+}. Absorption in Nd^{3+} bands region becomes higher as concentration of feed impurity increases: 0; 0.7; 1.5; 2 at%, respectively.

Electron radiation was applied, with high-current pulse accelerator as an exposure source; it generated electron flow with the following specifications: mean electron energy 0.2 MeV, pulse length -10 ns, energy absorbed by crystal per pulse 10^2 Gy. Pulse optical spectrometer was used for exposure and measurement of induced absorption, see the description in [5].

The following procedure was used to measure the spectra of radiation induced absorption in crystals at super-low temperatures. The sample was placed into the crystal holder of cryostat and was cooled down to the set temperature. Probing light from the exposure source was directed toward the sample. Light went through the sample, through monochromator and onto a CCD matrix, spectrum of the light that went through the sample was registered in PC. Then sample was exposed to a series of radiation pulses from small-scale pulsed electron accelerator. Then the radiation from the same source of probing light was directed toward the sample. The spectrum of the light that passed through the irradiated sample is different to the spectrum obtained when passing through unexposed sample. The result of measuring the spectrum that passed through the irradiated sample was also saved in PC. Ratio of spectrum values of probing light flows that passed through the sample before and after radiation exposure are obviously the spectral ratios of radiation induced transmission. Spectrum values of transmission rates were then converted into spectral values of transmission density of absorption factors. The corresponding spectra of radiation induced additional absorption in the sample were constructed based on calculated spectral values of transmission, optical density or absorption factors.

Typical spectra of extra absorption, induced by radiation by a series of pulses from electron flow from accelerator, in nominally pure $LiYF_4$ crystals and crystals with approx. 2at% Nd are represented in Fig. 1, a and b. Crystals have not been exposed before, and were transparent in the examined spectrum range. In the spectrum of induced absorption at 15K of pure and doped crystals, there are visible well-defined absorption bands at 2.1 and in 3.7 eV region, less defined band in 2.8 eV region, ill-defined band in 4.5 eV region. In doped crystals, 2.1 eV band is slightly widened into the short-wave region.

With increase of radiation dose, the size of additional absorption grows in the whole region of the spectrum, appearance of induced absorption spectra does not change visibly, but increment decreases with increase of dose. At radiation doses higher than 200 pulses, and corresponding ~$2\cdot10^4$ Gy, it tends to saturate the growth of induced absorption.

Pause in exposure, and pause in a series of sequential pulses results in diminution of the size of induced absorption. When $LiYF_2$ crystal + 2at% Nd, pre-exposed to a series of 200 subsequent pulses, is kept at constant temperature of 15K, decline of absorption in the whole spectrum range is observed, which reaches 25% relative to initial one, in 100 minutes. The decline of induced absorption when keeping the crystal at low temperature depends on the history of the crystal. Some crystals can show the decline of induced absorption up to 50%.

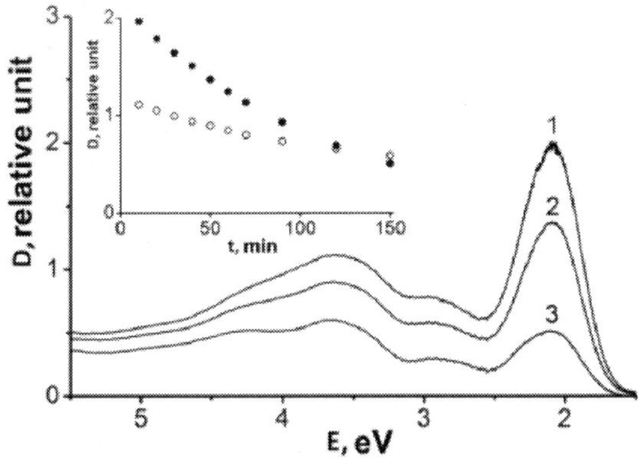

Figure 2. Spectra of induced and measured at 15K extra absorption in pure $LiYF_4$ crystals, exposed to radiation dose of $3{,}5\cdot10^4$ Gy and exposed to halogen lamp radiation for 10, 50, and 150 min. The inserted image shows the kinetic curve of crystal bleaching.

Further exposure of crystal after the pause in radiation in a series of consecutive pulses results in recovery of initial form of kinetics of growth of optical absorption in the whole spectral region.

Reduction of induced absorption size during the pause between series of radiation pulses in the band in 2.1 eV region is notably bigger that in the band in 3.7 eV region for all examined crystals.

Induced at 15K additional absorption is sensitive to the effect of radiation in optical spectrum. Exposure to radiation of halogen lamp of typical light of "Avantes" spectrometer of pure and doped $LiYF_4$ crystals, previously irradiated with series of electron pulses at 15K, results in decline of induced absorption in the spectral region. Figure 2 shows changes in spectrum of induced absorption with dose of $3.5\cdot10^4$ Gy at 15K in pure $LiYF_4$ crystal, with time of exposure of crystal to halogen lamp radiation. Partial bleaching of crystal was observed in the whole examined spectrum region. Appearance of spectrum remains unchanged, but the ratio of absorption in bands changes. Kinetics of decline of absorption is shown on inserted image on Figure 2. After 150 min of exposure to halogen lamp, absorption in the band of 2.1 eV region decreased ~ 4 times, band in 3.7 eV region – ~ 2 times. Such effect is observed when previously irradiated Nd doped crystals were exposed to halogen lamp radiation. For instance, photoexcitation of previously irradiated at 15K with a dose $2\cdot10^4$ Gy of $LiYF_4$ + 2 at% Nd crystal, absorption band in 2.1 eV region is reduced threefold, whereas band in 3.7 eV region only by 25%.

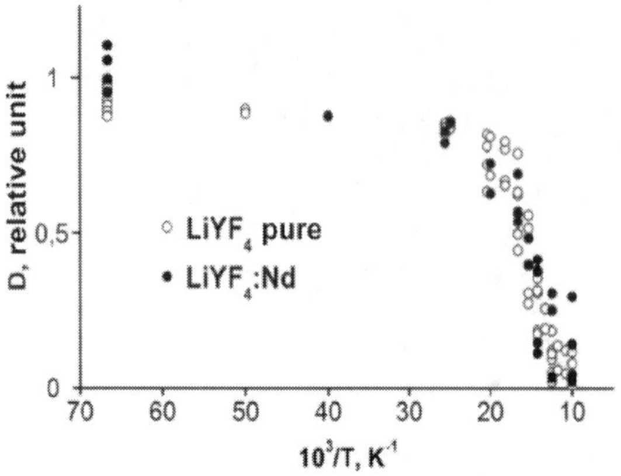

Figure 3. Temperature dependence of stability of radiation induced absorption bands at 15K in ranges 3.7 and 2.1 eV in pure and Nd doped $LiYF_4$ crystals.

Figure 3 represents the results of examining the thermal stability of radiation induced color centers at 15K in $LiYF_4$ crystals. Experiment was conducted as follows. The crystal was exposed to a series of electron pulses up to dose of 10^4 Gy, the spectrum of induced absorption was measured at this temperature. Then crystal was heated to certain temperature $T_i > 15K$, cooled to 15K, then induced absorption spectrum was measured again. Afterwards crystal was heated to $T_j > T_i$, cooled, spectrum measured at 15K, etc. This procedure was repeated until induced absorption disappeared. Resistance of induced color centers was examined for pure and Nd doped crystals. Figure 3 shows curves color center annealing, normalized for initial optical density at 25K, liable for bands in 3.7 and 2.1 eV regions. It should be noted that these bands are annealed absolutely likewise. The appearance of annealing curves is unaffected by the level of purity of crystal.

Research results show that when heating up to 45K, induced absorption remains almost unchanged. When temperature rises above 45K, absorption drops abruptly, centers start breaking. Maximum destruction speed occurs at 60K. By 80K induced centers and corresponding absorption disappears completely.

Preliminary irradiation of crystals at 15K, this is especially shown in doped crystals, and further annealing up to room temperature does not restore the initial defect structure of the sample completely. Exposure of crystals to doses up to 400 … 1000 pulses at 15K results in generation of characteristic spectra of additional absorption, given on Fig. 1. According to results given on Figure 3, further heating results in complete bleaching of crystals at 100K. Clarity of the crystal heated to room temperature can be compared to the initial one. Crystal remains clear when stored in darkness, from several days to one year. However, further exposure of such crystal (irradiated at 15K by a series of radiation pulses, annealed to room temperature and kept at this temperature for a long period) results in induction of additional absorption spectrum at 15K, which is visibly different to the one induced in crystal without preliminary radiation exposure.

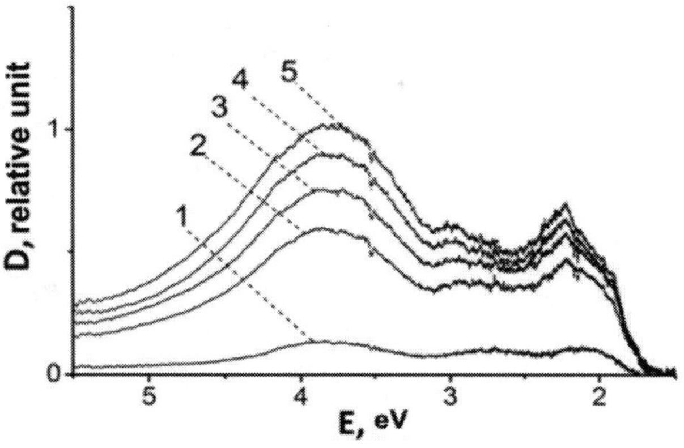

Figure 4. Spectra of absorption induced at 15K in $LiYF_4$ crystal doped with 2a at% of Nd, preliminarily exposed to $7.2 \cdot 10^4$ Gy radiation dose, then heated to room temperature and kept at this temperature for 15 days and exposed at 15K to radiation doses of 1 - $5 \cdot 10^2$, 2 - $3 \cdot 10^3$, 3 - $8 \cdot 10^3$, 4 – $1,2 \cdot 10^4$, 5 – $2 \cdot 10^4$ Gy.

Figure 4 shows results of research of absorption induced at 15K in crystal $LiYF_4$ with 2at% of Nd, previously exposed to 720 radiation pulses (7,2 $\cdot 10^4$ Gy), then heated to room temperature and kept at this temperature for 15 days. Absorption spectra were measured after exposure of such crystals to different doses of radiation pulses.

The band in 2.1 eV region of the induced absorption spectrum in such crystal has a significantly different shape and size from the one induced in the initial crystal. Band in 2.1 eV region grows slower with increase of radiation dose, compared to the band in 3.7 eV. Band in 3.7 eV is obviously dominating band in the spectrum.

Thus, doped crystal keeps the data about the previous exposure at 15K even after annealing, and storage at room temperature. Note that the described effect is not observed or is insignificant when annealing pure crystals, they do not save the data on previous exposure at low temperatures after annealing and storing at room temperature.

3. Conclusions

Exposure of pure and Nd doped $LiYF_4$ crystals to electron flows at 15K results in induced color centers in them, which emerge in absorption bands in 2.1 and 3.7 eV, and in ill-defined bands of 2.8 eV and 4.5 eV regions. The view of spectra of additional absorption, induced in pure and doped crystals is similar. Efficiency of coloring of doped crystals is slightly lower. When increasing the radiation dose, the absorption also increases in the whole examined spectral range – 1.5...5.5eV, growth speed tending to slow down with radiation dose. After termination of radiation exposure, even at 15K, absorption drops. In 100 minutes, absorption can drop to 25% from initial value. Influence of halogen lamp light results in significant acceleration of decline of induced absorption. In 150 minutes of influence of light flow onto irradiated at 15K with 2 $\cdot 10^4$ Gy $LiYF_4$ crystal + 2 at% Nd, there is threefold reduction of the absorption band in region 2.1 eV, when the band in 3.7 eV region – only by 25%. Within the same period of exposure to halogen lamp light, absorption in band in 2.1 eV of pure $LiYF_4$ crystal reduced by ~ 4 times, when band in 3.7 eV - ~ 2 times. Thus, radiation induced pure and $LiYF_4$ crystals color centers are unstable at 15K. Decomposition speed of centers responsible for bands 2.1 eV and 3.7 eV is different. This indicates that these centers are of different nature.

Heating crystals results in decomposition of induced color centers. When heating to 45K, induced absorption remains almost unchanged. When temperature rises above 45K, absorption drops abruptly, and center start to break. Maximum speed of decomposition of centers is at 60K. By 80K, induced centers and corresponding absorption disappear completely. Curves of decomposition of centers responsible for absorption in 2.1 eV and 3.7 eV regions are absolutely similar. This means that these centers decompose under the influence of same processes.

The following observation is of interest. If $LiYF_4$crystal with 2at% Nd, is exposed to high doses of radiation, and then heated to room temperature and kept at room temperature for a long time, crystal restores its clarity completely. But the reaction of such crystal to further exposure at 15K is absolutely different to the reaction to exposure of a crystal without pre-treatment. The band in 2.1 eV region in the induced absorption spectrum of such crystal is different in shape and size from the band induced in a crystal that has not been exposed. Band in 2.1 eV region grows slower with increase of radiation dose, compared to the band in 3.7 eV. Band in 3.7 eV is obviously dominating band in the spectrum.

Thus, doped crystal keeps the data about the previous exposure at 15K even after annealing, and storage at room temperature. Note that the described effect is not observed or is insignificant when annealing pure crystals, they do not save the data on previous exposure at low temperatures after annealing and storing at room temperature.

This effect can possibly be explained as follows. In the radiation field at low temperature, there is a transformation of the crystal structure in the region of impurity ion. The generated defective region is a competitive center of trapping carriers, involved in the formation of color center, responsible for band in 2.1 eV.

IOP Conf. Series: Materials Science and Engineering **38** (2012) 012039 doi:10.1088/1757-899X/38/1/012039

References

[1] Chinkov Y P, Krasnousov I V, Lisitsyna L A, Reyterov V M 1990 *Izv. Latv. Akad. Nauk.* **5** 60
[2] Kaczmarek S M, Bensalah A, Boulon G 2006 *Optical Materials* **28** 123
[3] Li Q E, Chen L F 2006 *Solid State Comm.* **140** 107
[4] Lisitsyn V M, Korepanov V I, Yakovlev V Y 1996 *Izv. Vuzov. Fizika* **11** 5
[5] Korepanov V I, Lisitsyn V M, Oleshko V I 200 *Izv. Vuzov. Fizika* **43** 22

IOP Conf. Series: Materials Science and Engineering **38** (2012) 012040 doi:10.1088/1757-899X/38/1/012040

Color centers and structural damage in LiF induced by 150 MeV Kr ions

A Rusakova[1], J Maniks[26], K Schwartz[3], A Dauletbekova[1], A Akilbekov[1], V.Lisitsin[4] and M Zdorovets[5]

[1]L.N. Gumilyov Eurasian national university, Astana, Kazakhstan
[2]Institute of Solid State Physics, University of Latvia, Riga, Latvia
[3]GSI Helmholtz Zentrum für Schwerionenforschung, 64291 Darmstadt, Germany
[4]Tomsk Polytechnic University, Tomsk, Russia
[5]Institute of Nuclear Physics, Kazakhstan

E-mail: manik@latnet.lv

Abstract Color centers and evolution of structure defects were investigated in LiF crystals irradiated at room temperature with 150 MeV ^{84}Kr ions with a beam current of 10nA/cm^2 in the fluence range 10^{11} - 10^{14} ions/cm^2 at the cyclotron accelerator DC-60 (Astana, Kazakhstan). At the fluence of 10^{11} ions/cm^2, SEM imaging revealed mainly formation of etchable ion tracks. Above this fluence, severe structural modifications in the irradiated layer were observed which include the ion-induced formation of dislocations and grains with nano-scale dimensions. The role of fluence in the concentration of electronic color centers and structural modifications is discussed.

1. Introduction

Energetic ions are an excellent tool for the creation of nano-structures on the surface and in the bulk of solids. This study is aimed to improve the understanding of mechanisms of ion-induced nano-structuring and continues our recent investigations of color centers and creation of structural damage in LiF under heavy ion irradiation ([1 – 5] and references within). The effect of fluence on formation of nano-structure in LiF irradiated with swift Kr ions is explored.

2. Experiments and discussion

2.1 Spectroscopy

LiF crystals were irradiated at room temperature with 150 MeV $^{84}Kr^{+14}$ ions (range R = 17.5 μm) at a beam current of 10 nA/cm^2 (corresponding to a flux $\varphi = 4.46\times10^9$ ions/cm^2.s) in the fluence range 10^{11} - 10^{14} ions/cm^2 at the cyclotron accelerator DC-60 (Astana, Kazakhstan). The experiments were performed in a period of two months after irradiation.
Using UV-VIS spectroscopy electron color centers (F and F_n) in the range of 1.4 – 6.5 eV (190 - 900 nm) were studied versus fluence (Φ). The absorption spectra for various fluences are presented in Fig.1. At $\Phi = 10^{11}$ ions/cm^2, the spectrum shows only the absorption of F and F_2 centers whereas at

[6] To whom any correspondence should be addressed

fluences $\Phi \geq 10^{12}$ ions/cm^2 besides absorption of F_2 centers also features corresponding to more complex F_3 and F_4 centers appear. The concentration of F centers was estimated using the Smakula - Dexter formula [6]:

$$n_F\left[cm^{-2}\right] = 9.48\times10^{15} D_{opt}(F) \quad (1)$$

where D_{opt} *(F)* is the optical density at absorption maxima for *F* centers (Table1). The volume concentration of *F* centers N_F (cm^{-3}) can be obtained as $N_F = n_F/R$ where *R* is the ion range. The distance between the ion impacts is equal to $d_{ion} = 2\times (1/\pi\Phi)^{1/2}$ and the average distance between the F centers is equal to $d_{F\text{-}F} = (1/N_F)^{1/3}$. The concentration of all F_n centers was estimated from the integral absorption (A_{Fn}) in the energy range from 1.5 to 4.2 eV [1].

The results from optical spectroscopy are presented in Fig. 1 and Table 1. A saturation of F centers takes place at $\Phi \sim 10^{13}$ ions/cm^2 where the volume concentration $N_F{\sim}10^{19}$cm^{-3} (Fig.1). At such fluence a strong interaction between neighbor tracks and their overlapping takes place (Fig.2). The distance of ion impacts d_{ion} is close to the average distance between the F centers ($d_{ion} \approx d_{F\text{-}F}$ =4.7 nm, Table1). As a result the concentration of single F centers decreases due to a strong aggregation of F centers according to F + F →F_2, F + F_2→ F_3, etc. These results are in a good agreement with other studies of aggregation processes in ionic crystals [6, 7].

The formation of F_n centers is most efficient in the fluence range between 10^{12} – 10^{13} ions/cm^2 where the strongest increase of the area A_{Fn} is observed (Table1). At fluences 10^{13} - 10^{14} ions/cm^2 $d_{ion} \approx d_{F\text{-}F}$ which leads to a stronger electron and hole center interaction (aggregate formation or recombination), as well as creation of structural damage [1, 3]. At such high fluences also ion induced stress plays an important role in the creation of structural damage [3].

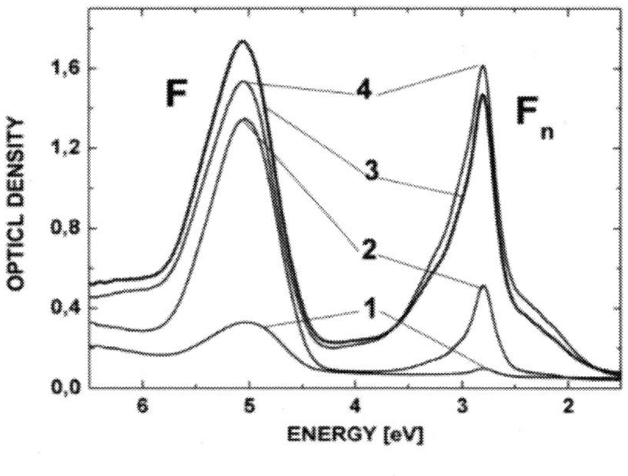

Figure 1. Absorption spectra for LiF crystals irradiated with 150 MeV Kr ions at a flux of φ = 4.46x10^9 ions/cm^2.s at fluences 1 – 10^{11}, 2 – 10^{12}, 3 – 10^{13}, 4 – 10^{14} ions/cm^2

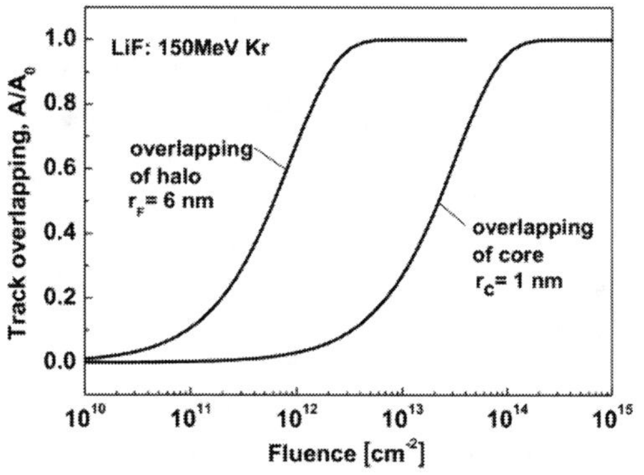

Figure 2. Calculated overlapping of halo and core regions of Kr tracks in LiF using radius of halo 6 nm and radius of track core 1 nm.

Table1. Parameters of color centers versus fluence

Φ, ions/cm^2	d_{ion}, nm	n_Fx10^{16}, cm^{-2}	N_Fx10^{18}, cm^{-3}	$d_{F\text{-}F}$, nm	A_{F2}, arbit.units
10^{11}	36	0.31	1.78	8.0	0.44
10^{12}	11	1.28	7.29	5.2	2.27
10^{13}	3.5	1.64	9.40	4.7	6.50
10^{14}	1	1.46	8.32	4.9	7.14

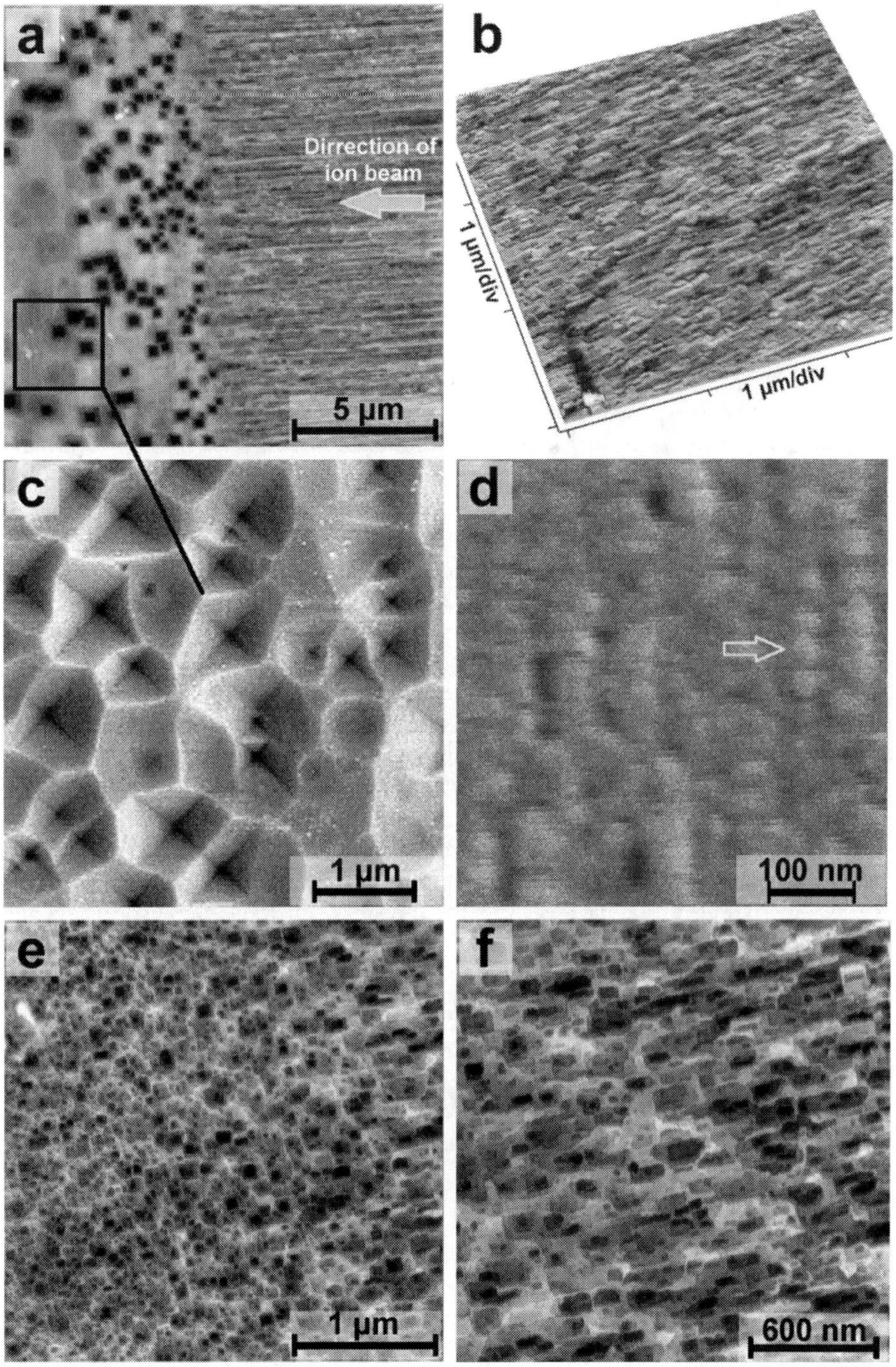

Figure 3. View of ion-induced structures on profile surfaces obtained by cleaving along the ion path: (a) –overall view (AFM image)), (b)- nanostructured zone, (c)- dislocation rich zone (SEM image), (d) – nanostructured zone at high magnification. Arrow shows nano-grains; (e) –dislocations in tail part of range, (f) - dislocation etch pits in the nanostructured zone. Irradiations were performed to fluences 10^{12} ions/cm^2 (samples *a* and *b*), 10^{13} ions/cm^2 (*c* and *d*), 10^{14} ions/cm^2 (*e* and *f*).

2.2 Structural study

Structural defects in the irradiated crystals were revealed by selective chemical etching in a saturated aqueous $FeCl_3$ solution. The surface topography was studied using a scanning electron microscope JSM-7500F (JEOL) and an atomic force microscopy CPII (Veeco). The investigation of structural modifications in the bulk of irradiated zone was performed on the cross-sections prepared by cleaving and subsequent etching the samples along the ion path (normal to the irradiated (100) surface).

An overview of the profile of irradiated layer is shown in Fig. 3 (a). A substantial change in the structure is observed up to 17.5 μm depth which coincides with the calculated ion range. Two zones exhibiting different types of structural features are clearly seen: (1) the nanostructured zone consisting of small columnar grains and (2) the zone with numerous square-based pyramidal etch pits which are typical for dislocations. Between the nanostructured and dislocation-rich zones we can see approximately 2 μm wide transitional region. At a higher magnification the grains with nanoscale dimensions (30 - 50 nm) (Fig. 3 (b) and (d)) and some amount of ion-induced dislocations (Fig.3 (f)) in the nanostructured zone are observed. Fig. 3 (c) and (e) present a detailed view of the dislocation-rich zone. The density of dislocations in this zone increases with fluence. A broadening of the transitional region with fluence is also observed. At the highest fluence the average density of dislocation etch pits in the transitional region reaches about 5×10^9 cm^{-2}.

It is well known that prismatic dislocation loops are formed in alkali halides through the aggregation of interstitials or vacancies that are generated by the irradiation. The observation of prismatic dislocation loops under irradiation with high-energy electrons inside a TEM was reported earlier in [8]. In irradiated crystals, prismatic dislocations of interstitial type are supposed to be dominating due to comparatively high room- temperature mobility of interstitials.

We compared the results of the structural modifications along the ion path with the calculated energy loss. The stopping of 150 MeV Kr ions in LiF is dominated by the electronic energy loss. The results show formation of nanostructure in a depth range where the electronic energy loss exceeds the threshold of 10 keV/nm for track core damage. At the tail part of the ion path, where the energy loss decreases below this limit, the formation of a dislocation-rich structure has been observed.

Structural studies were performed also on irradiated surface (Fig.4). After irradiation, the SEM images for all fluences show numerous pyramidal etch pits which are typical for ion tracks or dislocations. The energy loss of Kr ions on surface exceeds the threshold for track etching [9]. For that reason the etch pits are related mainly to ion impacts (tracks). However, the density of etch pits is consideredrably

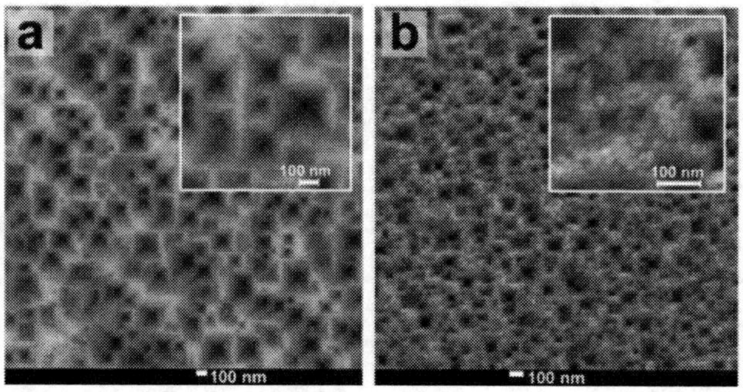

Figure 4. SEM image of LiF surface after irradiation with 150 MeV Kr ions at fluences (a) 10^{11} and (b) 10^{13} ions/cm^2. The inserts show the structures at a higher magnification.is about 5×10^9 cm^{-2}

lower than the applied fluence. Such difference can be explained by overlapping of a part of tracks. At fluences 10^{12} – 10^{14} cm^{-2}, the etching on irradiated surface reveals small (few tens of nm) grains that indicate ion-induced nanostructuring (Fig.3 (b), insertion). No features of nanostructuring were

observed on surface irradiated to fluence 10^{11} cm^{-2}. The results allow us to conclude that nanostructuring is observed above the threshold fluence of 10^{11} cm^{-2}.
The type of nanostructure formed under SHI irradiation was analyzed using XRD reciprocal space mapping [4]. In the nanostructured samples both the tilting of grains and variation of the interplanar distance were found to appear. The contribution of prismatic dislocation to the variation of interplanar distance is suggested. The estimates give low tilting angles that indicate formation of a mosaic-type nano-structure. The fragmentation in LiF crystals and reduction of the size of coherent domains was detected also by XRD measurements after neutron irradiation [10]. Low angle boundaries are known to be composed of dislocations. Therefore the role of dislocation processes in ion-induced nanostructuring is supposed to be significant. A major input in self-organized structural fragmentation in heavily irradiated LiF comes from ion-induced internal stresses [11]. The ion-induced defects, such as ion tracks and dislocations exhibit local stress field that determines their interaction. At fluences as high as 10^{13} ions/cm^{-2}, strong overlapping of halo regions and even core regions of tracks occurs (Fig.2) that facilitates the aggregation and interaction processes. The stress-driven ordering of dislocations and formation of boundaries are the main steps of nanostructuring.

3. Conclusion

The formation of a bulk nanostructure in LiF irradiated with 150 MeV Kr ions is confirmed. The nanostructure consists of LiF grains with nanoscale dimensions (30-50 nm). The results show that nanostructures appear at high fluence irradiation, when F centers reach saturation (volume concentration $>10^{19}$ cm^{-3}) and are observed only in regions where the ion energy loss has surpassed the threshold value (>10 keV/nm) for formation of the track core damage. The determined threshold fluence is between 10^{11} cm^{-2} and 10^{12} cm^{-2}.
The structural modifications include ion-induced formation of dislocations in all irradiated layer. The ordering of ion-induced dislocations in the stress field of tracks and other aggregates is a relevant stage of structural fragmentation and nanostructuring.

Acknowledgements: This work has been partly supported by the Latvian government grant No 09.1548

References

[1] Schwartz K, Sorokin M V, Lushchik A. 2008 *Nucl. Instr. Meth. B* **266** 2736
[2] Dauletbekova A, Akilbekov A, Zdorovets M 2010 *Phys. Status Solidi B* **247** 1227
[3] Dauletbekova A, Maniks J, Manika I, Zabels R, Aklibekov AT, Zdorovets MV, Bikert Y, Schwartz K 2011 *Nucl. Instr. Meth. B*, doi: 10.1016/j.nimb.2011.12.030
[4] Maniks J, Manika I, Zabels R, Grants R, Tamanis E, Schwartz K 2012 *Nucl. Instr. Meth. B* **282** 81
[5] Maniks J, Manika I, Zabels R, Schwartz K, Sorokin M and Papaleo R M 2011 *Applied Physics A* **104** 1121
[6] Perez A, Balanzat E and Dural J 1990 *Phys.Rev. B* **41** 3943
[7] Schwartz K, Trautmann C, El-Said A S, Neumann R, Toulemonde M and Knolle W 2004 *Phys. Rev. B* **70** 184104
[8] Kawamata Y 1976 *Journal de Physique* **37** C7-502
[9] Trautmann C, Schwartz K and Geiss O 1998 *J. Appl. Phys.* **83** 3560
[10] Kubo K, Hamaguchi Y, Motohashi H 1964 *J. Phys. Jap.* **19** 1105
[11] Kumar M, Singh F, Khan S A, Tripathi A, Avasthi D K, Pandey A C, 2006 *J Phys. D: Appl. Phys.* **39** 2935

IOP Conf. Series: Materials Science and Engineering **38** (2012) 012041 doi:10.1088/1757-899X/38/1/012041

Ab initio calculations of the *H* centers in MgF_2 crystals

F U Abuova[1,4], **A T Akilbekov**[1], **E A Kotomin**[2,3]

[1]*L. N. Gumilyov Eurasian National University, 3 Munaitpasova, Astana, Kazakhstan*
[2]*Institute of Solid State Physics, University of Latvia, 8 Kengaraga Str., LV-1063 Riga, Latvia*
[3]*Max Planck Institute for Solid State Research, Heisenbergstr.1, Stuttgart, Germany*
E-mail: fatika_82@mail.ru

Abstract. MgF_2 with rutile structure is important wide-gap optical material with numerous applications. We present and discuss the results of calculations for basic hole defects – interstitial F atoms (called also the colour *H* centres). This study is based on the large scale *ab initio* DFT calculations using hybrid B3PW exchange-correlation functional as implemented into CRYSTAL computer code. The electronic structure, atomic geometry, charge density distribution are calculated and discussed.

1. Introduction

Lenses and optical windows made of magnesium fluoride (MgF_2) are transparent over an extremely wide range of photon energies, from vacuum ultraviolet to infrared. It is also radiation-resistant material, the energy required to form a stable primary radiation defect known as the *F* centre (fluorine vacancy with trapped electron) between 5K and room temperature is much higher than in other alkali halides [1]. To our opinion, this could be related to a slow diffusion-controlled recombination of the primary radiation defects in MgF_2 - the complementary *F* and *H* centres (interstitial fluorine ions), partly due to defect diffusion and interaction peculiarities.

These two factors make MgF_2 an important material for space telescopes and similar applications. However, it is essential for materials scientists to understand at the atomistic level reasons of MgF_2 high radiation stability. Despite several *ab initio* calculations of basic MgF_2 properties (both bulk [2,3] and surfaces [4,5]), we are not familiar with defect studies in this material at the *ab initio* level (unlike other fluorites — CaF_2 and BaF_2 [6,7]). In this paper, we present results of *ab initio* calculations of the *H* centres in the MgF_2 bulk crystals.

2. Computational method

MgF_2 (rutile) has a tetragonal $P4_2/mnm$ structure with two formula units in the unit cell. As in previous calculations on the MgF_2 bulk and (001) surface [4,5], we used the CRYSTAL computer code [8] using localized, Gaussian-type basis sets. Our spin-polarized calculations were performed using the DFT-HF hybrid exchange-correlation B3PW functional [9] which gives better agreement of calculated and experimental MgF_2 optical gap than when applying other functionals: 9.5 eV (B3PW) *vs.* 19.65 eV (HF), 6.9 eV (GGA-PBE), (*cf.* with the experimental value of 12.5 eV – see Ref. [4]).

[4] To whom any correspondence should be addressed

The reciprocal space integration was performed by sampling the Brillouin zone with a 5 × 5 × 5 Pack-Monkhorst mesh [10]. To achieve high accuracy, large enough tolerances of 7, 8, 7, 7, and 14 were chosen for the Coulomb overlap, Coulomb penetration, exchange overlap, first exchange pseudo-overlap, and second exchange pseudo-overlap, respectively [8]. In our calculations for MgF_2 we applied the basis set developed by Catti *et al.* for the F atom [11] and by McCarthy and Harrison [12] for the Mg atom. The effective atomic charges were calculated using the Mulliken population analysis [13].

To simulate the *H* centres, we used a 2 × 2 × 2 times expanded unit cell (3 % defect concentration) containing 48 atoms. An interstitial F atom was then introduced therein and its chemical bonding with regular F ions was studied. It is well known that typical hole defects in alkali halides and fluorites are so-called split interstitials (dumbbells), *e.g.*, F_2^- ions in the singlet state (one unpaired electron) [14]. These defects are complementary *F* centres (fluorine vacancy). As a result of calculations, we obtained the relaxed lattice geometry, effective atomic charges and interatomic bond populations, the band structure, density of states (DOS) and defect formation energies.

3. Main results

We found two possible orientations of the *H* centres, with the interatomic bonds oriented along the (001) and (110) directions (Fig.1, a,b, respectively). The latter configuration is lower in energy (by 0.37 eV) than the former one, in agreement with the experiment [15]. The equilibrium interatomic distances in the (110) and (001) centres are 1.96 Å and 1.82 Å , respectively. Notice that the distance in the free F_2^- molecule is 1.98 Å, that is both molecular ions are compressed in a crystal, more in the (001) direction. For a comparison, the bond length in the *H* centre in fluorites BaF_2, SrF_2, CaF_2 are very similar, 1.98 [16].

The effective charges of the F and Mg ions in a perfect MgF_2 are -0.9 *e* and 1.80 *e*, respectively [4,5], which indicates a high ionicity of this material. When additional F-49 atom is introduced into the interstitial position in the (110) direction (Fig. 2), it shares the electron density almost equally with two nearest F ions (the effective charges are -0.59 *e* (F-49) *vs*. -0.68 *e* (*H1* and *H2*), respectively).

After forming a chemical bond with one of these F ions, the effective charges of the two atoms of the *H* centre along the (100) direction become -0.46 *e* each whereas the total spin on these fluorine atoms is 0.92 μ_B. That is, this defect is very close to the classical *H* centre model discussed above. This is well illustrated in the electron density maps in Fig. 3.a,b.

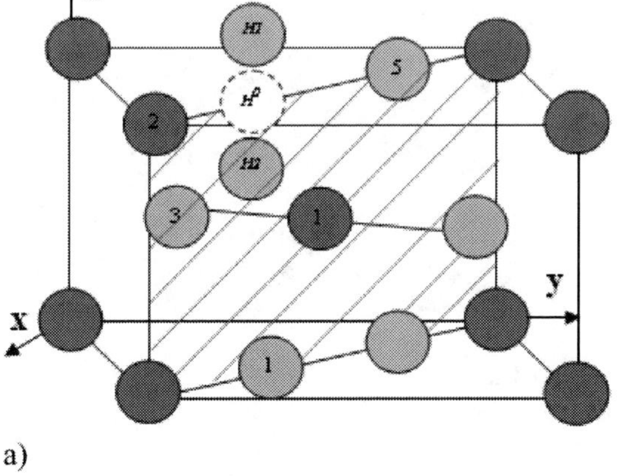

a)

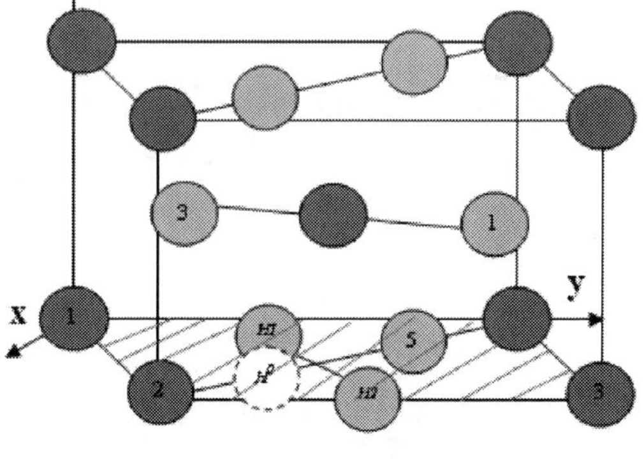

b)

Figure 1. Schematic view of the (001)- and (110)-oriented *H* centres (a and b, respectively). H^0 is a position of a regular F ion before defect formation.

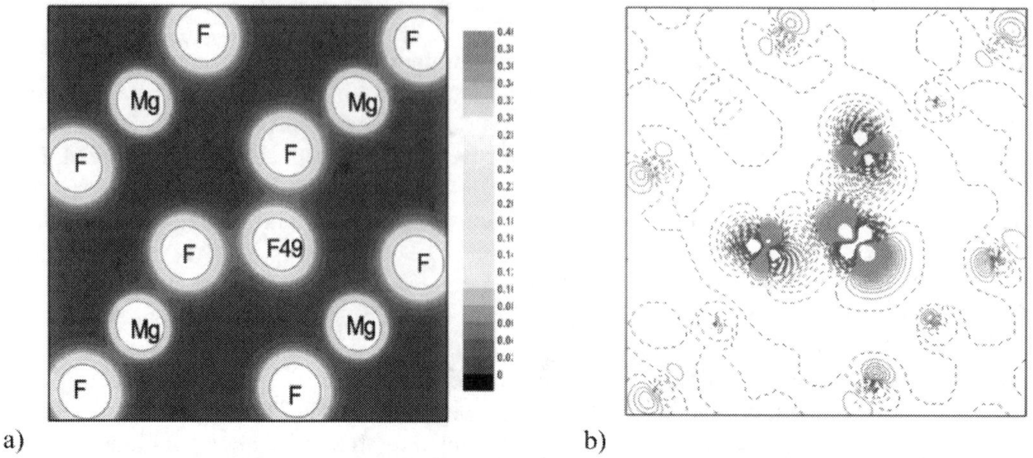

Figure 2. The total (a) and difference (b) electronic density maps for the additional F atom (F-49) incorporated into the 48 atom supercell of the perfect MgF_2. The red (full) and blue (dashed) colours indicate electron density deficiency and excess, respectively. The density range is from 0 to 0.4 *e*/bohr^3 with increment of 0.02 *e*/Bohr^3.

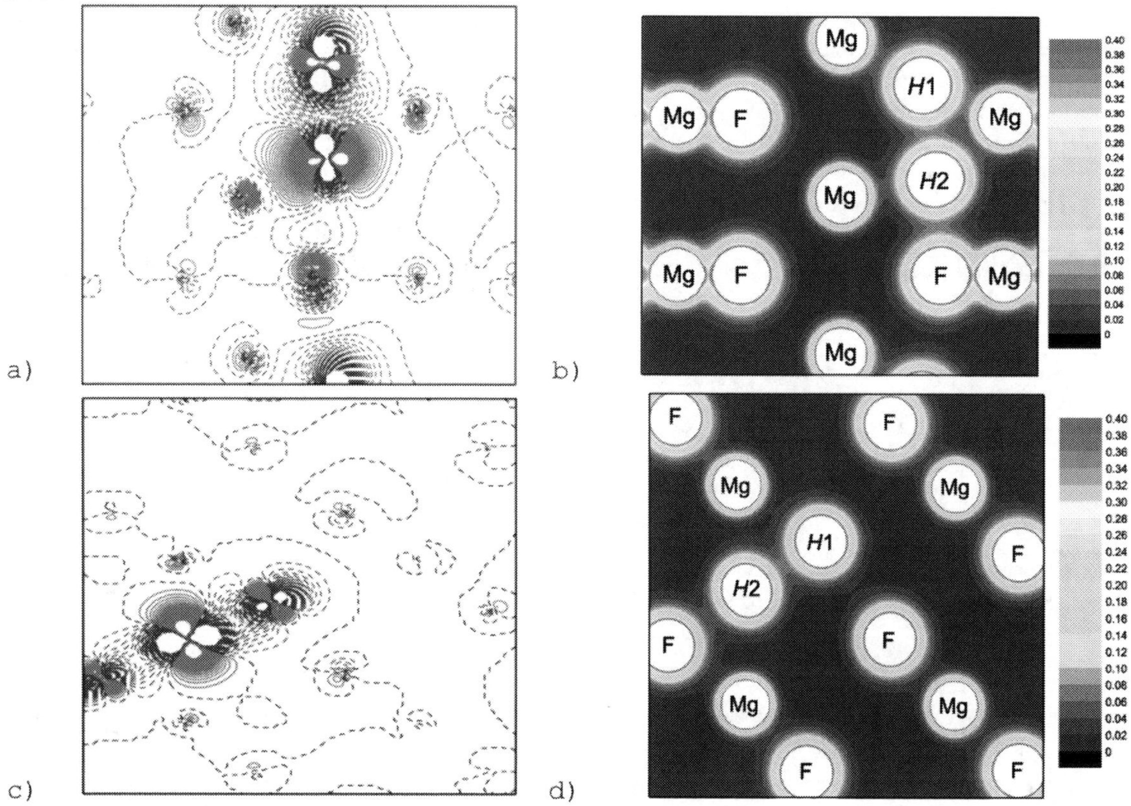

Figure.3. The difference and total electron density redistribution around the *H* centers in the (001) (a,b) and the (110) (c,d) planes. The density range is from 0 to 0.4 *e*/Bohr^3 with increment of 0.02 *e*/Bohr^3.

When the *H* centre is oriented along the (110) axis (Fig.3.c,d), atomic charges are distributed strongly asymmetrically: -0.31 e (*H1*) and -0.77 e (*H2*). This observation is in agreement with EPR experiments [17], where the atomic charge distribution was estimated as 0.4e – 0.6e. As mentioned in this paper, this asymmetry could be a reason for increased annealing temperature of these defects. The spins on these two atoms are also unequal: 0.62 μ_B and 0.23 μ_B, respectively. The deviation of the molecular axis from the (110) is 13° (to be compared with experimental estimate of 3° [15]). This discrepancy could be caused by computational difficulties in structure optimization for large number of atomic coordinates.

4. Conclusions

Our *ab initio* DFT calculations confirm the experimental conclusion [15,17] that the (110) oriented *H* centre is energetically more favourable in MgF_2 than the (001) one. Our calculations predict the difference in energies to be 0.3 eV and the deviation from the (110) axis by 13°. The (001) oriented *H* centres are symmetrical, with similar effective charges and spins on both atoms, unlike strongly asymmetric (110) *H* centres. Calculations of complementary *F* centres and radiation-induced *F-H* pairs of Frenkel defects are in progress.

Acknowledgements

Authors are indebted to V. N. Lisitsyn, A. K. Dauletbekova and S. Piskunov for many stimulating discussions.

References

[1] Blunt R F and Cohen M I 1967 *Phys. Rev.* **153** 1031
[2] Catti M, Pavese A, Dovesi R, Roetti C, Causa M 1991 *Phys. Rev. B* **44** 3509
[3] Babu K R, Lingam Ch B, Auluck S, Tewari S P, Vaitheeswaran G 2011 *J.Solid State Chem.*, **184** 343
[4] Vassilyeva A F, Eglitis R I, Kotomin E A, Dauletbekova A K 2010 *Physica B* **405** 2125
[5] Vassilyeva A F, Eglitis R I, Kotomin E A, Dauletbekova A K 2011 *Cent. Eur. J. Phys.* **9** 515
[6] Shi H, Eglitis R I and Borstel G 2005 *Phys. Rev. B* **72** 045109
[7] Shi H, Eglitis R I and Borstel G 2006 *J. Phys.: Condens. Matter* **18** 8367
[8] Saunders V R, Dovesi R, Roetti C, Causa M, Harrison N M, Orlando R and Zicovich- Wilson C M 2006 *CRYSTAL User Manual* (University of Torino, Italy)
[9] Becke A D, 1993 *J Chem Phys* **98** 5648
[10] Monkhorst H J and Pack J D 1976 *Phys. Rev. B* **13** 5188
[11] Catti M, Dovesi R, Pavese A and Saunders V R 1991 *J. Phys.: Condens. Matter* **3** 4151
[12] McCarthy and Harrison N M 1994 *Phys. Rev. B* **49** 8574.
[13] Catlow C R A and Stoneham A M 1983 *J. Phys.C: Solid State Phys.* **16** 4321
[14] Stoneham A M 1975 *Theory of Defects in Solids* (Cambridge Univ. press)
[15] Norman C D, Halliburton L E 1977 *Phys Rev B* **15** 5883
[16] Yue L, Jia R, Shi H, He X, Eglitis R I 2010 *J. Phys. Chem. A* **114** 8444.
[17] Ueda Y, 1976 *J. Phys Soc. Japan* **41** 1255

Fluctuations in an Inorganic Glass Forming System Capable of Liquid-Liquid Phase Separation

V Bogdanov [1], L Maksimov [2,4], A Anan'ev [2], S Nemilov [3], V Rusan [2]

[1] Saint Petersburg State University, Ulyanovskaya Ul. 1, Petrodvorets, Saint Petersburg, RU 198504 Russia

[2] Research and Technological Institute of Optical Material Science, Ul Babushkina 36, Bld.1 Saint Petersburg, RU 192171 Russia

[3] National Research University of Information Technologies, Mechanics and Optics, Kronverksky Prospekt 49, Saint Petersburg, RU 197101 Russia

E-mail: maksimov@interglass.spb.su

Abstract. Rayleigh and Mandel'shtam-Brillouin scattering (RMBS) spectroscopy and high temperature ultrasonic study (HTUS) are applied to PbO-Al_2O_3-B_2O_3 glass forming system characterized by over liquidus miscibility gap. Temperature dependences of ultrasonic velocity of glass melts were measured in 600-1200°C range. "Frozen-in" density fluctuations in two phase glasses were estimated from HTUS data by Macedo-Shroeder formulation. Landau-Placzek ratios were found from RMBS spectra of single phase glasses at room temperature. Results of RMBS and HTUS were compared with well-known SAXS data. It was found that contribution of "frozen-in" density fluctuations into light scattering by two-phase glasses is much smaller than the scattering on particles of the second glassy phase causing opalescence of the glasses. Abnormal "water-like" growth of ultrasonic velocity with melt temperature can be explained by coexistence of two types of packaging of structural elements.

1. Introduction

Fluctuation inhomogeneities in inorganic glass forming systems attract attention of glass scientists because they determine some glass properties important for designers of optical materials and components (glasses doped with heavy metals, glass ceramics, fibers, optical fibers, waveguides, etc.) that include Rayeligh scattering losses, Kerr coefficient, kinetic parameters of luminescence of rare-earth doped glasses, etc. [1, 2, 3]

It was found that combined application of Rayleigh and Mandel'shtam-Brillouin scattering (RMBS) spectroscopy for glass study and measurement of ultrasonic velocity of glasses and their melts opened the way to estimate separately various fluctuations as sources of light scattering by glasses [1]. Up to now RMBS and High Temperature Ultrasonic Study (HTUS) have been applied to glasses outside the composition area of liquid-liquid phase separation that ensured the possibility to perform RMBS measurements when light scattering followed the Rayleigh law and its intensity was sufficiently low [4, 5]. The paper is targeted on the extension of this approach to glasses of PbO-Al_2O_3-B_2O_3 system

---

[4] To whom any correspondence should be addressed

characterized by immiscibility of their melts. Attempts to determine composition areas with spinodal and nucleation and growth kinetics of phase decomposition were made in [6, 7]. Temperature dependences of mean dimensions of over-critical fluctuations in the melts have been found from SAXS data [6, 8]. It is expected that RMBS, HTUS, and SAXS results will allow estimating fluctuation inhomogeneities in both two- and single phase glasses.

2. Experimental

Two series of objects under study were synthesized in laboratory under conventional technique. Compositions of 1-8 glasses (see Table 1) suitable for HTUS coincide with the compositions of melts studied in [6, 8, 9] while glasses No 9-11 were found suitable for RMBS measurements. The glasses 1-8 were not transparent because of intensive opalescence while the glasses 9-11 were transparent. Measurements of RMBS spectra were performed by a piezo-electric scanned Fabry-Perot interferometer equipped with He-Ne laser and photon counting system. Scattering angle was 90°. HTUS of 1-8 glass melts in 600-1200°C range was carried out by ultrasonic interferometer. Ultrasonic velocities of two-phase glasses at room temperatures were measured, too. Details of RMBS and HTUS experimental procedures have been described in [10].

3. Results and discussion

Fig.1 shows the glass forming area of PbO-Al_2O_3-B_2O_3 system and compositions of synthesized glasses. Inhomogeneous structure of a single component glass determined by "frozen-in" density, ρ, and concentration, C, fluctuations can be characterized by Landau-Placzek ratio $R_{L\text{-}P} = I^R / 2I^{M\text{-}B}$ where I^R and $I^{M\text{-}B}$ are Rayleigh and Mandel'shtam-Brillouin intensities found from RMBS spectrum. $R_{L\text{-}P} \approx R_\rho + R_C$ where $R_\rho \approx (T_g / T)[(v/v_{0,Tg})^2 - 1]$ and R_C are the contributions of "frozen-in" density and concentration fluctuations into Landau-Placzek ratio, that is into Rayleigh scattering intensity, $v_{0,Tg}$ and v are the "equilibrium (low-frequency)" sound velocities at glass transition temperature, T_g, and ultrasonic velocity at room temperature, T, correspondingly [4].

Results of acoustic measurements and T_g values taken from [9] allow to determine R_ρ of two-phase glasses the light scattering of which is caused by scattering on glass particles caused by liquid-liquid phase separation of melts. Fig.3 illustrates the typical RMBS spectrum of a single phase PbO-Al_2O_3-B_2O_3 glass.

Comparison of the data in Table 1 evidences that lowering the temperature of glass melt leads to significant growth of dimensions of light scattering inhomogeneities of melts the compositions of which are found in the miscibility gap. When glass melt temperature approaches temperatures 35-103°C higher than the temperature of liquid-liquid phase decomposition it leads to increase of inhomogeneity dimensions from 30 to 100 Å while the further lowering temperature to the room one causes opalescence that makes RMBS spectroscopy inapplicable. Values of R_ρ found for the two-phase glasses are typical for inorganic single phase glasses [4]. Thus, the latter can be considered as an evidence of forming of a two phase structure through the development of over critical concentration fluctuations in melts. Sound velocity decreases with temperature practically in all solids and liquids except H_2O, liquid GeO_2 [10], B_2O_3, Al_2O_3, Te, Sb, SiO_2 where velocity increases with temperature. That may be explained in terms of the Hall model presupposing the existence of minimum two coexisting structures characterized by open "network-like" packing of structural elements (dominating at low temperatures) and more rigid, denser packaging (dominating at high temperatures) [10]. Fig.2 shows anomalous "water-like" insignificant growth of ultrasonic velocity in the PbO-Al_2O_3-B_2O_3 glass melts in high temperature range that corresponds to the tendency of diminishing the effect in multi component glass melts with the enhancement of ionic bonding in them.

4. Conclusion

It was found that liquid-liquid phase decomposition of glass melts did not lead to the enhancement of "frozen-in" density fluctuations that confirmed the main role of over-liquidus concentration fluctuations of glass melts in elastic light scattering by two-phase glasses. Coexistence of two types of packaging of structural elements in glass melts was confirmed by the results of high temperature acoustic study.

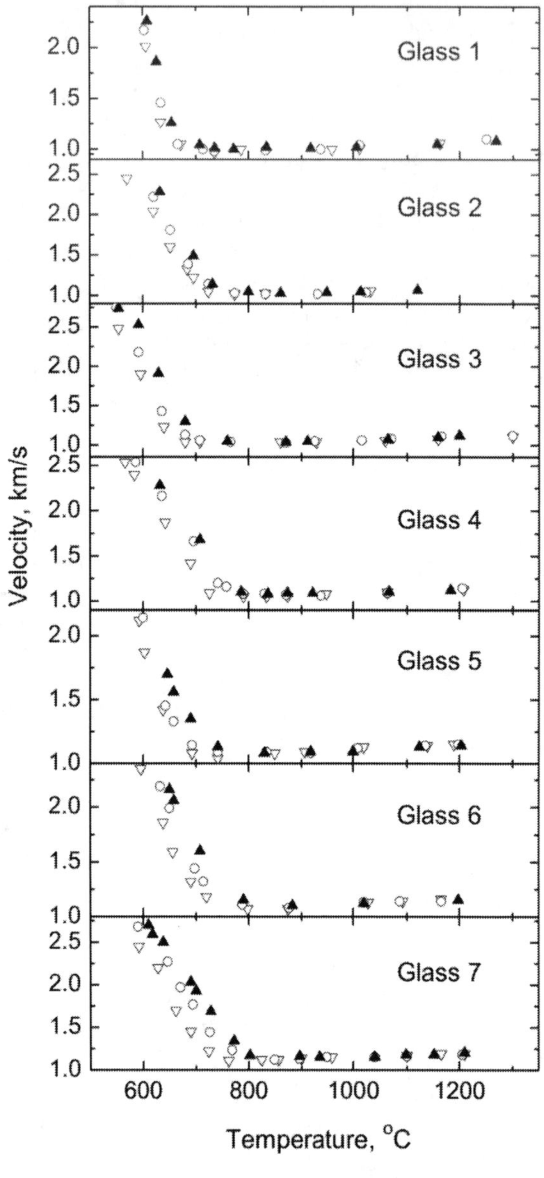

Figure 2. Temperature dependencies of ultrasonic velocity in PbO − Al_2O_3 − B_2O_3 melts at 1 (∇), 2 (○), and (▲) 4 MHz. Compositions of melts are listed in Table 1.

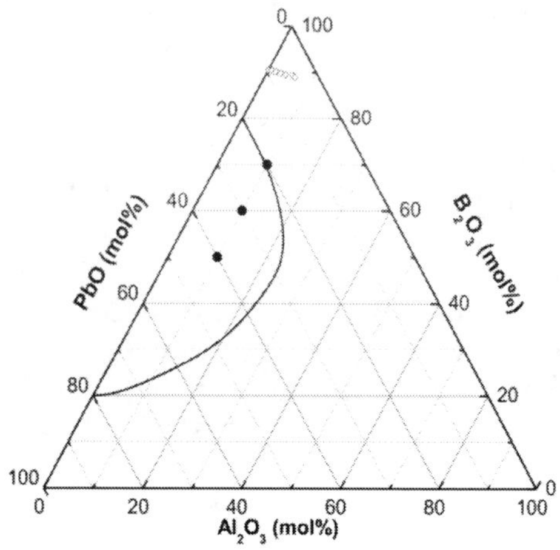

Figure 1. PbO − Al_2O_3 − B_2O_3 system. Glass forming area is outlined on the base of the data from [11]. Compositions of two phase glasses synthesized in [6, 8, and 9] and the present study are marked by ○. Single phase glass compositions are marked by ●.

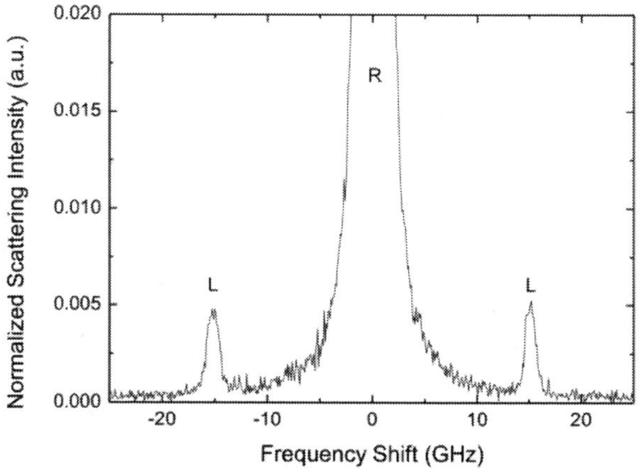

Figure 3. RMBS spectrum of 40PbO − $10Al_2O_3$ − $50B_2O_3$ glass. VV+VH polarization, λ = 632.8 nm. R is the Rayleigh scattering line, L is the lines of Mandel'shtam-Brillouin scattering caused by light scattering on longitudinal ultrasonic waves. $R_{L\text{-}P}$ = 105±5.

Table 1 PbO-Al_2O_3-B_2O_3 glass and melt compositions, Landau-Placzek ratios, R_{L-P}, of single phase glasses, contributions into Landau-Placzek ratios of 'frozen-in" density fluctuations of two-phase glasses, R_ρ, glass transition temperatures, T_g, temperatures of phase decomposition, T_L, fluctuation size in melts

Composition (mol. %)				Glass at room temperature: R_ρ (HTUS)	Glass at room temperature: R_{L-P} (RMBS)	Temperature, °C, of: Glass transition T_g ($\lg\eta$ =13) [9]	Temperature, °C, of: Liquid-liquid decom-position T_L	Melt: Fluctuation size (SAXS) [6]: 100 Å	Melt: Fluctuation size (SAXS) [6]: 50 Å	Melt: Fluctuation size (SAXS) [6]: 30 Å
№	PbO	Al_2O_3	B_2O_3							
1	9.47	0	90.53	17±2		350±5	785	796	804	830
2	8.82	0.82	90.36	18±2		350±5	756	764	779	826
3	8.35	1.63	90.14	21±2		410±5	724	733	751	803
4	7.39	2.84	89.77	36±4		420±5	685	694	720	787
5	6.53	3.97	89.50	29±3		425±5	675	675	709	778
6	6.27	4.35	89.15	27±3		420±5	675	675	708	778
7	5.74	5,11	89.32	24±2		430±5	669	682	715	782
8	4.90	6.21	88.89			430±5	703	712	733	788
9	20	10	70		120±6					
10	30	10	60		112±5					
11	40	10	50		105±5					

5. Acknowledgements
The study was performed under partial financial support of RFBR (Grant 10-03-00323-a).

References

[1] Maksimov L, Anan'ev A, Bogdanov V, Markova T, Rusan V and Yanush O 2011 *IOP Conf. Series: Materials Science and Engineering* **25** 1
[2] Denisov I A, Selivanov A G, Yumashev K V, Anan'ev A V, Maksimov L V, Ovcharenko N V, Bogdanov V N, Lipovskii A A and Vlasova A N 2007 *J. Appl. Spectroscopy* **74** 866
[3] Anan'ev A and Maksimov L 2008 *Advanced Materials Research* **39-40** 41
[4] Schroeder J 1977 Light Scattering of Glass *Treatise on Material Science and Technology. Vol 12 (Glass 1)* Ed. R H Doremus and M Tomozawa (New York N.Y: Academic) pp.157-222
[5] Cereyon A, Champagnon B, Martinez V, Maksimov L, Yanush O and Bogdanov V N 2006 *Optical Materials* **28** 1301
[6] Zarzycki J and Naudin F J 1971 *J. Non-Cryst. Solids* **5** 415
[7] Bras S, Craevich A, Sanchez J M, Williams C and Zanotto E D 1983 *Nuclear Instruments and Methods* **208** 489
[8] Zarzycki J and Naudin F 1970 *Disc. Faraday Soc. The Vitreous State* **50** 122
[9] Nemilov S V and Komarova N V 1977 *Fizika I Khimiya Stekla* **3** 568
[10] Anan'ev A, Bogdanov V, Champagnon B, Ferrari M, Karapetyan G O, Maksimov L V, Smerdin S N and Solovyev V A. 2008 *J. Non-Cryst. Solids* **354** 3049
[11] Mazurin O V, Streltsina M V and Shvaiko-Shvaikovskaya T P 1985 *Handbook of Glass Data Pt. B: Single-Component and Binary Non-Silicate Oxide Glasses: Physical Sciences Data* (Amsterdam: Elsevier) p 125

International Conference on Functional Materials and Nanotechnologies (FM&NT2012) IOP Publishing
IOP Conf. Series: Materials Science and Engineering **38** (2012) 012043 doi:10.1088/1757-899X/38/1/012043

Measurements of SiO_2 glass surface parameters by methods of microscopy

Eduards Gavars[1], Anda Svagere[1,*], Atis Skudra[1], Natalia Zorina[1], Raimonds Poplausks[2]

[1]Institute of Atomic Physics and Spectroscopy, University of Latvia, Skunu Street 4, Riga, Latvia, LV-1050
[2]Institute of Chemical Physics, University of Latvia, 4 Kronvalda Blvd. Riga, Latvia, LV-1010

E-mail: anda.svagere@gmail.com

Abstract. In this research we compare chemical and plasma treatment methods for surface of SiO_2 glass. For chemical treatment of surface tequila and alcohol were used but for plasma treatment - Ar+As and Ar+Se plasmas. Surface topography was analyzed using atomic force microscope. Comparison of chemical and plasma treatment methods shows that surface treated with plasma is smoother. Because of their various chemical compositions tequila and alcohol show different results.

1. Introduction

No one knows exactly when the glass as a material was first discovered but over the years number of its applications has significantly increased. Depending on purpose different types of glass are developed and produced and it is important to investigate how does chemical content and treatment change material's parameters.

SiO_2 glass is very popular material in different high-tech fields. In ideal conditions the surface of a glass is smooth and micro-structure free. Temperature, moisture and chemical compounds in the surrounding environment induce changes that lead to corrosion, blackening and other undesired effects [1, 2, 3].

Our work is concerned with the preparation and investigation of high-frequency electrode-less light sources (HFELS) that are widely used because of their narrow and intense spectral lines radiated in spectral region from VUV to IR. Lack of electrodes allows to use them in plasma investigation due to decreased amount of impurities. Light bulbs are made of SiO_2 glass and filled with different metal vapors and rare gases at low pressure. As the bulbs are of small size (diameter is about 5 millimeters) plasma interaction with wall material is significant. In order to improve our light sources' performance we have previously studied what impact does frequency and plasma temperature have on our lamps' inner surface [4]. The properties of light sources such as operating lifetime and emission stability are also strongly dependent on the wall material treatment methods used during the preparation phase as these methods can change the surface topography. In our laboratory we have used many different wall

---

* To whom any correspondence should be addressed.

material cleaning and pre-treatment methods, for example, cleaning with alcohol, plasma and thermal treatment. However, the modification of the surface topography during the treatment phase still needs to be investigated.

The goal of this work was to compare the modification of the surface of the SiO_2 lamp material after the cleaning with alcohol (96%) and tequila, as well as after the treatment with the discharge in Ar+Se and Ar+As plasmas. For the chemical treatment along with alcohol, well known as a cleaning solution, tequila was chosen because of its similar but yet different chemical content. In addition tequila was reported as precursor onto silicon for diamond film growth [5].

2. Experiment

2.1. Sample preparation

Five equal electrode-less lamps were made for the investigation. One was left untreated. Two chemically treated bulbs were rinsed with tequila and alcohol respectively. For plasma treatment lamps were filled with mixture of argon (gas pressure 3 Torr) and arsenic or selenium and then trained at 100 MHz frequency.

Afterwards all five lamps were broken in order to obtain samples for inner wall surface investigation.

2.2. Equipment

For wall topography studies we used an atomic force microscope (AFM) in a non-contact mode. The AFM tips were made of silicone nitride (Park Scientific instruments) and had a radius of curvature less than 35 nm. For image processing there were no filters used except for line-to-line leveling. The force constant of each cantilever was measured to be 0.1 and 0.16 N/m for cantilevers of two different dimensions.

3. Results

Table 1 and figures 1a-e represents the results obtained. It can be seen that surface without treatment is the smoother one and all four treated surfaces have different level of roughness. The most significant changes can be observed for surface treated with tequila – lots of relatively sharp spikes (Fig. 1a), reaching maximal value of 21 nm. Its spikiness is also characterized by large skewness (characterizes asymmetry of value distribution histogram) (9,81) and kurtosis (shows „peakiness" of the surface) (164) values. Because of different chemical content alcohol gives slightly different image (Fig. 1b) - spikes are lower (up to 15,5 nm) and flatter.

Table 1. Some characteristic parameter values for all five samples.

Treatment method	*Without treatment*	*Tequila*	*Alcohol*	*Ar+Se plasma*	*Ar+As plasma*
Standart deviation	172,708 pm	934,275 pm	2,276 nm	682,608 pm	288,829 pm
Maximal value	1,738 nm	20,908 nm	15,459 nm	7,571 nm	2,037 nm
Minimal value	-1,218 nm	-2,560 nm	-7,909 nm	-1,775 nm	-1,264 nm
Mean quadratic data value	172,707 pm	934,274 pm	2,276 nm	682,607 pm	288,829 pm
Average deviation	137,588 pm	489,536 pm	1,668 nm	487,223 pm	218,598 pm
Skew	0,0196	9,81	1,3	1,97	0,679
Kurtosis	0,272	164	4,0	9,37	2,57

Plasma treatment methods result in surface topography with no sharp spikes (maximal spikes values do not exceed 8 nm) but at the same time they are slightly different from one another – one can see that selenium plasma makes surface rougher but arsenic plasma – smoother. Experiments also show that surface without treatment still is the smoother one – difference between maxima and minimum is only approximately 3 nm.

By analysing obtained results and taking into account the mode of treatment used [6], we have built up such hypothesis:

a) surface modifications by plasma treatment are of both physical and chemical nature, first because of plasma properties such as temperature and pressure and the second one due to the impurities that occur during manufacturing process;

b) surface modifications from chemical treatment with alcohol/tequila are mostly of chemical nature – particles from these solutions are deposited on surface thus forming uneven coating.

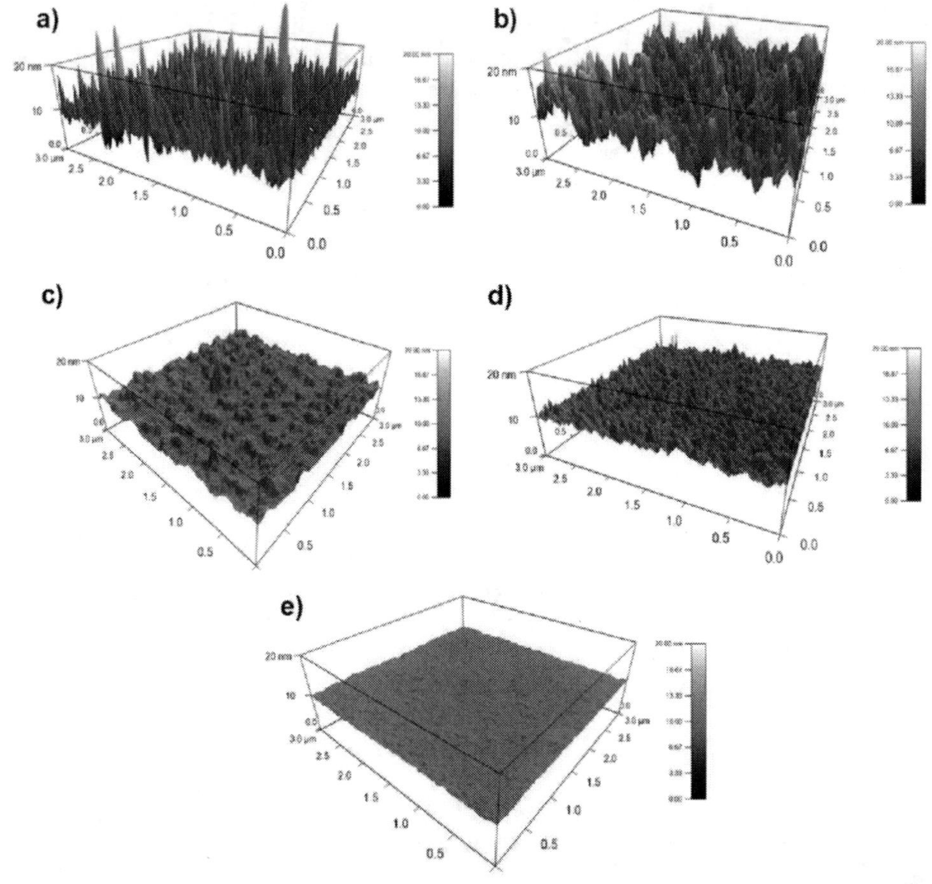

Figure 1. Examples of surface topography for samples treated with a) tequila, b) alcohol, c) Ar+Se plasma, d) Ar+As plasma and e) without treatment.

4. Conclusions

In this work, the surface topography of the SiO_2 glass, that is used for manufacturing of the spectral light sources, was investigated by means of atomic force microscopy after it was cleaned with tequila (40%), alcohol (96%) and plasma treatment (Ar+Se and Ar+As discharge).

The cleaning of SiO_2 surface with alcohol and especially tequila caused surprisingly large surface degradation that has to be considered more carefully in the future. The influence of the treatment with high frequency discharge plasma confirmed the previous results. Small but remarkable differences between modifications caused by argon plasma treatment in dependence of additive (Se or As) were observed. To understand the mechanisms of modification in detail, investigations need to be continued.

Acknowledgments
The work was partly supported by European Social Fund (ESF) project #2009/0210/1DP/1.1.1.2.0/09/APIA/VIAA/100 „Spectrometric techniques for detection of heavy metal contaminants" and National Research Program in Materialscience (Project #1 Multifunctional materials).

References

[1] Skudra A, Zorina N, Gavare Z, Berzins M and Erts D 2008 *phys. stat. sol.* (c) **5** No.4 915-7
[2] Arribart H and Abriou D 2000 *Ceramics Silikaty* **44** 121-8
[3] Witzke H D 2001 *Proc. of the 9th Internat. Symp. on the Science and Technology of Light Sources (Ithaca, New York, USA)* (Cornell University Press, Ithaca, NY) pp 357-368
[4] Skudra A, Gavare Z, Zorina N, Zinge M, Gavars E, Svagere A and Poplausks R 2011 *Journal of Materials Science and Engineering B* **1** 439-44
[5] Morales J, Apatiga L M, Castano V M 2009 *Rev. Adv. Mater. Sci.* **21** 134-138
[6] Prakash Sh, Karacor M B, Banerjee S 2009 Surface Science Reports **64** 233-254

IOP Conf. Series: Materials Science and Engineering **38** (2012) 012044 doi:10.1088/1757-899X/38/1/012044

Optical Properties of Irradiated Yttrium Aluminum Garnet

V. Skvortsova[1], **N. Mironova - Ulmane**
Institute of Solid State Physics, University of Latvia, 8 Kengaraga St., LV-1063, Riga, Latvia

E-mail: vera @cfi.lu.lv

Abstract. The results of investigation of the photoluminescence (PL) and optical absorption of crystals $Y_3Al_5O_{12}$(YAG) doped with different concentrations of manganese ions exposed to fast neutron irradiation and electron irradiation are presented. Photoluminescence spectra of YAG before neutron irradiation at T=80 K contain fine lines in orange region of spectrum, ascribed to Mn^{2+} ions in octahedral position. After irradiation band broadening is observed in the luminescence spectra of garnet crystals. Electron irradiation produced broad band with a complex structure related to Mn^{4+} ions. Exchange interaction between radiation defect and impurity ions during neutron irradiation and electron irradiation leads to appearance of additional lines and luminescence bands broadening in investigated crystals.

1. Introduction

Yttrium aluminum garnet ($Y_3Al_5O_{12}$, YAG) has been widely used in advanced optical technologies, tunable solid-state laser media and fluorescent materials for luminescent applications [1, 2]. Y^{3+} and Al^{3+} in YAG can be replaced by many kinds of other cations with different valency and size within a suitable range. The cubic cell of a general garnet compound $A_3B'_2B''_3O_{12}$ contains eight formula units, where A, B', .B" are metal ions occupying sites of different symmetry [3]. The crystal structure of YAG has a bcc structure (space group Ia3d or O^{10}_h) with 160 atoms in the cubic (primitive) cell. Y ions (A atoms) occupy 24 (c) sites and each of them is dodecahedrally coordinated to eight oxygen ions. Oxygen atoms occupy 96 (h) sites whose exact location depends on three structural parameters - x, y, and z and are different for different garnet oxides. There ate two different sites for Al: Al_{oct} (B' atoms) occupy 16 (a) sites with octahedral point symmetry (C_{3i}); and Al_{tet} (B") atoms occupy 24 (d) sites with tetrahedral point symmetry (S_4). The garnet structure can be considered as interconnected octahedrons, tetrahedrons, and dodecahedrons with shared oxygen atoms at the corners [3]. Each oxygen is a member of two dodecahedron, one octahedron, and one tetrahedron.

Hodges et al. [4] pointed out that Mn^{2+} in yttrium aluminum garnet can enter into any of three kinds of sites which are in the center of dodecahedron, octahedron and tetrahedron, respectively. The luminescence of different states of Mn ions in garnets was studied in the works [6-13].

Presence of point defects and their complexes in the crystalline structure effects a number of physical properties and makes impossible practical use of these crystals. The current understanding of the role of transition metal impurities in optical absorption and photoluminescence of YAG is not yet completely established, particularly in relation to the effect of irradiation in this material.

[1] Institute of Solid State Physics, University of Latvia, 8 Kengaraga St., LV-1063, Riga, Latvia

The photoluminescence (PL) and optical absorption of $Y_3Al_5O_{12}$ containing manganese ions and defects produced by fast neutron and electron irradiation are investigated.

2. Experimental details

The investigated samples were cut from single crystals grown by the Czochralski technique. The sample thickness was in the range of 1.05 –2.16 mm. YAG:Mn ($Y_3Al_{5-x}Mn_xO_{12}$, where "x"are the manganese concentration) crystals under study had the following composition: $Y_3Al_{4.988}Mn_{0.012}O_{12}$, $Y_3Al_{4.975}Mn_{0.025}O_{12}$, $Y_3Al_{4.963}Mn_{0.037}O_{12}$ and $Y_3Al_{4.95}Mn_{0.05}O_{12}$.
Photoluminescence (PL) spectra of Mn-doped yttrium aluminum garnet crystals have been measured at room and liquid nitrogen temperatures with a SPM-2 monochromator having diffraction grating of 651 lines per mm (Carl Zeiss Iena). The crystals were excited with a high pressure xenon lamp (DKSEL-1000) with 1000 W power connected to the monochromator (SPM-1) with a quartz prism. A signal was registered with a photomultiplier tube (FEU-119) by a synchronic detection method. The standard technique to measure absorption spectra has been based on a "Specord M-40" two-beam spectrometer operating in the 200-900 nm region.
The neutron irradiation was performed in the Latvian 5 MW water-water research reactor. The fluence of fast neutrons with energy > 0.1 MeV was in the range 10^{14}– $5{\cdot}10^{18}$ cm^{-2}. A cadmium filter was used for absorption of thermal neutrons. Irradiation by electrons with energy of 3.5 MeV of integral fluence up to 10^{16} cm^{-2} was carried out in accelerator ELU (Salaspils, Latvia). The sample temperature did not exceed 60°C during the electron irradiation and was below 40°C during the neutron irradiation. Optical measurements before and after irradiation were carried out at 293 and 80 K temperature.

3. Results and discussion

The ooptical absorption spectra of four garnet single crystals measured at room temperature before irradiation are similar. Figure 1.gives the absorption spectra of yttrium aluminum garnet single crystals (curve 1) before irradiation, after fast neutron irradiation (curve 2) and after electron irradiation (curve 3). Before irradiation bands at 370, 500 and 830 nm are observed. After neutron irradiation additional bands with maxima 549.5, 595.2 and 714.3 nm appear in absorption spectra of YAG containing manganese impurities. The 370 and 500 nm bands intensity increases but the 830 nm band intensity decreases after fast neutron irradiation. A weak band 370 nm corresponds to F-type centers [5].
Broad orange luminescence band in the range 580-620 nm appears at room temperature (Figure2, curve 1). Luminescence spectrum of $Y_3Al_{5-x}Mn_xO_{12}$ single crystal measured at liquid nitrogen temperature contains of the following main features: a sharp line at 587 nm and a series of weaker lines on the long wave side of the sharp line (curve 2, Figure 2).
After irradiation with fast neutrons at fluence $6{\bullet}10^{18}cm^{-2}$ the photoluminescence spectrum at liquid nitrogen temperature consists of a broad structureless band with a maximum at 607 nm (Figure 3). A little sharp line is seen also as a shoulder of the main band. Neutron irradiation produces appearance of anion and cation vacancies in the crystal. F-centre absorption band at 370 nm arises. After PL and optical measurement thermal annealing of the $Y_3Al_{4.988}Mn_{0.012}O_{12}$ and $Y_3Al_{4.975}Mn_{0.025}O_{12}$ crystal has been carried out. The crystal was heated in air at 473 K during 2 hours. The PL spectra measured after annealing at liquid nitrogen temperature is similar to the PL spectra before irradiation. It is suggested, that the reversibility of broadening of luminescence lines of Mn^{2+} ions is the evidence of the structure ordering reconstruction after annealing. The relatively low annealing temperature shows, that the structure changes after neutron irradiation concern only cation sublattice.
Photoluminescence spectra of $Y_3Al_{4.988}Mn_{0.012}O_{12}$ after electron irradiation at fluence $\Phi=10^{16}cm^{-2}$ is shown in Figure 4. PL measurements carried out at liquid nitrogen temperature have discovered a series of relatively narrow lines in 645-680 nm spectral regions. Mn^{2+} ions have a $3d^5$ electron configuration. All electron transitions for this configuration are forbidden by the selection rules of spin

and parity. The restrictions caused by the selection rule are removed owing to the exchange interaction. The emission with maxima around 590 nm is attributed to the $^4T_{1g} \rightarrow {}^6A_{1g}$ transition of Mn^{2+} ions in octahedral sites. Luminescence of Mn^{4+} ($3d^3$), which is isoelectronic to Cr^{3+} consists of the sharp R- lines peaked around 645- 675 nm, which are caused mainly by the $^2E_g \rightarrow {}^4A_{2g}$ transitions and partly by the $^4T_{2g} \rightarrow {}^4A_{2g}$ transitions. These bands are observed after electron irradiation of $Y_3Al_{5-x}Mn_xO_{12}$ together with the Mn^{2+} ion luminescence in octahedral sites. In contrast to paper [2] the green luminescence was not observed in our single crystals.

The broad band emission of Cr^{3+} with a maximum at 695 nm ascribed to the $^4T_{2g} \rightarrow {}^4A_{2g}$ transition, the sharp and narrow bands at 685 nm due to $^2E_g \rightarrow {}^4A_{2g}$, the diffused band centered at 675 nm (anti-Stokes vibronic sidebands) and two sharp diffused bands at 710 and 725 nm corresponding to the vibronic sidebands are observed in all investigated YAG crystals before and after irradiation (Fig.2, 4).

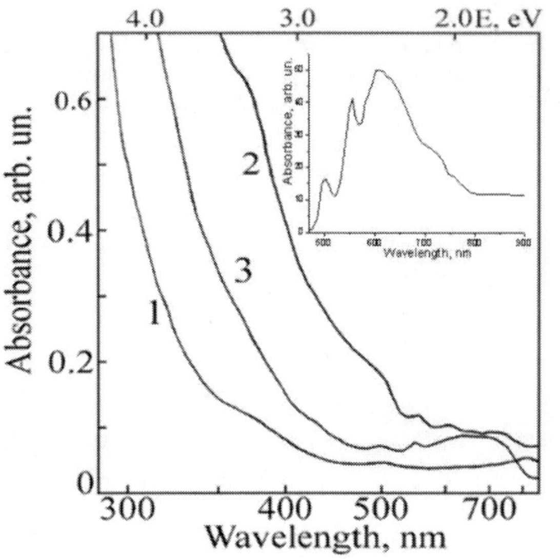

Figure 1. Absorption spectra of $Y_3Al_{5x}Mn_xO_{12}$: 1. before irradiation, 2. after neutron irradiation $\Phi = 10^{16} cm^{-2}$, 3.after electron irradiation $\Phi = 10^{16} cm^{-2}$. In the inset: additional absorption spectrum of neutron irradiated $Y_3Al_{4.963}Mn_{0.037}O_{12}$ crystal .

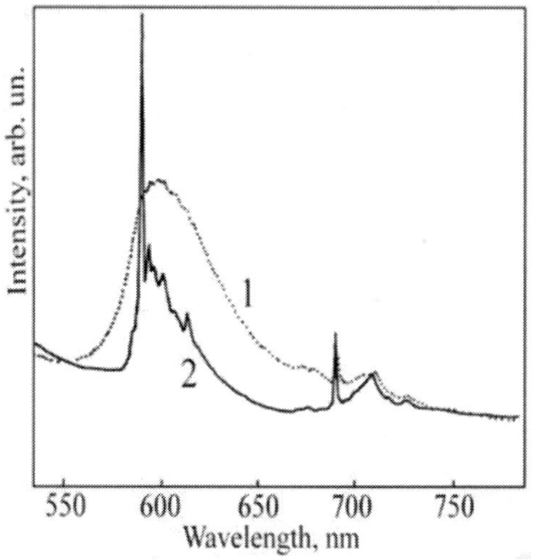

Figure 2. Photoluminescence spectra of $Y_3Al_{5-x}Mn_xO_{12}$: 1. at room temperature, 2. at 80 K, $\lambda_{ex} = 470$ nm.

The presence of a nearby F or F^+ center can modify the environment at the Mn^{2+} impurity site to produce an effective increase in the crystal field. In addition to the shift in energy, the perturbation induced by the F or F^+ center can increase the oscillator strength by introducing odd components to the crystal field, or by exchange coupling between the F or F^+ center and the Mn^{2+} impurity. The enhancement of the oscillator strength by the exchange interaction has been previously observed between F centers and transition metal impurity ions in MgF_2, $MgAl_2O_4$ and MgO [14, 15]. As the concentration of F and F^+ centers increased, the number of perturbed manganese sites also increased, so that the excitation peaks became more intense as a function of neutron dose. It is possible to assume that the band broadening is connected with a complex center "Mn^{2+}-F^+ (or F centre)". A part of holes generated after irradiation are trapped by Mn^{2+} which then produce Mn^{3+} and Mn^{4+} ions.

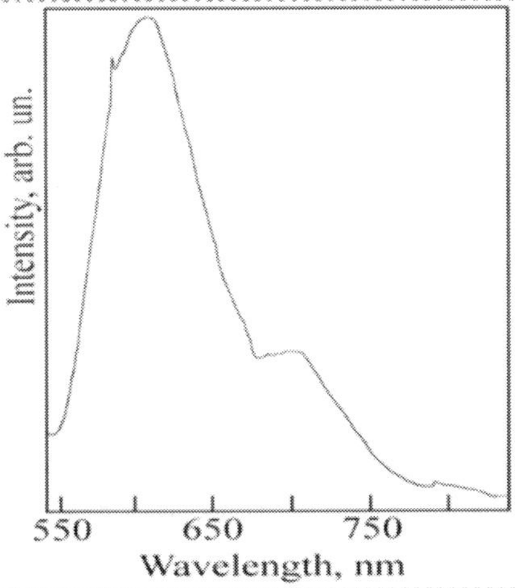

Figure 3. Photoluminescence spectra of $Y_3Al_{5-x}Mn_xO_{12}$ after neutron irradiation $\Phi = 10^{16} cm^{-2}$. T= 80 K, λ_{ex}= 470 nm.

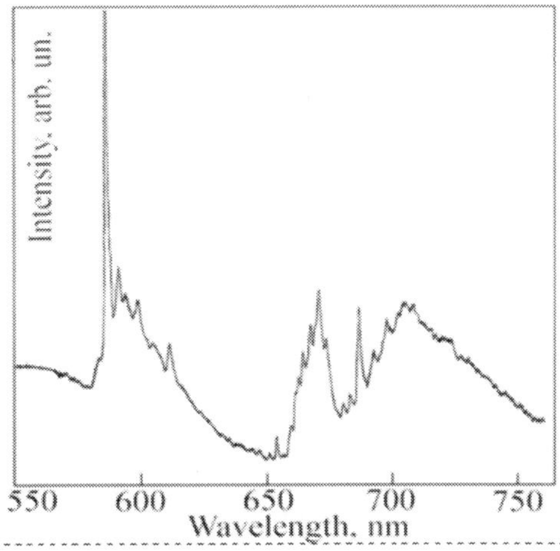

Figure 4. Photoluminescence spectra of $Y_3Al_{4.988}Mn_{0.012}O_{12}$ after electron irradiation $\Phi = 10^{16} cm^{-2}$. T= 80 K, λ_{ex}= 470 nm

4. Conclusion

The photoluminescence (PL) and optical absorption of yttrium aluminium garnet (YAG) crystals containing manganese ions and defects produced with fast neutron irradiation with fluence up to $5 \cdot 10^{18} cm^{-2}$ (E>0.1 MeV) and irradiated by electron fluence up to $5 \cdot 10^{18} cm^{-2}$ are investigated. Electron irradiation produces both Mn^{2+} and Mn^{4+} photoluminescence bands in $Y_3Al_{5-x}Mn_xO_{12}$. Exchange interaction between radiation defect and impurity ions during neutron irradiation leads to appearance of additional lines and luminescence band broadening in yttrium aluminum garnet crystals.

5. References

[1] Ter-Gabrielyan N, Merkle L D, Kupp E R, Messing G L and Dubinskii M 2010 *Opt. Lett.* **35**
[2] Singh V, Chakradhar R P S, Rao J L, Kwak H Y 2010 *J. Appl. Phys.B* **98**.407
[3] Galasso F S 1970 *Structure and Properties of Inorganic Solids* (Pergamon New-York) p 244
[4] Hodges J A 1974 *J. Phys. Chem. Solids* **35** 1385
[5] Chakrabarti K 1988 *J Phys. Chem. Solids* **49** 1009
[6] Gavrilovic P, Singh Sh 1994 *US patent* no. 22483
[7] Noginov M A, Loutts G B 1999 *J. Opt. Soc. Am.* **B 16** 3
[8] Zhydachevskii Ya, Suchocki A, Berkowski M, ZakharkoYa, 2007 *Radiat. Meas.* **42** 625
[9] Kuck S, Hartung S, Hurling S, Peterman K, Huber G 1998 *Phys. Rev.* **B 57** 2203
[10] Donegan F, Glynn T G, Imbusch G F, Remeika J P 1986 *J. Lumin.* **36** 93
[11] Yang Su-Hua., Lu Chinf-Yi 2007 *J. Electrochem. Soc.* **154** J397
[12] Riseberg L A, Weber M J 1971 *Solid State Communication* **9** 791
[13] Nijs A, Verweij M M, Blasse G 1992 *Mater.Chem. Phys.* **30** 199
[14] Cain L S, Pogatshnik, Chen Y 1988 *Phys. Rev.* **37** 2645
[15] Mironova N, Ulmanis U 1988 *Radiation defect and iron group metal ions in oxides* Zinatne, Riga (in Russian)

Acknowledgments
This work was supported by the grant of the Latvian Government (No 09.1580).

IOP Conf. Series: Materials Science and Engineering **38** (2012) 012045 doi:10.1088/1757-899X/38/1/012045

The luminescent properties of persistent strontium aluminate phosphor prepared by solar induced solid state synthesis

V Liepina[1,3], K Smits[1], D Millers[1], L Grigorjeva[1], C Monty[2]

[1]Institute of Solid State Physics, 8 Ķengaraga street, Riga, Latvia

[2]PROMES CNRS, Material and Solar Energy Laboratory, BP5 Odeillo, France

E-mail: virginija.liepina@gmail.com

Abstract. A novel method – solar induced solid state synthesis – for the synthesis of aluminate phosphor doped with Eu and Dy and the study of its luminescent properties are presented in this article. Two kinds of samples of persistent strontium aluminate phosphor have been prepared – ones via this new method and ones via conventional high temperature solid state reaction. Both kinds of samples were made using the same raw materials. The samples were tested for their chemical composition using X-Ray diffraction method. The luminescence properties and phase composition of the obtained samples have been compared. The behaviour of afterglow, thermally stimulated luminescence, excitation and emission spectra are presented.

1. Introduction

Phosphorescence materials have been widely studied by many researchers due to their great potential for device applications. The possible applications such as luminous paints, emergency signs, and more, demand phosphors with qualities like long lasting afterglow, high luminescence intensity and chemical stability [1,2]. Therefore long lasting phosphors such as Eu and Dy doped silicates and aluminates have attracted researchers' attention due to their good luminescent properties: long lasting afterglow, high quantum efficiency and good chemical stability. The luminescence decay time varies depending on the type of the aluminate, for example, $Sr_4Al_{14}O_{25}$: Eu, Dy is a phosphor with over 20 hour afterglow and the luminescence peak is at 490 nm [2,3].

The properties of strontium aluminates phosphors strongly depend on material preparation. Many articles describe the different possible synthesis methods of these phosphors, i.e. combustion method, co-precipitation method, sol-gel method etc [4]. The typical solid-state reaction process has been used intensively for phosphor synthesis, but this process has some major disadvantages. It often results in poor homogeneity due to unreacted components and requires high reaction temperature to get high diffusion rates as well as small particle sizes in the starting materials to decrease diffusion length. Furthermore, for some applications very small grain powders are necessary, but alas, the grain size of phosphor powders prepared through solid-state reaction method is in several tens of micrometres. Phosphors of small particles must be obtained by grinding the larger phosphor particles [5] . Those processes easily introduce additional defects and greatly reduce luminescence efficiency. Also, heating in air can lead to the development of unnecessary oxygen compounds, which leads to the need for reducing or neutral atmosphere for the reaction (usually about 1400° in $H_2 + N_2$ or N_2 atmospheres), which creates the need for specified expensive furnaces. However, it is the most

[3] To whom any correspondence should be addressed

common method for the synthesis of this phosphor due to the simplicity of the process. Another common method for synthesis is the sol - gel method, which excludes the need for very high temperatures, but it also has some major disadvantages such as large shrinkage during processing, long processing time and the outcome can be still amorphous. In this paper particularly we have tried a new method for the synthesis of strontium aluminate phosphors – solar induced solid state synthesis (SISSS) method. We compare the luminescent properties of RE doped strontium aluminate long afterglow phosphor with sample synthesized by the conventional high temperature solid state (SS) method to one synthesized by the SISSS method. It has some advantages compared to others, for example, it gives the possibility of high temperature SS reaction in very clean environment and the gas atmosphere and pressure can be controlled. The optical properties of the obtained samples have been studied and compared with those for similar composition phosphor synthesized by conventional SS reaction using the same raw material.

2. Experimental

2.1. Samples

In this paper we have tried a new method for SS reaction exploiting solar energy. Solar furnaces (Figure 1) (PROMES CNRS, France) are constituted by mobile plane mirrors ("heliostats") tracking the sun and reflecting the sunlight on a parabolic concentrator, thus it focuses to cooled sample holder. The sample are placed on holder and covered with glass balloons, thus allow prepare samples in vacuum or in different atmospheres. Peak power of such furnaces exceeds 2 kW and can heat the sample up to 3000°C. The main difference to convectional furnaces, is that the heat energy is transmitted in form of radiation, thus allow very fast heating of target, whereas other details of equipment were not heated. Two different samples were prepared - using conventional SS synthesis (Sample A) and SISSS (Sample B). For both methods the precursors were $SrCO_3$, Al_2O_3, $Eu(NO_3)_3$, Dy_2O_3.7%; H_2BO_3 was used as a flux agent. RE ions usually substitute Sr, that is why we chose 1 atomic % Dy and 1 atomic % Eu in order to obtain $Sr_{4-x}Al_{14}O_{25}$:Eu_x, Dy_x. For the SS synthesis the precursors were mixed, pressed and heated in N_2 atmosphere for 4 hours at 1300° C. The synthesis was done according to the N. Suriyamurthy et.al. article „Effects of non-stoichiometry and substitution on photoluminescence and afterglow luminescence of $Sr_4Al_{14}O_{25}$:Eu^{2+}, Dy^{3+} phosphor"[6]. For the SISS method the synthesis time was about 15 minutes approximately in 2000° C.

2.2. Measurements

The phase identification of the synthesized samples was carried out by an X-ray Diffractometer Bruker AXS GmbH DS Advance. The measurements of luminescence emission spectra were carried out using three different excitation sources – YAG laser FQSS266 (CryLas GmbH) 4th harmonic, 266 nm (4.66 eV), 2 ns pulses and X-ray excitation by X-ray lamp (W anode, 45 kV, up to 10 mA) at room temperature, and for excitation spectra measurments optical parametric oscillator (OPO) of type NT342/3UV (EXPLA).The same area of the samples was used for measurements and thus the comparison of luminescence intensities between the samples was possible. The luminescence spectra were recorded by Andor Shamrock B303-I spectrograph (spectral resolution ≈0.1nm) equipped with a CCD camera (Andor DU- 401A-BV) at exit port. Thermally stimulated luminescence (TSL) measurements were carried out using X-ray irradiation at liquid nitrogen temperature and heating the samples in 85-600K range.

3. Results and discussion

The XRD patterns of the samples are shown Figure 2. The XRD pattern analysis shows that both obtained samples consist of different complex aluminate oxides - mostly of $SrAl_2O_4$ and therefore the luminescent properties of the samples are comparable. $SrAl_2O_4$, is also a recognized and highly effective long lasting phosphor. In sample A besides the $SrAl_2O_4$, also the preferred strontium aluminate $Sr_4Al_{14}O_{25}$ was formed, whereas the sample B consists of $SrAl_2O_4$ and $SrAl_2O_3$.

Phases of the strontium aluminates change due to differences in heating speed, maximum temperature etc. The reason for the origination of resulting aluminates can then be explained by the fact, that the optimal heating speed and temperature was not considered. According to C. Chong, the $SrAl_2O_4$ conversion to $Sr_4Al_{14}O_{25}$ occurs at 1300°[7].

$$SrO + Al_2O_3 \xrightarrow{1200^0} SrAl_2O_4$$
$$SrAl_2O_4 + 5Al_2O_3 \xrightarrow{1250^0} SrAl_{12}O_{19}$$
$$7SrAl_2O_4 + 3SrAl_{12}O_{19} \xrightarrow{1300^0} 5Sr_4Al_{14}O_{25}$$

For sample prepared by SS method the reaction temperature was not high enough. As for the SISS method, the temperature was so high, that $Sr_4Al_{14}O_{25}$ was again reduced to $SrAl_2O_4$ and because of low atmospheric pressure in solar reactor, efficient oxygen reduction occurs and the $SrAl_2O_3$ are formed.

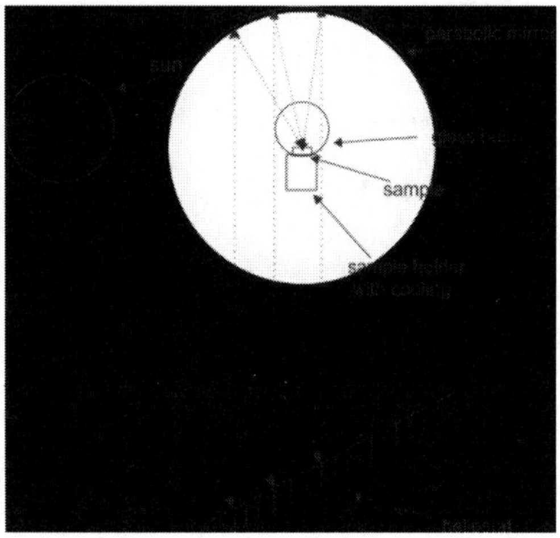

Figure 1. Solar furnace scheme

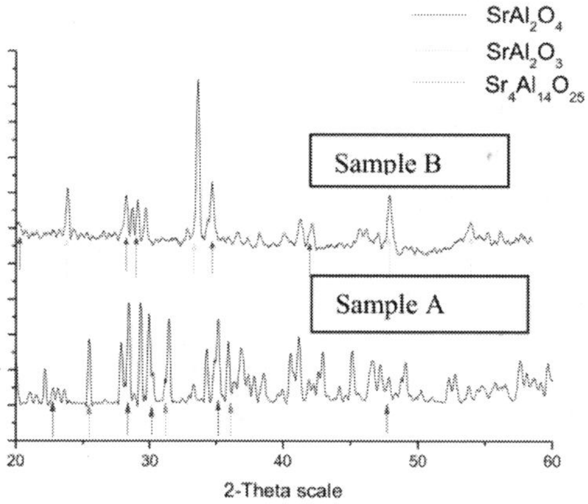

Figure 2. XRD Spectra

The excitation and laser (266 nm) excited luminescence spectra are shown in Figure 3. Similar excitation and emission spectra are observed for samples prepared via both processes. The excitation spectra (for emission at 520 nm) show a broad band from 260 nm to 400 nm. The emission for both phosphors is a symmetrical luminescence band at 520 nm due to the Eu^{2+} transition $4f^65d^1$-$4f^7$. As the peak position is strongly influenced by the surroundings of Eu^{2+}, a small displacement of the maxima of the emission and excitation spectra occurs and we can observe broadening in the excitation and emission spectra of Sample A. It could be explained by the changes in the crystal field around Eu^{2+} in different aluminates contents and thus the depth of the trapping centres.

Slightly different luminescence spectra could be obtained by X-Ray excitation. In case of Sample A we can observe that not all RE ions have been incorporated during the reaction, and in X-ray excitation they emit typical RE emission lines – Eu^{3+} ion emission lines 570 – 750 nm and Dy^{3+} ion emission lines 470 – 500 nm. In sample B the RE ions have been fully incorporated, however some reabsorbtion lines appears.

For sample B, we can assume, that rare earth ion migration to the surface in the very high temperature reaction occurs, because in case of photoluminescence the light penetration depth is relatively low and the luminescence is more from surface wheras X-rays penetrate deeply in the material deeply and that is why reabsorbtion is possible by the surface RE ions. X-ray radiation creates electron and holes which are migrating to luminescence centers, but in case of photoluminescence the energy is lower than the band gap (~5eV) and the selective excitation of defects and RE ions occurs. We can note that the luminescence intensity of sample A is 3 times higer the that of sample B.

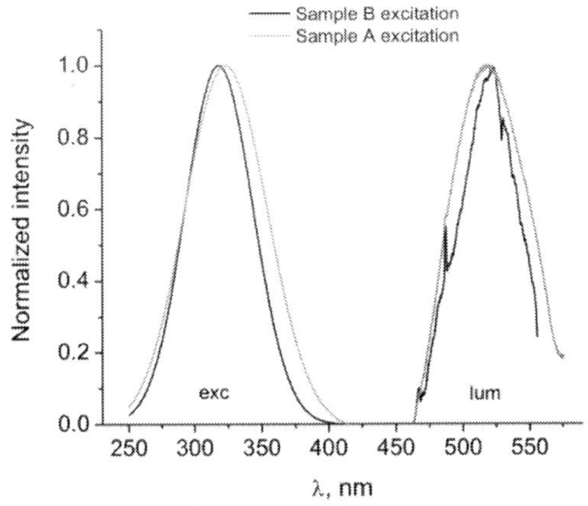

Figure 3. The excitation and emission spectra of Sample A and Sample B under UV irradiation

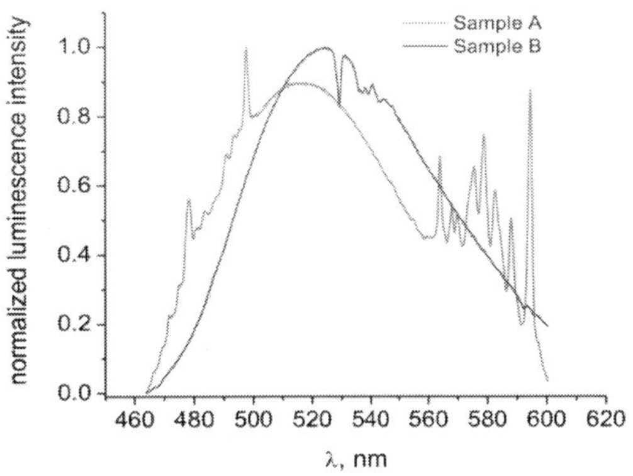

Figure 4. The emission of Sample A and Sample B under X-Ray irradiation

Figure 5 shows the luminescence intensity dependence of time after the termination of excitation light for Samples A and B. We can observe that the decay kinetics of Sample B is faster, which relates to the trap depth of the material – probably the traps are located deeper in the forbidden zone for Sample A. We can observe a long tail 5% the intensity of the initial luminescence 15 minutes after the termination of irradiation which descends very slowly.

The thermally stimulated luminescence (TSL) curve was recorded for both samples in order to obtain information about the trap depth in the material (Figure 6).

We can qualitatively tell from the spectrum that Sample B has only one major maximum in TSL curve at wavelength 536 nm (Eu^{2+} emission), which is a sign of an effective phosphor, but glow peak is below RT, which means that it's trapping centres are located close enough to the conduction band for the electrons them to be easily taken up by thermal energy and the recombination could occur. That is a reason for faster luminescence decay kinetic.

For the first order process the TSL spectra were taken with 2 different heating rates – 6°/min and 12°/min to determine the depth of the trapping centres by using the formula:

$$\frac{E_z}{kT_{maks}} = \frac{\omega_0}{v} * \exp(-\frac{E_z}{kT_{maks}}) \qquad (1)$$

where E_z – trap depth, k – Boltzmann constant, T – temperature, ω_0 – frequency factor, v – heating speed.

The evaluated trap depth for Sample A is 0.26 eV and for Sample B is 0.19 eV.

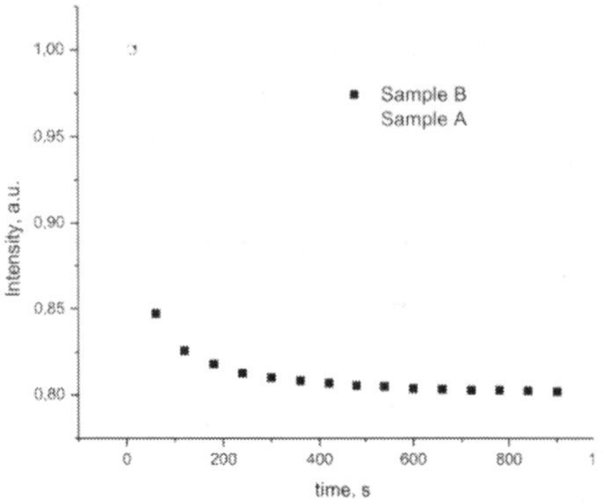

Figure 5. The fading of luminescence in time for Sample A and Sample B

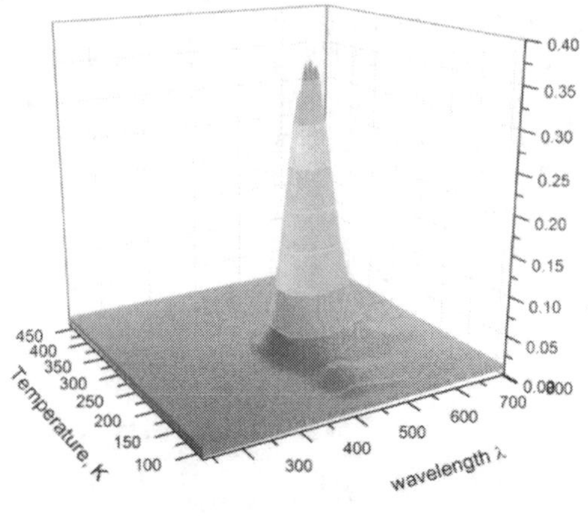

Figure 6. TSL Spectrum for Sample B at 12°/min.

To conclude, SISSS method proves itself to be a perspective method for phosphor synthesis. The synthesis temperature in case of SISSS sample was to high, but for SS to low. In perspective by changing the synthesis parameters it is expected to obtain one phase. Also for SISSS sample, all the RE ions had been incorporated. This method also reduces the possibility of unnecessary impurity incorporation due to the clean environment of the reaction. However, it also has some disadvantages, such as the unaccessability of the solar reactor and the problematic control of heating rate, and temperature, thus lead decomposition of aluminates. Small depth of the trapping centres in the obtained material, leading to a faster kinetic. Never less this is a promising method for phosphor synthesis and should be explored more.

Acknowledgements

The sample synthesis was supported by SFERA Access Programm SYNMAT (P10021900000067).
Virgīnija Liepiņa is supported by Nr2010/0253/ 2DP/2.1.1.0/10/ APIA/VIAA/079 10032 LZP grant.
Krišjānis Šmits is supported by EU SFP project 2009/0202/ 1D/1.1.1.2.0/09/APIA/VIAA/141

References

[1] Murayama Y, Takeuchi N, Aoki Y and Matsuzawa T 1995 *Phosphorescent phosphor, US Patent* 5,424,006
[2] Koen Van den Eeckhout, Philippe F. Smet and Poelman D 2010 *Materials* **3** 2536
[3] Gao F, Xiong Z, Xue H, Liu Y 2009 *Conference Series* **152** 012082
[4] Eeckhout K, Smet P, Poelman D 2010 *Materials* **3** 2536
[5] Dorenbos, P. *J. Electrochem. Soc.* 2005, **152,** H107–H110
[6] Suriyamurthy N, Panigrahi B S 2008 *Journal of Luminiscence* **128** 1809
[7] Chang C, Yuan Z and Mao D 2006 *Journal of Alloys and Compounds* **415** 220

EPR spectra of the Mn^{2+} ion in the oxyfluoride glass ceramics containing BaF_2 nanocrystalline phase

Dz Berzins, A Fedotovs[1], O Kiselova, A Sarakovskis

Institute of Solid State Physics, University of Latvia, Kengaraga str 8, Riga, LV-1063, Latvia

E-mail: andris.xyz@gmail.com

Abstract. Electron paramagnetic resonance (EPR) spectroscopy has been used to investigate structure of the oxyfluoride glass ceramics containing BaF_2 fluoride component. Mn^{2+} ion has been used as a paramagnetic probe. After successful material synthesis and thermal treatment at temperatures above first crystallization temperature transparent glass ceramics samples were obtained. For these samples superhyperfine (SHF) structure in the X-band EPR spectra taken at 77 K has been observed. Such SHF structure is characteristic to fluorine hyperfine interaction observed in the fluoride crystals. Obtained data allow analyzing the structure of the BaF_2 crystallites and possible location of the Mn^{2+} ion in the BaF_2 crystalline lattice.

1. Introduction

Oxyfluoride glass ceramics are known as excellent host material for rare earth ions. These materials possess high optical transparency, good mechanical and chemical properties. Crystalline component in the glass matrix provides low phonon energies thus energy transfer in rare earth ions could happen more efficiently than in the glass material alone [1-3]. The crystallites in the host glass matrix are created using thermal treatment. The size of the crystallites is dependent of the temperature and thermal treatment time and often is less than 100-200 nm for transparent glass ceramics [4,5]. It is very important to know the structural information of this crystalline component and its dependence of the thermal treatment temperature and time.

Paramagnetic probe method is useful for investigation of the coordination environment of the probe in the crystalline lattice. The Mn^{2+} ion are particularly useful for investigation of the disordered media like glasses and powders since it has practically isotropic hyperfine (HF) interaction constant [6,7,8]. Coordination of neighboring fluorine nuclei exhibits additional superhyperfine (SHF) interaction which renders the nearest environment of the paramagnetic probe in the EPR spectra as additional SHF structure [9]. Although, information of SHF interaction in disordered media remains obscured due to large distribution of EPR parameter values, if the SHF interaction is by large extent isotropic it is possible to observe spectral features characteristic to certain crystalline structures.

In this work we investigated EPR spectra of the oxyfluoride glass ceramics containing BaF_2 crystalline counterpart. Our previous studies have shown the usefulness of the paramagnetic probe

[1] To whom any correspondence should be addressed

method for the oxyfluoride glass ceramics containing CaF_2 and BaF_2 powders using Mn^{2+} ion as a probe [10, 11].

2. Experimental

The oxyfluoride glass ceramics sample was prepared with the following composition: $40SiO_2$-$25Al_2O_3$-$15Na_2CO_3$-$20BaF_2$-$0.5MnF_2$ (in mol%). Ingredients were mixed and melted at 1450 °C, then by pouring the melt on the solid steel block for rapid quench transparent glass has been obtained.
Glass samples were heated at temperatures close to and above first crystallization temperature T_c which was determined by differential thermal analysis (DTA) techniques. For comparison purposes pure BaF_2:Mn^{2+} powder was synthesized. We used pure (99.99%) BaF_2 and MnF_2 powders (99.5 and 0.5 mol% of particular fluorides, respectively), which were mixed in chloroform, and then dried. Powder sample were heated at 650 °C temperature for 30 minutes to achieve Mn^{2+} ions building into the BaF_2 lattice. Obtained samples were studied by conventional X band EPR spectroscopy using 50 mW klystron microwave radiation typically with 20 dB attenuation. The amplitude of the magnetic field modulation at 100 kHz was <1.0 G. All measurements were done at liquid nitrogen temperature.

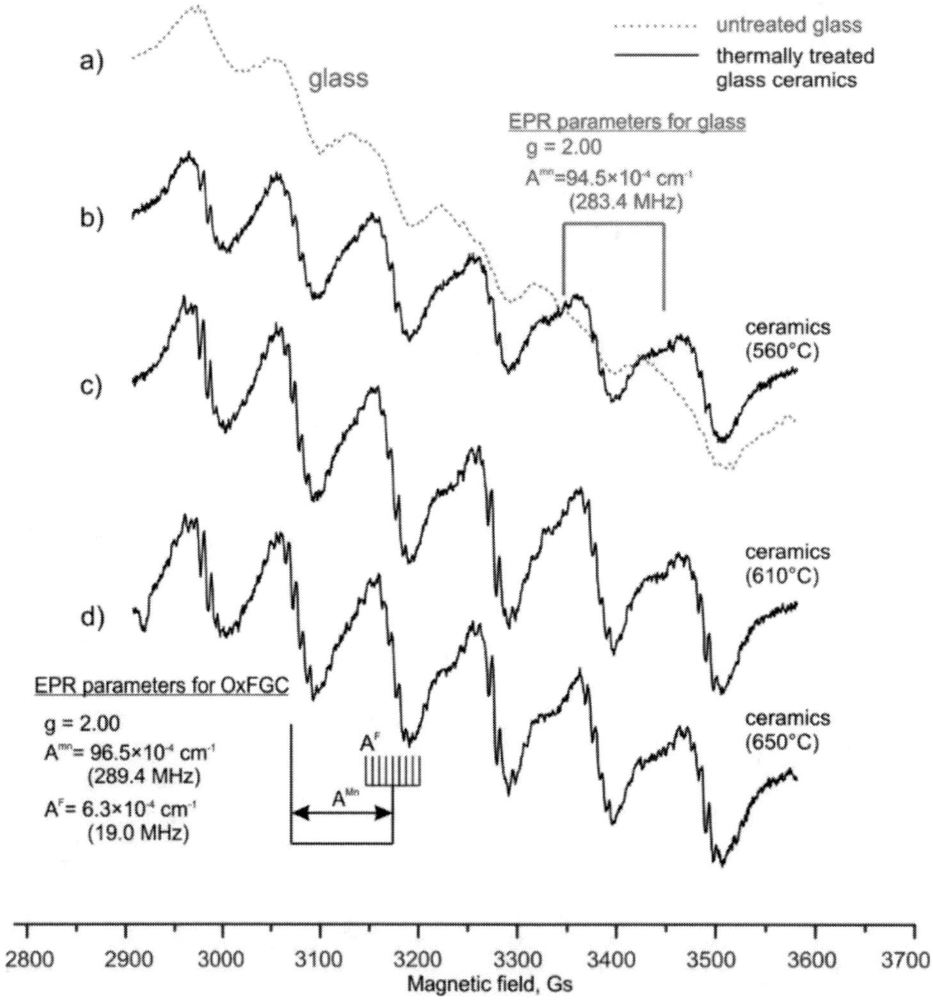

Figure 1. Curve a) depicts EPR spectrum of the BaF_2:Mn glass before thermal treatment. Curves b), c) and d) depict evolution of the EPR spectrum of BaF_2:Mn glass ceramics after thermal treatment at 560°C, 610°C and 650°C temperatures for 1h.

3. Results and discussion

The EPR spectra of the non-treated and thermally treated glass ceramics are shown in the Figure 1. The spectrum of non-treated material consists of characteristic manganese sextet at g=2.00 which occurs due to hyperfine interaction of the Mn^{2+} ions nuclei (I=5/2) with its electronic spin (S=5/2). The comparably large width of the lines is due to unresolved superhyperfine interaction of Mn^{2+} centers in the glass matrix and distribution in the fine structure parameters. Characteristic to glass structure is also the broad background line.

By means of DTA temperature of the first crystallization has been determined to be T_c = 590 °C. Structural changes during thermal treatment remain consistent until temperature of the second crystallization is reached. In Figure 1 the EPR spectra of the thermally treated glass ceramics are shown (curves b to d). After heating the samples at temperatures close to T_c pronounced SHF structure appears in the EPR spectra. However, further heating of the samples at temperatures higher than second crystallization temperature (~680 °C) degrades Mn^{2+} EPR spectra.

SHF structure is clearly visible on all six hyperfine structure lines. As temperature of the thermal treatment increases, spectrum part which contains SHF structure separates more and more from the other part of the spectrum which is due to remaining Mn ions in the glass matrix. The SHF structure consists of clearly resolvable 7 lines, although SHF structure spectrum of 9 lines is expected as will be discussed later. It is convenient to assume that some lateral lines are obscured due to low signal level.

From obtained EPR spectra it is clear that manganese ions tend to enter BaF_2 lattice upon thermal treatment. Such superhyperfine structure corresponds to the situation when manganese replaces

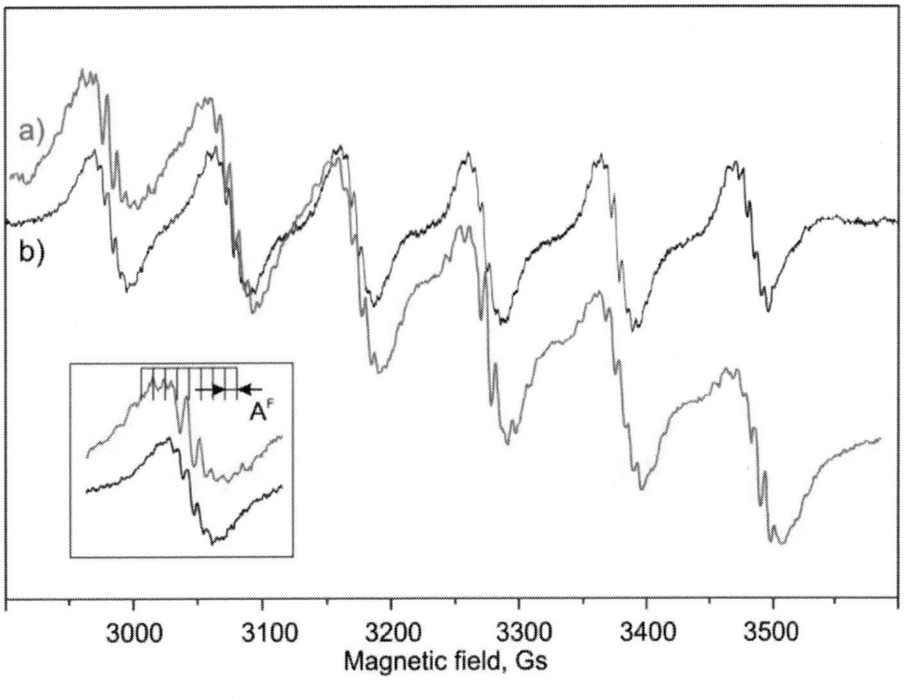

Figure 2. a) EPR spectrum of the oxyfluoride glass-ceramics sample heat treated at 650 °C, b) EPR spectrum of the BaF_2:Mn^{2+} powder spectrum heat-treated at 650 °C.

barium and is surrounded by eight equivalent fluorine ligands, thus 9-line SHF structure in the EPR spectra with binomially distributed intensities is created. Other types of manganese centers in the BaF_2 lattice are possible. Such so-called off-center or non-cubic Mn^{2+} defects have been reported in BaF_2 and in SrF_2 single crystals [12, 13]. Possibility for other crystalline structures to appear has been considered. Separate investigation of the EPR spectra of the Mn^{2+} ion in the pure BaF_2 shows similar picture after thermally treated powder samples. In Figure 2 correspondence of EPR spectra of Mn^{2+} ion in BaF_2 powder and in oxyfluoride glass ceramics is shown. Superhyperfine structure is practically the same in both cases, thus supporting hypothesis of the preferred crystalline environment of Mn^{2+} ions.

4. Conclusions

Oxyfluoride glass ceramics samples containing BaF_2 crystalline counterpart have been synthesized and crystallization effects in the EPR spectra has been observed after heating the samples at temperatures from 560°C to 650°C. The resolved SHF structure in the EPR spectra of thermally treated samples reveal cubic fluorine arrangement around the impurity ions as the most probable BaF_2 structure in the oxyfluoride glass ceramic materials.

Acknowledgements
This work was done with the support of the European Social Funds project No. 2009/0202/1DP/1.1.1.2.0/09/APIA/VIAA/141.

References
[1] Mortier M, Bensalah A, Dantelle G, Patriarhe G and Vivien D 2007 *Opt. Mater.* **29** 1263–70
[2] Chen D, Wang Y, Yu Y, Ma E and Hu Z 2005 *J. Phys.: Condens. Matter* **17** 6545–57
[3] Sarakovskis A, Dimitrocenko L, Misnevs A, Rogulis U and Springis M 2007 *J. Phys.: Conf. Ser.* **93** 012041
[4] Sung Y M 2012 *J. Non-Cryst. Sol.* **358** 36-39
[5] Shinozaki K, Honma T, Oh-ishi K, Komatsu T 2011 *Optical Materials* **33** 1350–1356
[6] Bogomolova L D, Grechko E G, Krasilnikova N A, Sakharov V V, 1985 *J. Non-Cryst. Sol.* **69** 299–307
[7] Bogomolova L D and Krasilnikova N A 1995 *Glass. Phys. Chem.* **21** 303
[8] Franco R W A, Lima J F, Magon C J, Donoso J P and Messaddeq Y 2006 *J. Non-Cryst. Sol.* **352** 3414–22
[9] Gehlhoff W, Ulrici W 1980 *Phys. Stat. Sol. (b)* **102** 11–59
[10] Fedotovs A, Berzins Dz, Sarakovskis A, Rogulis U, Doke G 2010 *IOP Conf. Series: Mater Sci. Eng.* **15** 012068
[11] Fedotovs A, Berzins Dz, Kiselova O, Sarakovskis A, Rogulis U 2011 *IOP Conf. Series: Mater Sci. Eng.* **23** 012018
[12] Orera V M, Alcala R, Alonso P J 1984 *Solid State Comm.* **50** 665-667
[13] Alcala R, Alonso P J, Cases R 1983 *J. Phys. C: Solid State Phys.* **16** 4693-4700

Characteristics of the Mn^{2+} EPR spectra in the oxyfluoride glass ceramics containing SrF_2 nanocrystals

A Fedotovs[1], Dz Berzins, O Kiselova, A Sarakovskis

Institute of Solid State Physics, University of Latvia, Kengaraga str 8, Riga, LV-1063, Latvia

E-mail: andris.xyz@gmail.com

Abstract. Characteristics of the electron paramagnetic resonance (EPR) spectra of the Mn^{2+} ion in the oxyfluoride glass ceramics containing SrF_2 crystalline counterpart has been investigated. Explicit superhyperfine structure (SHF) has been observed in the X-Band EPR spectra taken at 77 K after thermal treatment of the glass precursor. Such SHF structure is characteristic to Mn^{2+} ion in the fluorine coordinated nearest environment which is characteristic to cubic structure of the alkaline earth fluorides. Manganese replaces strontium in the SrF_2 lattice where it has pronounced SHF interaction with surrounding fluorine nuclei. Thus Mn^{2+} ion provides information about crystalline structure of the particular oxyfluoride glass ceramics.

1. Introduction

Nowadays significant number of investigations has been done seeking the compromise between complex single crystal technology and cost effective, more simpler and flexible glass-ceramics technology. Main goal is to find effective host material for rare earth ions which serve as energy transformers in proper optical radiation in many optical and optoelectronic applications. Oxyfluoride glass ceramics are composite materials which possess high durability of the glass and transparency and low-phonon properties of the crystalline media [1]. The essential optical properties are defined by the structure of the crystalline counterpart. Its creation is strongly dependent on the relative ratios of the ingredients and dopant ions [2,3].

Incorporation of alkaline earth fluorides in the glass matrix is of particular interest recently [3]. Oxyfluoride glass ceramics containing SrF_2 nanocrystalline phase has been studied in [3, 4]. Common method of structure analysis is XRD techniques. However, in some cases more subtle structure sensitive methods could be used for structure analysis of materials containing glass and crystalline counterparts could be used.

In this work we used Mn^{2+} as indicating impurity ions to investigate structure of the crystalline counterpart of the oxyfluoride glass ceramics containing SrF_2 crystalline phase. Successful investigations using Mn^{2+} ion as a paramagnetic probe in other alkaline earth fluorides have been done previously [5-7].

[1] To whom any correspondence should be addressed

2. Experimental

The samples of the oxyfluoride glass ceramics containing SrF_2 were prepared with the following composition: $50SiO_2$-$10Al_2O_3$-$20ZnO_2$-$20SrF_2$-$0.25MnF_2$ (in mol%). Ingredients were stirred in chloroform and melted at 1450 °C. The melt was poured on the solid steel block for rapid quenching which resulted in 2 mm thick transparent glass samples. For SrF_2:Mn^{2+} powder synthesis we used pure (99.99%) SrF_2 and MnF_2 powders (99.5 and 0.5 mol% of particular fluorides, respectively), which were mixed in chloroform, and then dried. Additional synthesis of ZnF_2:Mn^{2+} powder was done using same type ingredients and synthesis techniques. Powder samples were thermally treated under air atmosphere at various temperatures. A home-built EPR spectrometer working in X-band was used for the EPR measurements. The microwave frequency was ~9.1 GHz. All measurements were done at 77K by immersing samples in liquid nitrogen.

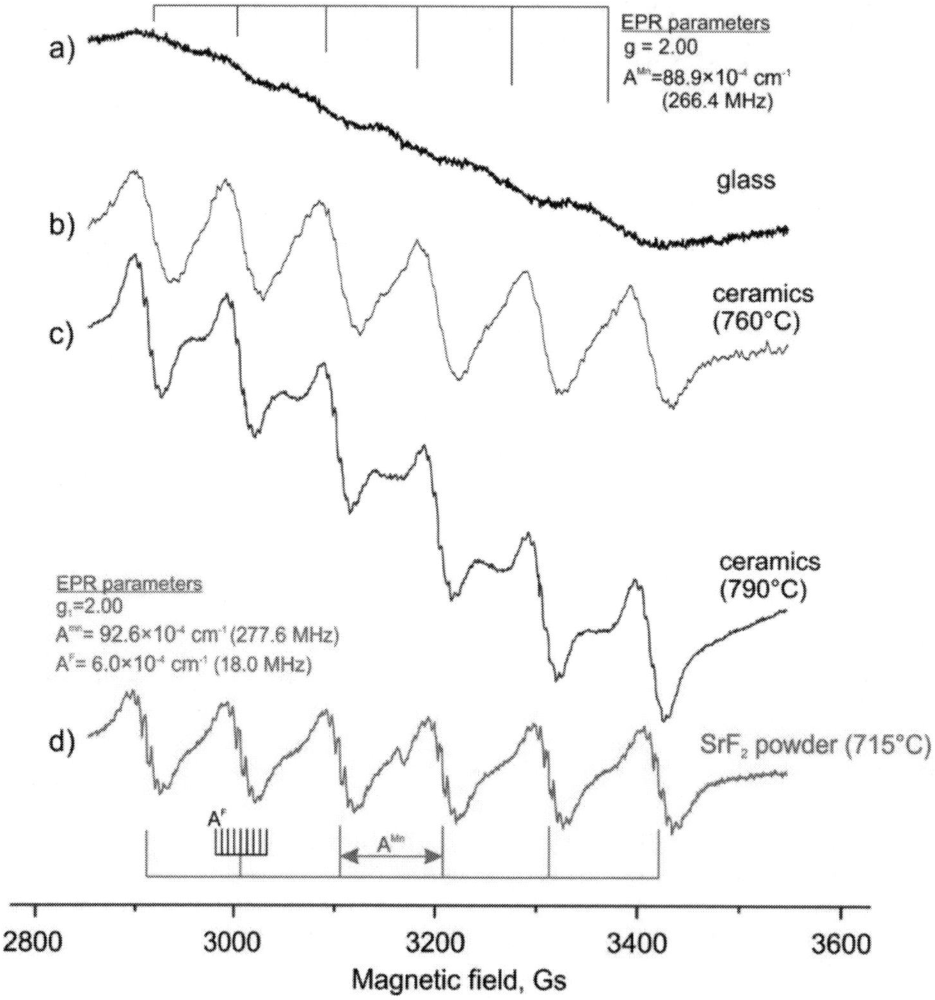

Figure 1. Curves a)-c) oxyfluoride glass ceramics with SrF_2:Mn^{2+} EPR spectra before and after heat treatment at various temperature. d) EPR spectrum of the SrF_2:Mn^{2+} powder after heat treatment at 715 °C taken at 9.1 GHz.

3. Results and discussion

Figure 1 shows the EPR spectra of the heat treated glass-ceramics and untreated glass material. A very weak manganese hyperfine structure sextet located at g~2.0 could be detected in the untreated sample (curve a). After thermal treatment EPR spectra have been changed considerably (curves b-d). At first, hyperfine structure lines have gained intensity and additional superhyperfine structure appears. At second, shift of two manganese sextets becomes evident: one corresponding to Mn^{2+} spectrum in the glass remains at its position while sextet with SHF structure is slightly shifted towards higher fields. This shift is due to slightly different g-factor and zero field splitting parameters for Mn^{2+} ions in distinct environments. As could be seen in the spectra the HF interaction constant is larger for Mn^{2+} ions in the crystalline lattice (see Figure 1 for corresponding experimental EPR parameters).

Each hyperfine structure line is overlaid with SHF structure which consists of resolvable 9 lines. Enlarged structure is shown in the inset in Figure 2. According to qualitative conventional spin Hamiltonian analysis the nine-line structure with binomial intensity distribution corresponds to hyperfine interaction with 8 nearest equivalent nuclei with nuclear spin I=1/2 [6]. In SrF_2 case these ligands are fluorine nuclei which are located at the vertices of the cube.

The possibility of other types of fluoride phases in particular glass matrix has been reported [8-10]. In this case other probable fluoride could be ZnF_2. In contrast to SrF_2 EPR spectra of manganese ions in ZnF_2 powders do not render additional SHF structure (see Figure 2 curve b), although manganese tends to build into this structure also. EPR spectra of $ZnF_2:Mn^{2+}$ and $SrF_2:Mn^{2+}$ powder mix (Figure 2 curve a) show that SHF structure is particularly characteristic to SrF_2 crystalline phase.

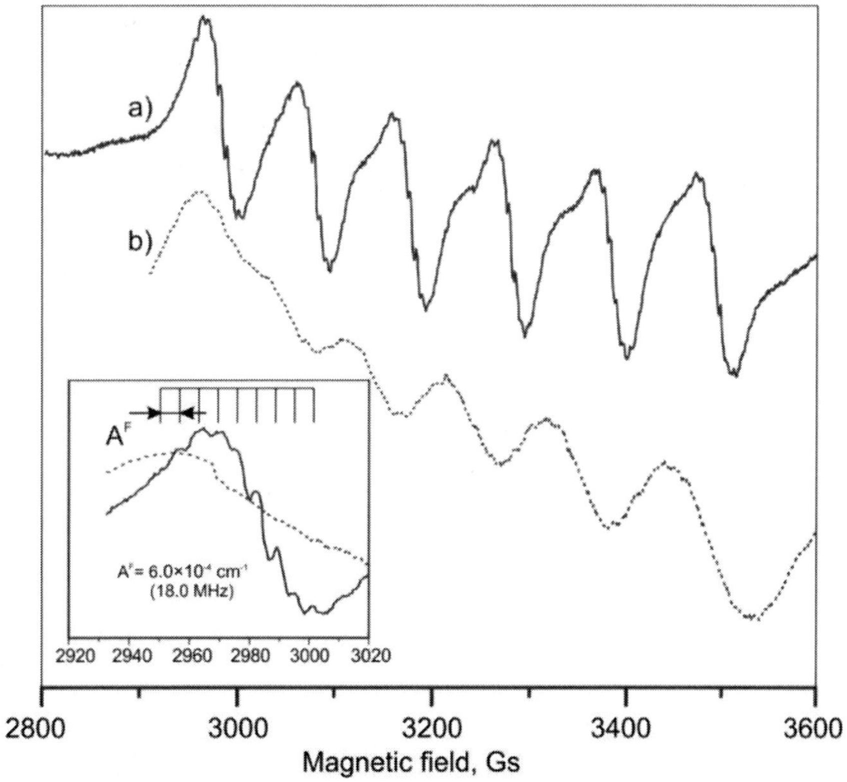

Figure 2. Curve a) EPR spectrum of the Mn^{2+} ion in the SrF_2 and ZnF_2 powder mix. Curve b) EPR spectrum of the Mn^{2+} ion in the ZnF_2 powder. Both spectra were obtained after heating powder samples at 650 °C. In the inset the first peak of the Mn^{2+} hyperfine structure in the SrF_2 with clearly resolved SHF structure of nine lines is shown.

4. Conclusions

Heat treatment of the glass with composition $50SiO_2$-$10Al_2O_3$-$20ZnO_2$-$20SrF_2$-$0.25MnF_2$ creates SrF_2 nanocrystals in still transparent glass matrix. In this process Mn^{2+} impurity ion enters into the SrF_2 lattice where it has SHF interaction with eight equal fluorine ^{19}F nuclei. This interaction is rendered in the SHF structure of the EPR spectra of Mn^{2+} ion.

Acknowledgements
This work was done with the support of the European Social Funds project No. 2009/0202/1DP/1.1.1.2.0/09/APIA/VIAA/141.

References

[1] Shinozaki K, Honma T, Oh-ishi K, Komatsu T 2011 *Optical Materials* **33** 1350–1356
[2] Nazabal V, Poulain M, Olivier M, Pirasteh P, Camy P, Doualan J L, Guy S, Djouama T, Boutarfaia A, Adam J L 2011 *J. Fluor. Chem.* in press Bogomolova L D and Krasilnikova N A 1995 *Glass. Phys. Chem.* **21** 303
[3] Sung Y M 2012 *J. Non-Cryst. Sol.* **358** 36-39
[4] Reben M 2011 *J. Non-Cryst. Sol.* **357** 2653-2657
[5] Franco R W A, Lima J F, Magon C J, Donoso J P and Messaddeq Y 2006 *J. Non-Cryst. Sol.* **352** 3414–22
[6] Fedotovs A, Berzins Dz, Sarakovskis A, Rogulis U, Doke G 2010 *IOP Conf. Series: Mater Sci. Eng.* **15** 012068
[7] Fedotovs A, Berzins Dz, Kiselova O, Sarakovskis A, Rogulis U 2011 *IOP Conf. Series: Mater Sci. Eng.* **23** 012018
[8] Sroda M and Olejniczak Z 2011 *J. Mol. Struc.* **1001** 78-82
[9] Sroda M and Olejniczak Z 2011 *J. Non-Cryst. Sol.* **357** 1696-1700
[10] Hill R G, Law R V, O'Donnell M D, Hawes J, Bubb N L, Wood D J, Miller C A, Mirsaneh M, Reaney I 2009 *J. Europ. Ceram. Soc.* **29** 2185-2191

IOP Conf. Series: Materials Science and Engineering **38** (2012) 012048 doi:10.1088/1757-899X/38/1/012048

Work function of colloidal semiconducting nanocrystals measured by Kelvin probe

P Kulis[1], J Butikova, B Polyakov, G Marcins, J Pervenecka, K Pudzs, I Tale

Institute of Solid State Physics, University of Latvia
Kengaraga Str. 8, Riga Lv-1063, Latvia

E-mail: p.kulis@cfi.lu.lv

Abstract. Colloidal suspensions of CdS, PbS, Cu_2S, Ag_2S and ZnO nanocrystals were prepared by chemical route in presence of organic capping ligands. The formation of nanocrystals was studied by using UV-visible absorption spectroscopy. Kelvin probe measurements of work function were performed on nanocrystals thin film on ITO (indium tin oxide) coated glass.

1. Introduction

Synthesis and characterization of nanocrystals (NCs) of semiconducting metal sulphides and metal oxides has been in research focus due to their interesting properties and potential applications [1]. For example, NCs are widely used in the solar cell technologies. A significant advantage of NC-based solar cell technology is synthesis from the solution, which is much cheaper in comparison with conventional semiconductor technologies used for silicon solar cells and easily can be scaled up. Several NC based solar cell architectures were demonstrated in recent time. Main architectures are Schottky type, heterojunction all NC or NC/polymer cells, and Grätzel type solar cell [2]. For example, an efficiency of 5.1% was reported for a single junction PbS/TiO_2 NC solar cell [3].

Proper arrangement of NC electronic band offsets is critical for all NC based solar cell architectures. Due to self-purification effect of impurity, atoms are expelled to the surface of nanocrystals for thermodynamic reasons [4]. However, stoichiometry and surface chemistry influence electronic properties of NCs and determine Fermi level position [5, 6].

Kelvin probe is a technique producing relative work-function measurement between a reference probe and the sample in a noninvasive fashion, compatible with vacuum or ambient operating conditions. It has been utilized, e.g., for thin-film studies, characterization of oxides and thin films, semiconductor doping profiling and chemical processed monitoring on sample surface [7]. In this work we used Kelvin probe for measuring work function of different NCs.

2. Experimental

Nanocrystal syntesis. PbS NCs were synthesized from PbO in mixture of oleic acid and octadecene, using hexamethyldisilanthiane as sulphur source [8]. When NCs were dispersed in oleylamine and stored at 60°C for 24 hours for ligand exchange to proceed, then precipitated by methanol and washed by toluene. Finally NCs were dispersed in chloroform. CdS NCs were synthesized from $CdCl_2$ in

[1] To whom any correspondence should be addressed

oleylamine and elemental sulphur, washed and redispersed in chloroform [9]. ZnO NCs were synthesized from zinc acetate in methanol using KOH [9], then washed and dispersed in oleylamine for ligand exchange. Later ZnO NCs were precipitated by methanol, washed and redispersed in chloroform. The synthesis of Cu_2S and Ag_2S NCs was adopted from [10]. NCs were synthesized from copper acetate (0.5 mmol) and silver nitrate (0.5 mmol) in mixture with octadecene (10 mL) oleylamine (1 mL), oleic acid (1 mL) and dodecanthiol (0.25 mL) at 200°C for 30 min under N_2 flow under gentle stirring, until the colour of the solution had changed to dark brown. The NCs were naturally cooled to room temperature, separated from the solution by addition the mixture of methanol and acetone. Then NCs were redispersed in toluene, washed by methanol, precipitated and redispersed in chloroform.

Optical absorption measurements. The optical absorption of NC solution diluted in toluene was measured in UV-VIS SPECORD spectrophotometer (M40, *Carl Zeiss*) in a quartz cell.

Kelvin probe measurements. Kelvin probe (KP) measurements were performed by SKP5050 instrument (*KP Technology*) using Au probe. As a substrate, ITO (indium tin oxide) coated glass slide (*PGO*, 40 Ohm/sq) was cleaned in acetone in ultrasound bath for 5 min. NCs capped with oleylamine were deposited onto the substrate by dipping ITO-coated glass slide in NC solution several times, followed by soaking in 1% ethanedithiol solution in acetonitrile and drying in air. PbS NCs samples were finally dried at 90°C for 5 min [8], CdS NCs samples dried at 150°C for 5 min [11], Cu_2S NCs samples were dried at 150°C for 10 min [11]. ZnO NCs without oleylamine were coated onto ITO glass slide and annealed at 260°C for 10 min [8].

3. Results and discussion

Optical absorption can be used to determine the optical band gap and extinction of NCs. Band gap can be found from optical absorption data using following equation [12]:

$$\alpha h\nu = A(h\nu - E_g)^n,$$

where α is the absorption coefficient measured as a function of photon energy $h\nu$, A is a constant factor determined by the transition probability, E_g is the energy gap, and n is a number that depends on the type of the transition process, where n=1/2, 2, 3/2, or 3 for allowed direct, allowed indirect, forbidden direct, and forbidden indirect electronic transitions, respectively. CdS, Ag_2S, ZnO are known as direct band gap materials having E_g ~2.4-2.6 eV, 0.95-1.3 eV, 3.2 eV, respectively [13-15]. Cu_2S is an indirect band gap material with significant influence of direct optical transitions, having band gap 1.2-1.6 eV in NCs [16]. We found for CdS NCs optical band gap ~2.4 eV, for Ag_2S 1.3 eV, for ZnO NCs 3.25 eV, and 1.5 eV for Cu_2S NCs. These values are in good agreement with the literature, confirming successful synthesis of NCs.

Additionally, if average diameter of NCs is smaller than Bohr exciton radius, blueshift of the absorption edge can be used to estimate the diameter of NCs. For example, optical absorption spectra of PbS NCs have a well defined excitonic peak centred at about 765 nm (1.62 eV) as it is shown in figure 1. An average diameter of the NCs of about 2.4 nm was calculated using formula of band gap dependence E_g on particle size from [17]. Bohr radius of CdS, ZnO and Cu_2S is only 3 nm [18], 2.3 nm [19] and 3-5 nm [5], respectively (data for Ag_2S are not available), and almost no manifestation of quantum confinement effects can be seen in their absorption spectra.

We used Kelvin probe technique to determine the Fermi level E_f of the PbS, CdS, Cu_2S, Ag_2S and ZnO NC layers deposited on ITO coated glass slide. Kelvin probe measurements determine the contact potential difference (CPD) or position of the Fermi level relative to the reference surface, which was a gold electrode in our case. To find the exact value of gold electrode, we used freshly cleaved HOPG having work function of 4.45 eV [20].

Kelvin probe measurements were carried out in two regimes. In the first one, the NCs-coated/uncoated border of the ITO glass slide was scanned by the Kelvin probe. The CPD potential graphs are presented in figure 2. The diameter of the Kelvin probe is 2 mm, and the CPD signal gradually changes during the probe crossing the border of NC coated / uncoated regions. In the second regime, Kelvin probe was positioned over NC coated ITO glass slide and measured CPD signal during some time to accumulate statistics and obtained the time-averaged value. Measurement results for PbS, CdS, Cu_2S and ZnO coated samples are presented in table 1.

In table 1, we compare our results with the literature data for corresponding bulk material and NCs. Unfortunately, the information on work function value or Fermi level position in NCs can be found in the literature only for few NCs materials. Pattantyus-Abraham et al. reported Fermi level $E_f = 4.8$ eV and 4.9 eV for 5.5 nm and 3.7 nm PbS NCs [3]. Fermi level of ZnO NCs was measured by Lakhwani et al. $E_f = 4.7$-4.8 eV [21], indicating n-type material (E_c =4.4 eV). Our measurement for the Fermi level for PbS NCs was 4.7 eV, and for ZnO – 4.8, which is in a good agreement with the literature data.

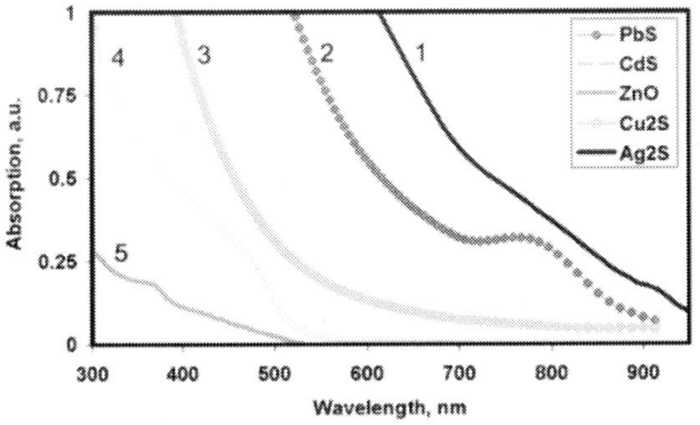

Figure 1. Optical absorption spectra of Ag_2S (1), PbS (2), Cu_2S (3), CdS (4) and ZnO (5) NCs measured in suspension.

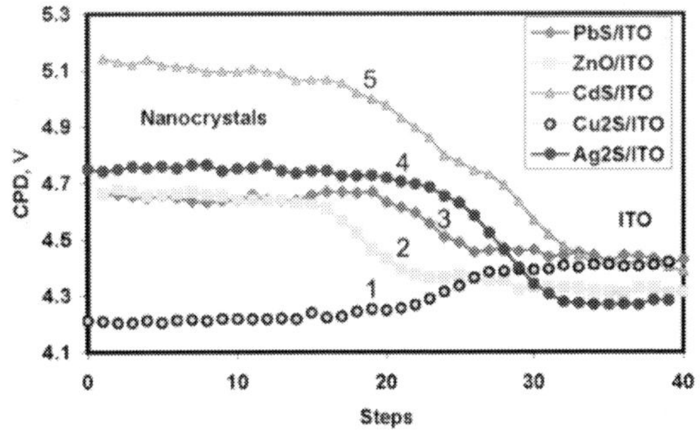

Figure 2. Contact potential difference between Cu_2S (1), ZnO (2), PbS (3), Ag_2S (4), CdS (5) NCs coated and bare ITO slide.

CdS NCs are supposed to be n-type semiconducting material [22], however no direct data about their work function are available in the literature. The Fermi level E_f=5.3-5.7 eV for bulk CdS is (E_c =4.5 eV) [23]. We found E_f=5.2 eV, which indicates n-type property of CdS NCs.

Bulk Cu_2S is known to be a p-type semiconductor with shallow Cu vacancy acceptors, electron affinity of 3.6 eV and work function 4.2 eV; Lee et al. used bulk values of electron affinity and work function for NCs [24]. Depending on the number of Cu vacancies, the number of charge carriers in Cu_2S NCs varies [5]]. We found E_f=4.2 eV, which correspond to nearly stoichiometric i-type material.

Bulk Ag_2S, having electron affinity of 3.66 eV [14], is mainly referred as a n-type semiconductor. However, we found E_f=4.8 eV, which correspond to the p-type material.

Table 1. Comparison of the literature values of work function for bulk and NC of PbS, CdS, ZnO, Cu_2S and experimental values measured by Kelvin probe.

	Work function, eV (experimental data)	Work function, eV (literature data)	
Substance	NCs	Bulk	NCs
ZnO	4.8	3.6-5.1 [25]	4.7-4.8 [21]
PbS	4.7	4.8 [26]	4.8-4.9 [3]
CdS	5.2	5.3-5.7 [23]	No data
Cu_2S	4.2	4.2 [24]	No data
Ag_2S	4.8	4.3 [14]	No data

As it is shown in figure 2, work function of ITO is in range of 4.3-4.4 eV. This is in a good agreement with values of work function referred by Park et al [27]. They found ITO $E_f = 4.4$-4.5 eV depending on annealing temperature. We believe that the slight dissimilarities in the values of ITO work function for ZnO and other NCs appeared due to the heating at different temperatures, and possible contamination by NCs of bare ITO surface during rinsing whole slide after NC deposition.

4. Conclusions

Colloids of CdS, PbS, Cu_2S, Ag_2S and ZnO nanocrystals stabilized by oleic acid or oleylamine ligands were synthesized in solution and deposited on ITO glass by dip coating using ethanedithiol for cross linking. Work function of CdS, PbS, ZnO, Cu_2S and Ag_2S nanocrystals thin films was measured by Kelvin probe, resulting in 5.2, 4.7, 4.8, 4.2 and 4.8 eV, respectively.

Acknowledgements

P.K. and J.B. were supported by the ESF Grant Nr. 2009/0202/1DP/1.1.1.2.0/09/ APIA/VIAA/141. B.P. and I.T. were supported by the ERAF Nr.2010/0252/2DP/2.1.1.1.0/10/ APIA/VIAA/009.

References

[1] Talapin V D, Lee J-S, Kovalenko V M and Shevchenko V E 2010 *Chem. Rev.* **110** 389
[2] Alivisatos P A 1996 *Science* **271** 933
[3] Pattantyus-Abraham G A, Kramer J I, Barkhouse R A, Wang X, Konstantatos G, Debnath R, Levina L, Raabe I, Nazeeruddin K M, Gratzel M and Sargent H E 2010 ACS Nano **4** 337
[4] Norris J D, Efros L A and Erwin C S 2008 *Science* **319** 1776
[5] Luther M J, Jain K P, Ewers T A and Alivisatos P A, 2011 *Nature Mater.* **10** 361
[6] Lee J-S, Kovalenko V M, Huang J, Chung D S, Talapin V D 2011 *Nature Nanotechnol.* **6** 348
[7] Baikie I D, Peterman U, Lagel B and Dirscherl K 2001 *J. Vac. Sci. Technol.* A **19** 1460
[8] Luther J, Gao J, Lloyd M T, Semonin O, Beard M and Nozik A 2010 *Adv. Mater* **22** 3704
[9] Yong K-T, Sahoo Y, Swihart T M and Prasad N P 2007 *J. Phys. Chem. C* **111** 2447
[10] Zhuang Zh, Lu X, Peng Q and Li Y 2011 *Chem. Eur. J.* **17** 10445
[11] Wu Y, Wadia C, Ma W, Sadtler B and Alivisatos P A 2008 *Nano Lett.* **8** 2551
[12] Yu P and Cardona M. 2002 *Fundamentals of Semiconductors*, 3rd ed. (Springer)
[13] Khallaf H, Oladeji O I, Chai G and Chow L 2008 *Thin Solid Films* **516** 7306
[14] Barman S , Shanthi N, Shukla A and Sarma D D 1996 *Phys. Rev.* B **53** 3746
[15] Shim M and Guyot-Sionnest M 2001 *J. Am. Chem. Soc.* **123** 11651
[16] Klimov I V and Karavanskii A V 1996 *Phys. Rev.* B ***54*** 8087
[17] Moreels I, Lambert K, Smeets D, Muynck D, Nollet T, Martins C J, Vanhaecke K, Vantomme A, Delerue C, Allan G and Hens Z 2009 *Asc Nano* **3** 3023
[18] Tiwari A, Mishra K A, Kobayashi H and Turner P F A 2012 *Intelligent Nanomaterials* (New Jersey: John Wiley and Sons)
[19] Gu Y, Kuskovsky L I, Yin M, O'Brien S and Neumark G F 2004 *Applied Phys. Lett.* **85** 3833
[20] Hansen N W and Hansen J G 2001 *Surface Science* **481** 172
[21] Lakhwani G, Roijmans F H R, Kronemeijer J A, Gilo Jt, Janssen A J R and Meskers C J S 2010 *J. Phys. Chem.* C **114** 14804
[22] Shim M and Guyot-Sionnest Ph 2000 *Nature* **407** 981
[23] Hu H, Kung Sh, Yang L, Nicho M and Penner R 2009 *Solar Energy Materials & Cells* **93** 51
[24] Lee H, Yoon W S, Kim J E and Park J 2007 *Nano Lett.* **7** 778
[25] Ellmer K, Klein A and Rech B 2008 *Transparent Conductive Zinc Oxide: Basics and Applications in Thin Film Solar Cells* (Heidelberg: Springer-Verlag)
[26] Knapp A R 1963 *Phys. Rev.* **132** 1891
[27] Park Y, Choong V, Gao Y, Hsieh B R, and C. W. Tang C W 1996 *Appl. Phys. Lett.* **68** 2699

IOP Conf. Series: Materials Science and Engineering **38** (2012) 012049 doi:10.1088/1757-899X/38/1/012049

Deposition of Cu Nanoparticles on the Surface of Metallic Aluminum

A Lescinskis[1], J Katkevics [1,2], D Erts[2], A Viksna [1,2,3]

[1]Faculty of Chemistry, University of Latvia, Kr. Valdemara 48, Riga, LV-1013, Latvia
[2]Institute of Chemical Physics, University of Latvia, Kronvalda blvd. 4, Riga, LV-1586, Latvia

e-mail: arturs.viksna@lu.lv

Abstract.
Deposition of Cu particles by electrolysis at constant electrode potential and by internal electrolysis methods was investigated. The composition of deposited material was confirmed by optical and scanning electron microscope methods. Combination of electrolysis at constant electrode potential with internal electrolysis method was found most effective for fabrication of nanoparticle arrays. Single crystalline Cu particles are fabricated by internal electrolysis, while polycrystalline ones obtained by combined chronopotentiometric and internal electrolysis methods. The formation mechanism of Cu nanoparticles is described.

Introduction

Cu nanoparticles are perspective materials for application as catalysts, photocatalysts, adsorbents, and sensors. Cu nano and microparticles have been deposited on a ultrathin polypyrrole film [1], on a surface of boron doped diamond electrode by constant current electrolysis [2], and on Al surfaces by internal electrolysis methods [3]. Early stages of formation of nanoparticles on Al surfaces have not been investigated in detail.
In the present study, the deposition of Cu nanoparticles on Al surface were performed by internal electrolysis (IE) method in the acidic electrolyte media and by electrolysis at constant electrode potential in acidic electrolyte media combination with internal electrolysis.

2. Experimental

0.10 mm thick Al foil with 99.9 % purity was used in all experiments. Al foils were treated with acetone and then annealed at 400 °C for 20 minutes, electropolished in a perchloric acid - ethanol (1 : 3) bath at 15 V dc and 10 °C for 5 min. After anodizion at a constant voltage of 15 V in 0.3 M oxalic acid Al foils were etched 3.5 hours at room temperature in 0.2 M $K_2Cr_2O_7$ / 0.4 M H_3PO_4 solution mixture to remove an oxide layer. Etching process was controlled by Electrochemical impedance spectroscopy (EIS). Cu was deposited on Al surface from electrolyte containing 0.30 ÷ 0.03 M $CuSO_4$ in 10% H_3PO_4 at 40 °C, but 0.3 M concentration was used for further investigations as optimal. Deposition of copper was performed in two ways. In the first experimental part internal electrolysis at different concentration of the

[3] To whom any correspondence should be addressed

copper solution was investigated. In the second experimental part chronopotentiometric method at constant potential from -0.9 to -1.9 V with potential step -0.2 V and electrolysis time 30 s at each potential was combined with internal electrolysis. Electrolysis time of internal electrolysis was applied from 1 to 15 minutes. EIS was used to control Al oxide dissolution process. Electrodes were characterized by optical and scanning electron microscope methods. A computer controlled frequency response analyzer (FRA) with module Autolab PGSTA–30 was used in experiments.

3. Results and discussion

Copper nanoparticles have to be deposited on pure metallic aluminum surface in electrolyte, which does not oxidize aluminum. Therefore an electropolished and anodized Al foil was etched in 0.2 M $K_2Cr_2O_7$/0.4 M H_3PO_4 solution mixture and during the process EIS was collected. In the previous studies [3] it was found that the oxide layer was removed completely after 50 min of aluminum surface treatment in 0.2 M $H_2Cr_2O_7$/0.4 M H_3PO_4 electrolyte, but dissolution process in this electrolyte is too fast to get good quality of Al surface. In the current study the oxide removal process was controlled directly by EIS measurement (Fig.1a). Impedance modulus was used to characterize the oxide dissolution process. As seen from Figure 1b, oxide dissolution process continues up to 135 minutes, after which EIS shows pure metallic Al surface and the value of impedance modulus decreased from $|Z|_\omega$ = 10.4 kΩ.cm^2 (10 min treatment) to $|Z|_\omega$ = 0.014 kΩ.cm^2 (135 min treatment). Impedance modulus is practically invariable in the treatment interval from 135 min to 24 hours (Fig.1b) and impedance spectra (Fig. 1a, curves 5;6) are similar, and therefore the Al surface is ready for copper electrodeposition.

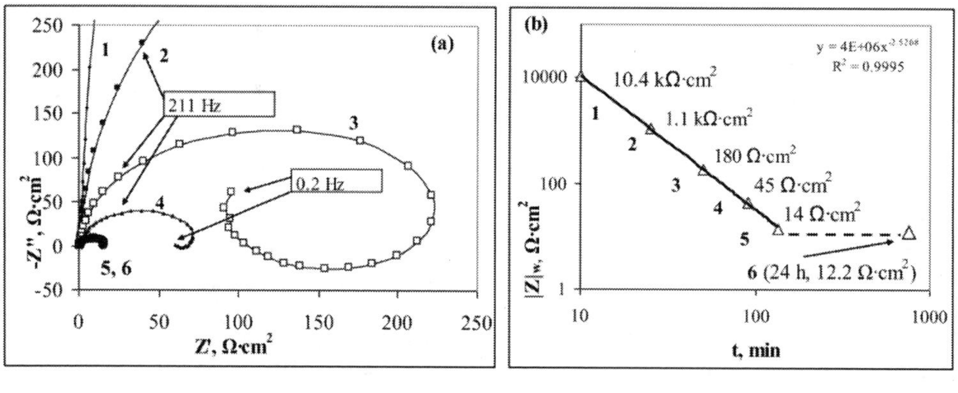

Figure 1. Electrochemical impedance spectra (a) and impedance modulus values (b) after etching of Al foil in 0.2 M $K_2Cr_2O_7$ / 0.4 M H_3PO_4 at different treatment times: 1 – 10 min; 2 – 25 min; 3 – 50 min; 4 – 90 min; 5 – 135 min; 6 – 24 hours.

During previous studies [3] the optimal conditions of internal electrolysis for the deposition of Cu micro size particles on the Al surface were developed. In current study we investigate initial stages of electrolysis. SEM images revealed that the smallest globule size is around 100 nm in diameter (Fig. 2a) with the internal electrolysis time less than 1 minute. At the optimal internal electrolysis conditions (0.3M $CuSO_4$ in 10% H_3PO_4 at 40°C) the density of particles of around 5×10^3 particles/cm^2 are achieved (Fig. 2b). The obtained particles exhibit monolithic crystal structure (Fig. 2c). After 14 minutes of internal electrolysis the Cu globules can reach from 1.5 to 40 μm in diameter in size (Fig. 2b,c).

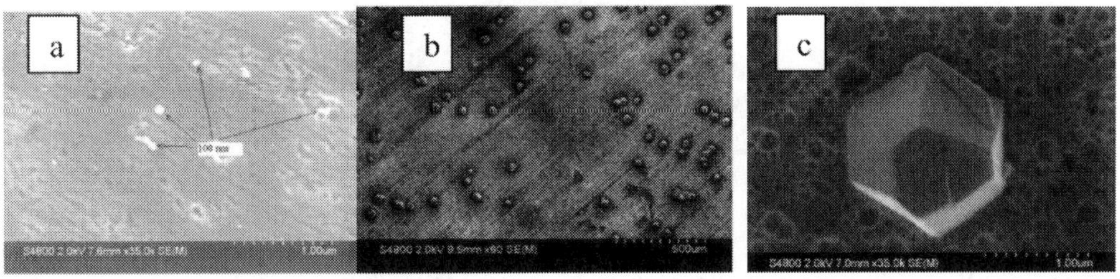

Figure 2. SEM images of Al/Cu-Al2O3 electrode after deposition of copper by internal electrolysis of 0.3 M $CuSO_4$ in 10% H_3PO_4 solution at 40 °C with electrolysis time 1 min (a) and 14 min (b, c).

Initiation of nanoparticle formation can be performed also by electrolysis at constant current before internal electrolysis. The formation process of nanoparticles depends on dynamic current flow in the electrochemical cell and on the potential on the electrode. As shown in Figure 3 the highest current density (0.5 ÷ 11 mA/cm^2) was observed at the beginning of applied constant potential and in time interval from 0.3 seconds at -1.1 V to 9.0 seconds at -1.9 V the current density decreases to 0.01 till 7.5 mA/cm^2 at -1.1 V and -1.9 V respectively and it was stable for the next 21 to 29 seconds. Different sizes of Cu particles (Fig. 4) were observed by chronopotentiometric deposition of Cu in five different potential cycles (Fig. 3). To stabilize the diffusion process, 30 s of electrolysis time was used.

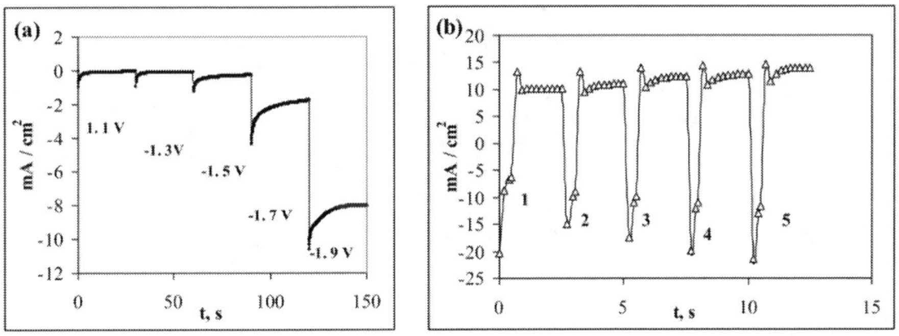

Figure 3. Chronopotentiometric mode of copper deposition on Al surface. Applied potential range from -0.9V to -1.9 V with potential step -0.2 V and electrolysis time 30 sec (a); Cyclic copper deposition regime at -1.9 V with 0.5 s and 2 s internal electrolysis (b).

As can be seen in Figure 4, the Cu particles are polycrystalline in contrast to single crystalline Cu particles obtained with internal electrolysis (Fig. 2c). SEM image shows granules of different diameters: 100, 300, 1000, 1600 and 2000 nm. The granularity is higher for larger particles. It can be stated that new crystallization centers are initiated by the first current pulse. In following current pulses new Cu particles are formed not only on the surface of metallic Al, but also on the existing Cu particles. The observed density of nanoparticles on the metallic Al surface (i.e. the Cu depositions centers) is about 5.2×10^6 particles/cm^2 which is 3 orders magnitude higher in comparison to internal electrolysis.

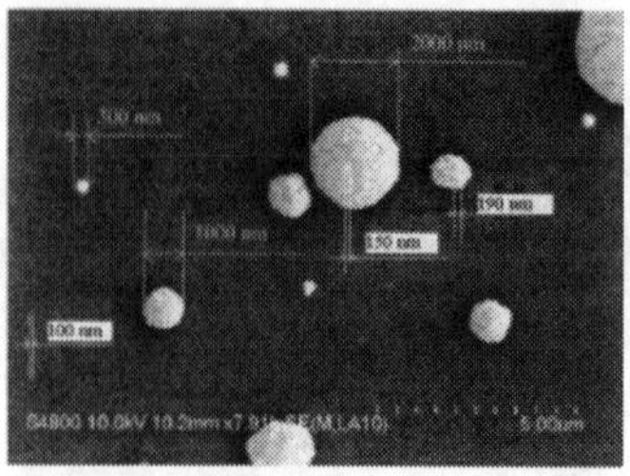

Figure 4. SEM images of copper nanoparticles and microparticles obtained from 0.30 M $CuSO_4$ in 10% H_3PO_4 electrolyte at 40 °C using combination of chronopotentiometric method with internal electrolysis (1 minute electrolysis time).

In order to obtain higher particle density, the cyclic electrodeposition at -1.9 V for 0.5 s combined with 2 s internal electrolysis was used. As seen from Figure 3a the highest current pulse is formed at -1.9 V. The cyclic regime is presented in Figure 3b. The observed density of nanoparticles on metallic Al surfaces, which was obtained by cyclic regime combining electrodeposition at constant potential with short internal electrolysis, is about 5.2×10^9 particles/cm^2 (Fig. 5). The obtained Cu particle density on the Al surface is six orders higher compared to the case of internal electrolysis method.
The electric double layer (EDL) is formed on the Al surface by immersing aluminum electrode into a copper sulphate solution. The formation of copper globule centers is initiated by current pulse. After applying a voltage on the EDL, the nanoparticle centers are initiated and the formation process is continued by internal electrolysis.

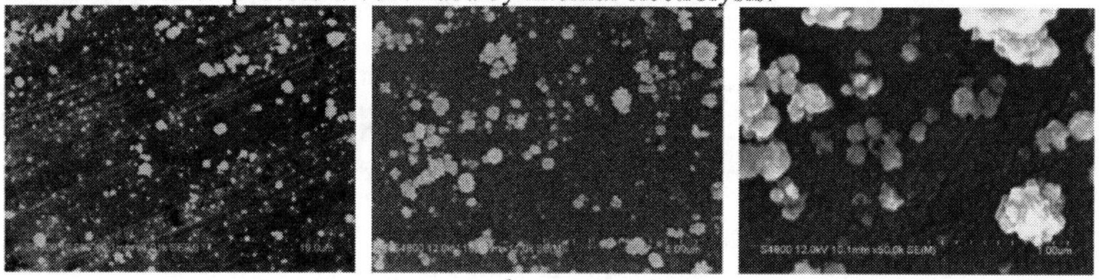

a b c

Figure 5. SEM images of copper nanoparticles obtained from 0.03 M $CuSO_4$ in 10% H_3PO_4 electrolyte at 40 °C using cyclic copper deposition for 0.5 s at -1.9 V and 2 s internal electrolysis.

The mechanism of Cu nanoparticles formation is demonstrated in Figure 6 in static equilibrium (a), electrolysis at constant potential (b) and internal electrolysis (c) regimes. At static regime the formation of Helmholtz and diffusion layers occurs, but after applying certain constant potential the formation of crystallization centers is initiated and further particle growth process proceeds by internal electrolysis. It is necessary to note that additional dissolution process of Al electrode occurs during internal electrolysis.

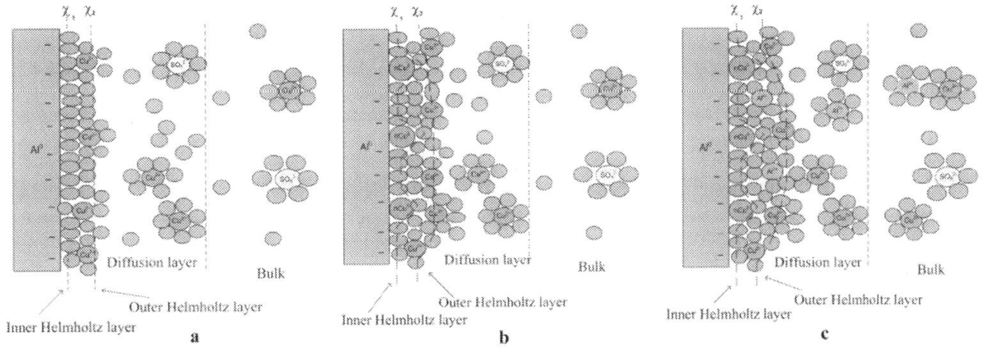

Figure 6. Mechanism of Cu nanoparticles formation process: static equilibrium (a), short current pulse (b), internal electrolysis (c).

Conclusions

We have demonstrated that applied electrochemical methods are useful for the fabrication of Cu particles on the Al surface. The oxide barrier layer of aluminum dissolves completely within 135 minutes using $K_2Cr_2O_7$ solution in H_3PO_4. This process can be easily controlled by electrochemical impedance spectroscopy. The transport of copper ions from electric double layer to the Al surface depends on applied potential. The structures of Cu nanoparticles depend on the electrochemical method. Cu particles fabricated by internal electrolysis exhibit monolithic crystalline structure. The surface structure of Cu particles deposited combining electrolysis at constant potential with internal electrolysis is grainy consisting as aggregate of smaller particles. The density of deposited Cu particles on the Al surface by using combined method is at least 6 orders higher compared with internal electrolysis method. The developed method is perspective for the fabrication of Cu particles on Al surface with high density.

Acknowledgements

The work was supported by ERAF project No 2010/0251/2DP/2.1.1.1.0/10/APIA/VIAA/096.

References

[1] Zhou X J, Harmer A J, Heinig N F, Leung K T 2004 *Langmuir* **20** 5109
[2] Welch C M, Hyde E H, Banks CE, Compton R G 2005 *Anal. Sci.* **21** 1421
[3] Denisova J, Katkevics J, Erts D, Viksna A 2011 *IOP Conf. Series: Materials Science and Engineering* **23** 012040

IOP Conf. Series: Materials Science and Engineering **38** (2012) 012050 doi:10.1088/1757-899X/38/1/012050

Zirconia nanocrystals as submicron level biological label

K Smits[1], J Liepins[2,3], M Gavare[2], A Patmalnieks[2], A Gruduls[1] and D Jankovica[1]

[1] Institute of Solid State Physics, University of Latvia, 8 Kengaraga Str., Riga, Latvia LV-1048

[2] Institute of Microbiology and Biotechnology, University of Latvia, Kronvalda Blvd. 4, Riga, Latvia LV-1586

E-mail: janis.liepins@lu.lv

Abstract. Inorganic nanocrystals are of increasing interest for their usage in biology and pharmacology research. Our interest was to justify ZrO_2 nanocrystal usage as submicron level biological label in baker's yeast *Saccharomyces cerevisia* culture. For the first time (to our knowledge) images with sub micro up-conversion luminescent particles in biologic media were made. A set of undoped as well as Er and Yb doped ZrO_2 samples at different concentrations were prepared by sol-gel method. The up-conversion luminescence for free standing and for nanocrystals with baker's yeast cells was studied and the differences in up-conversion luminescence spectra were analyzed. In *vivo* toxic effects of ZrO_2 nanocrystals were tested by co-cultivation with baker's yeast.

1. Introduction

A great variety of measurement methods are used currently in biology and pharmacology; however, the demand for more measurement possibilities is continuous. Many of the new methods include usage of inorganic nanocrystals [1,2].

Zirconia nanocrystals are promising candidate for biosensing applications due to their chemical stability and mechanical and optical properties [2,3]. On one hand, zirconia optical properties are oxygen sensitive [4] while on the other hand, it is possible to use up-conversion luminescence properties for temperature measurement [5]. Due to infrared excitation, up-conversion luminescence applications in biology have several advantages over the "classical" one-photon Stokes process, including: lack of autofluorscence, larger penetration depth and less harmful for cells [6, 7].

Information on the nanostructured ZrO_2 effects on organisms *in vivo* is scarce. Thus far, ZrO_2 has been proposed in medical implants, assuming that the material is biologically inert [8]. In a comparable study of metal oxide nanocrystal toxicity in mammalian cell lines (human mesothelioma and rodent fibroblasts), ZrO_2 revealed medium toxicity [9, 10]. Thus, we conclude that zirconia nanoparticle effects on live organisms are loosely covered, especially regarding traditional eukaryotic model organisms such as baker's yeast *Saccharomyces cerevisiae*.

In the work presented here, we synthesized zirconia nanocrystals, measured up-conversion luminescence from nanocrystals in biological media, and checked the impact of different ZrO_2 nanocrystal concentrations on baker's yeast *S. cerevisiae* growth. An attempt to simultaneous submicro scale imaging of *S. cerevisiae* with ZrO_2 nanocrystals was also made.

[3] To whom any correspondence should be addressed

2. Materials and methods

2.1. Zirconia nanocrystal synthesis

The ZrO2 nanocrystals with 1 mol% Er2O3 and 2 mol% Yb2O3 were prepared by the sol-gel combustion technique [11, 12]. After synthesis, the samples were calcinated at 900°C for 2 hours. Phase composition of the prepared sample was determined by X-ray diffraction analysis (XRD) (D8 Advance, Bruker AXS). The determined grain size was 40 nm. The crystallite size was calculated from the broadening of diffraction maxima using Topas 3 software (Bruker AXS). ZrO2 nanocrystal transmission electron microscopy (TEM) was done by Jeol JEM-1011 instrument.

2.2. Yeast strains and growth conditions

The *S. cerevisiae* haploid reference strain BY4741 (EUROSCARF, Germany) was used for toxicity studies. The yeasts were grown in the presence (1, 2, 5 or 20 mM) or absence of ZrO_2. Cells were cultivated as reported elsewhere [13]: in media containing 2% **P**eptone, 1% **Y**east extract, 2% **D**extrose (further abbreviated as YPD), +30°C, on orbital shaker with agitation rate 180 revolutions per minute. Yeasts were inoculated at OD600=0.2 and grown for 22 hours. 2 ml samples for biomass measurements were harvested and frozen for further analyses.

2.3. Biomass concentration and glucose measurements

The yeast growth was measured spectrophotometrically at 600 nm (**o**ptical **d**ensity at **600** nm, further abbreviated as OD600); one absorbance unit of OD600 equals approx. 2 X 10^7 yeast cells/ ml. OD600 measurements were done on the Ultraspec 2100pro (GE Healthcare Life Sciences) instrument using empty YPD broth as a reference.

Since ZrO_2 nanocrystals increase the absorbance of cell suspension at 600 nm, thus giving erroneous estimation of cell growth dynamics, an alternative Fourier transform infrared spectroscopy (FT-IR)s based method for biomass measurements for samples containing ZrO_2 was developed. FT-IR analysis was carried out using the HTS-XT microplate reader (Bruker, Germany) over the range 4000-600 cm^{-1} and data processed using OPUS 6.5 software. The spectral region specific for lipids (3002-2797 cm^{-1}) [14] was chosen to linearly correlate OD600 curve. No absorption of ZrO_2 particles over the chosen wavelength region was observed.

Glucose concentration was estimated by sulphuric acid anthrone method [15].

2.4. Luminescence measurements

The luminescence measurements were prepared using the L975P1WJ 975 nm laser diode (Thorlabs) as the excitation source. The luminescence spectra were recorded using the Andor Shamrock B-303i spectrograph equipped with a CCD camera (Andor DU-401A-BV) at exit port. Nanocrystal luminescence imaging with baker's yeast cells was done with a modified Carl Zeiss microscope. The image was captured using a Canon 5D Mark II COMOS sensor coupled with C3C21 colour filter (LOMO) in camera input, to cut off the IR radiation.

3. Results and discussion

3.1. Zirconia nanocrystal up-conversion characteristics

At first, it should be noted that zirconia nanocrystal up-conversion luminescence spectra were environment dependent (figure 1 A). Not only did the luminescence spectral distribution changed, but also the luminescence intensity weakened strongly in aqueous solutions. Therefore we concluded, that large specific surface areas were responsible for strong zirconia nanocrystals interaction with environment.

However, the ZrO_2 nanocrystal up-conversion luminescence did not change if measured in various aqueous media (water, empty YPD broth, ethanol, YPD broth with yeasts). The nanocrystal up-conversion intensity changed in aqueous media, but the spectral distribution was the same (results not shown).

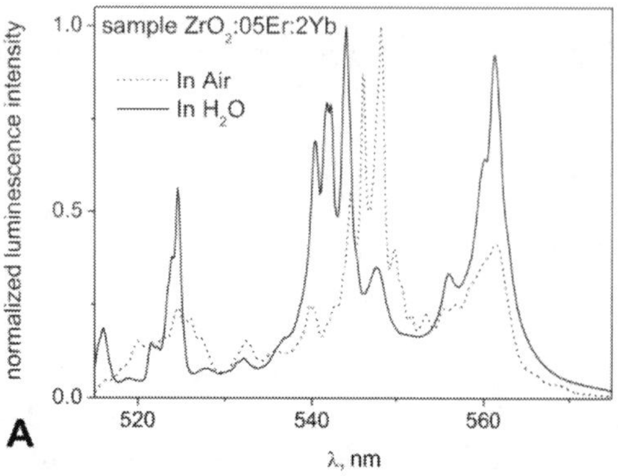

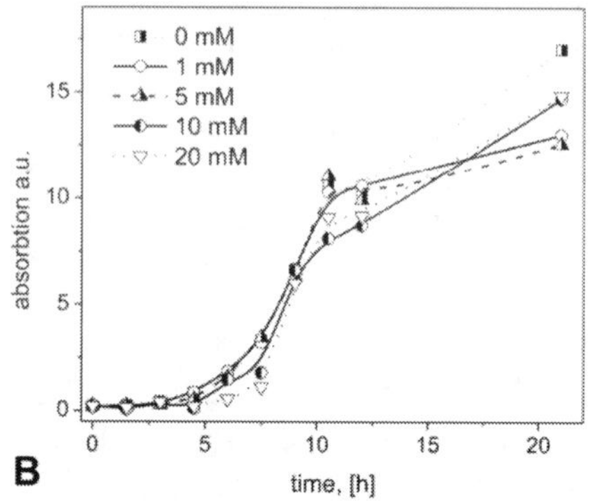

Figure 1. Zirconia nanocrystal up-conversion luminescence (A) and *S. cerevisiae* BY4741 strain growth in broth with different (0-20 mM) concentrations of zirconia nanocrystals (B)

3.2 ZrO_2 nanocrystal toxicity measurements

The toxicity of zirconia nanocrystals on eukaryotic cells was assayed by monitoring baker's yeast growth in YPD broth containing different concentrations (0-20 mM) of zirconia nanoparticles. Similarly to previously reported toxicity assays [16]- OD600 was used to estimate biomass growth in the presence of nanoparticles. Results were plotted as OD600 dynamics over time (figure 1 B). We observed no significant impact of ZrO_2 nanocrystals on *S. cerevisiae* fermentative growth (till 12 hour) over all of the concentrations applied. Similar results in the same concentration range were reported previously with ZnO, TiO_2 and CuO nanoparticles [17]. However, ZrO_2 nanocrystals slowed *S. cerevisiae* growth at concentrations of 1 and 5 mM after glucose exhaustion (after 12 hours of cultivation). Other authors noted that nanoparticles tend to form aggregates in water and microbial cultivation broth; therefore, their toxic effects were not pronounced due to the fact that cells were not exposed to small sized nanoparticles, but rather to their comparatively large aggregates instead [18].

It seemed that ZrO_2 nanocrystals were biologically inert while the cell's respiratory metabolism was suppressed by glucose, and therefore might be used as a biosensor during fermentative growth of yeast. Further investigations should be done to clarify whether rare earth doped zirconia nanocrystals have any specific effects on baker's yeast respiratory growth.

3.2. Up-conversion and TEM imaging

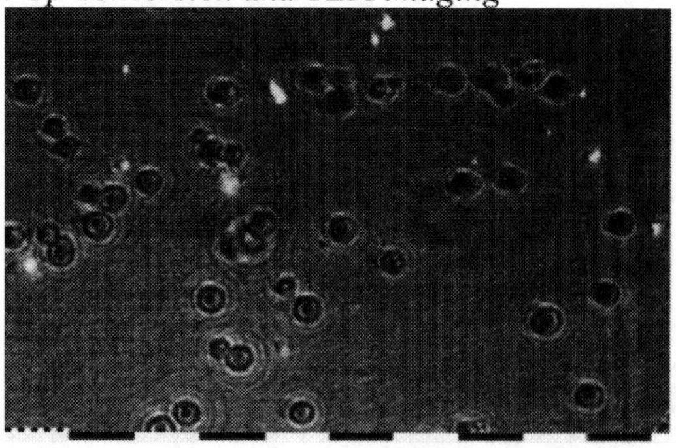

A

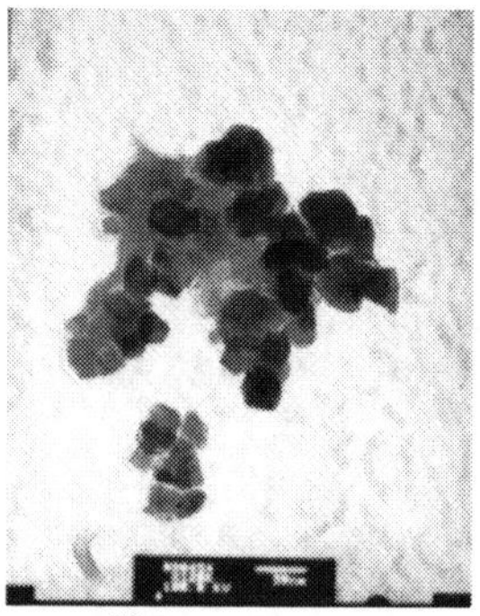

B

Figure 2. ZrO_2 nanocrystal IR upconversion microscopy (scale bar 10 μm) image with *S. cerevisiae* cells (A) and TEM micrography (scale bar 50 nm) (B)

Zirconia nanocrystals prepared by the Sol Gel method tend to form agglomerates. The IR up-conversion luminescence optical microscope image (Fig. 2 A) showed that there were particles of size under 1 μm, mostly agglomerates; however, it is not possible to determine the agglomerate size at the sub micron level with optical microscopy methods. When the particle size decreases the luminescence intensity decreases too.

The smallest ZrO_2 nanocrystals were possible to observe by TEM (Figure 2 B). Therefore, in order to detect smaller particles by upconversion imaging - the sensitivity of registration system should be increased. Due to low intensity and low resolution we were not able to observe whether uptake of nanocrystals had taken place in yeast cells.

4. Conclusions

The zirconia up-conversion luminescence peak positions were sensitive to the environmental surrounding. However, the presence of yeast cells did not show any significant influence on the spectral composition of the luminescence. Rare earth doped zirconia nanocrystals did not had any toxic effects on yeast cell fermentative growth; besides, it was possible to observe nanocrystals in biological media via upconversion process. Thus, we concluded that it is possible to use zirconia nanocrystals as micro size markers in baker's yeast based bioassays.

5. Acknowledgments

Authors highly acknowledge Joseph Spanier for language correction.
The present investigation was supported by European Social Fund.
Krisjanis Smits and Dzidra Jankovica contract Nr.2009/0202/1DP/1.1.1.2.0/09/APIA/VIAA/141
Janis Liepins, Marita Gavare contract Nr.2009/0202/1DP/1.1.1.2.0/09/APIA/VIAA/128
Arturs Gruduls contract Nr. 2011/0015/1DP/1.1.2.1.1/11/IPIA/VIAA/008

6. References

[1] Rad A G, Abbasi H and Afzali M H 2011 *Physics Procedia* **22** 203–8
[2] Chen G Y, Zhang Y G, Somesfalean G, Zhang Z G, Sun Q and Wang F P 2006 *App. Phys. Lett.* **89** 163105
[3] Smits K, Grigorjeva L, Millers D, Sarakovskis A, Opalinska A, Fidelus J D, and Lojkowski W 2010 *Opt. Mater.* **32** 827–31
[4] Fidelus J D, Lojkowski W, Millers D, Grigorjeva L, Smits K and Piticescu R R 2007 *Solid State Phenom.* **128** 141–50
[5] Sarakovskis A, Grube J, Doke G and Springis M 2010 *Opt. Mater.* **32** 832–5
[6] Li C, and Lin J 2010 *J. Mater. Chem.* **20** 6831–47
[7] Kim D H and Kang J U 2010 *Microscopy: Science, Technology, Applications and Education* (Extremadura, Formatex) 571-82
[8] Chevalier J, 2006 *Biomaterials* **27** 535–543
[9] Brunner T J, Wick P, Manser P, Spohn P, Grass R N, Limbach L K, Bruinink A and Stark W J 2006 *Environ. Sci. Technol.* **40** 4374–81
[10] Yamamoto A, Honma R, Sumita M and Hanawa T 2004 *J. Biomed. Mater. Res. A*, **68** 244-56
[11] Lamas D G, Lascelea G E and Walsoe de Reca N E 1998 *J. Eur. Ceram. Soc.* **18** 1217-21
[12] Singh K A, Pathak L C and Roy S K 2007 *Ceram. Int.* **33** 1463-8
[13] Corbacho I, Teixido F, Velazquez R, Hernandez L M and Olivero I 2011 *Antonie Leeuwenhoek* **99** 591–600
[14] Corte L, Antonielli L, Roscini L, Fatichenti F, and Cardinali G. 2011 *Analyst* **136**, 2339-2349
[15] Rossignol T, Dulau L, Julien A and Blondin B 2003 *Yeast* **20** 1369–85
[16] Jones N, Ray B, Ranjit K T and Manna A C 2008 *FEMS Microbiol. Lett.* **279** 71–76
[17] Kasemets K, Ivask A, Dubourguier H C and Kahru A 2009 *Toxicol. In Vitro* **23** 1116–22
[18] García-Saucedo C, Field J A, Otero-Gonzalez L and Sierra-Álvarez R 2011 *J. Hazard. Mater.* **192** 1572–79

IOP Conf. Series: Materials Science and Engineering **38** (2012) 012051 doi:10.1088/1757-899X/38/1/012051

Structure of nanoporous carbon materials for supercapacitors

A Volperts[1], N Mironova-Ulmane[2], I Sildos[3], D Vervikishko[4], E Shkolnikov[4], G Dobele[1,5]

[1] Latvian State Institute of Wood Chemistry, Riga, Latvia
[2] Institute of Solid State Physics, University of Latvia, Riga, Latvia
[3] Institute of Physics, University of Tartu, Tartu, Estonia
[4] Institute of High Temperatures, Russian Academy of Sciences, Moscow, Russia

E-mail: gdobele@edi.lv

Abstract. Activated carbons with highly developed porous structure and nanosized pores (8 – 11 Å) were prepared from alder wood using thermochemical activation method with sodium hydroxide. Properties of the obtained activated carbons were examined by benzene and nitrogen sorption, X-Ray diffraction and Raman spectroscopy. Tests of activated carbons as electrodes in supercapacitors were performed as well. It was found that specific surface area of above mentioned activated carbons was 1800 m^2/g (Dubinin – Radushkevich). Raman spectroscopy demonstrated the presence of ordered and disordered structures of graphite origin. The performance of activated carbons as electrodes in supercapacitors have shown superior results in comparison with electrodes made with commercial carbon tissues.

1. Introduction

Carbon porous materials are an important element of the human life-support; therefore the perfection of their structure, search for cheap sources of raw materials and new technological solutions, are urgent tasks of many investigations. At present, the studies of active carbons on the basis of plant biomass have assumed new perspectives in connection with the possibility of obtaining carbon with a developed nanoporous structure. Lignocellulosic materials (wood sawdust, nutshells, fruit stones, rice husks, etc.) are widely used as precursors and, actually, account for around 45% of the total raw materials used for the manufacture of activated carbons, materials with a high added value, suitable, for example, for use in electric double layer capacitors (EDLC), also known as supercapacitors. However, commercial activated carbons, having a low surface area and poor adsorption properties, cannot be applied to these fields. Therefore, it is necessary to prepare activated carbons with high surface area. Technical large scale implementation of super capacitors is being hindered by high costs of electrodes, which is an important factor in their acceptance at the world markets. Many researches use activated carbon electrodes due to their lower costs, wide working conditions range and high potential comparing to electrodes made from other carbon materials (fullerenes, nanotubes, etc.) [1].

Aims of the current study were obtaining of nanoporous carbon materials on the lignocellulosic base applying thermochemical activation and investigation of their porous structure and applicability for supercapacitor electrodes.

---

[5] To whom any correspondence should be addressed

2. Experimental

Active carbons on the basis of alder wood mechanical processing residues (fraction 0.16-0.2 mm) were prepared using thermochemical activation with NaOH. Activation took place at the temperature range 500 – 700 °C. Activated carbons were subsequently washed with hot water, 1M HCl solution and hot water again until neutral pH has been reached.

Porous structure of the activated carbons was evaluated from benzene (limited evaporation method [2]) and N_2 sorption/desorption isotherms using Brunauer–Emmet–Teller (BET) and Dubinin – Radushkevich (DR) methods for high (>0.2) (mesopores) and low (<0.2) (micropores) pressure regions respectively [3].

X-Ray spectra were taken using DRON-3 (Cu Kα irradiation (λ=0,154 nm), U=30 kV, I=20 mA), rate 1 degree per minute.

Raman spectra were collected using 50x microscope objective using Renishaw inVia micro-Raman spectrometer equipped with argon laser (514.5 nm, max cw power 10 mW). The spectral signal was dispersed by the 2400 grooves/mm grating onto Peltier-cooled (-60 °C) CCD detector.

Supercapacitor testing was performed as follows: two electrodes were assembled in a Swagelok system. The electrode was composed of carbon, H_2SO_4 media. Potentiostat-galvanostat Autolab-Ecochemie PGSTAT 30 (1-100 mA/cm^2, scan rate 20 mV/s) was used.

3. Results and discussion

Comparison of NaOH activated carbons obtained at the temperatures 500, 600 and 700 °C shows that their main parameters increase with the temperature increase. Increase of alkali content leads to intensive volatiles formation, probably due to redox reactions and hydrogen and carbon monoxide formation. Pore size distribution, specific volume and specific area were evaluated, showing considerable increase of two latter parameters in favor of 700 °C activation.

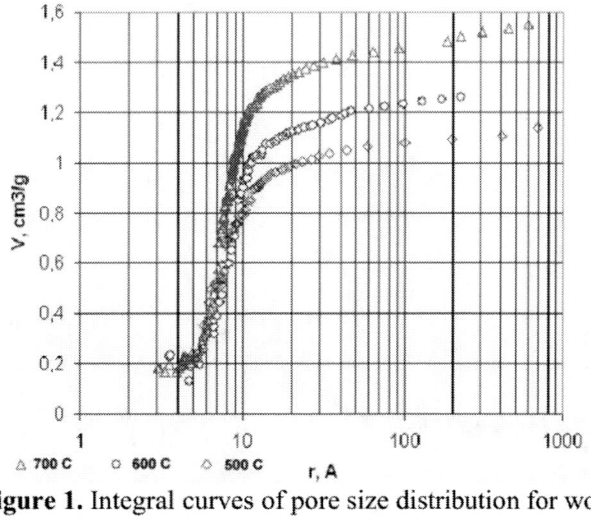

Figure 1. Integral curves of pore size distribution for wood based activated carbons (calculation on benzene desorption)

It was experimentally established that temperature 500 °C is not sufficient (S_{BET} 1200 m^2/g, S_{DR} 700 m^2/g) to reach the maximum porous structure of carbon materials. At 700 °C specific surface maximum (S_{BET} 3570 m^2/g, S_{DR} 1800 m^2/g) is being achieved after 90 minute synthesis. At the figure 1 the dependence of pore volume vs. activation temperature can be observed – it is quite clear that higher activation temperature increases pore volume due to higher micropores (L=8-11 Å) formation.

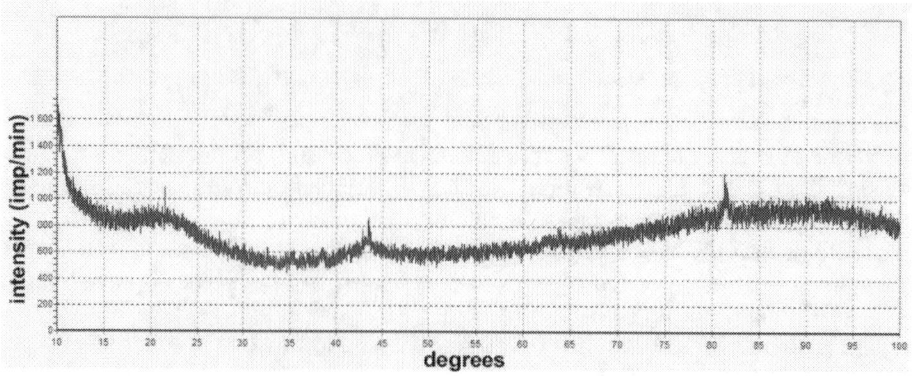

Figure 2. X-Ray scattering plot of wood based activated carbon (700 °C activation)

It is a well known fact that at temperatures below 1000 °C wood is not subject to graphitization. In the chemical activation environment at the lower temperatures the appearance of nanoporous structure can be explained by the formation of pores represented as gaps between carbon layers, while the carbon layers can be considered as graphene sheet stacks [4]. To prove it the X-ray and Raman scattering tests were performed.

On the X-ray scattering plots (figure 2) of the obtained carbon materials two diffraction maximums can be observed at 21.7 and 42.8 degrees, which is characteristic for the materials with turbostrate structure and presence of graphite-like disordered carbon layers. Average coherent scattering areas sizes L=12 and L=19 Å can be attributed to the average crystallites sizes.

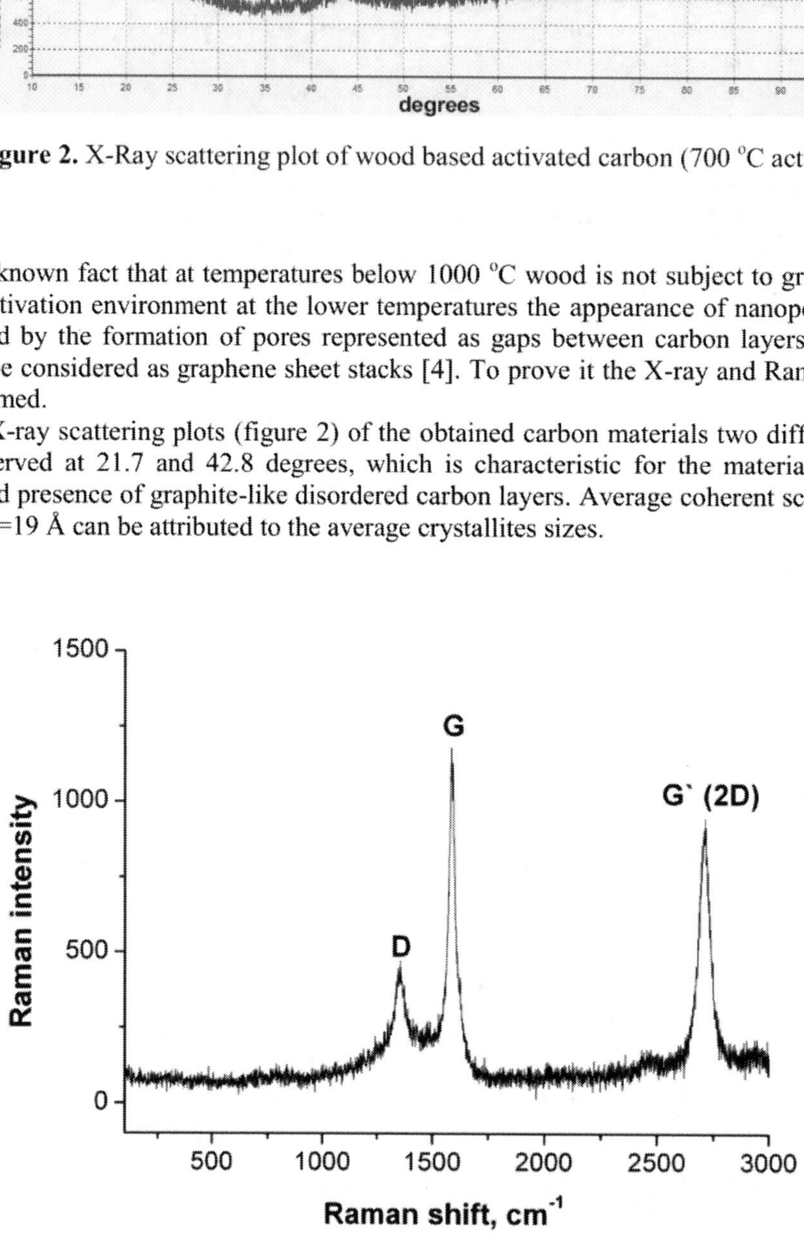

Figure 3. Raman spectra of wood based activated carbon (700 °C activation)

Raman spectroscopy is very sensitive to carbon bond, so it is an important method for the identification of diamond, amorphous carbon and graphite structures. We performed micro-Raman spectroscopy in order to investigate the vibrational properties of the synthesized nanoporous carbon materials. Three Raman scattering bands were observed (figure 3). The peak at approximately 1333 cm^{-1} is considered to represent a more disordered structure and is labeled as the D (disordered) band. D-band is present in Raman spectra (1250 -1450 cm-1) of carbon materials, including amorphous carbon. The intensity of this band characterizes the degree of deficiency of perfect symmetry of the graphite layer. Note that in a perfect graphite crystal the first-order vibrational mode of the D band is forbidden. Band at 1585 cm^{-1} corresponds to crystalline graphite and second order overtone of D band - sharp G` (or 2D) band (2685 cm^{-1}), corresponds to the graphite and graphene [5].

Since XRD analysis did not reveal presence of grapehens, Raman spectroscopy tests demonstrate the appearance of ordered and disordered structure of graphite origin. It should be noted that there is a number of works devoted to use of graphenes as materials for electrodes for suprecapacitors showing their benefits and properties [6-8].

To further investigate properties of the obtained nanoporous activated carbons they were tested in two electrode system. Charge/discharge cyclic voltammograms were recorded and compared to voltammograms obtained for electrodes based on commercial carbon materials (figure 4). All three materials exhibit a regular box-like shape with a steep current change at the switching potential characteristic of the behavior of an ideal capacitor. It is clear that nanoporous activated carbons show higher specific capacitance. The other benefit of disperse carbon material is the ability to control thickness of carbon layer.

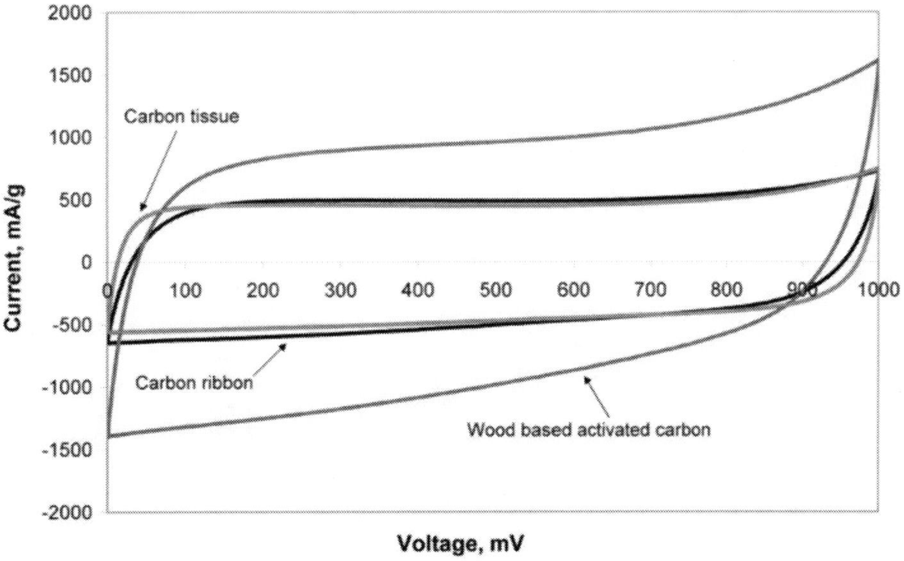

Figure 4. Cyclic voltammograms (two electrode cells) of the capacitors built from wood based activated carbons electrodes, commercial carbon tissue electrodes and commercial carbon ribbon electrodes (H_2SO_4 aqueous solution)

4. Conclusions

Nanoporous wood based carbon materials with pores sizes ranging 8 - 11 Å and specific surface area 1800 m^2/g (Dubinin-Radushkevich) were obtained. The X-Ray diffraction analysis testifies formation of turbostrate structure and presence of graphite-like disordered carbon layers. Raman spectroscopy tests showed the presence of ordered and disordered structure of graphite origin. The obtained nanoporous carbon materials can be recommended for the use as selective sorbents as well as for storage and transmission of energy as supercapacitor electrodes.

5. References

[1] Béguin F and Frackowiak E 2009 *Carbons for Electrochemical Energy Storage and Conversion Systems* (London: Taylor & Francis)
[2] Shkolnikov E and Vitkina D, 2010 *Teplofizika Vysokikh Temperatur* **48** 854–861
[3] Bansal R and Goyal M 2005 *Activated Carbons Adsorption* (London: Taylor & Francis)
[4] Gurianov V, Petukhova G and Dubinina L. 2007 *Izvestia Rossiyskoy Akademii Nauk. Seria Khimicheskaya* **1** 40–43.
[5] Ferrari A 2007 *Solid State Communications* **143** 47
[6] Park S and Rouff R 2009 *Nature* **4** 217
[7] Stoller M, Park S, Zhu Y, An J and Rouff R 2008 *Nano Letters* **8**
[8] Wang Y, Shi Z, Huang Y, Ma Y, Wang CY, Chen M and Chen Y 2009 *J. of Phys. Chem.* **C 113** 13103

Acknowledgements

The research leading to these results has received funding from Latvian Budget (Grant 1546), Latvian National Programme VPP-5,2,2.4. This study was supported by the European Union through the European Regional Development Fund (Centre of Excellence "Mesosystems: Theory and Applications", TK114).

Synthesis of Silica-Carbon Nanotube Composite Materials and Their Application for Laser Systems

J Pilipavicius, D Sakalauskas, A Beganskiene, A Kareiva

Vilnius University, Department of General and Inorganic Chemistry,
Naugardukas str. 24, Vilnius, Lithuania LT-03225

E-mail: Jurgis.pilipavicius@chf.vu.lt

Abstract. Currently, fabrication of CNT and polymer composites for saturable absorbers is very popular. But due to nature of organic polymer matrix, these composites are unstable on higher temperatures or in high energy light exposition. Therefore a sol-gel process for the fabrication for inorganic (i.e. silica) or semi-organic (organic-modified silica) matrix can be a promising way to produce photo-chemically and thermally stable CNT composites. In this work we investigated SWCNT's solubility in ethanol using different surfactants – Triton, polyvinylpyrrolidone, and Poly(4-vinylpyridine). Most stable colloidal solution was obtained using a P4VP as surfactant – solutions were stable up to 6 weeks. Therefore this surfactant was used for the fabrication of SWCNT/Silica composite coatings. SWCNT's was directly dispensed in a colloidal silica sol and coatings made by spin-coating technique. Samples were investigated under optical, atomic force microscopy methods, and optical non-linearity was measured. Thermal stability was investigated using Raman spectroscopy.

1. Introduction

Carbon nanotubes (CNT) have been widely investigated by researchers to improve understanding of the unique electric mechanism and optical properties [1]. Currently researchers are interested in unique carbon nanotubes non-linear optical properties [2-4]. Now saturable absorption of light is applied in lasers manufacturing to produce femto-second laser systems. Semiconducting, single-walled carbon nanotubes (SWCNT) is a promising material for saturable absorbers in laser mode locking [5] because of the fast recovery time, which covers a broad spectral range in the near infrared. Furthermore, SWCNT-based saturable absorbers can be fabricated by simple and economical methods, such as spray, spin coating [6] or horizontal evaporation methods[7]. However, application of those techniques requires dissolution of CNT's. Unfortunately, CNTs are insoluble in any solvents and forming bundles due to strong van der Waals interactions that tightly hold them together. CNT's insolubility decrease the overall yield of usable material and interfere with most of the desired properties of the CNT's[8].

In last decade a lot of intense research were on solubilisation and chemical modification of CNT's [8]. Better CNT solubility can be reached via chemical covalent or non-covalent modification of tube surface. The use of various surfactants and polymers can highly increase solubility of CNT in polar and non-polar media. Most stable solutions were with a nitrogen containing surfactants, like polyvinylpirrolidone (PVP) [9] and Poly(4-vinylpyridine) (P4VP) [10], and some non-polar interacting surfactants as sodium dodecylbenzene sulfonate (SDS) or Triton x-305 [9].

First approach to make SWCNT composites as saturable absorbers included application of various polymers as matrix [11]. Unfortunately as noticed later, polymer matrix decomposing on higher

IOP Conf. Series: Materials Science and Engineering **38** (2012) 012052 doi:10.1088/1757-899X/38/1/012052

energy light exposure [11, 12]. Moreover, due to reaction with oxygen carbon nanotubes starts to oxidize. So, the application of silica matrix can be promising solution for saturable absorbers [13]. Therefore the aim of this research – make SWCNT/Silica composites via sol-gel method, using Raman spectroscopy investigate their chemical stability and measure optical non-linearity.

2. Experimental

Aqueous colloidal solutions of semiconducting (IsoNanotubes-S 0,01 mg/mL) and unsorted (PureTubes 0,25 mg/mL) single walled carbon nanotubes were purchased from Nanointegris, Polyvinylpirrolidone (PVP 360 000M), Poly(4-vinylpiridine) (P4VP 60 000M), Tetramethoxysilane (TMOS 98%), Ammonia (NH_4OH 33%) ordered form Sigma-Aldrich. Ethanol was ordered form Car-Roth (EtOH 96%) and dried by boiling with CaO and then distilled. BK7 glass (1 inch diameter) substrates for spin-coating were obtained from Altechna (Lithuania).

Coatings were made using SCS inc. SpinCoater P6700. In all samples coatings obtained using 100 μL of colloidal solution on substrate and spinning 30 s at 1000 rpm for each layer. Raman spectra were registered with confocal Raman spectrometer/microscope LabRam HR 800 (Horiba JobinYvon Inc.) using 632,8 nm laser for excitation. Non-linear absorption curves of composites were registered with light conversion laser "Pharos" which has wavelength at 1030 nm and 500 kHz pulse frequency. First collimated light was transmitted through $\lambda/2$ plate and polarizing cube to adjust laser output power. Then light was focused by 10 cm focusing lens. Sample was placed at focal length, which position along light beam adjusted with translator. Using two power meters (sample is between power meters) laser power ratio was measured. AFM images obtained with Atomic force microscope Veeco Bioscope 2 using tapping mode.

3. Results and Discussions

3.1. SWCNT's solubilisation in Ethanol

For formation of composite coatings via spin-coating technique, solutions must be non-aqueous. So first of all, solubility of SWCNT's were investigated by varying surfactants – PVP (360 000M), P4VP (60 000M), Triton X-100 and Triton X-305. For that purpose a drop (about 100 μL) unsorted SWCNT's solution dripped into 10 ml 1% v/v surfactant solution in Ethanol. Solutions were sonified 15 minutes in ultrasonic bath.

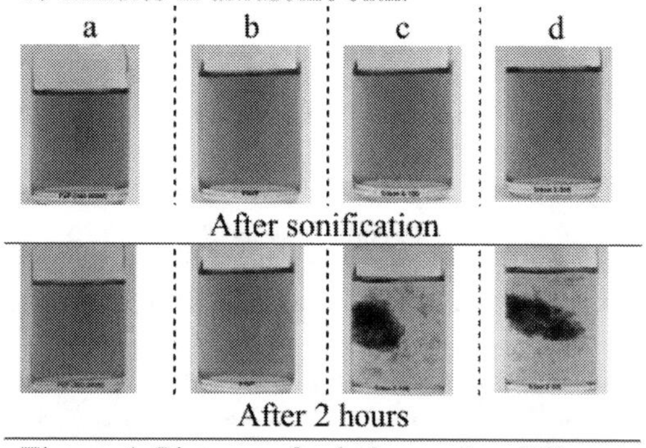

Figure 1. Pictures of solutions using different surfactants and aging time: a – PVP, b – P4VP, c – Triton X-100, d – Triton X-305

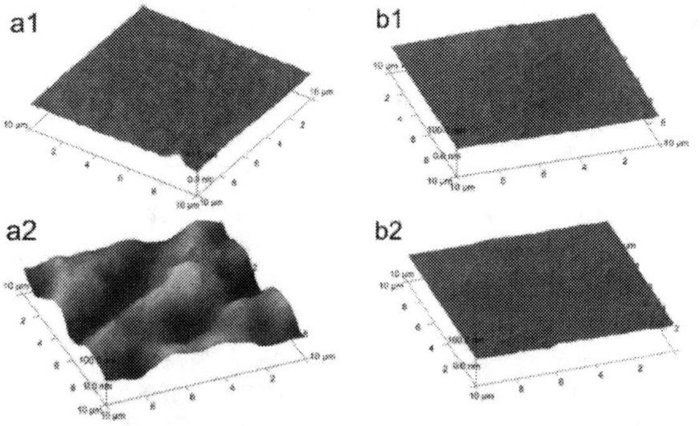

Figure 2. AFM images of SWCNT films made with different surfactants and aging time: SWCNT/PVP after sonification (a1) and after 3 days (a2); SWCNT/P4VP after sonification (b1) and after 3 days (b2)

As seen from figure 1, colloidal solutions with surfactants PVP and P4VP were stable, but in solutions with Triton X-100 and Triton X-305 SWCNT's aggregates were visible already after few hours. Moreover must be mentioned, that colloids containing P4VP and PVP was stable up to six weeks.

For thorough investigation of colloids stability, PVP and P4VP surfactants were used with higher SWCNT concentration. In this case, a 1 ml of aqueous unsorted SWCNT solution (0,25 mg/ml) was freeze dried and added 1 ml Ethanol containing 0,1 mg/ml of surfactant. Samples were sonified for 30 min, a part of solution separated and aged for 3 days. Form all solutions using spin-coating method thin films was produced, and AFM images where registered (fig. 2).

As seen from figure 2, solutions with PVP right after sonification forms uniform film, but after 3 days aging obtained films was very rough and non-uniform. However, P4VP/SWCNT solutions form uniform films right after sonification and after aging for 3 days. So, using P4VP as surfactant is promising way to produce a silica/SWCNT's coatings.

3.2. Non-linear measurements of SWCNT's

In this work we used two types of SWCNT's – unsorted and semiconductive type. Therefore, before making SWCNT/silica composites, the differences in non-linear absorption must be evaluated. For this purpose, two colloidal solutions with different type SWCNT were prepared. For first solution 0,25 mg/ml unsorted SWCNT was added, and second – 0,01 mg/ml semiconductive SWCNT. In both solutions 0,1 mg/ml of P4VP were added to make stable suspensions. Solutions were used for spin-coating on BK7 glass substrates up to 8 layers and non-linear absorption was measured (figure 3).

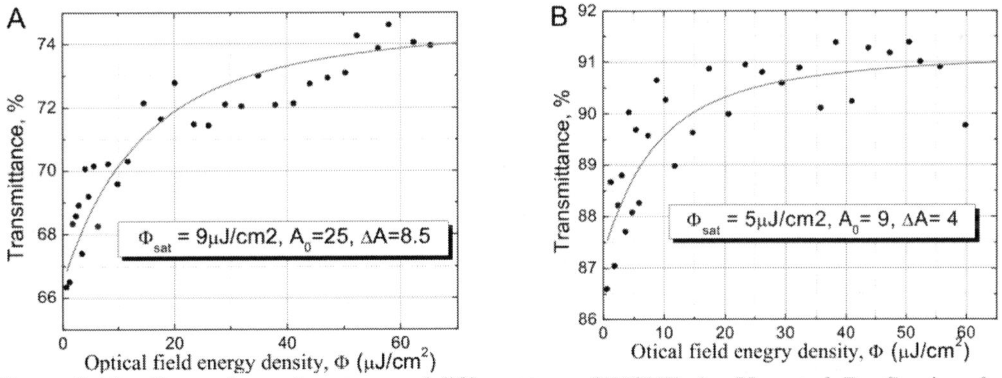

Figure 3. Non-linear absorption curves of different types SWCNT: A – Unsorted; B – Semiconductive

As seen from figure 3, unsorted SWCNT's has higher saturation energy (Φ_{sat}), and higher non-saturated absorption (A_0), than semiconductive. Thus semiconductive has better non-linear absorption values than unsorted. However, due to better solubility of unsorted SWCNT's, their concentration in film is higher and as result, the ratio between absorbed and transmitted light (ΔA) are higher.

3.3. SWCNT/Silica composites – investigation of thermal stability

As mentioned, single walled carbon nanotubes are not stable in high intensity light exposure, due to reaction with oxygen. So we presumed that processes in photo-oxidation are very similar to thermal oxidation. To investigate silica ability to preserve carbon nanotubes structure, using sol-gel process SWCNT (semicond.)/Silica composite coatings were formed.

For this purpose semiconductive SWCNT using P4VP was suspended in ethanol (concentrations are 0,01 mg/ml and 0,1 mg/ml respectively) and TMOS and NH4OH introduced into colloidal solution. Solution was aged for 1 h and then spin-coated on BK7 glass. SWCNT (semicond.)/P4VP coatings were made for comparison, with same method like in section 3.2. Samples were heated at 200, 300, 400, 450 C^0 (two samples remain unheated) and Raman spectra were recorded (figure 4).

In Raman scattering spectrum (figure 4) clearly seen, that carbon nanotubes without silica layer starts to change his structure already in 200 ^{0}C – decreasing RBM (radial breathing mode) peak intensity and G (sp^2 stretching) peak shape. Also, no distinctive SWCNT peaks were detected in sample heated at 400 ^{0}C. In spectra of SWCNT/Silica composite coatings heated at 200 ^{0}C, RBM and

D peaks remain unchanged, small deformations of peaks are visible only in samples heated up to 300 ^{0}C. But no distinctive SWCNT peaks were detected in sample heated at 450 ^{0}C. However Silica layer has positive impact to thermal stability of SWCNT's.

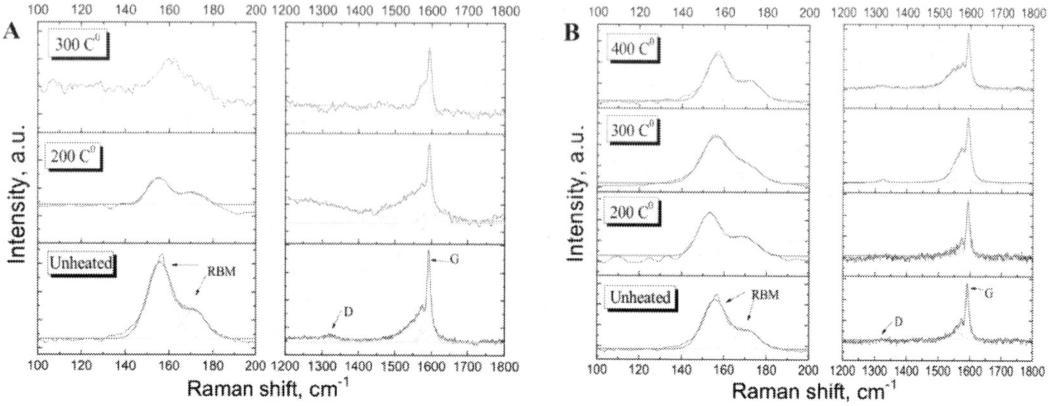

Figure 4. Raman scattering spectrums of semicondutive SWCNT films heated in different temperature: A – SWCNT/P4VP; B – SWCNT/Silica

4. Conclusions

1. Good solubility of SWCNT in ethanol was obtained using P4VP - colloid was stable up to 6 weeks.
2. Semiconductive SWCNT has better non-linear absorption properties, but due better solubility of unsorted SWCNT, possible to obtain more concentrated composites.
3. Silica/SWCNT composites have higher thermal stability than polymer embedded composites and accordingly they can be used in higher intensity laser systems.

References

[1] S. Iijima and T. Ichihashi 1993 *Nature,* **363** 603-605.
[2] S. Lefrant, J. P. Buisson, J. Y. Mevellec, M. Baibarac, and I. Baltog 2010 *Mol Cryst Liq Cryst,* **522** 472-479.
[3] A. M. Da Silva, G. M. A. Junqueira, H. F. Dos Santos, and A. C. M. Carvalho 2009 *Quim Nova,* **32** 315-U61.
[4] L. Vivien, E. Anglaret, D. Riehl, and F. Hache 2000 *Mater Res Soc Symp P,* **597** 257-262.
[5] Y. Sakakibara, A. G. Rozhin, H. Kataura, Y. Achiba, and M. Tokumoto 2005 *Jpn J Appl Phys,* **44** 1621.
[6] J. H. Yim, W. B. Cho, S. Lee, Y. H. Ahn, K. Kim, H. Lim, G. Steinmeyer, V. Petrov, U. Griebner, and F. Rotermund 2008 *Appl. Phys. Lett.,* **93** 161106-3.
[7] M. A. Solodyankin, E. D. Obraztsova, A. S. Lobach, A. L. Chernov, A. V. Tausenev, V. I. Konov, and E. M. Dianov 2008 *Opt Lett,* **33** 1336-1338.
[8] N. Karousis, N. Tagmatarchis, and D. Tasis 2010 *Chem. Rev.,* **110** 5366-5397.
[9] A. J. Blanch, C. E. Lenehan, and J. S. Quinton 2010 *J Phys Chem B,* **114** 9805-9811.
[10] J. H. Rouse 2004 *Langmuir,* **21** 1055-1061.
[11] T. Schibli, K. Minoshima, H. Kataura, E. Itoga, N. Minami, S. Kazaoui, K. Miyashita, M. Tokumoto, and Y. Sakakibara 2005 *Opt. Express,* **13** 8025-8031.
[12] F. Wang, A. G. Rozhin, V. Scardaci, Z. Sun, F. Hennrich, I. H. White, W. I. Milne, and A. C. Ferrari 2008 *Nat Nanotechnol,* **3** 738-742.
[13] H.-J. Kim, H.-J. Choi, S.-M. Nam, and Y.-W. Song 2011 *Opt. Express,* **19** 4762-4767.

The effect of nanosize ZnO on the properties of the selected polymer blend composites

A Grigalovica[1], I Bochkov[1], R Merijs Meri[1], J Zicans[1,4], J Grabis[2], R Kotsilkova[3] and I Borovanska[3]

[1]Institute of Polymer Materials, Riga Technical University, Latvia
[2] Institute of Inorganic Chemistry of Riga Technical University, Riga, Latvia
[3] Institute of Mechanics, Bulgarian Academy of Sciences, Sofia, Bulgaria

E-mail: zicans@ktf.rtu.lv

Abstract. In the current research the effect of ZnO nanoparticles on the structure and properties of common thermoplastic polymers (polyoxymethylene (POM), polypropylene (PP), ethylene-α-octene copolymers (EOC)) and their binary blends is investigated. EOC content in the composites varies from 0 to 50 wt. %. The amount of nanostructured ZnO filler in the composites is changed in the interval from 0 to 5 wt. %. Tensile and frictional properties of ZnO modified nanocomposites are investigated. Results of the investigation show that ZnO additions cause increment in stiffness and strength as well as coefficient of friction of the investigated nanocomposites. The effect of ZnO modifier is the highest at low EOC content. The effect of ZnO is strongly dependent on the compatibility and crystallinity of the investigated nanocomposites.

1. Introduction

In different sectors of national economy polymer materials are increasingly exploited. Polymers commonly are used as a direct replacement for other materials because they are lightweight, corrosion resistant and can be easily processed in different shapes. Nowadays, great attention is paid to polymer composites, because in this way one can obtain new materials with advanced or even new properties. For example, improved impact behaviour in wide temperature range can be imparted to POM or PP modified with elastomers, rubbers or polyurethanes [1-3].

Another way of polymer modification, that has recently achieved a great interest, is development of polymer nanocomposites. In several studies it is reported about modification of broad range of thermoplastic and thermosetting polymers with various inorganic and organic nanofillers. Considerably less effort, however, is devoted to the development of polymer blend nanocomposites. Therefore, in the current project thermoplastic polymer composites are investigated as potential candidates for developing of specific nanocomposites.

Thermoplastic elastomer ethylene-α-octene copolymer is used as high molecular weight modifier of POM and PP. It is expected that addition of certain amounts of EOC will increase impact behaviour of POM and PP based binary composites, allowing their use in various applications, where tough superelastic materials are required. On the other hand, addition of EOC could compensate some property drop for PP and POM based nanocomposites.

[4] To whom any correspondence should be addressed

Nanostructured ZnO filler is used in variety of applications due to its photocatalytic reactions, self-purification, antibacterial and other properties. It has been shown that ZnO increases thermal stability [4,5], mechanical stiffness and strength [5] of certain composites, as well as decreases their coefficient of friction [6]. In these studies, however, pure polymers are used as matrices. Meanwhile, there is practically no information about polymer blend usage as matrix for ZnO modified polymer nanocomposites.

2. Methods

In this research following commercial products were used: polyoxymethylene copolymer (POM), *Kolon Plastics Inc., Kocetal K300*; homopolymer of isotactic polypropylene (PP), *Lyondell Basell, Moplen HP456J*; thermoplastic elastomer of ethylene-α-octene copolymer (EOC) with alpha octene content 38 %, *The Dow Chemical Company, Engage 8200.*

Nanostructured ZnO nanofiller with average crystallite sizes ~ 70 nm was synthesized at the *Institute of Inorganic Chemistry of Riga Technical University* (Salaspils, Latvia).

Polymer composites with various EOC and ZnO contents (0, 10, 30, 50 wt. % of EOC and 0, 2, 5 wt. % of ZnO) were prepared by roll milling. POM based blends were processed at 170^{0}C, while PP based blends - at 180^{0}C. Blending time was 7min. Suitable melt processing method was used to prepare specimens corresponding to the testing requirements.

Structure was characterized by using *Mettler Toledo DSC1/200W* differential scanning calorimeter (DSC). All measurements were made under N_2 flow; samples of ~10 mg were heated from -100^{0}C up to 200^{0}C at a scanning rate of 10^{0}C/min.

Tensile properties were investigated by using *Zwick/Roell BDO FB-020FN* universal testing machine (test temperature 22+/-2^{0}C; test speed for modulus determination – 1 mm/min.; test speed for determination of other stress-strain characteristics - 50 mm/min.).

Coefficient of friction (COF) was determined by using *CETR Universal Material Tester UMT*, Pin on Flat mode. Stainless steel metallic pin (D=6.35; L=25.4) contacted with the surface of flat polymer sample. All measurements were made under 20 N load, force 10 mm/s, 10 mm distance.

3. Results and discussion

Tensile stress-strain characteristics of the investigated compositions are shown in the figure below. As already expected, addition of EOC lowers stiffness (Fig. 1) and strength (Fig. 2) of the investigated compositions and increases their relative elongations to certain extent (Fig. 3).

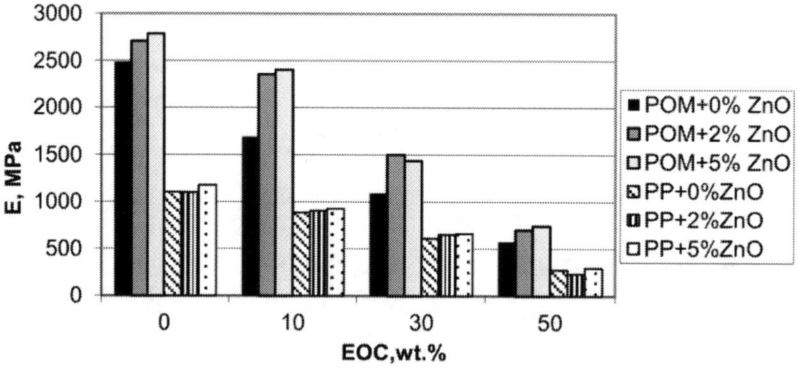

Figure 1. Modulus of elasticity E of ZnO modified POM/EOC and PP/EOC blends as functions of EOC and ZnO weight contents.

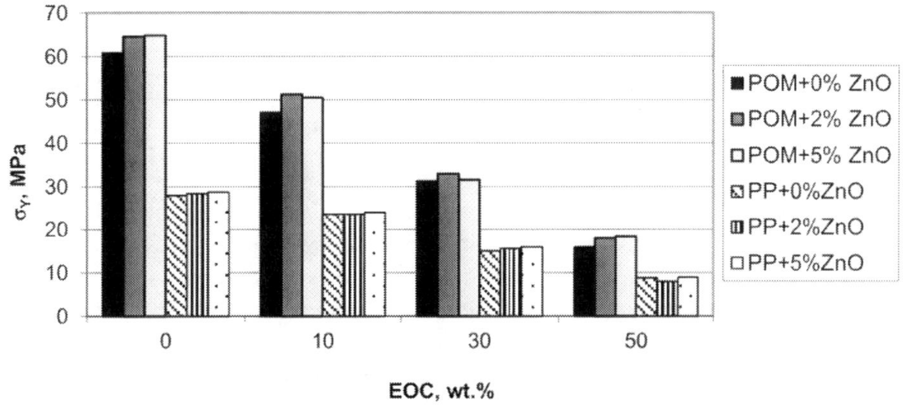

Figure 2. Yield strength σ_Y of ZnO modified POM/EOC and PP/EOC blends as functions of EOC and ZnO weight contents.

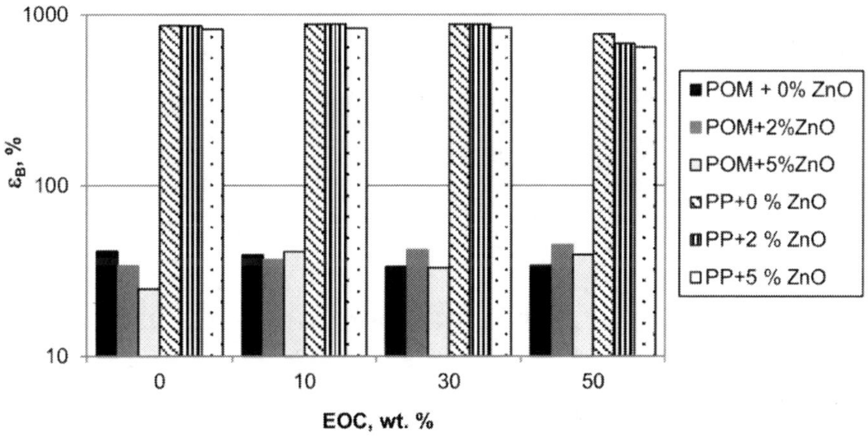

Figure 3. Relative elongation at break ε_B of ZnO modified POM/EOC and PP/EOC blends as functions of EOC and ZnO weight contents.

Limited increase in relative elongation values of the compositions along with addition of EOC (observed only up to 30 wt. %) most probably is connected with incompatibility of the blend components, which is especially pronounced in the case of POM/EOC blends. Probably this could be explained with less similar macromolecular structure of POM and EOC. Considering the effect of ZnO on the mechanical properties of thermoplastic matrices, it is reported [7] that the extent of stiffness and strength increase usually is highly dependent on the dispersion degree of the nanofiller. Similar statements evidently are true also in the case of the investigated blends. For an example, the modulus of elasticity of POM90/EOC10/ZnO5 is increased by ~40 %, while that of POM70/EOC30/ZnO5 by ~30 % in respect to the modules of corresponding binary blends (no EOC, no ZnO added). These changes can be explained also because of the effect of ZnO on the crystallinity degree χ of the above mentioned compositions; with raising nanofiller content up to 5 wt. % χ is raised by 15 %. As in the case of PP based composites, these changes are smaller (~ 1 % and 12 %, respectively for analogous compositions), smaller is the effectiveness of ZnO addition. It should be, however, mentioned that addition of 2 wt. % of ZnO nanofiller is optimal for modification of both investigated polymer systems. At higher nanofiller content, reinforcing efficiency is somewhat reduced, especially in the case of POM based systems. This could be connected with greater agglomeration of ZnO in the matrix of POM based systems. Another factor that evidently influences the effectiveness of ZnO modifier is incompatibility of the blend components. Consequently, reinforcing effectiveness of ZnO is higher for the blends with lower EOC concentrations.

Apart from evaluating stress-strain characteristics of the investigated POM/EOC and PP/EOC nanocomposites, the effect of ZnO on lowering of the coefficient of friction (COF) of EOC modified thermoplastics is also assessed. It is demonstrated that addition of ZnO nanofiller allows successfully decrease COF of the materials investigated, starting from POM, already having excellent tribological properties (COF decreased by ~1.5 % at maximum ZnO content), and ending with EOC, showing the highest COF (COF decreased by ~16 % at maximum ZnO content). It has been also observed that addition of 2 wt. % of ZnO ensures most rapid decrease of the values of COF; at higher ZnO content modifying effectiveness of nanofiller is evidently reduced, most probably due to increasing agglomeration trends, as it has been reported elsewhere [5].

Conclusions

In the current research the effect of ZnO nanoparticles on the tensile and frictional properties of POM, PP and their binary blends with EOC are investigated. Results of the investigation show that addition of ZnO increases stiffness and strength, while the coefficient of friction of the investigated compositions decreases. It should be mentioned, however, that the effect of ZnO nanofiller is higher at lower EOC content, being the function of the compatibility of the polymer blend matrix. The effect of ZnO depends also on the crystallinity of the investigated nanocomposites. Potential applications of the investigated blends could be casing for household equipment, sanitary wares, medical goods (such as inhalers etc).

Acknowledgements

This research has been supported by the COST Action MP0701

References

[1] Gao X Qu C Fu Q Peng Y and Zhang Q 2004 *Macromol.Mater.Eng.* **289** 41
[2] Nalini Uthaman R Pandurangan A and Abdul Majeed S S M 2007 *J Polym Res.* **14** 441
[3] Bai S-L Wang G-T Hiver G-M G'Sell S 2004 *Polymer.* **45** 3063
[4] Liufu S C Xiao H N and Li G P 2005 *Polym. Degrad. Stabil.* **87** 103
[5] Wacharawichanant S Thongyai S Phutthaphan A and Eiamsamang C 2008 *Polym. Test.* **27** 971
[6] Carrión F J Sanes J and Bermúdez M-D 2007 Wear **262** **11**, 1504
[7] Wacharawichanant S Thongbunyoung N Churdchoo P Sookjai T 2010 *Sci. J. UBU* **1** 20

Carrier dynamics in epilayers and nanocolumns of ternary AlGaN with a tunable bandgap

R Aleksiejūnas[1,4], P Ščajev[1], S Nargelas[1], S Miasojedovas[1], K Jarašiūnas[1], L Trinkler[2], J Grigorjeva[2], B Berzina[2], K H Chen[3], Y T Chen[3], M W Chen[3], and L C Chen[3]

[1]Institute of Applied Research, Vilnius University, Sauletekio Ave. 9 - III, Vilnius LT-10222, Lithuania

[2]Institute of Solid State Physics, University of Latvia, 8 Kengaraga, LV-1063 Riga, Latvia

[3]Center for Condensed Matter Sciences, National Taiwan University, Taipei 106, Taiwan

E-mail: ramunas.aleksiejunas@ff.vu.lt

Abstract. We apply several optical time-resolved techniques to investigate the dynamics of excess carriers in $Al_xGa_{1-x}N$ nanocolumns grown on (111) silicon substrates and in thick $Al_xGa_{1-x}N$ epitaxial layers deposited on sapphire. We demonstrate that carrier lifetime drops in nanocolumns by several times if compared to epilayers with similar Al content; in addition, recombination rate displays a strong nonlinearity on excitation. On the other hand, carrier localization effects observed in the epilayers completely disappear in nanocolumns.

1. Introduction

Quasi one-dimensional group-III nitride nanostructures (nanowires, nanocolumns or nanorods) attract increasing amount of attention as they offer solutions to some fundamental problems that hinder development of planar nitride devices. Probably most important of all, the dislocation density can be greatly reduced in III-nitride nanocolumns (NC) grown on foreign substrates like sapphire or silicon. Several groups reported the growth of fully-relaxed, dislocation-free monocrystalline GaN NC [1, 2]. Quasi one-dimensional structures also feature a large surface-to-volume ratio which is desirable for applications like biosensing [3] or solar water splitting [4]. Large area of NC sidewalls, oriented along the non-polar directions, also allow expecting higher internal quantum efficiency of NC-based LEDs. In addition, emission wavelength of NC LEDs can be tailored and shifted towards deeper UV by changing both composition and diameter of NC [5].

On the other hand, growth of NC involves many challenges making it difficult to manufacture NCs with well controlled parameters like thickness, doping, composition etc. (for a review see Ref. 6). Therefore, it is important to know how these parameters influence optical and electrical NC properties. Unlike quantum wells, NCs are very difficult objects for electrical characterization; this is why contactless optical techniques gain a big advantage for investigations of carrier recombination in NCs.

[4] To whom any correspondence should be addressed

In the given paper, we present an optical study of carrier recombination in MBE grown $Al_xGa_{1-x}N$ NC samples on Si (111). We combine and analyze data obtained by differential transmission (DT), light-induced transient grating (LITG), and photoluminescence (PL) optical techniques. The data for $Al_xGa_{1-x}N$ NCs is then compared to that in the $Al_xGa_{1-x}N$/sapphire layer of similar x.

2. Samples and experimental details

We studied two $Al_xGa_{1-x}N$ NC samples grown on Si(111) substrates using a home-made MBE system. After chemical and thermal cleaning of Si substrates, $Al_xGa_{1-x}N$ NCs were grown at 10^{-9} torr background pressure. Samples NC1 and NC2 were grown at 1000°C temperature of Ga and Al sources and constant N_2 flow rate of 3 sccm. Different Al content was achieved by varying the power of N plasma generator (350 W and 450 W for samples 2 and 3, correspondingly). This resulted in Al content of 1.5% (NC1) or 6.6 % (NC2), as determined by XRD. The typical nanocolumn diameter was 55-60 nm. For comparison, we investigated carrier dynamics in 0.5 μm thick $Al_xGa_{1-x}N/Al_2O_3$ (0001) layers labeled NZ1-NZ3 with x = 6.7%, 7.9% and 13%, correspondingly. PL spectra peaked at 3.51 eV, 3.56 eV and 3.64 eV for NZ1–NZ3 samples or 3.42 eV and 3.44 eV for NC samples. We note that Al content in NC2 sample appears to be smaller (~3 %) if determined from the PL peak position.

The experimental setup for time-resolved DT was based on a commercial Ti:Sapphire femtosecond laser (SuperSpitfire, Spectra Physics) delivering 800 nm pulses of 120 fs duration at 1 kHz repetition rate. The output of the laser was split into two equal parts. One was used to pump the optical parametric amplifier (TOPAS, Light Conversion) that provided 120 fs pulses set to 330 nm wavelength (3.75 eV). The second beam was delayed and used to generate white light in a CaF_2 window. DT technique provided means to observe evolution of state occupancy in AlGaN NCs with high temporal and spectral resolution.

Unlike the conventional pump-probe, LITG for excitation uses an interference field of two coherent beams [7]. This results in recording of transient diffraction grating, which can be probed by diffraction of a delayed probe beam. Such an approach has several advantages, but in the given case we exploited mostly the fact that LITG signal – diffraction efficiency – was proportional to the square of free carrier density $N^2(t)$, integrated over the excited volume.

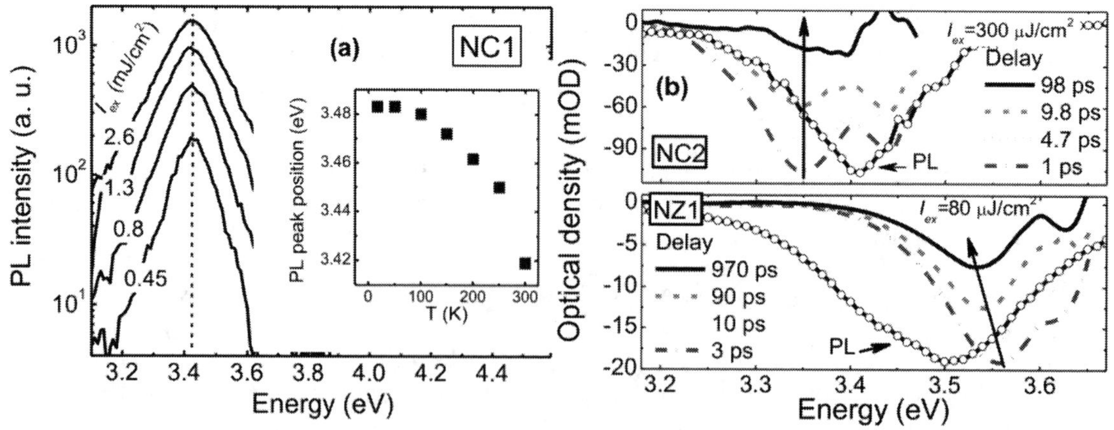

Figure 1. (a) Time-integrated PL spectra recorded in NC1 sample at various excitations. The inset shows PL peak position at various temperatures; (b) DT spectra at various delay times (lines) and time-integrated PL spectra (dotted lines) recorded in NC2 (upper) or NZ1 layer (lower) samples.

This provided a convenient way to measure instantaneous carrier density. LITG setup was based on a YLF:Nd^{3+} 10 Hz repetition rate laser (PL2243, Ekspla) emitting 8 ps duration pulses at 1053 nm. For excitation, we used the frequency quadrupled beam at 263 nm (4.7 eV) wavelength, which produced the transient grating on the sample using a holographic beam splitter. LITG measurements were performed within 80 – 600 K temperature range when photoexcited carrier density varied within ~10^{18} – 10^{20} cm^{-3} range.

The time-integrated photoluminescence (PL) spectra of the samples were measured using a home-made system consisting of solid-state laser DTL-389QT with pulse energy 4.7 μJ at 1 kHz emitting at 263 nm (4.7 eV) and grating monochromator ANDOR SR-303i-B with CCD camera DV420A-BU2. The sample was put into the close-cycle cryostat CCS-100 Janis Research Co enabling a temperature range from 8 K up to 300 K. Time-resolved PL kinetics were measured using another setup under pulsed excitation at 266 nm (PL2143, Ekspla) with 25 ps time resolution achieved by using a Kerr shutter with toluene.

3. Results and discussion

NC1 sample displayed a rather broad PL line (FWHM of ~150 meV) peaked at 3.42 eV at 300 K (3.48 eV at 17 K). The peak spectral position (i) didn't shift with excitation up to ~2.6 mJ/cm^2; (ii) exhibited Varshni type of red shift with temperature; no s-shape type of behavior has been observed (Fig. 1 (a) with inset). These two facts confirmed an absence of free carrier localization effects in the nanocolumn structures. PL behavior was found qualitatively similar in NC2.

DT spectra were recorded at various delays and excitations in NC2 sample (Fig. 1 (b), upper part). Unexpectedly, DT peak was shifted towards the lower energy with respect to PL line indicating that radiative transitions in NC occur between higher energy states. DT peak spectral position (3.35 eV) was constant with delay and injection, i.e. it did not depend on carrier density. The situation appeared to be different in the AlGaN layers on sapphire (Fig. 1 (b), lower part). DT spectra were shifted towards higher energies from PL peak and they suffered a noticeable shift with decreasing free carrier density, which was attributed to contribution of carrier localization.

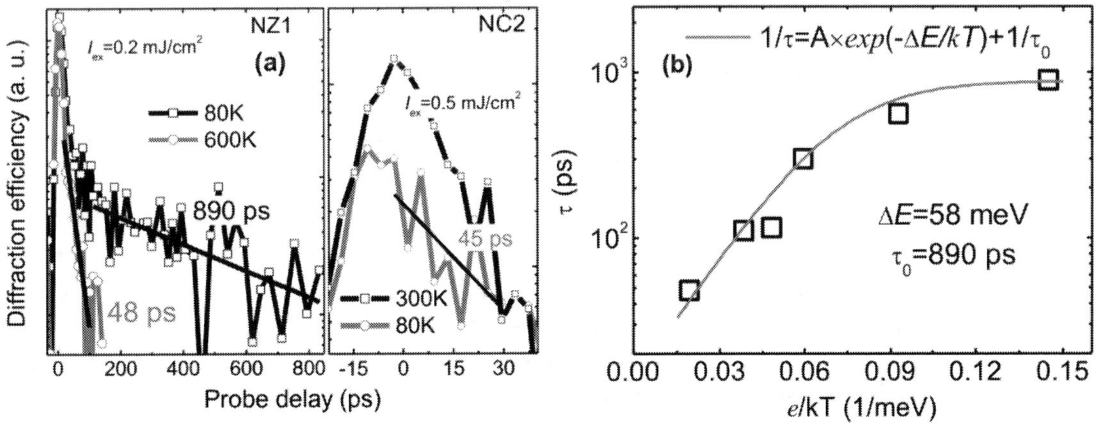

Figure 2. (a) LITG kinetics recorded in NZ1 layer (left) and NC2 nanocolumns (right) at different temperatures; (b) Arrhenius plot of free carrier lifetime τ vs. temperature *T* obtained from LITG kinetics

Distinct dynamics of free and localized carriers were revealed in LITG kinetics (i.e. dependence of diffraction efficiency vs. probe delay time) recorded at various temperatures (Fig. 2 (a), left part). LITG traces in AlGaN layers displayed two components, a fast and a slow one. The slow component was present in the layers but not in the NCs and was attributed to the signature of carriers in localized states. It can be speculated that localization in AlGaN layers most probably is related to Al composition fluctuations. The experimental data suggests that Al distribution is more homogeneous in the NCs. The measured decay rate of the slow component was thermally activated with activation energy of 58±8 meV (Fig. 2 (b)).

The fast component represents recombination of free carriers via defect states. DT provided free carrier lifetime shortening from 170 ps in NZ1 down to 80 ps in NZ3 with increasing Al content due to increase of Al related defect states. In NCs sample, the carrier lifetime was noticeably shorter than in the layers and varied within 14 – 50 ps range (Fig. 3 (a)). Short free carrier lifetime was confirmed by time-resolved PL that provided values of ~ 20 ps. It has been found that the lifetime strongly depended on carrier density. In Fig. 3 (b) we show LITG exposure characteristics, i.e. diffraction efficiency as a function of pump fluence for several probe delays. Extremely fast decline of the slope value with excitation ($m = 1.5$ if measured synchronously with the excitation pulse and $m < 1$ at later delays) indicates the presence of some efficient nonlinear recombination pathway, which cannot be ascribed to band-to-band or defect recombination. The exact origin of this phenomenon is not clear yet. It can be speculated, though, that at high carrier densities the potential barriers that are present near the surface of nanorods [8] get screened, which increases a probability of carriers reaching the surface and recombining at surface states.

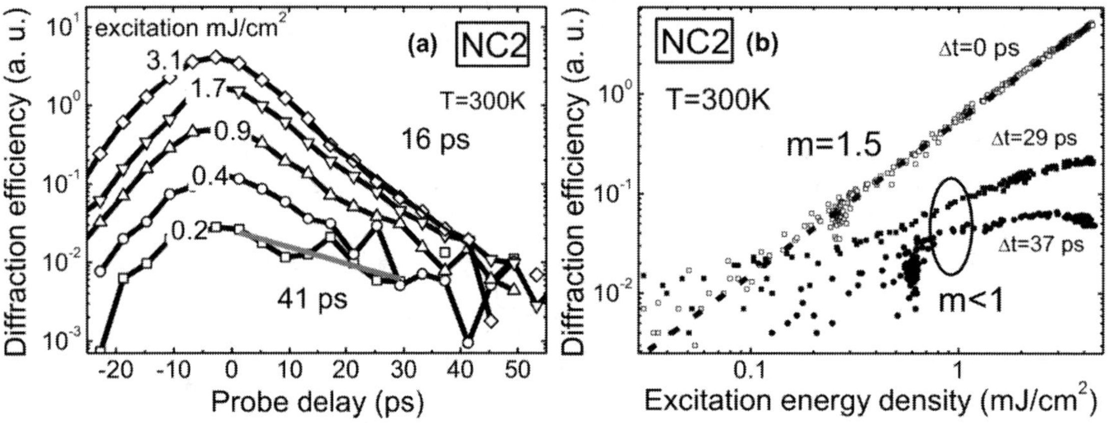

Figure 3. (a) LITG kinetics at various excitation pump fluence and (b) LITG exposure characteristics at various delay times in NC sample at room temperature (m are the slope values of the dependences in log-log plot).

4. Conclusions.

We have performed the time-resolved optical study of carrier recombination in $Al_xGa_{1-x}N$ nanocolumns and thick layers. We show that carrier localization effects play a prominent role in carrier recombination in (Al,Ga)N epilayers, while it is not present in nanocolumn samples; this can be the indication of more homogeneous Al distribution in (Al,Ga)N nanocolumns. The 2–8 fold decrease in free carrier lifetime in nanocolumn samples is attributed to higher concentration of defects and fast carrier recombination at surface defect states.

Acknowledgement Financial support of Taiwan National Science Council and Research Council of Lithuania (grant TAP LLT 05/2012) is kindly acknowledged.

References

[1] Kim J H 2012 *Cryst. Growth Des.* **12** 1292
[2] Tchernycheva M 2007 Nanotechnology **18** 385306
[3] Sofikiti 2009 *Appl. Phys. Lett.* **95** 113701
[4] Hwang Y J, Wu Ch H, Hahn C, Yang H E J, and P 2012 *Nano Lett.* **12** 1678

[5] Sekiguchi H, Kishino K, and A. Kikuchi A 2010 *Appl. Phys. Lett.* **96** 231104
[6] Li S and Waag A 2012 *J. Appl. Phys.* **111** 071101
[7] Jarašiūnas K 2007 *Rev. Sci. Instrum.* **78** 033901
[8] Calarco R, Marso M, Richter T, Aykanat A I, Meijers R, Hart A, Stoica T, and Luth H 2005 *Nano Lett.* **5** 981

International Conference on Functional Materials and Nanotechnologies (FM&NT2012) IOP Publishing
IOP Conf. Series: Materials Science and Engineering **38** (2012) 012055 doi:10.1088/1757-899X/38/1/012055

Investigation of Electric Properties of Nanocomposites in Frequency Domain – Compact Measurement Data Representation

I Matīss, Ē Pentjušs, O Krievs and J Zicāns
Institute of Polymer Materials, Riga Technical University, Latvia

E-mail: oskars.krievs@rtu.lv

Abstract. The paper presents a procedure for mapping large amounts of experimental measurement data in a compact and representable form, preserving information on the properties of the research object for further investigations. The raw data set is compacted using mathematical description of dielectric relaxation. This approach provides a possibility to recover the dielectric properties for any frequency value, even extrapolating outside the actual measurement range. It also favours the understanding the causes for the measured phenomena and identification of imperfections within the measurement procedure.

1. Introduction

The investigated properties of nanocomposites must be characterized over a wide frequency range, providing a possibility to transform the results for different applications - for example to describe the characteristics expressed in dielectric terms with conductivity terms and vice versa [1]. The problem is - the larger the experimental data set, the more incomprehensive and inconvenient is it's analysis and further research. In practice, the amount of measurement data points for just a single nanocomposite test piece can reach several thousands, which in a raw form is hardly applicable for the main research objectives - description of the material structure, optimization of the production technology or the comparison of different materials.

2. Theoretical background for a model based representation of raw measurement data

In electrical engineering it is accustomed to describe materials with dielectric permittivity or the conjugated term – electric conductivity. These quantities are complex and frequency dependant [2]:

$$\varepsilon(\mathrm{i}\omega) = \varepsilon'(\omega) + \mathrm{i}\varepsilon''(\omega) = \frac{\sigma(\mathrm{i}\omega)}{\mathrm{i}\omega\varepsilon_0} = \varepsilon''(\omega)\omega\varepsilon_0 + \mathrm{i}\varepsilon'(\omega)\mathrm{i}\omega\varepsilon_0; \qquad (1)$$
$$\sigma(\mathrm{i}\omega) = \sigma'(\omega) + \mathrm{i}\sigma''(\omega),$$

where ε' and σ' are real parts of the complex dielectric permittivity and electric conductivity respectively; ε'' and σ'' are imaginary parts of the complex dielectric permittivity and electric conductivity; ω is the angular frequency and $\varepsilon_0 = 8{,}854\cdot10^{-12}$ – vacuum permittivity.

The frequency-dependant nature of dielectric properties of materials is the consequence of phenomenon of dielectric relaxation, which can be described with mathematical models such as Debye, Havrilaki-Negami, Cole-Cole [3],[4] etc. Several relaxation processes are characteristic to nanocomposites. Furthermore, a correlation has been observed between the real part of the complex dielectric permittivity and the concentration of inclusions within ferrite, as well as increase of the real component and shift of imaginary component's resonance spike towards higher frequencies [5]. However, there are examples in literature where nanocomposites do not behave this way [3]. So measurement results should be interpreted with precaution, avoiding elimination of separate measurement data points just because they do not agree with "classic" behavior and seem faulty at the first glimpse. In such cases it must be clarified whether the declination of experimental coherencies from the possible model is the consequence of a new physical property or experiment faults.

In order to map large amounts of experimental measurement data in a compact and representable form, two models of dielectric relaxation have been chosen - Debye, Havrilaki-Negami. The following advantages can be expected of such representation:

- Compact description of the measurement results;
- Possibility to recover the frequency characteristics for any argument value, even extrapolating outside the actual measurement range to some extent;
- Contribution in explaining the causes of the measured phenomena;
- Opportunity to identify the imperfections within the measurement procedure;
- Possibility to transform the measurement results from dielectric terms to conductivity terms.

2.1. The Debye and Havrilaki-Negami models

The Debye and Havrilaki-Negami models of dielectric permittivity can be expressed with the following equations respectively:

$$\varepsilon(\mathrm{i}\omega) = \varepsilon'_{\infty} + \sum_{k=1}^{n} \frac{\Delta\varepsilon_k}{1+\mathrm{i}\omega\tau_k}, \tag{2}$$

$$\varepsilon(\mathrm{i}\omega) = \varepsilon'_{\infty} + \frac{\Delta\varepsilon}{\left[1+\left(\mathrm{i}\omega\tau_0\right)^{1-\alpha}\right]^{\beta}}, \tag{3}$$

where ε'_{∞} is the real component of dielectric permittivity at infinite frequency $\omega \to \infty$; $\Delta\varepsilon_k$ is the dielectric increment - the difference between values of dielectric permittivity at $\omega = 0$ and $\omega \to \infty$ for the kth relaxation mechanism; τ_k is the relaxation constant of the kth relaxation mechanism; α,β - the dynamic parameters of the dielectric relaxation characterizing the width and symmetry of dielectric spectrum respectively.

As evident from the equation (2), the Debye model is a superposition of separate relaxation mechanisms, each characterized by a constant relaxation time – thus the model considers a discrete distribution of relaxation times. This model has proven to provide a more adequate representation of measurement data of nanocomposite test pieces, tracking the experimental coherencies of real and imaginary components of dielectric permittivity relatively precisely (e.g. imaginary component of dielectric permittivity on figure 3.).

The Havrilaki-Negami model (3), in turn, is based on a continuous distribution of relaxation time. The curves of this model are explicitly monotonous – the model is more applicable in cases when integral characteristics of the material or its components are of interest.

3. Results of the compacted measurement data representation

The experimental data comprise measurements of the complex dielectric permittivity of nanocomposite test pieces, carried out over frequency range of 100Hz to 15MHz. The measurements were performed using impedance/gain-phase analyzer "HP 4194A" with the measurement sensor "HP 16451B".

As an example of compact measurement data representation with Debye model (n=2), the real and imaginary components of dielectric permittivity of test pieces with different $NiZnFe_4O_8$ admixture concentration (2p0N-0%, 2p2N-2%, 2p5N-5% and 2p10N-10%) are shown on figure 1. and figure 2. Two relaxation transitions can be identified on figure 2., one of them occurring at low frequencies.

The possibility to fit the synthesized curve to the complicatedly shaped experimental curve is illustrated with a particularly selected nanocomposite test piece on figure 3. Correcting action of the model based (2) approach can be observed in this figure – the measurement faults (negative imaginary component values) of in the low and high ends of the frequency range are omitted and the curve is smoothened altogether. The synthesized frequency curve can be recovered and, if necessary, extrapolated with just five parameters of dielectric relaxation: $\varepsilon'_\infty = 1{,}29$; $\Delta_1 = 0{,}013$; $\tau_1 = 0{,}015ms$; $\Delta_2 = 0{,}024$; $\tau_2 = 0{,}0003ms$. Again, two relaxation transitions can be identified in the curve.

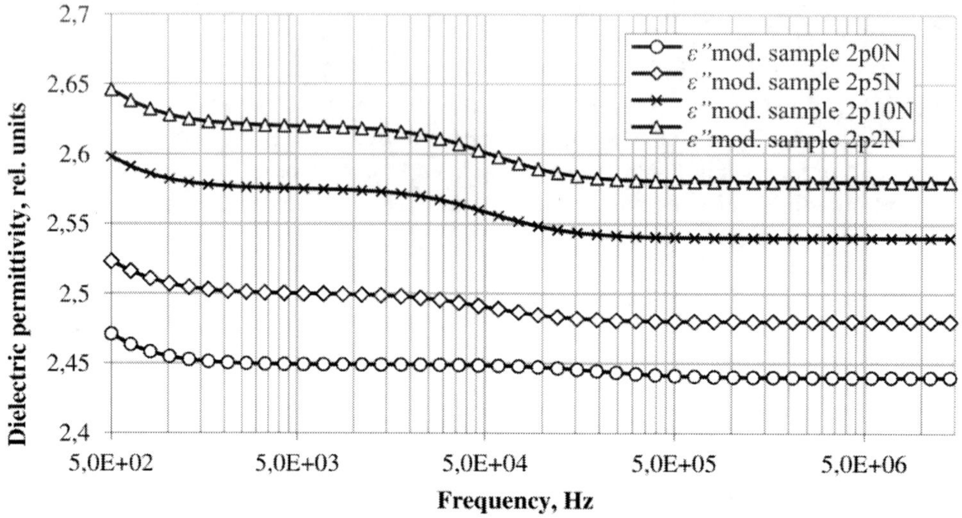

Figure 1. The real components of dielectric permittivity of the Debye model applied to measurement results of nanocomposite test pieces with different concentration of $NiZnFe_4O_8$ admixture.

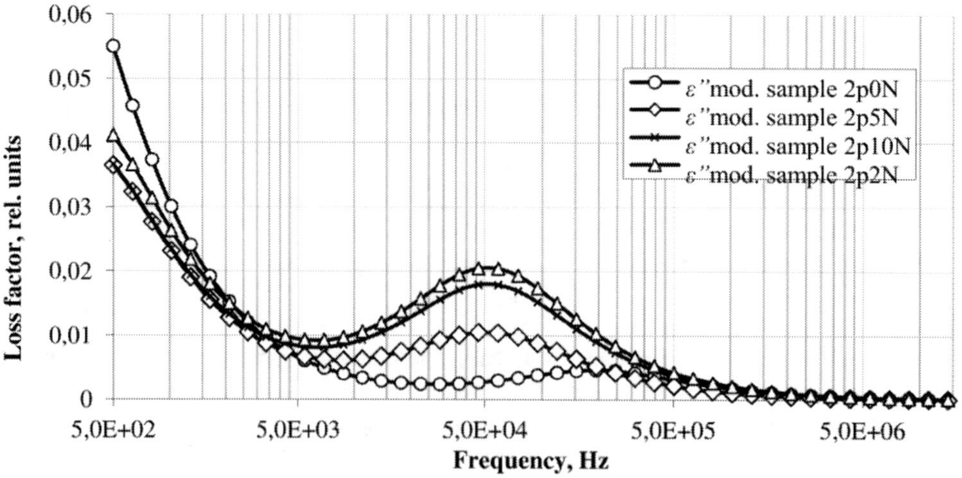

Figure 2. The imaginary components of dielectric permittivity of the Debye model applied to measurement results of nanocomposite test pieces with different concentration of $NiZnFe_4O_8$ admixture.

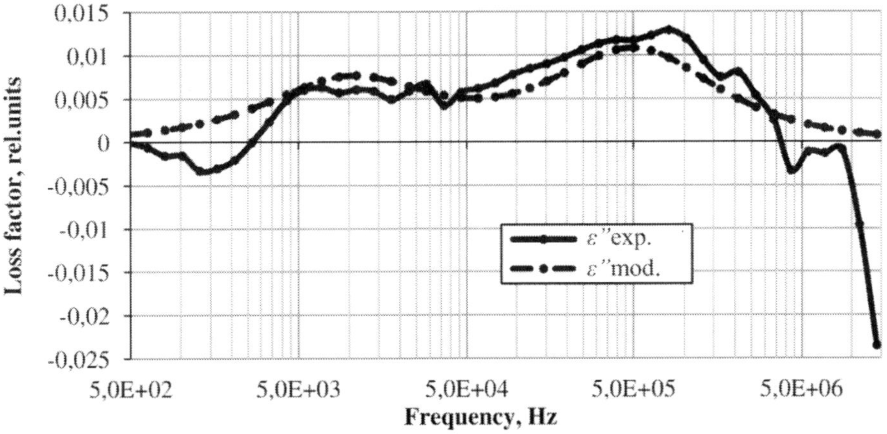

Figure 3. The experimental and Debye model frequency curves of imaginary components of dielectric permittivity.

4. Conclusions

A procedure for mapping large amounts of experimental measurement data in a compact and representable form, preserving information on the properties of the research object, is presented.

Before synthesizing the frequency curve, the experimental data have to be evaluated for conformity with physical properties of dielectric relaxation. Any declination of experimental coherencies from the possible model has to be carefully checked for experiment faults and measurement tolerance errors. An appropriate frequency range has to be selected for further synthesizing of the mathematical model.

The dynamic parameters of dielectric relaxation are very sensitive – even one per cent deviation of these parameters has a notable effect.

The research results show that the real component of the dielectric permittivity can be evaluated more precisely (with a higher statistical significance). Consequently, it can be used as a criterion in selecting the usable frequency range from the measurement data.

It must be noted, that, in contrary to analysis, the problems of synthesis do not necessary have a single correct solution. Nevertheless, the main advantage of the offered approach is the possibility to recover the dielectric properties for any frequency value, even extrapolating outside the actual measurement range to some extent.

Acknowledgment

The research was carried out within the project No.2010/0209/2DP/ 2.1.1.1.0/ 10APIA/VIAA/028 financially supported by ERDF.

References

[1] Abdeen, A M 1998 Electric conduction in Ni–Zn ferrites *Journal of Magnetism and Magnetic Materials* vol 185 Issue 2 pp 199-206

[2] Matiss, I Purvinsh A 2006 Measurement of Frequency Dependent Dielectric Properties by the Capacitance Technique *Industrial Electronics, IEEE International Symposium on Industrial Electronics)* vol 4 pp 2743-2749

[3] Rao, P B C, Setty S P 2010 Electrical properties of Ni-Zn nano ferrite Particles *International Journal of Engineering Science and Technology* vol 2(8) pp 3351-3354

[4] Mohan G R et. al. 1999 Dielectric properties of polycrystalline mixed nickel–zinc ferrites *Materials Letters* 40 pp 39–45

[5] Kambale R C. et. al. 2009 Dielectric properties and complex impedance spectroscopy studies of mixed Ni-Co ferrites *Smart Materials and Structures* vol 18 issue 8 pp 085-114

IOP Conf. Series: Materials Science and Engineering **38** (2012) 012056 doi:10.1088/1757-899X/38/1/012056

Structure analysis and size distribution of particulate matter from candles and kerosene combustion in burning chamber

M Baitimirova, A Osite, J Katkevics, A Viksna[1]
Department of Chemistry, University of Latvia, Riga, Valdemara Street 48, Latvia, LV-1013

E-mail: arturs.viksna@lu.lv

Abstract. Burning of candles generates particulate matter of fine dimensions that produces poor indoor air quality, so it may cause harmful impact on human health. In this study solid aerosol particles of burning of candles of different composition and kerosene combustion were collected in a closed laboratory system. Present work describes particulate matter collection for structure analysis and the relationship between source and size distribution of particulate matter. The formation mechanism of particulate matter and their tendency to agglomerate also are described. Particles obtained from kerosene combustion have normal size distribution. Whereas, particles generated from the burning of stearin candles have distribution shifted towards finer particle size range. If an additive of stearin to paraffin candle is used, particle size distribution is also observed in range of towards finer particles. A tendency to form agglomerates in a short time is observed in case of particles obtained from kerosene combustion, while in case of particles obtained from burning of candles of different composition such a tendency is not observed. Particles from candles and kerosene combustion are Aitken and accumulation mode particles.

1. Introduction

The particle diameter (size) is one of the most important parameters for the aerosols' behavior characterizing, as nearly all aerosol properties depend on particle size. Different particle sizes have different penetration efficiencies into a human respiration system. The particle size distribution is valuable for the indoor air quality assessment and control strategies.

Candles burning produce fine particles of black carbon [1]. In our previous research, the black carbon as a component of particulate matter was investigated by electrochemical impedance spectroscopy [2, 3]. As mentioned in literature [1, 4], the candle burning produce many fine particles. In turn, kerosene combustion produces many larger particles. So, candles with different waxes composition were used for the burning experiments for fine particles formation and kerosene - for larger particles formation.

Solid fine particles easily can agglomerate into coarse particles. In the atmosphere fine particles easily coagulate and condense with water vapor, becoming bigger in size. Coarse particles, because of larger mass, precipitate due to the influence of gravitation. Airborne particles mainly are in the size range from 0.1 to 10 μm. The typical airborne particle size distribution includes three modes: nucleation or Aitken mode (particle size 0.001-0.1 μm), accumulation mode (0.1 – 2 μm) and coarse particles (2 – 100 μm) [5].

The main aim of current studies is to prove laboratory built particulate matter sampling system for the applications in the further investigations of combustible aerosols.

2. Experimental

Candles of different waxes compositions were used for the burning to obtain fine particulate matter. A stearin candle, a stearin-paraffin candle (mixed in ratio 1:2) and paraffin candle was used in the burning experiments. Kerosene was used in the burning experiment for the larger particles formation. The closed system for different candles and kerosene combustion was laboratory built (Figure 1).Through a tube with

[1] To whom any correspondence should be addressed

filter on the end *(1)* the purified air came to the burning chamber. The pressure in the chamber was controlled by manometer *(7)*. The sticky tape *(4)* was used the particulate matter collection for sample analysis by optical microscope. The Laser Aerosol Particle Size spectrometer (LAP) *(5)* with the software *PASWin (6)* was used for the particle size distribution measurements. Obtained particles from candles and kerosene combustion for analysis by the scanning electron microscope (SEM) were collected on filters, using the particle sampler which consists of filter holder *(8)*, pump *(9)*, rotameter *(10)* and flowmeter *(11)*.

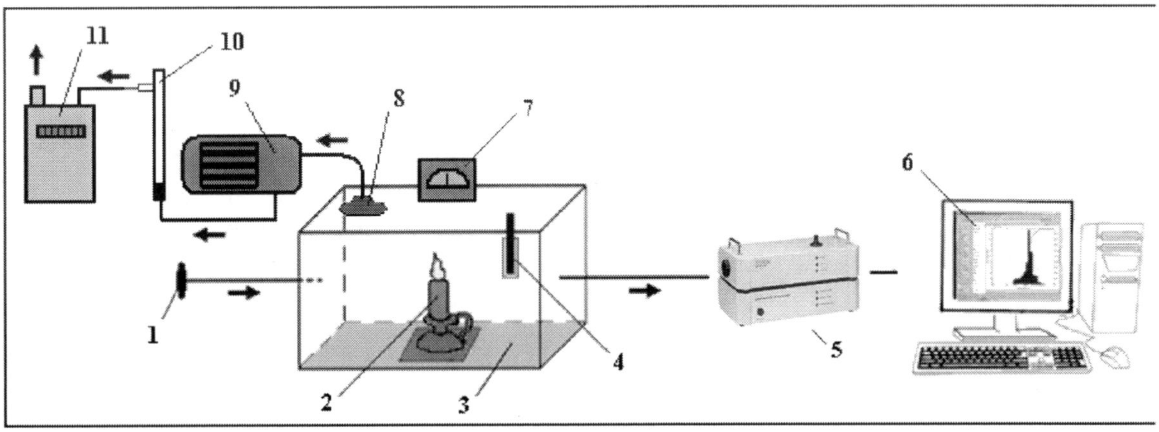

Figure 1. The burning system for the obtaining particulate matter from combustion process (1- filter, 2 – candle, 3 – burning chamber, 4 - sticky tape, 5 – Laser Aerosol Particle Size spectrometer (LAP), 6 – software PASWin, 7 – manometer, 8 - filter holder, 9 – pump, 10 – rotameter, 11 – flowmeter).

3. Results and discussion

Optical microscope image of particulate matter obtained from the burning of stearin candle is shown on Figure 2. Optical microscope images of particles from burning candles of different composition and kerosene combustion were similar. We can see coarse and fine particles on the microscope image. The particle size distribution (Figure 3) was obtained by analyzing such images. As mentioned in the literature [1; 4], the fine particles are created mostly by burning combustibles. Such observation was in our experiment – the fine particles (smaller than 3 μm) were mostly in all emissions of candles burning. The difference between size distribution of particles from burning of candles of different composition and kerosene combustion was not observed using the optical microscope.

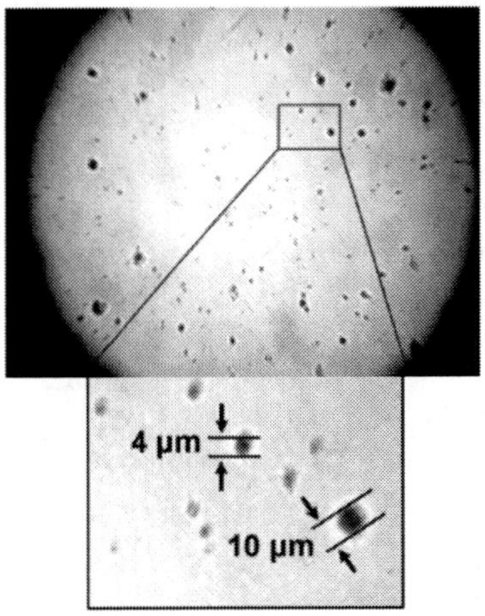

Figure 2. An optical microscope image of particulate matter from the stearin candle burning.

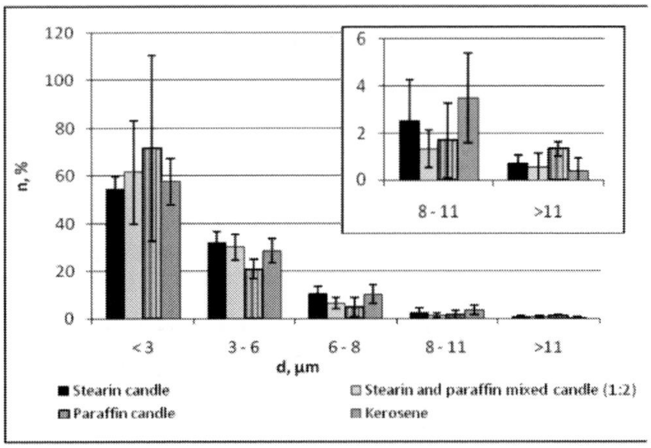

Figure 3. The particle size distributions of particulate matter from burning of candles of different composition and kerosene combustion (data from the analysis of optical microscope images).

On the other hand, the differences between particle size distributions were observed during analysis of the LAP spectrometer data (Figure 4). Particles obtained from the kerosene combustion have normal size distribution. Whereas, particles obtained from the stearin burning have distribution shifted towards finer particle size range. A tendency to form agglomerates in a short time is observed in case of particles generated from kerosene combustion, while in case of particles obtained from burning of candles of different composition such a tendency is not observed. If an additive of stearin to paraffin candle is used, particle size distribution is observed in range of towards finer particles. It was assumed that the simultaneous burning of a stearin candle and kerosene would produce particulate matter with size distribution shifted towards finer particle size range. In Figure 4 we can see that the particle size distribution was not shifted towards finer size range, so the hypothesis was not confirmed.

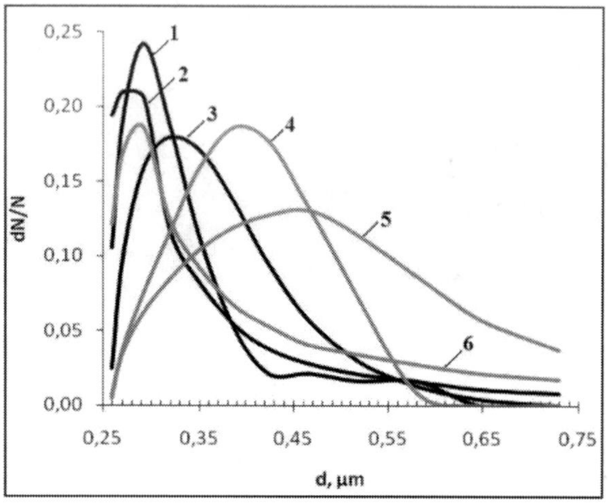

Figure 4. Particle size distributions from the burning of candles of different composition and kerosene combustion analyzed by LAP spectrometer (1 – stearin candle, 2 – stearin-paraffin candle (mixed in ratio 1:2), 3 – paraffin candle, 4 – kerosene, 5 – simultaneous burning of kerosene and stearin candle, 6 – airborne particulate matter).

Average particle sizes of obtained particulate matter and their confidence interval, analyzed by LAP spectrometer, are shown in Table 1. The average particle size ($\bar{d}$) was calculated from number of particles *(n)* in particles size class with the class width Δd_i by the equation (1):

$$\bar{d} = \frac{1}{N}\sum_{i=1}^{k} d_i n_i ; \qquad (1)$$

where k is number of particle size classes (LAP spectrometer has 64 classes). N is total number of particles that was calculated by the equation (2):

$$N = \sum_{i=1}^{k} n_i \qquad (2)$$

Obtained particles size distributions was asymmetric distribution. So, the confidence interval was calculated by the equation (3) in conformity with asymmetric trust evaluation:

$$sz_1 < \sigma < sz_2 \; ; \qquad (3)$$

where s is standard deviation, that was calculated by the equation (4):

$$s = \left(\frac{1}{N}\sum_{i=1}^{k}(\bar{d} - d_i)^2\right)^{1/2} \qquad (4)$$

In the equation (3) z is coefficient, which depend on confidence interval (α=0.05) and degree of freedom.

Table 1. An average particle sizes from the burning of candles with different compositions and kerosene, obtained by LAP spectrometer (d', µm) and SEM micrograph (d'', µm)

Source of particulate matter	d', µm	Confidence interval (α=0.05)	d'', µm	Particulate mode
Airborne particulate matter	0.33	$0.06 < \sigma < 0.13$	0.3 ÷ 18.0	accumulation, coarse
Stearin candle	0.33	$0.05 < \sigma < 0.12$	0.15÷ 0.35	accumulation
Stearin and paraffin mixed candle	0.32	$0.24 < \sigma < 0.50$		accumulation
Paraffin candle	0.35	$0.05 < \sigma < 0.11$	0.05÷ 4.0	accumulation, coarse
Kerosene	0.39	$\sigma \pm 0.11$		accumulation, coarse
Kerosene and stearin candle	0.45	$0.09 < \sigma < 0.20$		accumulation, coarse

Particulate matter from candles burning was analyzed by the SEM. The size of fine particles in SEM micrographs is under 100 nm (Figure 5). Mainly aggregates of particulate matter are seen on the SEM micrographs. It proves that the coarse particle is formed from several fine particles. Summarizing the obtained particle size data, it can be concluded that solid particles obtained both from burning of candles of different composition and kerosene combustion are aggregates of Aitken and accumulation mode particles according to literature data [1, 5, 6].

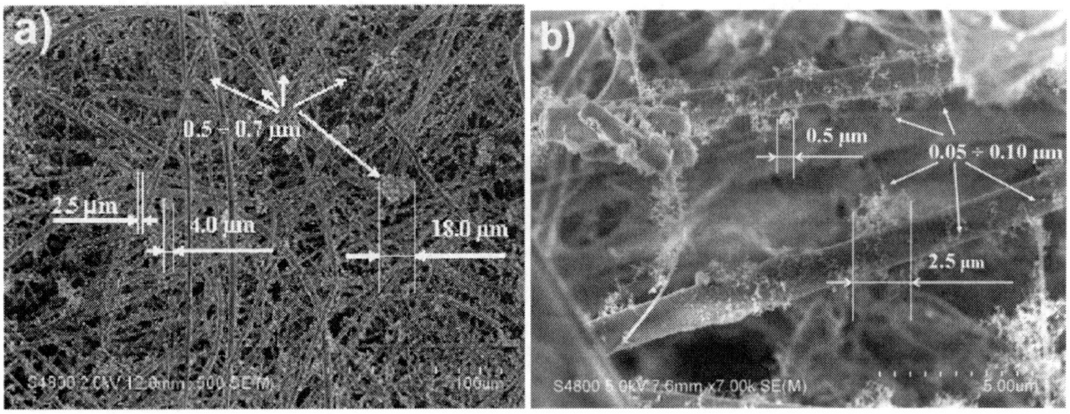

Figure 5. SEM micrographs of airborne particulate matter (a) and particulate matter from paraffin (b) candles burning.

4. Conclusions

Analysis of particles generated from candles of different composition and kerosene burnt in the closed burning chamber shows sampling system is usable for the combustible aerosol investigations. The obtained results are in a good agreement with literature data. Particulate matter obtained from the kerosene combustion has normal particle size distribution, and a tendency to form agglomerates in a short time is observed in case of these particles. Whereas, particulate matter from burning candles of different composition have the size distribution shifted towards finer particle size range. In case of these particles a tendency to form agglomerates in a short time is not observed. Size of particles, obtained from candles and kerosene burning, is in the range from 0.01 to 10 µm. From this it can be concluded that solid particles obtained from burning of candles of different composition and kerosene combustion are agglomerates of Aitken and accumulation mode particles.

Acknowledgments

Work was supported by ESF project No. 2011/0015/1DP/1.1.2.1.1/11/IPIA/VIAA/008.
Margarita Baitimirova is grateful for financial support from the European Social Fund (ESF).

References

[1] Pagels J, Wierzbicka A, Nilsson E, Isaxon C, Dahl A, Gudmundsson A, Swietlicki E, Bohgard M 2009 *Aerosol Science* **40** 193-208
[2] Osite A, Katkevich J, Viksna A, Vaivars G 2011 *IOP Conf. Ser.: Materials Science and Engineering* **23** 012029
[3] Baitimirova M, Osite A, Katkevich J, Viksna A 2011 *Enviromental and Climate Technologies* **7** 19-26
[4] Tumolva L, Park J Y, Kim J, Miller A L, Chow J C, Watson J G, Park K 2010 *Aerosol Science and Technology* **44**, 202-215
[5] Harrison R M. Pollution: Causes, Effects and Control, Fourth Edition. UK, The Royal Society of Chemistry, 2001, 171-178
[6] Zai S, Zhen H, Jia-Song W 2006 *Aerosol Science* **37** 1484-1496

Ab initio modeling of sulphur doped TiO_2 nanotubular photocatalyst for water-splitting hydrogen generation

O Lisovski[1,5], S Piskunov[234], Y F Zhukovskii[4] and J Ozolins[1]
[1]Faculty of Materials Science and Applied Chemistry, Riga Technical University, Latvia
[2]Faculty of Computing, University of Latvia, Latvia
[3]Faculty of Physics and Mathematics, University of Latvia, Latvia
[4]Institute for Solid State Physics, University of Latvia, Latvia

E-mail: oleg.lisovski@gmail.com

Abstract. In order to construct an efficient visible-light-driven TiO_2 photocatalyst for water splitting applications, one has to perform improvements of its electronic structure. In this theoretical study we consider single-walled anatase TiO_2 nanotubes having following morphologies: (101) 3-layered wall with chirality indexes $(n,0)$ and (n,n), (101) 6-layered wall with $(n,0)$ and $(0,n)$, (001) 6-layered wall with $(n,0)$ and $(0,n)$, and (001) 9-layered wall with $(n,0)$ and $(0,n)$. The latter configuration occurs to be the most energetically stable, due to possessing negative strain energy. In our study the most stable 9-layered anatase (001) $(0,n)$ nanotube has been doped with sulphur. According to obtained results sulphur dopant creates the mid-gap states making the TiO_2 nanotube to be a good candidate for efficient photocatalyst working under day light irradiation.

1. Introduction

Issues of alternative energy sources and environment protection belong to the important nowadays topics. Hydrogen is one of the most environment-friendly energy sources; moreover, it is widely used in industry. Currently, hydrogen is generally produced from fossil fuels, and the disadvantage of the method is a presence of CO_2 as a side product. One of the ways how to get rid of this drawback is a hydrogen generation by photocatalytic water splitting [1].

There are several requirements toward the efficiency of photocatalysts. First, it should be a semiconductor, *i.e.*, it should possess a band gap. Second, band gap width should be approximately 2.0 eV, [1] in order to allow the catalyst to use not only UV light (which consists only approximately 4% of total solar radiation [2,3]), but also a significant amount of visible light spectrum. Third, positions of band edges should be adjusted, in order to generate excited electrons and electron holes at appropriate energy levels where hydrogen cations and hydroxyl group anions, respectively, are reduced and oxidized [4]. Fourth, the photocatalyst's structure must be free from impurities, in order to block recombination of excited electrons and electron holes. Such recombination occurs at impurities [4].

TiO_2 samples (band gap ~3.18 eV for the bulk phase of anatase) are one of the most attractive photocatalysts proposed for water-splitting applications [4]. TiO_2 is relatively cheap, chemically stable and non-toxic material. Its top valence band is lower than the energy level which hydroxyl groups are

[5] To whom any correspondence should be addressed

oxidized at, while its bottom of conduction band is higher than the energy level where H^+ are reduced, so energetic requirements are fulfilled.

Nevertheless, the band gap of TiO_2 bulk is too wide, and its crystalline structure normally contains a great number of impurities meaning that it cannot serve as a visible-light-driven photocatalyst. To surmount these disadvantages one can use nanotubes made of titania and dope them with various atoms. Comparing with conventional catalysts, titania nanotubes (NTs) possess larger surface area and, consequently, have better crystallinity, better adsorption capacity, and higher photocatalytic activity. It means that impurity concentration in nanotubes should be lower than in the bulk.

Table 1. Lattice constants and band gaps of TiO_2 (bulk phase) as calculated by means of various E_{xc} [6].

E_{xc}	a, Å	c, Å	$\Delta\varepsilon_g$, eV
B3PW	3.785	9.580	3.667
PBE0	3.778	9.545	4.109
B3LYP	3.800	9.647	3.641
SOGGAXC	4.449	9.539	0.669
LDA	3.766	9.353	1.999
PBE	3.821	9.597	2.019
PWGGA	3.817	9.595	2.023
HF	3.777	9.643	12.357
Experiment	3.784	9.507	3.18

Thus, it is also easier to introduce dopants in the NT structure. It is known that doping by S shifts the absorption spectrum of titania to a lower energy and effectively leads to visible light photocatalytic activity [5]. Therefore, in the current study we concentrate on titania nanotubes doped with sulphur.

Although pristine nanotubes tend to have larger band gap ($\Delta\varepsilon_g$) as compared to the bulk, low-dimensional structures of TiO_2 with $\Delta\varepsilon_g > 3.2$ eV are better starting point for doping which leads to the formation of the induced mid-gap states diminishing the band gap to the required interval. In this study we consider single-walled titania nanotubes of 8 different morphologies: built of 3-layered anatase (101) nanosheet with chirality indexes (n,0) and (n,n), 6-layered anatase (101) (n,0) and (0,n), 6-layered anatase (001) (n,0) and (0,n), 9-layered anatase (001) (n,0) and (0,n) and define which of them is most stable and prospective for further doping. Then we study effects of S atom (as substitute for O atom) which is being used as a dopant too.

2. Computational details

Our *ab initio* calculations have been performed using linear combination of atomic orbitals (LCAO) method with atom-centered Gaussian-type functions (GTF) as basis set (BS), according to implementation in computer code *CRYSTAL* [6]. We have used hybrid exchange-correlation functional B3LYP [6] within the framework of density functional theory (DFT). We have considered two BSs for Ti (ECP-411*sp*-311*d*(G) [7] and ECP-5*s*-6*sp*-5*d* [8]). We have figured out that the latter, containing uncontracted GTFs would require irrational amount of computational time for calculations of NTs containing hundreds atoms per unit cell, so we chose the former. The full electron BS for oxygen has been taken in the form 6*s*-311*sp*-1*d*. Reciprocal space integration was performed by sampling the Brillouin zone (BZ) with the 4 k-points uniformly distributed along the line segment of the irreducible BZ. The cutoff threshold parameters of *CRYSTAL* for Coulomb and exchange integrals evaluation (ITOL1–ITOL5) have been set to 7, 8, 7, 7, and 14, respectively.

To perform reliable calculations, we have compared three hybrid exchange-correlation functionals (E_{xc}), one LDA and three GGA, which allow us to performed Hartree-Fock (HF) calculations too within hybrid functional formalism. We have calculated the band gap and lattice constants of bulk TiO_2 in anatase phase using different E_{xc} too. We have compared the obtained results with experimental data (results are present in the Table 1). According to Table 1, the best agreement with experiment is achieved for hybrid B3LYP E_{xc} (band gap 3.64 *vs*. 3.18 eV). Nevertheless, the result was not accurate enough for further calculations as far as band gap is the key feature for a photocatalyst, so, we modify an admixture coefficient for non-local HF exchange in B3LYP. The result provided in the Table is obtained with default 20% of HF exchange. Lowering HF exchange contribution down to 14% we have been able to reproduce experimental $\Delta\varepsilon_g$ (3.16 *vs.* 3.18 eV) while TiO_2 lattice constants remain practically unchanged.

3. Results and discussion

In our study, we have considered four different anatase nanosheets to roll-up NTs (see Fig. 1). These are 3-layered slab cut parallel to (101) anatase face, 6-layered slab parallel to (101), 6- and 9-layered

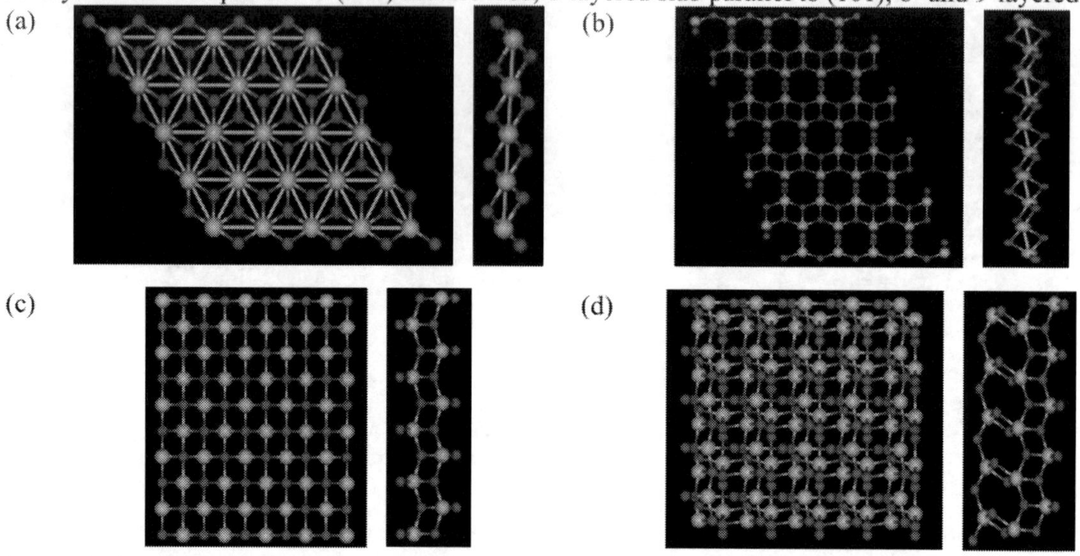

Figure 1 (color online). Schematic representation (top view and side view) of nanosheets considered in current study to roll-up NTs: (a) 3-layered anatase (101) nanosheet, (b) 6-layered anatase (101) nanosheet, (c) 6-layered anatase (001) nanosheet and (d) 9-layered anatase (001) nanosheet.

slabs parallel to (001). It is possible to form two types of titania nanotubes from the two-dimensional structure since the nanotube configuration depends on how chirality vector is oriented [7]. For 3-layered anatase (101) of hexagonal morphology [7], any (n,m) chirality indexes are possible, however, in current study, we have considered only achiral structures having (n,n) and $(n,0)$ chirality indexes. For the rest NTs of rectangular morphology, only $(n,0)$ and $(0,n)$ chirality indexes are possible [7].

We have calculated energetics for NTs folded from nanosheets mentioned above and having diameters ranging from 0.6 to 4 nm. The purpose of that was to define the most energetically stable NT. According to our calculations, 9-layered (001) NT with $(0,n)$ chirality indexes possess negative strain energy, *i.e.* it is energetically more favorable to form nanotube rather than to keep original 2D structure. Dependence of calculated strain energy on NT diameters is shown in Figure 2. It is worth mentioning that our prediction is in good agreement with earlier theoretical study performed by Ferrari *et al.* [8]. Based on our strain energy calculations (Figure 2), for further doping with sulphur, we have chosen (0,36) 9-layered anatase (2×2) (001) NT with internal diameter of 3.47 nm, wall thickness of 0.67 nm, and having 648 atoms in the unit cell. Sulphur has replaced oxygen in six possible configurations yielding defect concentration of ~8%. The notation of dopant positions is the following: S1 – dopant placed in the NT's outer wall, S2 and S3 – dopant placed in two irreducible positions of outer subsurface NT layer, S4 and S5 – dopant placed in two irreducible positions of inner subsurface NT layer, and S1 – dopant placed in the NT's inner wall.

Fig. 3 shows band edge positions calculated for TiO_2 nanosheet, pristine and S-doped NTs. For five configurations of S-doped NT, their top valence bands are essentially shifted in a beneficial position,

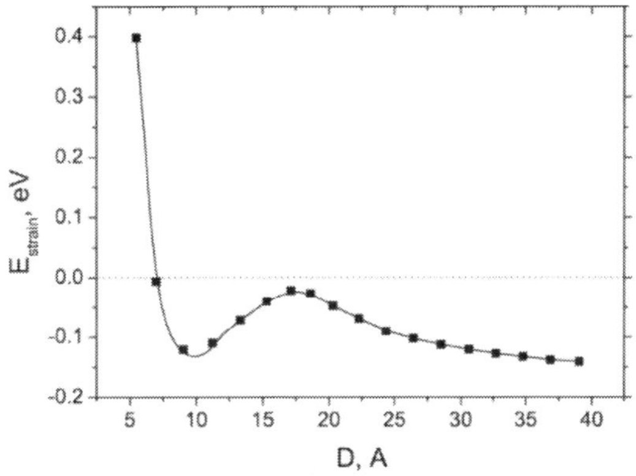

Figure 2. Calculated strain energies, E_{strain}, *vs.* NT diameters for (0,*n*) 9-layered (001) NT.

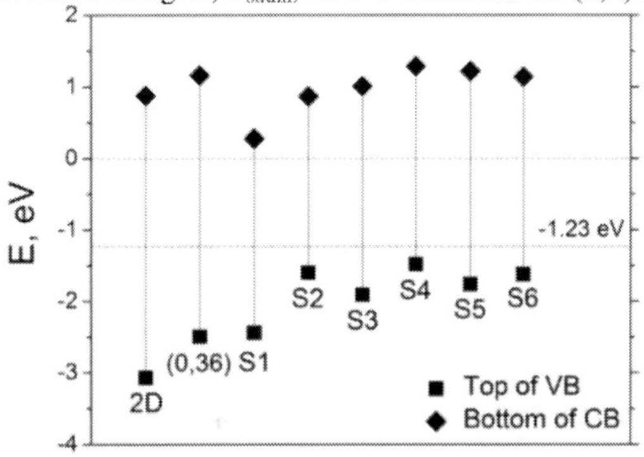

Figure 3. Calculated positions of band edges for TiO_2 nanosheet (2D), pristine (0,36) NT and S-doped (0,36) NTs denoted as S1 – S6 (see text). Energy scale shown with respect to standard hydrogen electrode (SHE, -4.44 eV *vs.* vacuum level). Dash-dotted line corresponds to the oxygen redox potential (-1.23 eV).

while the bottoms of conduction bands are positioned substantially above the standard hydrogen electrode (SHE, -4.44 eV *vs.* vacuum level). However, there is a single configuration (S1) where bottom of conduction bands lies close to SHE. In this last case dopant is introduced into the outer NT surface layer, and it occurs to be the most energetically favorable position. The band gap of this structure is 2.72 eV (narrower with respect to the bulk TiO_2) which is close to interval required for construction of efficient visible-light-driven photocatalysts.

4. Conclusion

In current study, we have defined the most energetically stable nanotube. It can be folded from 9-layered anatase (001) nanosheet with chirality indices (0,*n*). We have performed calculations for the S-doped (0,36) NT. Doping of nanotubular TiO_2 by sulphur results in narrowing of its band gap with the band edges positions, close to the limits required for efficient H_2O splitting. Therefore, S-doped NT can be predicted as suitable photoanode for day-light photocatalytic applications. The energetically favorable position of S dopant is in the outer surface layer of the NT.

Acknowledgements. S.P. is thankful for the financial support through the ESF project No. 2009/0216/1DP/1.1.1.2.0/09/APIA/VIAA/044.

References

[1] Cook T R, Dogutan D K, Reece S Y, Surendranath Y, Teets T S and Nocera D G 2010 *Chem Rev* **110** 6474-6502
[2] Thimsen E, Biswas S, Lo C S and Biswas P 2009 *J Phys Chem C* **113** 2014-2021
[3] Valdes A 2012 *Phys Chem Chem Phys* **14** 49-70
[4] Kudo A and Miseki Y 2009 *Chem Soc Rev* **38** 253–278

[5] Periyat P, Pillai S C, McCormack D E, Colreavy J and Hinder S J 2008 *J. Phys. Chem. C* **112** 7644–7652.
[6] Dovesi R, Saunders V R, Roetti C, Orlando R, Zicovich-Wilson C M, Pascale F, Civalleri B, Doll K, Harrison N M, Bush I J, D'Arco Ph and Llunell M. *CRYSTAL-2009 User Manual*: University of Torino: Turin 2009
[7] Piskunov S, Heifets E, Eglitis R I and G. Borstel 2004 *Comput Mater Sci* **29** 165-178
[8] Evarestov R A, Bandura A V, Losev M V, Piskunov S and Zhukovskii Yu F 2010 *Physica E* **43** 266-278
[9] Ferrari A M, Szieberth D, Zicovich-Wilson C M and Demichelis R 2010 *J Phys Chem Lett* **1** 2854–2857

IOP Conf. Series: Materials Science and Engineering **38** (2012) 012058 doi:10.1088/1757-899X/38/1/012058

Ab initio modelling of UN grain boundary interfaces

E A Kotomin[1], Yu F Zhukovkii[1], D Bocharov[2,3,4], D Gryaznov[1]
[1]Institute of Solid State Physics, University of Latvia, Kengaraga Str. 8, LV-1063, Riga, Latvia.
[2]Faculty of Computing, University of Latvia, Raina blvd. 19, LV-1586, Riga, Latvia.
[3]Faculty of Physics and Mathematics, University of Latvia, Zellu Str. 8, LV-1002, Riga, Latvia.

E-mail: bocharov@latnet.lv

Abstract. The uranium mononitride (UN) is a material considered as a promising candidate for Generation-IV nuclear reactor fuels. Unfortunately, oxygen in air affects UN fuel performance and stability. Therefore, it is necessary to understand the mechanism of oxygen adsorption and further UN oxidation in the bulk and at surface. Recently, we performed a detailed study on oxygen interaction with UN surface using density functional theory (DFT) calculations. We were able to identify an atomistic mechanism of UN surface oxidation consisting of several important steps, starting with the oxygen molecule dissociation and finishing with oxygen atom incorporation into vacancies on the surface. However, in reality most of processes occur at the interfaces and on UN grain boundaries. In this study, we present the results of first DFT calculations on O behaviour inside UN grain boundaries performed using GGA exchange-correlation functional PW91 as implemented into the *VASP* computer code. We consider a simple interface (310)[001](36.8°) tilt grain boundary. The N vacancy formation energies and energies of O incorporation into pre-existing vacancies in the grain boundaries as well as O solution energies were compared with those obtained for the UN (001) and (110) surfaces.

1 Introduction

Uranium mononitride (UN) is a compound with metallic properties (metallic colour and low electrical resistivity) [1], possessing rock-salt *fcc* structure over a wide temperature range. UN is considered nowadays by the Generation IV International Forum of nuclear reactors [2] as one of the most promising future nuclear fuels alternative to currently used UO_2. However, UN reveals unwanted oxidation in air which could affect the fuel fabrication process and fuel performance [3]. Due to a considerable amount of chemically aggressive oxygen impurities in any UN samples, it is necessary to understand the atomistic mechanism of O adsorption and further oxidation process.

The detailed study of single-crystalline UN surface including their interaction with oxygen have been performed by us recently [4] using the DFT method as implemented in the VASP computer code

[4] To whom any correspondence should be addressed.

[5]. We performed such the calculations for a pure UN bulk and surface, nitride and uranium vacancies on the surface, atomic and molecular adsorption as well as oxygen diffusion on the UN (001) surface (we studied mainly the (001) surface since according to the traditional Tasker analysis [6] it has the lowest surface energy).

The results obtained for interaction of O atoms and oxygen molecules with the UN (001) surface [4] demonstrate strong chemisorption, typical for metallic adsorbents. The possibility for spontaneous dissociation of the adsorbed O_2 molecules upon the perfect UN surface, analogous to the O_2 dissociation on metallic surfaces, has been demonstrated. After molecular dissociation, O adatom (O_{ads}) forms a strong chemical bond with U atom in the outermost surface plane (U_{surf}) which can be considered as one-center surface complex. High mobility of O_{ads} atoms along the surface due to relatively low migration barriers (<0.5 eV) has been found. The possibility of low-barrier (~0.5-1 eV) for O_{ads} incorporation (from the nearest adsorption site atop U_{surf}) into pre-existing N vacancy has been also shown as well as energetic stability of UN surface containing incorporated O atoms [4]. To increase validity of results, we have additionally performed the (110) surface calculations. We have compared the energies of N vacancy formation, oxygen atom adsorption upon U or N surface atom as well O atom incorporation into the N vacancy evaluated for both the UN(001) and (110) surfaces [4].

Nevertheless, many questions on UN oxidation remain open. Taking into account the fact that polycrystalline UN fuel powder contains nanoparticles with differently oriented crystallographic facets and a wide variety of interfaces between the grains [7], we consider here a role of grain boundaries (GB) in UN oxidation. This makes theoretical simulation of grain boundaries between different UN facets and their interaction with defects and O impurities very important for a realistic description of actinides. It is also important to study similarities and differences in oxygen incorporation processes energetic on both single surface and grain boundary.

In this paper, we consider a simple grain boundary interface (310)[001](36.8°) model and present first results for the energetics of N vacancy formation in the grain boundaries as well as oxygen incorporation into pre-existing N vacancies.

2 Computational details

The results were obtained using the plane wave calculations with *VASP* computer code, employing a plane-wave basis set combined with *PAW* pseudopotentials for U, N and O atoms (containing 14, 5 and 6 valence electrons, respectively) [5].

In this study, we have generated *k*-points using Monkhorst-Pack's technique [8] whereas the electron populations were determined following the method of Methfessel and Paxton [9] as implemented in the *VASP* code. For each series of calculations, we have found the optimal k-point mesh that provides convergence of the results: the 4×4×4 mesh for grain boundary calculations (the 4×4×1 and 8×8×1 meshes was used for previous surface calculations).

We have performed spin-relaxed calculations of ferromagnetic UN. The smearing parameter of 0.2 eV has been found to be optimal for reasonable convergence suggesting the electronic entropy contribution of the order of 10 meV. The optimal cut-off energy has been found to be equal to 520 eV. In this study, we present results of first DFT calculations on oxygen behaviour between UN grain boundaries.

3 Model

We consider a model of (310)[001](36.8°) tilt grain boundary used previously for describing of electron-trapping in rock-salt compounds like MgO, NaCl and LiF [10] a slab model for surface modeling. The periodic supercell with 15.40 Å × 4.87 Å × 34.13 Å linear dimensions contains 160 atoms (or 159 for N vacancy calculations). We have considered three sites for the N vacancies in positions of N atoms enumerated as 1 to 3 in Fig. 1. The lattice constant of UN slabs is fixed at 4.868 Å, taken from the lattice relaxation of UN bulk [11]. In all calculations we performed a complete structure optimization inside the supercell of fixed linear dimensions using criterion of the total energy minimization.

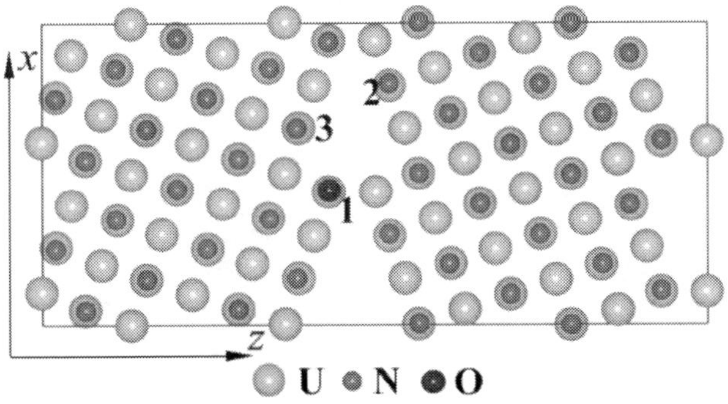

Figure 1. Cross-section of the (310)[001](36.8°) tilt grain boundary supercell (15.40 Å × 4.87 Å × 34.13 Å) for UN slab model with oxygen incorporated into one of possible positions (here N atom in position 1 is substituted by O atom). Other calculations show either formation of N vacancies in positions (1)-(3) or N atom substitution by O atom in positions (2) and (3).

4 Results and discussion

For three considered positions (1 to 3 in Fig.1) we estimated the N vacancy formation energy as well as oxygen incorporation and solution energy, respectively.
The formation energy $E_{form}^{N\_vac}$ of N vacancy was calculated as

$$E_{form}^{N\_vac} = E_{GB(N\_vac)} + \mu_{N\_atom} - E_{GB}, \qquad (1),$$

the oxygen incorporation energy E_{inc} for the first time was suggested by Grimes and Catlow [12]:

$$E_{inc} = E_{GB(O\_inc)} - E_{GB(N\_vac)} - \mu_{O\_atom}, \qquad (2),$$

and oxygen solution energy E_{sol} was calculated as

$$E_{sol} = E_{inc} + E_{form}^{N\_vac}, \qquad (3)$$

where $E_{GB(N\_vac)}$ is the total energy of a fully relaxed grain boundary containing one N vacancy in positions (1), (2) or (3) as depicted in Fig. 1, $E_{GB(O\_inc)}$ is the total energy for the grain boundary with O atom incorporated into the N vacancy, E_{GB} is the same for grain boundary without defect, and, finally, μ_{N_atom} (μ_{O_atom}) is the energy of an atom in the N_2 (O_2) molecule.

Table 1. Formation energies $E_{form}^{N\_vac}$ (in eV) for N vacancy in the grain boundary for three different positions (Fig. 1) and on the UN (001) and (110) surface (outermost) and central layers and as well as the average spin magnetic moment μ_{av} of U atoms in these systems.

Grain boundary				Surface layer				Central layer			
				(001) surface		(110) surface		(001) surface		(110) surface	
positions in Fig. 1.	$E_{form}^{N\_vac}$	$\mu_{av}(\mu_B)$	Number of layers in slab	$E_{form}^{N\_vac}$	$\mu_{av}(\mu_B)$	$E_{form}^{N\_vac}$	$\mu_{av}(\mu_B)$	$E_{form}^{N\_vac}$	$\mu_{av}(\mu_B)$	$E_{form}^{N\_vac}$	$\mu_{av}(\mu_B)$
(1)	3.47	1.46	7, 2×2	3.70	1.55	3.03	1.59	4.43	1.49	4.52	1.51
(2)	3.34	1.46	9, 2×2	3.70	1.45	3.04	1.51	4.42	1.41	4.35	1.49
(3)	3.48	1.46	11, 2×2	3.71	1.39	3.03	1.45	4.42	1.36	4.42	1.43
			7, 3×3	3.65	1.49	2.97	1.50	4.42	1.49	4.56	1.47

The N vacancy formation energies $E_{form}^{N\_vac}$ are summarized in Table 1. The results for the N vacancies in grain boundaries are compared with previous results for the N vacancies on the UN (001) and (110) surfaces [4]. We clearly see spin-changes similar for both UN surface and grain boundary. The formation energies $E_{form}^{N\_vac}$ in the GB are 3.3-3.5 eV. These values are comparable with analogous values for UN (001) (3.6-3.7 eV) and (110) (~3.0 eV) surfaces but are smaller than those in the bulk material (~4.4 eV) or in the (001) or (110) slabs central layer (4.3-4.6 eV). It indicates a clear trend for segregation of vacancies towards the grain boundaries and surfaces.

Table 2. Incorporation energy E_{inc} and solution energy E_{sol} (in eV), average spin magnetic moment of U atom μ_{av} as well as effective charge of incorporated O atoms into the UN (001) and (110) surface (outermost) atomic layer, central layer of UN(001)slab as well as grain boundary (Figure 1).

	O incorporation into grain boundary				O incorporation into central layer of (001) slab				
positions in Fig. 1.	E_{inc}	E_{sol}	$\mu_{av}(\mu_B)$	$q_{eff}(e^-)$	Number of layers	E_{inc}	E_{sol}	$\mu_{av}(\mu_B)$	$q_{eff}(e^-)$
(1)	-5.92	-2.45	1.45	-1.41	7, 2×2	-6.61	-2.18	1.47	-1.42
(2)	-5.67	-2.33	1.44	-1.38	9, 2×2	-6.61	-2.19	1.39	-1.38
(3)	-5.92	-2.44	1.44	-1.36	7, 3×3	-6.60	-2.18	1.45	-1.42

	O incorporation into surface layer of slab							
	(001) surface				(110) surface			
Number of layers	E_{inc}	E_{sol}	$\mu_{av}(\mu_B)$	$q_{eff}(e^-)$	E_{inc}	E_{sol}	$\mu_{av}(\mu_B)$	$q_{eff}(e^-)$
7, 2×2	-6.18	-2.48	1.495	-1.36	-5.82	-2.79	1.52	-1.29
9, 2×2	-6.19	-2.48	1.41	-1.36	-5.82	-2.78	1.47	-1.29
11, 2×2	-6.195	-2.48	1.365	-1.35	-5.82	-2.79	1.42	-1.29
7, 3×3	-6.13	-2.48	1.46	-1.36	-5.75	-2.78	1.47	-1.28

Oxygen incorporation energy E_{inc} and solution energy E_{sol} are summarized in Table 2. These energies for the vacancies in the grain boundaries (-5.6÷-5.9 vs. -2.3÷-2.5 eV, respectively) are close to those for the UN surface (-5.7÷-6.6 eV for O atom incorporation and -2.4÷-2.8 for O atom solution).

5 Conclusion

Ab initio DFT method was used to analyze the role of point defects and O interaction with UN grain boundaries using (310)[001](36.8°) tilt model. We obtained for the first time the energies of nitrogen vacancy formation and O atom incorporation into N vacancy in the GB and compared these results with previously obtained results for both UN(001) and (110) surfaces. We have observed a clear evidence for vacancy segregation at the interfaces, open surfaces and grain boundaries. Overall, defects energetics of the UN grain boundary and surface very well correlate with each other.

Acknowledgements: The authors thank A.L. Shluger and K. McKenna for help in constructing the grain boundary model, as well as A. Kuzmin, Yu.A. Mastrikov and S. Piskunov for a technical assistance. D.B. is grateful to European Social Fund project No. 2009/0216/1DP/1.1.1.2.0/09/APIA/VIAA/044.

References

[1] Curry N, 1965 *Proc Phys Soc.* **86** 1193-1198
[2] www.gen-4.org
[3] Arai Y, Morihira M and Ohmichi T 1993 *J Nucl Mater* **202** 70-78
[4] Zhukovskii Yu F, Bocharov D, Gryaznov D and Kotomin E A 2012 Chapter in *Advances in Nuclear Fuel* (Ed Shripad T Revankar, InTech Open Acess Publishers) 95-122
[5] Kresse G and Hafner J 2007 *VASP the Guide* (University of Vienna)
[6] Tasker P W 1979 *J Phys C: Solid State Phys* **12** 4977-4984
[7] Chinthaka Silva G W, Yeamans Ch B, Ma L, Cerefice G S, Czerwinski K R and Sattelberger A P 2008 *Chem Mat* **20** 3076-3084
[8] Monkhorst H J and Pack J D 1976 *Phys Rev B* **13** 5188-5192
[9] Methfessel M and Paxton A T 1989 *Phys Rev B* **40** 3616–3621
[10] McKenna K and Shluger A 2008 *Nature Mater* **7** 859-862
[11] Evarestov R A, Bandura A V, Losev M V, Kotomin E A, Zhukovskii Yu F, and Bocharov D 2008 *J Comput Chem* **29** 2079
[12] Grimes R W and Catlow C R A 1991 *Phil Trans Roy Soc A* **335** 609-634

IOP Conf. Series: Materials Science and Engineering **38** (2012) 012059 doi:10.1088/1757-899X/38/1/012059

Preparation and photoactivity of electrophoretic TiO_2 coating film

I Liepina[1], G Bajars, J Gabrusenoks, L Grinberga, J Kleperis, A Lusis

Institute of Solid State Physics, University of Latvia, Latvia

E-mail: liepina.ineta@gmail.com

Abstract. TiO_2 thin films have been obtained by a sol-electrophoretic deposition method on metallic Ti and Pt substrates. X-ray diffraction, Raman spectroscopy and scanning electron microscopy were used to investigate the structure and morphology of plated materials. Annealed TiO_2 films mainly are formed in anatase structure with different morphology depending on substrate material. Light induced potential measurements indicate that the morphology substantially have an impact on photoactivity of TiO_2 thin films.

1. Introduction

The solar energy utilisation needs different thin film coatings and deposition technologies for different types of devices (solar energy collectors, photovoltaic, photo-catalysis convertors, etc.). As the most promising photo-catalyst, titanium dioxide (TiO_2) is expected to play a key role for effective utilization of solar energy in photovoltaic devices and for hydrogen generation due to its excellent resistance to chemical and photochemical corrosion in aggressive aqueous environments [1,2]. It is also less expensive than other photo-sensitive materials.

Many efforts have been made to improve their properties by preparing porous films with high surface area. Traditionally, the photoactive coatings are deposited by sol-gel, spin coating, spray pyrolysis, chemical vapor deposition, laser ablation, sputtering, electron beam evaporation methods which are suitable for flat substrates [3-5].

For coating of irregular conductive surfaces the electrochemical methods are most eligible. The advantages of electrophoresis over other coating methods are the homogenity of deposition for complex shapes, reduction of material usage and high production rates, low contamination level and negligible initial capital investment, as well as easy transfer from the research laboratory to industry [3,6,7].

The purpose of current work is the optimization of parameters for the sol-electrophoretic deposition process to produce nanostructured TiO_2 thin films on various substrates. Obtained TiO_2 thin films were characterized by phase composition, morphology and their microstructures using X-ray diffraction, Raman spectroscopy, as well as scanning electron microscopy (SEM). Since titanium dioxide is a semiconductor under the form of thin films and enables to absorb only in the near ultraviolet region [8], the experimental investigation of their photocatalytic properties has been performed exactly in the near ultraviolet region, because there is no necessity for additional treatment of TiO_2 films to make them sensitive to visible light.

[1] To whom any correspondence should be addressed

2. Experimental

The first step was the preparation of TiO_2 sol. Titanium tetrabutoxide (TTB) was used as a precursor, hydrochloric acid as a catalyst for the peptization and deionized water as a dispersion media. A water-acid mixture was stabilized at 50 °C with continuous stirring. An appropriate amount of TTB was added next forming the white precipitate that gradually peptized forming a clear sol.

For the electrophoresis growth, the Pt anode and cathode are placed parallel in TiO_2 sol with a distance 1 cm in between. The cathodes used were metallic Ti and thin Pt layer on silicon. A constant voltage of -1 V was applied by a potentiostat/galvanostat EG&G PAR Model 173 between the electrodes and held for 1 – 5 h. The charge was registrated by a digital coulometer EG&G PAR Model 179. The as-deposited thin films were first dried at 100 °C for 24 h and then heated at a rate of 10 °C/min and were finally annealed at 500 °C for 2 h.

The morphology of TiO_2 films was characterized by scanning electron microscope FEI SEM "Quanta 200". The X-ray diffraction patterns (XRD) were recorded on a Philips X'Pert Pro MPD and Raman spectra by using spectrometer Spex Ramalog.

Measurements of photoactivity were carried out in two electrode cell with 1 M Na_2SO_4 water solution as an electrolyte and Pt foil as reference electrode and counter electrode [9]. The samples were exposed to light source (120 W mercury lamp) placed in distance of 25 cm from quartz window of cell. The open-circuit voltage and photocurrent density were measured by potentiostat VoltaLab PGZ 301.

3. Results and Discussion

The X-ray diffraction patterns from as-deposited and annealed TiO_2 thin films on Ti substrate are shown in Figure 1. No diffraction peaks were detected for the as-deposited films due to its low crystallinity. An intensive peak at $2\theta \sim 35°$ is associated with metallic Ti substrate because the thickness of oxide film is only 600 nm. After thermal treatment at 500 °C for 2 h the film show an increase of crystallinity. Two diffraction peaks are observed for this sample characteristic of anatase (2θ – 25.3°) and rutile (2θ – 27.4°).

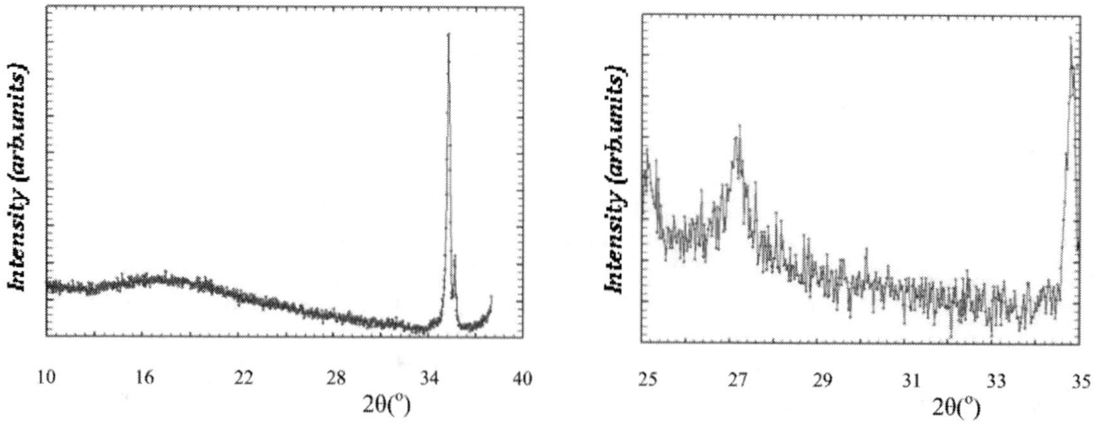

Figure 1. XRD patterns of TiO_2 thin films on Ti substrate as-deposited (left) and annealed at 500 °C for 2 h (right).

XRD usually reveals the long-range order of materials and gives average structural information. Raman scattering as a local probe is very sensitive to crystallinity and microstructures of material. Figure 2 shows the Raman spectrum of annealed TiO_2 thin films on different Ti and Pt/Si substrates.

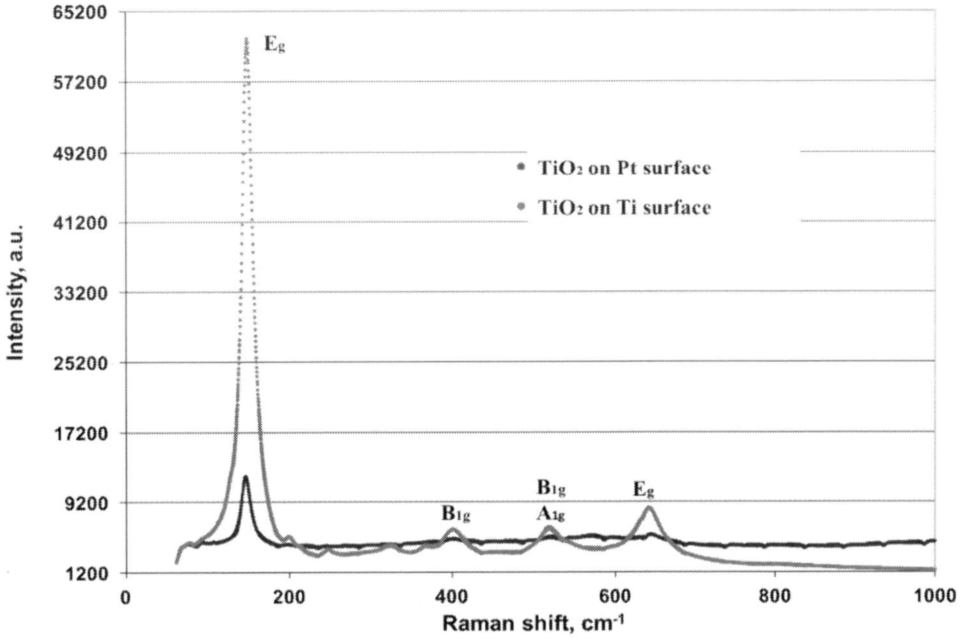

Figure 2. Raman spectrum of annealed TiO_2 thin films on Ti and Pt/Si substrates.

The Raman lines at 150, 405, 515, and 633 cm^{-1} are assigned to E_g, B_{1g}, A_{1g} or B_{1g}, and E_g, respectively. The strongest E_g mode at 150 cm^{-1} arising from the external vibration of the anatase structure is clearly explicit, which indicates that TiO_2 thin films obtained by sol-electrophoretic deposition mostly are formed in anatase structure.

Figure 3 shows the typical SEM images of as-deposited TiO_2 thin film with its corresponding annealed film on Ti substrate (average thickness 0.6 μm) and annealed film on Pt/Si substrate (average thickness 0.8 μm).

Figure 3. SEM images TiO_2 thin films: as-deposited on Ti substrate (left), annealed at 500 °C for 2 h on Ti substrate (middle) and Pt/Si substrate (right).

According XRD analysis as-deposited TiO_2 film is amorphous with disordered structure what is confirmed by SEM micrograph. The annealed film has homogeneous structure. The TiO_2 surface is comprised of a number of fine cracks. The length of cracks is about 1 – 5 μm. Annealed TiO_2 on Pt/Si substrate is formed by crystallites with an average size 0.5 – 1 μm. This crystallite grain structure is formed when pure anatase is present in the thin films. TiO_2 thin film on Ti substrate does not show such crystallites indicating the presence of rutile form as admixture.

An increase of photopotential of crystalline TiO_2 thin films on Ti and Pt/Si substrates demonstrates a similar character, but essentially different values (Figure 4). The change of photopotential in the case uniform TiO_2 films (~ 250 mV) is almost ten times higher than that for the films on Pt/Si substrate (~ 30 mV). The presence of both TiO_2 forms on Ti substrate gives higher photoactivity than for pure anatase on Si/Pt substrate. Microscopic observations of the TiO_2 thin films deposited on different metals revealed that the morphologies of the films are quite different what substantially have an impact on photoactivity of investigated photosensitive materials. It is expected that by increasing of thin film porosity and having expanded surface area the photoactivity of TiO_2 thin films would also increase.

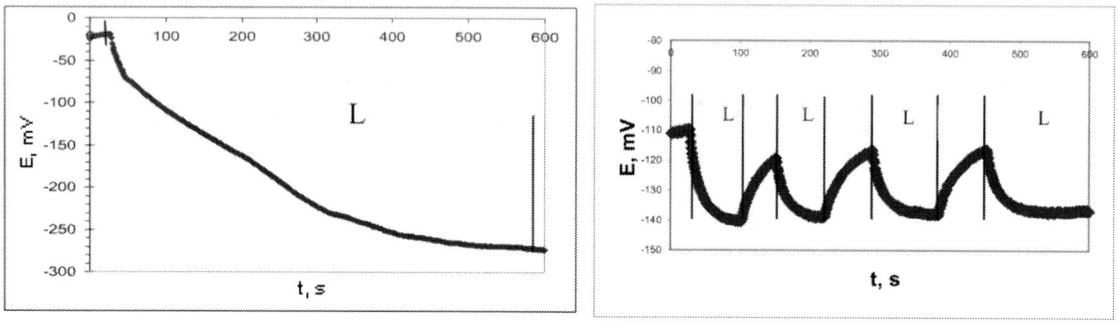

Figure 4. Photopotential changes of annealed TiO_2 thin films on Ti (left) and Pt/Si (right) substrates, where L – light, unmarked section - dark

4. Conclusions

In summary, highly uniform TiO_2 films have been successfully obtained by a sol-electrophoretic deposition method on metallic substrate. XRD and Raman spectra analyses confirmed that the crystallization of TiO_2 thin films annealed at 500 °C occurred mainly in the form of anatase phase. The morphologies of the TiO_2 thin films substantially have an impact on their photoactivity.

Acknowledments

This work was mainly financially supported by European Regional Development Fund project No.2010/0243/2DP/2.1.1.1,0/10/APIA/VIAA/156.

L.G. acknowledges European Social Fund project Nr. 2009/0202/1DP/1.1.1.2.0/09/APIA/ VIAA/141 for financial support.

The authors thank G. Bebris, Latvian State Forensic Science Bureau, for his technical support for SEM observation.

References

[1] Ye M, Vennerberg D, Lin C and Lin Z 2012 *J. Nanosci. Lett.* **2** 1
[2] Yoong L S, Chong F K and Dutta B K 2009 *Energy* **34** 1652
[3] Abdel Aal A M, Barakat A and Mohamed R M 2008 *Applied Surface Science* **254** 4577
[4] Lopez A, Acosta D, Martinez A I and Santiago J 2010 *Powder Technology* **202** 111
[5] Mohammadi M R, Odrikhani F O, Fray D J and Khomamizadeh F 2011 *Particuology* **9** 161
[6] Toh A G G, Cai R and Butler D L 2009 *Wear* **266** 585
[7] Hamadanian M, Jababari V, Gravand A and Asad M 2012 *Surface & Coatings Technology* doi: 10.1016/j.surfcoat.2012.03.017
[8] Mechiakh R and Bensaha R 2006 *C. R. Physique* **7** 464
[9] Linitis J, Kalis A, Grinberga L and Kleperis J 2011 *IOP Conf. Serie: Material Science and Engineering* **23** 012010

Hydrogen Adsorption in Zeolite Studied with Sievert and Thermogravimetric Methods

P Lesnicenoks[1,3], A Sivars[2], L Grinberga[2], J Kleperis[2]

[1]Faculty of Materials Science and Applied Chemistry, Riga Technical University
Āzenes iela 14/24, Riga, Latvia
[2]Institute of Solid State Physics, University of Latvia
Ķengaraga iela 8, Rīga, LV-1063, Latvija

E-mail: peteris.lesnicenoks@rtu.lv

Abstract. Natural clinoptilolite (mixture from clinoptilolite, quartz and muscovite) is activated with palladium and tested for hydrogen adsorption capability at temperatures RT – 200°C. Thermogravimetric and volumetric methods showed that zeolite activated with palladium (1.25%wt) shows markedly high hydrogen adsorption capacity – up to 3 wt%. Lower amount of adsorbed hydrogen (~1.5 wt%) was found for raw zeolite and activated with higher amount of palladium sample. Hypothesis is proposed that the heating of zeolite in argon atmosphere forms and activates the pore structure in zeolite material, where hydrogen encapsulation (trapping) is believed to occur when cooling down to room temperature. An effect of catalyst (Pd) on hydrogen sorption capability is explained by spillover phenomena were less-porous fractions of natural clinoptilolite sample (quartz and muscovite) are involved.

1. Introduction

Nowadays zeolites are the most important catalysts in petrochemical refineries [1]. Synthetic zeolites mostly are exploited because around 200 different new zeolites with specific pore geometries have been synthesized, but millions of hypothetical structures have been generated on the basis of energy minimization [2]. Zeolites as hydrogen storage materials are investigated broadly [3-5] and it is found that as small molecules such as molecular hydrogen and ammonia can be easily absorbed into a flexible network of zeolites. Previous researches found that hydrogen storage in zeolites can follow either an encapsulation mechanism or an adsorption regime [6,7].

At elevated temperatures, encapsulation (trapping) is believed to occur, when cooling to room temperature and release of excess pressure causes the hydrogen molecules to become encapsulated (or trapped) within the pores [7-9]. When the zeolite is reheated the hydrogen molecules are released [9]. Storing hydrogen molecules in porous media based on a physisorption mechanism is one possible approach to reach the US Department of Energy (DOE) targets (9 wt% 2015) for on-board hydrogen storage [10]. Some of the results reported on hydrogen storage in zeolites this far show that metal–organic frameworks can meet most of the 2010 DOE targets (6 wt%) on a materials basis when operating at 77 K, nevertheless researchers are optimistic when faced with the challenge of increasing

[3] To whom any correspondence should be addressed

the H_2 binding energy in zeolites to produce a hydrogen storage system that will ultimately function near ambient temperature [11].

Nowadays most developed materials for hydrogen storage are metal hydrides [12] and stable chemical hydrides [13], which are ready to be commercialized in transport sector and autonomous power supply systems, replacing tanks with compressed gas. They are characterized by slow kinetics and non-reversibility accordingly. Fast kinetics and reversibility can be expected for zeolites, since physisorption is a non-activated process. Weak van der Waals forces (dispersion, orientation and induction) are responsible for hydrogen physisorption in porous zeolites [14].

Our research is focused on natural zeolite – clinoptilolite and methods used in the research are volumetric (Sievert) and thermogravimetric.

2. Experimental setup and process

Natural clinoptilolite (origin in Ukraine) is used as raw material, grinded to 5 – 7 mm large grains. Pyrolytic-extraction method [15] is used to activate clinoptilolite with palladium nanopowder (≤20 nm, Aldrich). Two types of modified zeolite were prepared – with 1.25 wt% Pd and 5wt% Pd.

Structure and morphology of samples are studied by X-ray diffraction (XRD) diffractometer D-8 Advance (Bruker AXS) with CuKα radiation (λ=1.5418 Å) at Faculty of Chemistry, University of Latvia. The specific surface area of the powders is measured at Institute of Wood Chemistry with Sorptiometer KELVIN 1042 (COSTECH Instruments) by BET method using nitrogen gas as an adsorbate.

Sievert (volumetric) sorption method is realized by PCT–Pro2000 SETARAM Instrumentation) at hydrogen atmosphere under pressure 10 bars [15]. This instrument determines the amount of gas sorbed by measuring pressure changes in a reservoir of known volume and temperature. By knowing the initial and final pressure and volume of a sample chamber the quantity of gas absorbed or desorbed by different samples like metals, ceramics or polymers, is calculated. Pressure-composition isotherm (PCT) and cycle life measurements can be performed where the tolerance of the sample to various atmospheres can be determined. The PCTPro-2000 uses advanced software for automation of the experimental measurements. An experimental range of pressure lies between 0.01 and 200 bars and temperatures in a sample chamber can be set up from RT to 200°C.

Thermogravimetric hydrogen sorption method is elaborated in this work using commercial instrument thermogravimeter SHIMATZU TG–TDA at constant flow 50 ml/min of two different test gases: argon – while heating, and hydrogen – while cooling the samples. In this method the samples are first heated to 200°C at rate 10°C/min in Ar atmosphere and then cooled down in hydrogen atmosphere to RT (room temperature). Each sample went through at least two cycles of heating, the first one being the cycle of cleaning from adsorbed gases and the loading with hydrogen during the cooling. Next cycle started with second cleaning procedure in argon atmosphere to make sure – whether we can release the same amount of gas which was previously adsorbed by the sample. Palladium nano-powder was chosen as reference material (Figure 1).

As it is seen from figure 1, the hydrogen desorption from palladium in argon atmosphere (heating) and hydrogen absorption in hydrogen atmosphere (cooling) are described with understandable results. Previously (1st cycle) charged with hydrogen palladium is releasing about 0.7 wt% hydrogen gas when heated, with characteristic heat release at 85 °C, and is absorbing the same hydrogen amount, when cooled in hydrogen atmosphere with characteristic heat absorption peak. In the next measurement cycle the palladium has been replaced with activated and raw zeolites (see Results).

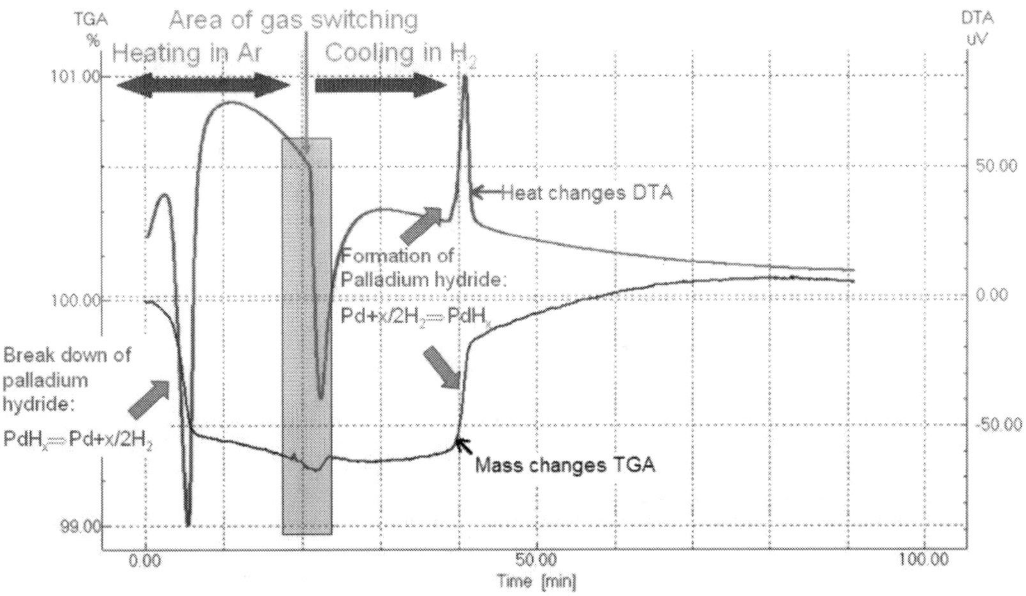

Figure 1. The test of developed measurement methodology with palladium nanopowder sample (2nd cycle).

3. Results

Structural investigations by XRD analysis method gave the following results - the sample consists of at least three crystalline phases – clinoptilolite $[Na,K,Ca)_{2-3}Al_3(Al,Si)_2Si_{13}O_{36}\bullet 12(H_2O)]$ ~ 60%; quartz (SiO_2) ~ 25% and muscovite $KAl_3Si_3O_{10}(OH)_2$ ~ 15%.

The specific surface area of the powders measured using BET method and nitrogen gas as an adsorbate gave the following results: BET Surface Area: 10.63 m^2/g; Langmuir Surface Area: 14.60 m^2/g; Total pore volume: 30.25 mm^3/g.

Zeolite activated by palladium (1.25%wt) show markedly high hydrogen adsorption capacity – up to 3 wt% (Figure 2). Similar results were measured for other zeolite samples – raw zeolite adsorbed ~1.5 wt% hydrogen, same amount that was adsorbed by 5 wt% activated clinoptilolite.

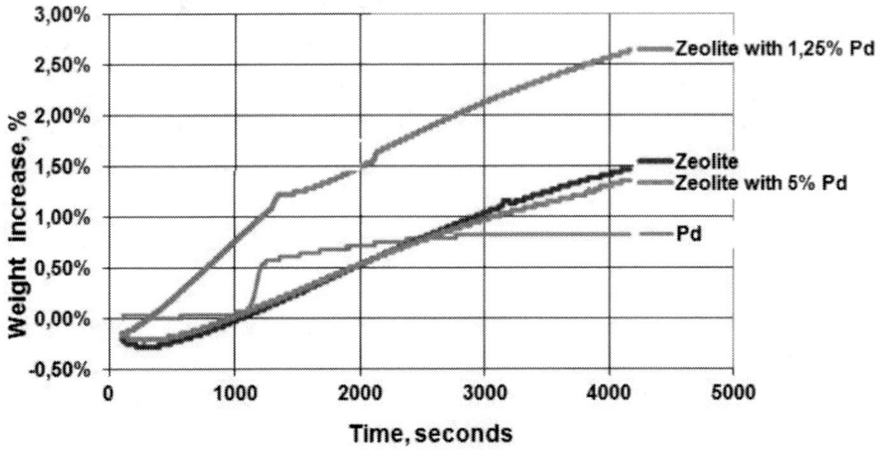

Figure 2. Results of thermogravimetric measurements of clinoptilolite/Pd 1,25wt% sample (2nd cycle).

Volumetric (Sievert) adsorption method was realized by PCT–Pro2000 equipment for the same samples and results presented in figure 3.

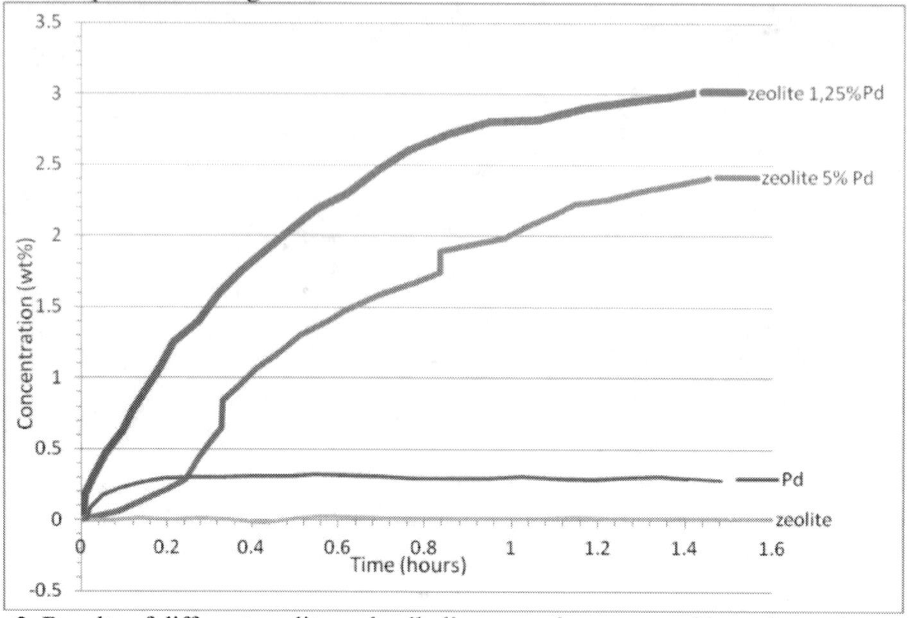

Figure 3. Results of different zeolite and palladium samples measured by volumetric method at hydrogen pressure 10 bars.

As it is seen from figure 3 volumetric method gives similar results as the gravimetric method – most efficient sample is zeolite with 1.25wt% palladium catalyst; next is with 5wt% palladium, and smallest amount of adsorbed hydrogen is measured for raw clinoptilolite.

4. Discussion

Experiments with hydrogen have shown that it is possible to encapsulate gases in the structure of zeolite, representing the material of certain molecular sieve type [6-9]. This method was investigated broadly in the 1980ies with aim to screen appropriate materials for temporarily storing and disposing of tritium. The method may also prove safer, and may enable isotope separation. Initial experiments were performed with hydrogen and interesting results found. Typical zeolites are crystalline, hydrated aluminosilicate compounds (known some 100 synthetic and some 34 naturally occurring types) with extremely uniform pore openings (3 to 10 angstroms) [6]. This fine and precise structuring is in contrast with large-pore adsorbents like silica gel (100 to 200 angstroms) and activated charcoal (20 to >10,000 angstroms).

Zeolites have a three-dimensional interconnecting network structure of SiO_4 and AlO_4 tetrahedra [6]. This structure permits an orderly sharing arrangement from tetrahedra bonded with corners in 3D structure forming form an array of truncated octahedral with sodalite-type cages. Water molecules which are bonded in zeolite structure as crystallization water, can be removed by heating, leaving voids that are refillable with other molecules. Therefore a zeolite can be used to selectively adsorb only those molecules that can enter the pores while excluding molecules whose larger diameters prohibit them from passing through the pores.

Regarding our results, the hypothesis may be raised that the heating of zeolite in argon atmosphere forms and activates the pore structure in material where hydrogen encapsulation (trapping) is believed to occur when cooling to room temperature. It should be noted that we did not observe this effect, when air was used instead of an argon atmosphere. It is therefore concluded that the Argon is special role in pore formation process in zeolite for hydrogen adsorption at elevated temperatures. The release of coupled hydrogen occurs in next heating procedure. Much smaller amounts of bonded hydrogen in zeolites at room temperature and atmosphere pressure are reported [9] - about 1.5 and 2.0 μmol/g, but so far no one has made hydrogen desorption - absorption experiments in zeolite as we are.

We must not forget the role of the Pd catalyst in hydrogen adsorption process, and potential explanation can be related with the spill-over effect [16]. The spill-over of hydrogen involves splitting of gaseous H_2 molecule on palladium nanocrystal, spill of hydrogen atoms to substrate and adsorption in the form of atomic hydrogen (Figure. 4). In such process a previously inactive material can be activated for hydrogen physisorption. The spill-over of atomic hydrogen is highly dependent on the chemical bridges formed at the interface; thereby less porous material can provide better media for atomic hydrogen transport and storage. In our case such less porous materials are quartz and muscovite, which present in natural clinoptilolite investigated in this work. From this point of view the mixture of quartz and muscovite with zeolite contributes to the higher amount of adsorbed hydrogen.

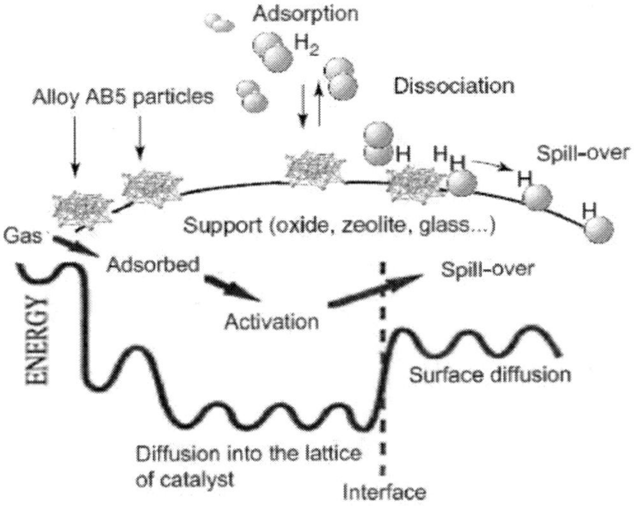

Figure 4. Explanation of spillover effect in hydrogen adsorption on inert substrates [15].

4. Conclusions

Original Thermogravimetric method is developed to measure hydrogen adsorption capability in porous materials – commercial instrument is adjusted to constant flow (50 ml/min) of two different test gases: argon – while heating, and hydrogen – while cooling the samples. In this method the samples are first heated to 200°C (at rate 10°C/min) in Ar atmosphere and then cooled down to RT in hydrogen atmosphere. The method was checked by using palladium powder and reliable results have been obtained.

Both thermogravimetric and volumetric methods showed that zeolite activated with palladium (1.25%wt) show markedly high hydrogen adsorption capacity – up to 3 wt%. Lower amount of adsorbed hydrogen (~1.5wt %) was found for raw Zeolite sample and activated with higher amount of palladium sample. Hypothesis is proposed that the heating of zeolite in argon atmosphere forms and activates the pore structure in zeolite material, where hydrogen encapsulation (trapping) is believed to occur when cooling down to room temperature. An effect of catalyst (Pd) on hydrogen sorption capability is explained by spillover phenomena, where less-porous fractions of natural clinoptilolite sample (quartz and muscovite) are involved.

Acknowledgement
Authors acknowledge financial support of ERDF project Nr. 2010/0188/2DP/2.1.1.1.0/10/APIA/VIAA/031.

IOP Conf. Series: Materials Science and Engineering **38** (2012) 012060 doi:10.1088/1757-899X/38/1/012060

References

[1] Degnan T F 2000 *Topics in Catalysis* **13** 349
[2] Sartbaeva A, Wells S A, Treacy M M J and Thorpe M F 2006 *Nature Materials* **12** 962
[3] Langmi H W and McGrady G S 2007 *Coordination Chemistry Reviews* **251** 925
[4] Logar N Z and Kaučič V 2006 *Acta Chim. Slov* **531** 117
[5] Broom D P 2011 *Hydrogen Storage Materials: The Characterisation of Their Storage Properties* (London: Springer-Verlag)
[6] Dyer A 1988 *Introduction to Zeolite Molecular Sieves* (Chichester: Wiley)
[7] Fraenkel D and Shabtai J 1977 *J. Am. Chem. Soc.* **99** 7074
[8] Weitkamp J, Fritz M and Ernst S 1995 *Int. J. Hydrogen Energy* **30** 967
[9] Efstathiou A M, Borgstedt E R, Suib S L and Bennett C O 1992 *Journal of Catalysis* **135** 135
[10] http://www1.eere.energy.gov/hydrogenandfuelcells/mypp/
[11] Dinca M and Long J R 2008 *Angew. Chem. Int. Ed.* **47** 6766
[12] Sakintuna B, Lamari-Darkrim F and Hirscher M 2007 *Intern. J. of Hydrogen Energy* **32** 1121
[13] Van den Berg A W C and Arean C O 2008 *Chem. Commun* **2008** 668
[14] Xiaoming D, Jing L and Erdong W 2010 *Progress in Chemistry* **22** 248
[15] Grinberga L, Sivars A, Kulikova L, Serga V and Kleperis J 2011 *IOP Conf. Ser.: Mater. Sci. Eng.* **23,** 012009
[16] Grinberga L, Kleperis J, Bajars G, Vaivars G and Lusis A 2008 *Solid State Ionics* **179** 42

Threshold character of temperatures on deuterium thermal desorption in Mg-V composite grown atom-by-atom

I M Neklyudov, O M Morozov[1], V G Kulish, V I Zhurba, A G Galitskiy, N S Lomino, A S Kuprin, V D Ovcharenko, E N Reshetnyak

National Science Center "Kharkov Institute of Physics and Technology", 1 Akademicheskaya St., 61108 Kharkov, Ukraine

E-mail: morozov@kipt.kharkov.ua

Abstract. The plasma evaporation-sputtering method was applied to make composite materials of the Mg-V system. The ion-implanted deuterium desorption temperature variations as a function of the component concentration were studied. It has been established that, by introducing vanadium into magnesium, the deuterium desorption temperature can be appreciably decreased (to 300-330 K) in comparison with the case of deuterium desorption from magnesium (~800 K). A step-like form of the curve of deuterium desorption temperature evidences on the presence of two different structure states of the Mg-V system depending on the ratio of components. The deuterium temperature decrease can be caused by filamentary inclusions formed, in the process of composite making and annealing, by the insoluble component (vanadium) atoms providing the deuterium diffusion from the sample at a lower temperature (channels for deuterium diffusion through the surface barrier). A necessary high diffusion mobility of deuterium is provided by the amorphous state of $Mg_{83+x}V_x$ samples. The deuterium desorption data obtained on the example of Mg-V and Mg-Zr composites provide support for further research into hydrogen storage materials containing low-soluble chemical elements in the alloy components.

1. Introduction

Magnesium-based alloys are promising in the view of present-day requirements to the metal-hydride hydrogen storage systems. Behavior of hydrogen in the magnesium-based alloys is of scientific and applied interest that is confirmed by many publications [1-10]. However, the use of such alloys presents some difficulties because of the high hydrogen desorption temperature (550-600 K).

Particular attention is given to materials in a nanocrystalline state. An exclusive position of nanocrystalline materials as hydrogen storage materials is determined by their unique structural properties that can, probably, provide a high sorption capacitance and potentially high concentrations for hydrogen storage. Therefore it is reasonable that reports about investigations on the hydrogen behavior in the nanocrystalline materials become more and more frequent in the literature [8-10].

One of the methods for obtaining materials in the nanocrystalline state is introduction of nanoformative elements. These are chemical elements with low solubility or non interacting with other elements in the composite being formed. By making nonequilibrium alloys, to introduce in them low-soluble chemical elements, one, usually, uses a high-energy ball mill or magnetron sputtering. The literature data confirm that investigations in this direction are promising [5-7]. The purpose of this

[1] To whom any correspondence should be addressed

work is to make Mg-V composites using the method of atom-by-atom mixing of components and to study mechanisms of hydrogen storage and thermoactivated release from the samples obtained depending on the component concentration. Note, that, practically, Mg and V are not alloy able, therefore, in the system the liquid-and-solid phase immiscibility is observed [11].

2. Experimental

To manufacture Mg-V composites the plasma evaporation-sputtering method was used enabling the atom-by-atom component growth. Thus, the composites with a wide range of the ratios of insoluble components were obtained [12]. The Mg-V composite layer of 1-5 μm thickness was deposited on the molybdenum foils (0.2 mm thickness, 10 mm width, 250 mm length) placed between the cathode assemblies in the facility. In parallel with the molybdenum foils there were arranged the copper plates (10×10 mm^2) as standards for determining the thickness of obtained samples by the gravimetric method and composite component concentrations – by the X-ray fluorescent method. Figure 1 presents the magnesium and vanadium concentration in the Mg-V composite samples as a function of the distance between the magnesium and vanadium cathodes.

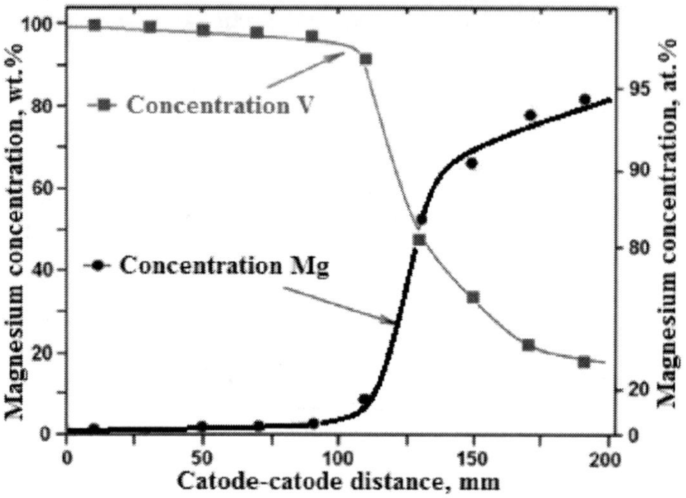

Figure 1. Magnesium and vanadium concentration in the Mg-V composite samplesversus the distance between the magnesium and vanadium cathodes.

Deuterium introduction into the samples was performed by the ion implantation method. The deuterium desorption temperature ranges and the deuterium storage levels were determined by the thermal desorption spectroscopy. The temperature was measured with the chromel-alumel thermocouple fastened to the heater.

3. Results

The most characteristic spectra of ion-implanted deuterium thermal desorption from the Mg-V composites are shown in Figure 2. The Figure shows the thermal desorption spectrum dependence on the both the sample's composition and implanted deuterium dose, as well as the corresponding temperature of the desorption peaks in the spectrum against the Mg content.

A single-peak character of the deuterium thermal desorption spectrum observed for magnesium concentration values ranging from 1 to 20 at.% evidences on the homogeneity of composite structural state in this range (Figure 2a). The temperature range of deuterium desorption from the magnesium-vanadium composites with a high vanadium content on the temperature scale locate the region of the deuterium solid solution in vanadium.

As the magnesium content increases, the maximum temperature peak in the thermal desorption spectrum moves in the direction of temperature increasing. Besides, in the spectra of deuterium desorption from the $Mg_{65}V_{35}$ and $Mg_{83}V_{17}$ composites an additional low-intensity lower-temperature broad peak arises near the mean peak, beginning from the temperature of ~300 K (see Figure 2b and

2c), that is evidence of changes which occur in the structural state of the magnesium-vanadium composites.

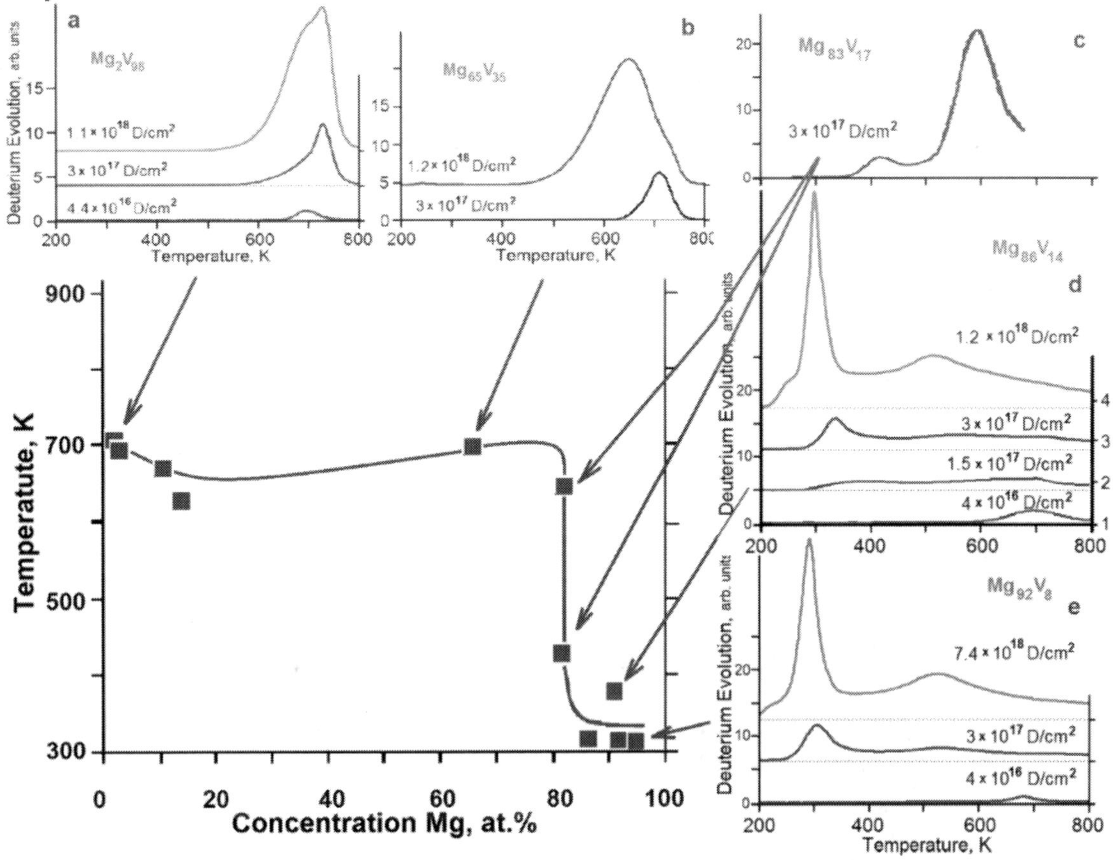

Figure 2. Deuterium desorption temperature versus the Mg-V composition (■) for a deuterium dose of ~3×10^{17} D/cm^2 ($T_{irr.}$~100 K) and thermodesorption spectra of deuterium from composites: (a) $Mg_{02}V_{98}$; (b) $Mg_{65}V_{35}$; (c) $Mg_{83}V_{17}$; (d) $Mg_{86}V_{14}$; (e) $Mg_{92}V_{08}$

A further increase of magnesium concentration in composites leads to the qualitative changes in the deuteron desorption spectrum. For example, in the case of the $Mg_{86}V_{14}$ composite of stoichiometry, as compared with the $Mg_{65}V_{35}$ composite of stoichiometry, the intensity in the low-temperature part of the deuteron desorption spectrum (T_m ~300 K) is increasing and in the high-temperature part of the spectrum it is decreasing (see in Figure 2d the low-intensity flat plateau of the deuterium desorption in the temperature range between 400 and 600 K). It should be noted that in the case of $Mg_{86}V_{14}$ the deuterium thermal desorption spectrum structure depends on the implanted deuterium dose. At a low dose (~4×10^{16} D/cm^2) there is only one high-temperature peak in the spectrum. The increase of the implanted deuterium dose results in the deuterium desorption temperature range extension towards the temperature decreasing (Figure 2d, curves 3 and 4). According to the data obtained the implanted deuterium atoms occupy at first the composite traps having a highest temperature.

Further decrease of the vanadium concentration in the composites do not produce qualitative changes in the deuteron desorption spectrum. There are observed only insignificant variations of the maximum temperature values for the low-temperature peak at ~300 K (Figure 2d and 2e). An additional low-temperature peak in the thermal desorption spectrum, which becomes prevailing with the implanted deuterium dose increasing and reaches the maximum desorption temperature of the main low-temperature peak ~300 K, indicates on the significant temperature decreasing during deuterium desorption from the Mg-V composites with the vanadium concentration less than 17 at.%.

An important feature of the spectra of deuterium thermal desorption from $Mg_{86+x}V_{14-x}$ composites is the fact that the main amount of implanted deuterium is releasing at a room temperature. In addition, there is a high-temperature low-intensity wide region of deuteron desorption comprising indistinct peaks. To all appearance, the composites with a vanadium concentration below 17 at.% are in the amorphous state that is characterized by the implanted deuterium desorption spectrum in the form of a continuous wide temperature region of deuterium desorption between 400 K and 800 K. X-ray structure analysis has shown that $Mg_{83+x}V_x$ composites are amorphous. The X-ray diffraction patterns demonstrate only intense lines from the molybdenum substrate and the background.

The deuterium temperature decrease can be caused by filamentary inclusions formed, in the process of composite making and annealing, by the insoluble component (vanadium) atoms providing the deuterium diffusion from the sample at a lower temperature (channels for deuterium diffusion through the surface barrier). A necessary high diffusion mobility of deuterium is provided by the amorphous state of $Mg_{83+x}V_x$ samples.

3. Conclusions

The present paper shows that by introducing vanadium into magnesium the deuterium desorption temperature can be significantly decreased (to 300-330 K) as compared to the case of magnesium samples (~800 K).

A step-like form of the curve of thermal activated deuterium desorption temperature evidences on the presence of two different structure states of the Mg-V system depending on the ratio of components.

The deuterium temperature decrease can be caused by filamentary inclusions formed, in the process of making and annealing, by the insoluble component (vanadium) atoms providing the deuterium diffusion from the sample at a lower temperature (diffusion channels for deuterium through the surface barrier). A necessary high diffusion mobility of deuterium is provided by the amorphous state of $Mg_{83+x}V_x$ samples.

The deuterium desorption data obtained using Mg-Zr [12] and Mg-V composites provide support for further research into the hydrogen storage materials containing low-soluble chemical elements in the alloy components.

Acknowledgement

This study has been done with financial support to the program at the National Academy of Sciences of Ukraine "Hydrogen in the alternative energy and new technologies" (Project #27-12).

References

[1] Jain I, Lal C, Jain A 2010 *Int. J. Hydrogen Energy* **35** 5133
[2] Sakintuna B, Lamari-Darkrim F, Hirscher M 2007 *Int. J. Hydrogen Energy* **32** 1121
[3] Sohn H, Emami S 2011 *Int. J. Hydrogen Energy* **36** 8344
[4] Chen M, Yang X-B, Cui J, Tang J-J, Gan L-Y, Zhu M, Zhao Y-J 2012 *Int. J. Hydrogen Energy* **37** 309
[5] Kwon S, Baek S, Mumm D, Hong S, Song M 2008 *Int. J. Hydrogen Energy* **33** 4586
[6] Checchetto R, Bazzanella N, Miotello A, Mengucci P 2007 *J. Alloys Compd.* **446-447** 58
[7] Ershova O, Dobrovolsky V, Solonin Yu, Khyzhun O, Koval A 2008 *J. Alloys Compd.* **464** 212
[8] Varin R, Czujko T, Wronski Z 2009 *Nanomaterials for Solid State Hydrogen Storage* (Springer)
[9] Jurczyk M, Nowak M, Szajek A, Jezierski 2012 *Int. J. Hydrogen Energy* **37** 3652
[10] Chen X, He Y, Zhao Y and Wang X 2010 *Nanotechnology* **21** 8
[11] 1999 *State diagrams of double metal systems (in Russian)* vol 3-1 Ed. N.P. Lyakishev (Moscow: Mashinostroyenie) pp 340-341
[12] Neklyudov I, Morozov O, Kulish V, Zhurba V, Lomino N, Ovcharenko V, Kuprin O 2011 *IOP Conf. Ser.: Mater. Sci. Eng.* **23** 012028

IOP Conf. Series: Materials Science and Engineering **38** (2012) 012062 doi:10.1088/1757-899X/38/1/012062

EPM Fine-Disperse Platinum Coating on Powder Carriers

V. Serga, L. Kulikova, A. Cvetkov and A. Krumina
Institute of Inorganic Chemistry, Riga Technical University, Latvia
vera_serga@inbox.lv

Abstract. In the reported investigation the extractive-pyrolytic method of fine-disperse platinum coating on powder carriers was applied. Nanopowders of Al_2O_3, γ- AlO(OH), Y_2O_3, CeO_2, SiO_2 were used as carriers. Investigations on the effect of synthesis parameters on the mean size of platinum crystallites in the produced composites (metal content 4.8 wt%) have revealed that the increase of the pyrolysis temperature, annealing period, metal concentration in the precursor $[(C_8H_{17})_3NH]_2PtCl_6$ in toluene as well as the decrease of the specific surface area result in growth of the mean size of platinum crystallites. Microscopic studies show the formation of platinum spherical particles sized 5 to 35 nm as a results of the pyrolysis of the platinum-containing precursor in a water-soluble carrier (fine-disperse NaCl).

1. Introduction

Platinum has been widely used in many applications, especially for catalysis and fuel cell technology. To coat platinum nanoparticles on various carriers, different methods of synthesis, such as impregnation [1] and incipient wetness impregnation [2], metal-organic chemical vapour deposition [3], micro-emulsion [4], surfactant-stabilized colloidal [5] and others are used. In this paper, a novel route–extractive-pyrolytic method (EPM) for fine-disperse platinum coating on powder carriers is proposed. Advantages of the EPM for the synthesis of oxide materials of various applications are demonstrated for metal extraction systems with monocarboxylic acids [6]. With reference to the extraction systems with anion-exchange extractant tri-n-octylamine $(C_8H_{17})_3N$, we studied the possibilities of the EPM to produce palladium nanoparticles on α-Al_2O_3 micro granules [7] and platinum nanocrystalline films on glass substrates [8].

2. Experimental

A solution of tri-n-octylamine $(C_8H_{17})_3N$ in toluene was used to produce an organic extract of platinum (tri-n-octylammonium hexachloroplatinate $[(C_8H_{17})_3NH]_2PtCl_6$ in toluene - precursor). The method for the production of metal-containing organic extract is described in [9]. Thermogravimetric and thermal measurements have shown [8] that the process of thermal decomposition of the precursor has three stages at temperatures of 210, 230 and 270 ^{0}C and that the mass losses cease at 280 ^{0}C.

In order to coat platinum particles, a carrier sample was dispersed in a solution of the precursor, then it was dried and thermally treated by heating from room temperature until 300 ^{0}C and by annealing for 5 min in air. As carriers, we used plasma processed nanopowders of Al_2O_3 and Y_2O_3, commercial nanopowders of γ-AlO(OH), CeO_2, SiO_2 and nanopowder of γ-Al_2O_3 produced by thermal treatment of γ-AlO(OH) at t = 700 ^{0}C within 2 hours. The platinum content in the composites was determined using an X-ray spectrometer S4 Pioneer (Bruker AXS GmbH).

The produced composites were analyzed by X-ray diffraction (XRD) using a D-8 Advance (Bruker AXS) diffractometer with CuK_α radiation (λ = 1.5418 Å) in a wide range of Bragg angles ($10^o < 2\theta < 75^o$) with a scanning rate of 0.02 ^{o}/s at room temperature. In order to get more accurate Pt crystallite size information, slow scans (0.001 ^{o}/s) were performed in the range of 2θ corresponding to one of the most intensive diffraction peaks of platinum [(111), (200),

(220)], which has been selected with reference to the carrier X-ray image. TEM measurements were made using JEM-1230 operating at 100 kV.

3. Results

The XRD phase analysis of the platinum-containing composites produced on the nanopowder of Al_2O_3 has shown that the peaks for Pt overlap, at least partially, with the carrier at all diffraction angles that makes the definition of accurate value of the mean size of platinum crystallites (d_{Pt}) at a small content of the metal in the composite ($\leq$ 2.4 wt.%) complicated (Fig. 1, curves 1, 2, 3). Therefore, for the mentioned carrier, d_{Pt} was defined only for the composites with the 4.8 wt.% Pt content using the most intensive peak (111).

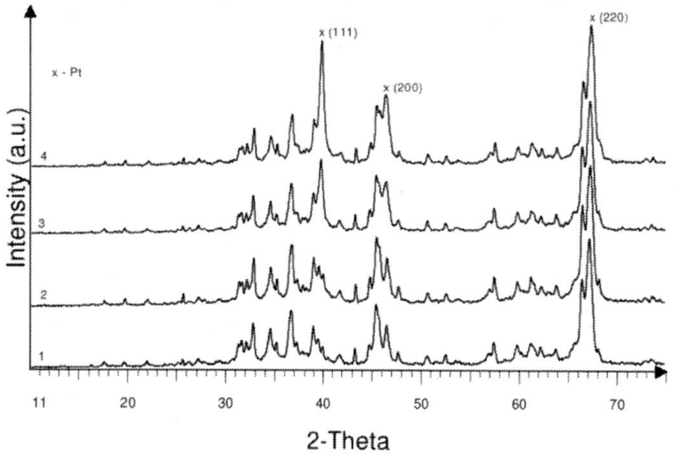

Figure 1. XRD patterns : 1 - Al_2O_3; 2 - Al_2O_3/ Pt 1.2 wt.%; 3 - Al_2O_3/ Pt 2.4 wt.%; 4 - Al_2O_3/ Pt 4.8 wt.%.

While processing X-ray images of the composites produced with Y_2O_3, we used the peak (220) and the peak (111) with CeO_2. The X-ray phase analysis has revealed that the selected peaks do not overlap with the carrier peaks, but they exhibit a low intensity, which does not allow to define d_{Pt}, with the metal content being less than 4.8 wt.%.

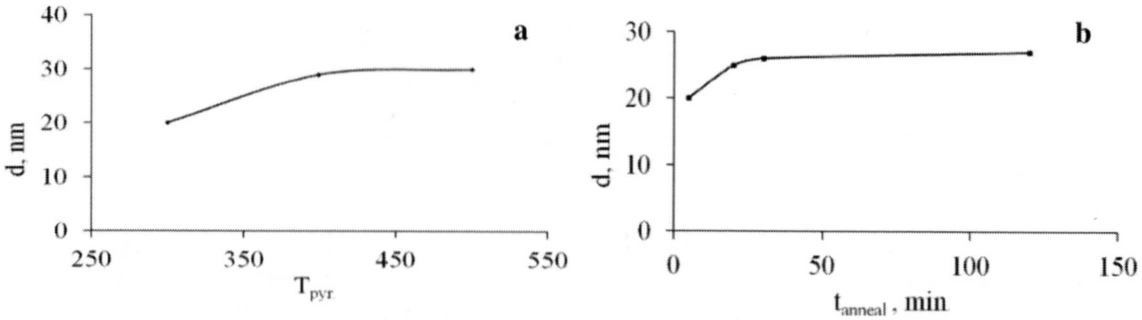

Figure 2. Dependence of d_{Pt} in Al_2O_3 /Pt 4.8 wt.% composites on: a) pyrolysis temperature (annealing duration 5 min), b) annealing period (pyrolysis temperature 300 °C).

The influence of the pyrolysis temperature (Fig. 2a) and annealing period at a constant temperature of pyrolysis (Fig. 2b) for platinum coating on the Al_2O_3 nanopowder was studied. The presented data evidence that the pyrolysis temperature rise from 300 ^{0}C to 500 ^{0}C results in a more considerable growth of d_{Pt} than the increase of the annealing period from 5 to 120 min at the pyrolysis temperature 300 ^{0}C.

With reference to the obtained results and thermogravimetric measurements [8], the further investigations were performed under the following pyrolysis conditions: t_{pyr} = 300 ^{0}C, the annealing period 5 min.

Amorphous SiO_2 (the specific surface area (SSA) is 190 m^2/g) was used as a carrier when investigating the influence of the Pt concentration in the precursor solution on d_{Pt}. With the concentration of the metal being reduced, the observed diffraction peak (111) broadening (Fig. 3a) evidences of the d_{Pt} decrease (Fig. 3b).

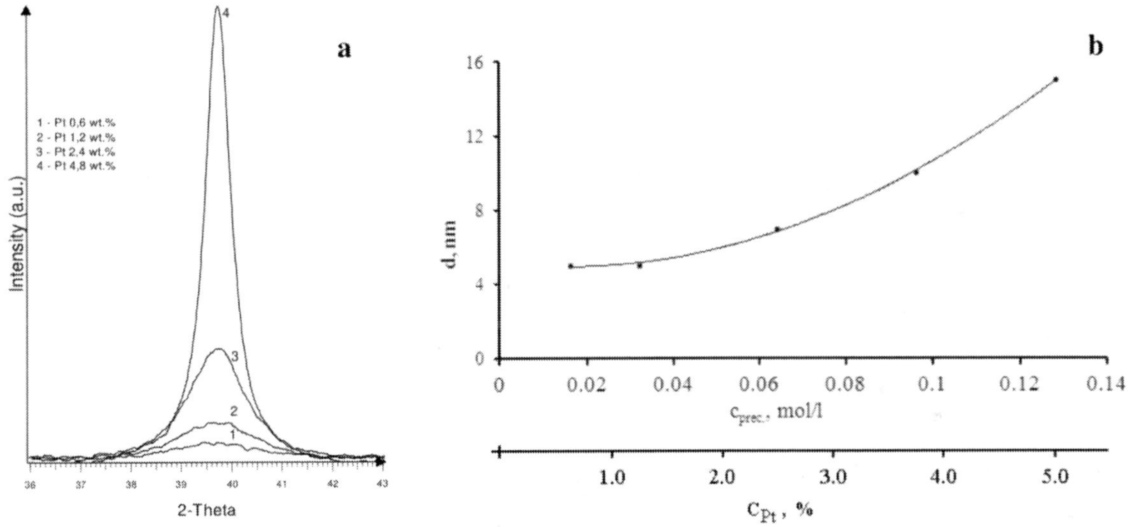

Figure 3. Diffraction peak (111) of SiO_2 /Pt composites with different metal concentrations (a), and the influence of metal content in the composite on d_{Pt} (b).

As a rule, carriers with a larger SSA are used to produce nano-sized platinum particles [1-5]. We have shown (see Table 1) that when applying the EPM, even if the metal concentration is relatively high (4.8 wt.%), a considerable decrease of d_{Pt} can be achieved by enlarging the SSA of the carrier.

Table 1. Dependence of the mean size of platinum crystallites on the carrier SSA in composites with the 4.8 wt.% platinum content.

Carrier	Al_2O_3 *	γ-Al_2O_3 **	γ-AlO(OH) *
SSA, m^2/g	41	122	145
d_{Pt}, nm	20	10	5

To calculate d_{Pt}, * peak (111) and ** peak (220) were used.

On the other hand, the SSA of the nanopowders of CeO_2 and Y_2O_3 is relatively small, respectively, 26 m^2/g and 21 m^2/g; small particles of the metal are also formed (d = 9-10 nm). Probably, in this case, some part of the metal interacts with the carrier that results in decrease of the metal concentration on the surface.

The microscopic studies of the samples have shown that it is impossible to correctly define the dispersion of the platinum particles in the presence of the carrier nanoparticles, since the metal particles and the carrier particles are difficult to distinguish. Therefore, we conducted the pyrolysis of the platinum-containing precursor on a water-soluble carrier (NaCl), after removal of which separate particles of the metal were obtained. The presented micrograph (see Fig. 4) shows that the particles are spherically-shaped and their size ranges from 5 to 35 nm. Thus, the size of the platinum particles on this carrier varies in a wide enough range.

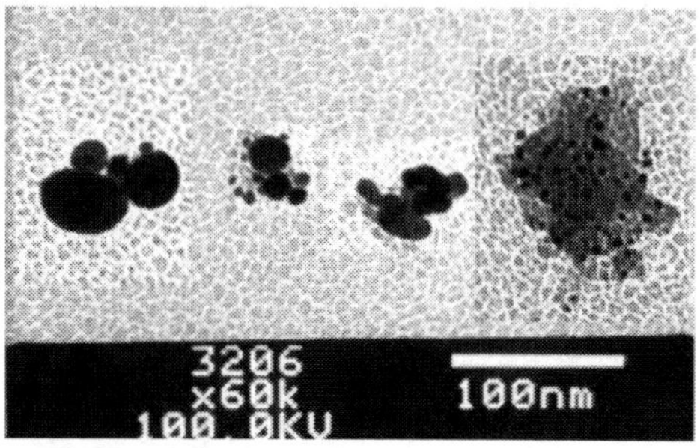

Figure 4. Micrographs of Pt nanoparticles produced after removal of the carrier from the composite NaCl/Pt 4.8 wt.% (SSA_{NaCl} = 9 m^2/g). XRD analysis has provided d_{Pt} = 24 nm.

4. Conclusions

The investigations on the influence of the EPM parameters on the mean size of platinum crystallites in the produced composites have shown that the increase of the pyrolysis temperature and the longer annealing period as well as the increase of metal concentration in the precursor solution and the reduction of the carrier SSA, with the metal content in the composite being high (4.8 wt.%), result in growth of d_{Pt} in the composites. Microscopic studies have revealed the formation of spherical platinum particles sized 5-35 nm during the pyrolysis of a platinum-containing precursor on a water-soluble carrier (NaCl).

Acknowledgments. This work was supported by ERAF project No. 2DP/2.1.1.1.0/10/APIA/VIAA/087.

References

[1] Job N, Pereira M F R, Lambert S, Cabiac A, Delahay G, Colomer J F, Marien J, Figueiredo J L and Pirard J P 2006 *J. Catal.* **240** 160-171
[2] Min M, Park C, Kim H, Kwak C, Serov A A, Kweon H and Lee S 2006 *Electrochim.Acta* **52** 1670–1675
[3] Lashdaf M, Lahtinen J, Lindblad M, Venäläinen T and Krause A O I 2004 *Appl.Catal. A: General* **276** 129–137
[4] Brimaud S, Coutanceau C, Garnier E, Léger J-M, Gérard F, Pronier S and Leoni M 2007 *J. Electroanalyt.. Chem.* **602** 226–236
[5] Lim D-H, Lee W-D, Choi D-H, Park D-R and Lee H-I 2008 *J. Power Sources* **185** 159–165
[6] Kholkin A I and Patrusheva T N 2006 *Extractive pyrolytic method* (Com Book, Moscow) (in Russian)
[7] Serga V, Kulikova L, Palcevskis E, Maiorov M and Krumina A. 2009 *Proc. Int. Conf. Functional Materials and Nanotechnologies* (Riga: Latvia), p. 158
[8] Serga V, Kulikova L, Grekhov V, Kalnacs J, Murashov A and Vilken A. 2011 *Proc. Int. Conf. Functional Materials and Nanotechnologies* (Riga: Latvia), p. 245
[9] Serga V, Kulikova L, Maiorov M, Krumina A and Heidemane G. 2009 *Latv. Kim. Zurn.* **4** 288-296.

The measurement of impurities in silicon for solar cell production

G Kolosovska[1], A Viksna[1,3], G Chikvaidze[2], A Osite[1], A Opalais[1]

[1]Faculty of Chemistry, University of Latvia, Kr. Valdemara Str. 48, Riga, LV-1013, Latvia

[2]Institute of Solid State Physics, University of Latvia, Kengaraga Str. 8, Riga, LV-1063, Latvia

E-mail: arturs.viksna@lu.lv

Abstract. The power conversion efficiency of solar cells is largely dependent on impurity levels in silicon. In the current investigations two sample preparation methods of silicon had been used before analysis – acid dissolution at atmospheric pressure and pressured microwave assisted technique. Quantification of impurities in solar silicon was done by inductively coupled plasma mass spectrometry and electrothermal atomic spectrometry. Microwave assisted dissolution of solar silicon was more effective compared to the traditional dissolution method on the hot plate, but complete dissolution of impurities and silicon matrix was achieved with a nitric and hydrofluoric acid mixture. The mass of solar silicon for the microwave dissolution is about 0.5 g and optimal ratio of HF and HNO_3 acids mixture is 3 to 5.

1. Introduction

Solar energy has the potential to provide most of the primary energy for the global requirements. It is an alternative source to fossil fuels, but this clean energy system requires high initial costs related to the solar cells. Silicon is used as a raw material for 95% of the solar cell module. Polycrystalline silicon is used in 60% of all cases. About 25 to 50% of the solar cell cost comes from the silicon. Since the power conversion efficiency is largely dictated by the impurity levels, there is a high demand for silicon and emphasis is being directed to cheaper processes for supplying silicon for the growing solar cell market.

Several investigations have been performed for determining the impurities in solar grade purity silicon during the last decade. Silicon is a complicated material, which locks up impurities in its crystal lattice, so the assessment of impurities can be difficult. Some volatile compounds can be produced during sample preparation and this provides a problem for determining the purity of silicon. The actual mass concentration of the impurities is hard to determine due to the low concentration level and the effect of the silicon matrix on analyzed elements using advanced analysis methods, such as inductively coupled plasma mass spectrometry (ICP-MS).

The main aim of current study is to evaluate the dissolution behavior of impurities in solar silicon for the fractionation studies.

[3] To whom any correspondence should be addressed

The impurities of solar purity grade silicon

The type and content of impurities in silicon largely depends on the choice of method for producing silicon. Metallic impurities found in silicon ingots can cause different defects such as a bad coating, gaps, leaks; also fragility and overload. It is very important to distinguish the primary impurities and those arising from the treatment of silicon. Metallic impurities affect the electrical properties, the diffusion length of the conductor and the efficiency of the solar cells. For instance, rapid temperature increase, which is followed by rapid refrigeration, can cause the deliquescence of metal clusters, which reduces the efficiency of batteries. The impurities can further lower the solar cell efficiency by up to 20% that no longer makes the process economically viable [1, 2].

Phosphorus and boron affects the properties of solar silicon more than other chemical elements. The segregation coefficients of these elements are rather similar (k_B=0.7-0.8 and k_P=0.35), which means that these elements are collected in silicon crystals and are not removed during crystallization. It is known that Mo, Nb, Zr, W, Ti, V, Cr, Mn, Fe and Co have a major effect on the solar cell efficiency, while Al, Ni and C have very small effect [3, 4].

Methods of silicon purification

Electron beam melting of silicon can effectively purify silicon from impurities. It completely excludes trichlorosilane technology from the production process of solar silicon that reduces the cost of production of solar silicon and makes its production safe for humans and for the environment.

Metallurgical silicon with a purity of 99.95 % was received from "KazSilicon" (Kazakhstan).

The analyzed silicon samples were purified at "Poltava Quartz Glass" plant, Poltava (Ukraine) using the electron-beam method. Silicon was remelted in a vacuum furnace with the help of powerful electron beam; afterwards it was crystallized with a casting, which provides a possibility to obtain renewed polycrystalline silicon.

Two methods of silicon purification were applied. In the first method samples were purified by electron-beam remelting in high vacuum (vacuum refining - LG). The alternative method involved purification by oxidative refining (in oxygen - AK-10-F) as well as electron-beam melting. The concentrations of impurities in all samples are unknown in the current investigations. It is necessary to note that reference materials for solar silicon are not available. It is an additional problem for the development of quantification method. Preliminary work showed that solubility of silicon depends on the purification method. This can be attributed to the polycrystalline nature of silicon. It was observed that silicon used in the current study is not soluble at room temperature in the nitric and hydrofluoric acid mixture.

2. Experimental

Two methods have been applied for the analysis of silicon samples – acid dissolution at atmospheric pressure and acid dissolution using the closed microwave assisted technology. For many analysis methods silicon samples should be dissolved. Since silicon is soluble only in HF acid, two different methods of quantification of impurities were used.

Acid dissolution at atmospheric pressure.

Initial dissolution of 0.5 g of silicon powder samples in 5 mL 65 % HNO_3 (Suprapur, Merck) was performed at 130°C on the hot plate in the clean box. After 24 hours 5 mL 65 % HNO_3 and 30 % H_2O_2 (Suprapur, Merck) acid mixture (1:1) was added and dissolution was performed again. The dissolution of impurities in the silicon matrix was done in a semi-closed quartz container for 48 hours.

Dissolution with microwave assisted technology.

To 0.5 g of silicon powder samples 5 mL 65 % HNO_3 and 2 mL 30 % H_2O_2 was added (first method). To dissolve the silicon matrix (second method), 3 mL of 40 % hydrofluoric acid (Suprapur, Merck) was added. An Anton Paar 'Multiwave 3000' microwave digestion system was used for dissolution of samples. The temperature was set up to 130 °C and maximum microwave power was applied. Process was done in the pressure resisted 100 mL teflon (PTFE) vessels.

3. Results and Discussion

The silicon matrix influences the quantification of different samples. Most of analytical quantification methods require converting solid samples into liquid state to get representative results.

Hydrofluoric acid is the only reagent which completely dissolves silicon. There are two possibilities to quantify impurities in solar silicon samples. The first approach dissolves the impurities in nitric acid and ignores the matrix effects for the quantification by ICP-MS or any other possible analytical technique. Second method completely dissolves the solar silicon samples, but in this case it is necessary to use a nitric and hydrofluoric acid mixture. In the second case it is necessary to take into account the possible interference from the matrix, Si from solar silicon and F from hydrofluoric acid.

The experimental results show that dissolution of impurities in nitric acid is unsuitable for the total quantification of impurities in the solar silicon due to formation insoluble silicon compounds with trace elements or incorporation of impurities in the silicon crystalline lattice during purification. Experimental results of solar silicon dissolution process in nitric acid at atmospheric pressure on the hot plate and pressured microwave system are presented in Fig. 1 and 2 respectively. Comparing dissolution efficiencies it is seen that dissolution using a closed microwave method is more effective. After repeated dissolution with nitric acid the obtained normalized mass concentrations using closed microwave technique were 2 to 10 time smaller compare to dissolution at atmospheric pressure on the hot plate.

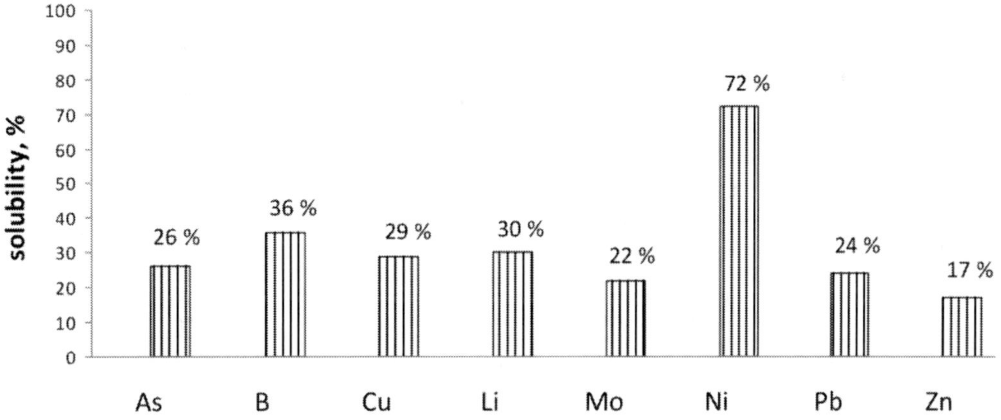

Fig. 1. Repeated extraction of impurities from solar silicon (LG) in nitric acid at atmospheric pressure on a hot plate at 130 °C.

The dissolution process in the microwave system after immersion in nitric acid is not complete. To check the extent of dissolution, hydrofluoric acid was added to nitric acid to further dissolve the silicon lattice.

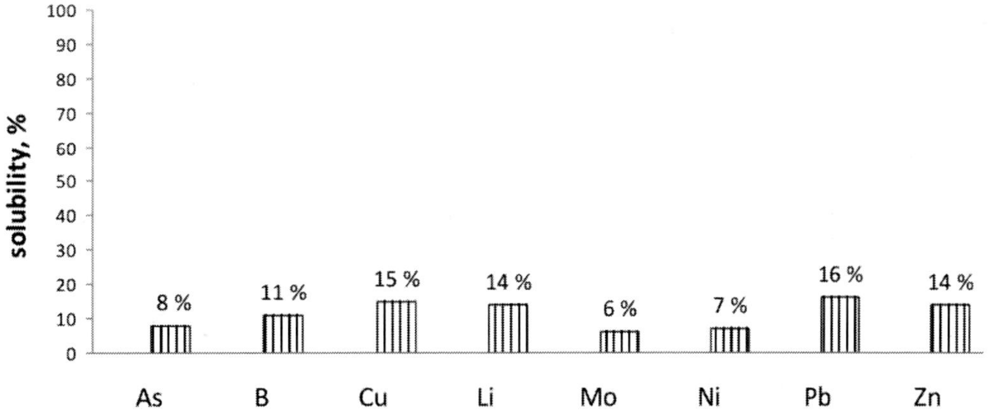

Fig. 2. Repeated extraction of impurities from solar silicon (LG) in nitric acid under pressure in a closed microwave system.

The optimal combination of nitric and hydrofluoric acid for the complete dissolution of 0.4 - 0.6 g of solar silicon is 5 to 3. Analysis results are summarized in Figure 3. The accuracy of ICP-MS method was confirmed by the results of electrothermal atomic absorption spectrometry (ETAAS), flame atomic absorption spectrometry (FAAS), flame photometry and spectrophotometry.

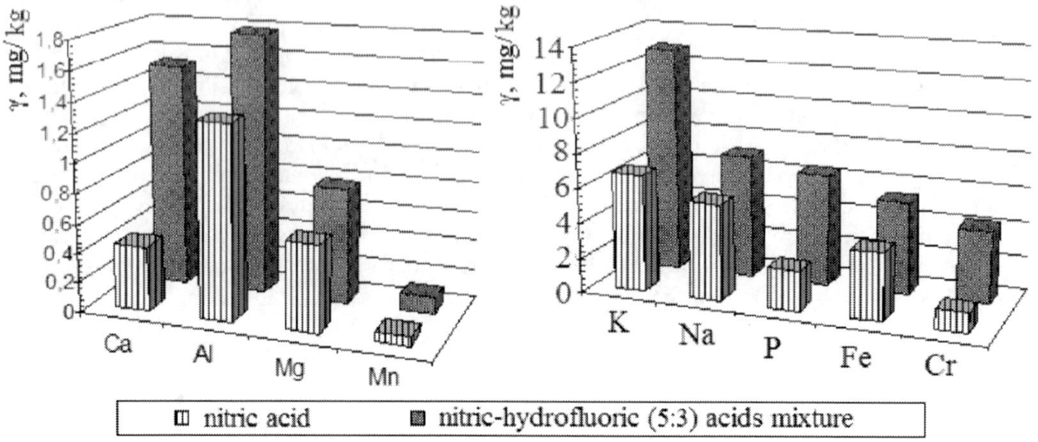

Fig. 3. Comparison of mean mass concentrations in solar silicon (AK-10-F) samples after dissolution in nitric acid and nitric-hydrofluoric (5:3) acids mixture in a closed microwave system.

4. Conclusions

The ability to determine the concentration of impurity elements with ICP-MS depends on sample preparation method. The first method preferentially dissolved the impurities from solid silicon in nitric acid. Use of pressured microwave assisted dissolution method is more effective than immersion into nitric acid heated on a hot plate. A comparison of the extraction and complete dissolution of silicon reveals that a part of the impurities remain locked in the silicon lattice or silicon compounds, but another part is found in the fraction that can be dissolved by nitric acid. The concentration of impurity elements depends on the process of silicon manufacturing.

References

[1] Buonassisi T, Heuer M, Istratov A A, Picket M D, Weber E R 2006 *Mater. Sci. Eng.*1-5
[2] Rein S, Kwapil W, Geilker J, Emanuel G, Spitz M, Reis I, Weil A, Biro D, Glatthaar M, Soiland A K, Enebakk E, Tronstad R 2009 *European Photovoltaic Solar Energy Conference and Exhibition* **8**
[3] Hanazawa K, Yuge N, Kato Y 2004 *The Japan Institute of Metals* 2004 **3** 6.
[4] Giannattasio A, Giaquinta A, Porrini M 2011 *Phys. Status Solidi A* **208** 564

Acknowledgements

This work was supported by ERAF Project Nr.2010/0245/2DP/2.1.1.1.0/10/APIA/VIAA/114

IOP Conf. Series: Materials Science and Engineering **38** (2012) 012064 doi:10.1088/1757-899X/38/1/012064

Properties of the Nafion membrane impregnated with hydroxyl ammonium based ionic liquids

Valeriy Garaev[1], Janis Kleperis[2], Sanita Pavlovica[1] and Guntars Vaivars[1,2,3]

[1] Department of Chemistry, University of Latvia, Riga, Kr. Valdemara Str. 48, Latvia, LV-1013

[2] Institute of Solid State Physics, University of Latvia, Riga, Kengaraga Str. 8, Latvia, LV-1063

E-mail: Guntars.Vaivars@cfi.lu.lv

Abstract. In this work, the Nafion 112 membrane impregnated with nine various hydroxyl ammonium based ionic liquids have been investigated. The used ionic liquids were combined from hydroxyl ammonium cations (2-hydroxyethylammonium/HEA, bis(2-hydroxyethyl)ammonium/BHEA, tris(2-hydroxyethyl)ammonium/THEA) and carboxylate anions (formate, acetate, lactate). The membranes are characterized by conductivity and thermal stability measurements. It was found, that almost all composites have 10 times higher ion conductivity than a pure Nafion 112 at 90 °C in ambient environment due to the higher thermal stability. The thermal stability of Nafion membrane was increased by all studied nine ionic liquids. In this work, only biodegradable ionic liquids were used for composite preparation.

1. Introduction

The Nafion membrane is one of the most common and commercially available proton conducting membrane. Also, it is characterized by significant thermal, mechanical and chemical stability. Due to its good properties, Nafion membrane is widely used in fuel cells and other electrochemical applications. However, the pure Nafion membrane' working temperature does not exceed 100 °C since the ion conductivity is very sensitive to the water content changes. There are several options to solve the problem, e.g., it is possible to produce the composite membranes using zirconium phosphate [1], heteropolyacids or inorganic oxides [2] and stabilize the water content. Recently, due to their excellent chemical, electrical and thermal properties the room temperature ionic liquids are used to impregnate the polymer membranes. The application of ionic liquids containing heterocyclic cations, such as imidazolium cations [3] and pyrrolidinium cations [4], with fluorinated anions is described by other authors. These ionic liquids are not very safe from the toxicology view [5].

In this work, we are using hydroxyl ammonium based ionic liquids with the carboxylate anion. In previous researches it was found that these ionic liquids are highly biodegradable and practically non-toxic [6, 7].

[3] To whom any correspondence should be addressed

2. Experimental

2.1. Materials

Nafion 112 (equivalent weight 1100 g/mol, thickness 50.8 μm) was purchased from Ion Power. The ionic liquids 2-hydroxyethylammonium formate (HEA-F), acetate (HEA-Ac), lactate (HEA-L); bis(2-hydroxyethyl)ammonium formate (BHEA-F), acetate (BHEA-Ac), lactate (BHEA-L); tris(2-hydroxyethyl)ammonium formate (THEA-F), acetate (THEA-Ac), lactate (THEA-L) (for chemical structures see figure 1) were synthesized by reaction of hydroxyethyl amines with formic, acetic or lactic acids by the method described elsewhere [6, 7]. Hydroxyethyl amines and acids (Aldrich) were distilled in vacuum before use. The ionic liquids purity is ≥98% for lactates and ≥99% for formates and acetates (by HPLC/MS).

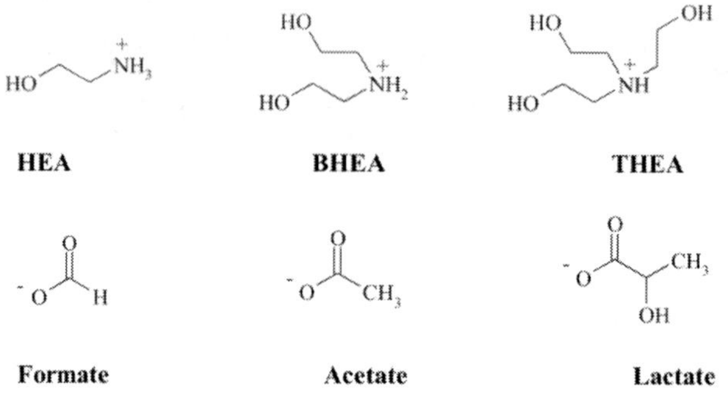

Figure 1. Chemical structures of used ionic liquids cations and anions

2.2. Impregnation with ionic liquids

Nafion samples with dimensions of 1.5 cm x 2.0 cm were used for composite fabrication. These samples were pretreated by sequential immersion in boiling 0.5 mol L^{-1} H_2SO_4 solution and then in boiling deionized water. After that membrane was blotted with a tissue to remove any residual liquid. Impregnation with ionic liquids was performed by immersing the membrane samples into the corresponding ionic liquid at 50-55 °C temperature for 90 minutes. After impregnation the membranes surface was blotted with a tissue. Membranes were afterwards dried few days at a room temperature. Thermal analysis showed that about 1 – 3 wt. % of water is left in structure. The amount of absorbed ionic liquid per sulfonic acid group λ can be calculated by equation (1) [4]:

$$\lambda = \left(\frac{m_{IL}}{M_{IL}}\right) \times \left(\frac{M_{eq}}{m}\right) \quad (1)$$

where m_{IL} – absorbed ionic liquid weight; M_{IL} – the molecular weight of ionic liquid; M_{eq} – the molecular weight of the Nafion 112; m – the weight of the pure Nafion membrane.

2.3. Conductivity measurements

Conductivity measurements of membranes were made by in-plane method with four platinum electrodes in BekkTech conductivity cell at temperature range from 30 °C to 90 °C. During measurements the current was kept constant and controlled by Voltalab PGZ 301 (Radiometric Analytical). Voltage was determined between two inner electrodes with DVM850BL (Velleman) multimeter. The conductivity can be calculated by equation (2):

$$\sigma = \frac{L \times I}{U \times W \times T} \quad (2)$$

where L – the distance between two inner electrodes, which is fixed at 0.425 cm; I – the amperage; U – the voltage between two inner electrodes; W – the membrane sample width; T – the membrane thickness.

2.4. Thermogravimetry
The thermal analyses of modified membranes have been performed by Shimadzu (DTG-60) in argon atmosphere with a heating rate of 10 °C min^{-1} and a gas flow rate of 50 mL min^{-1} at temperature range from 30 °C to 400 °C. Analyses of ionic liquids were performed at temperature range from 30 °C to 300 °C.

3. Results and Discussion

3.1. Impregnated membranes
The impregnated Nafion membranes are characterized by uptake of ionic liquid (Fig. 2). The uptake of ionic liquids with HEA cations has the same values – 2.20-2.32 mmol IL/mmol SO_3H. The uptake with DHEA cations is higher and it is about 3 mmol IL/mmol SO_3H. The highest value of impregnated ionic liquids has THEA-F. The uptake value increases in HEA-DHEA-THEA formates with the increase of cation geometric size. On the other hand, the uptake of ionic liquids with lactate anions does not increase and are found to be between 2.2 and 2.3 mmol IL/mmol SO_3H.

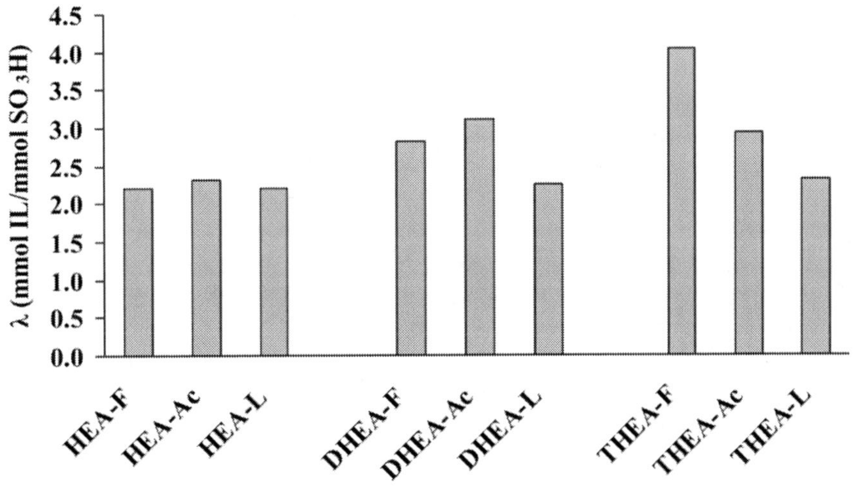

Figure 2. Impregnated amount of hydroxyl ammonium based ionic liquids into Nafion 112 at 50-55 °C for 90 minutes.

3.2. Conductivity
The ion conductivities of pure Nafion 112 and those of the impregnated with ionic liquid are shown in Arrhenius plot (Fig. 3). The conductivity of pure Nafion 112 decreases by 10 times increasing temperature from 33 °C to 92 °C due to the water desorption from the membrane. All ionic liquid containing composites increase their ion conductivity 3-10 times in the same temperature range. Except DHEA-Ac, all composites show higher conductivity than pure Nafion at 90 °C. Composite with HEA-F shows conductivity increase by two orders due to the high activation energy. Schmidt et al. [4] reported that the conductivity increases with increasing amount of the ionic liquid in the membrane, therefore, it is difficult to compare the membranes conductivity, because the ionic liquid uptake is slightly different for different membranes. The activation energy of the ion conducting process was determined from Arrhenius plot and it is in a range from 0.11 eV to 0.36 eV, which is characteristic range for solid electrolytes. The composite membrane with HEA-F showed higher activation energy (1.06 eV) and it might be related to different conductivity mechanism.

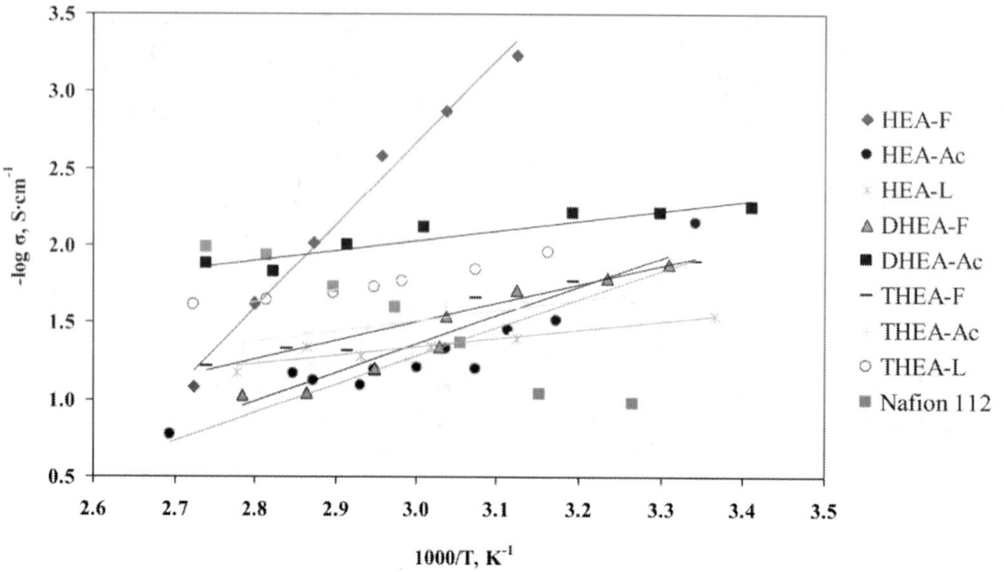

Figure 3. The ion conductivities of pure Nafion 112 and of Nafion 112 impregnated with hydroxyl ammonium based ionic liquids in Arrhenius plot.

3.3. Thermal analysis

The thermogravimetric curves of pure Nafion 112, pure HEA-F and composite of Nafion + HEA-F are shown on Fig. 4. The HEA-F starts to evaporate at 140 °C. The other ionic liquids used in our work have similar evaporation temperatures. The first weight loss is between 30 °C and 110 °C, which corresponded to the loss of water. Water content in composite is from 1.5% to 3%, but this content is smaller than in pure membrane. The second weight loss in composite is from 140 °C to 210 °C. This temperature range corresponds with the HEA-F evaporation. After this weight loss from composite remains only Nafion matrix. Pure Nafion membrane begins degradation process at 300 °C but composite heavily loss weight at 350 °C. Figure 5 shows comparison between pure Nafion and composites of Nafion and other ionic liquids. All composites have higher thermal stability versus pure Nafion and degrade at 350 °C. So ionic liquids decrease the water content in Nafion membrane and affect as plasticizers, increasing the thermal stability of Nafion matrix.

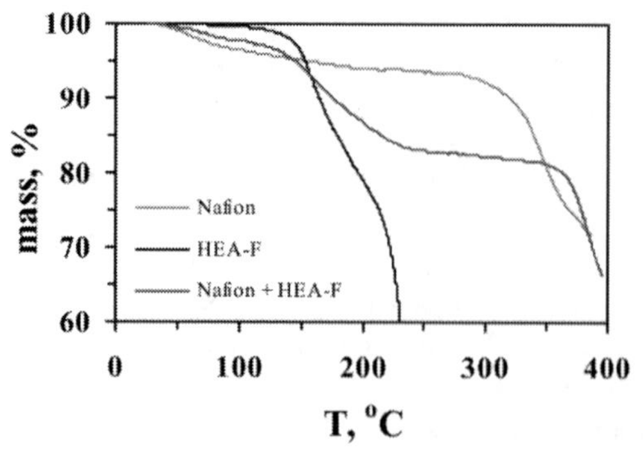

Figure 4. TGA curves of Nafion 112, Nafion

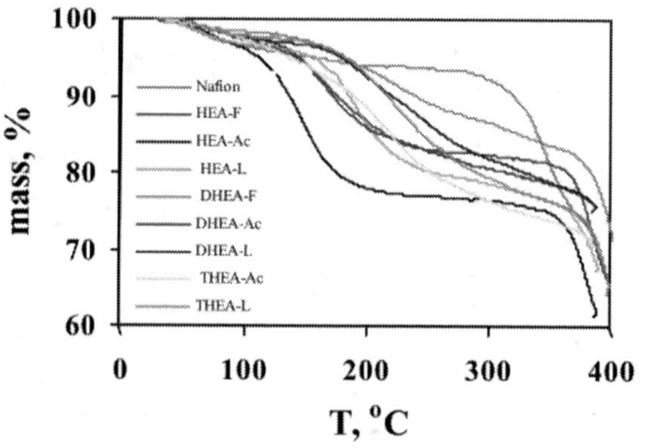

Figure 5. TGA curves of Nafion 112 and

112 + HEA-F, HEA-F (IL). modified membranes.

4. Conclusions

All studied composite Nafion 112 membranes impregnated with different hydroxyl ammonium ionic liquids showed similar behavior. The conductivity decrease was not observed in a temperature range from 30 °C to 90 °C in ambient environment due to the potential water loss. Almost all composites have higher ion conductivity than a pure Nafion 112 at 90 °C. All ionic liquids increased the thermal stability of Nafion membrane approximately by 50 °C. The composite membrane of Nafion with ionic liquids has potential to be used in ionic devices at increased temperatures, especially acetates. However, the conductivity in a temperature range over 90 °C also should be studied. At the same time, different ionic liquids with higher temperature stability should be selected for composite preparation. Such materials will find application in medium range PEM fuel cells. In this work, only biodegradable ionic liquids were used for composite preparation.

References

[1] Yang C, Costamagn P, Srinivasan S, Benziger J and Bocarsly A B 2001 *J. Power Sources* **103** 1
[2] Shao Z, Xu H, Li M and Hsing I 2006 *Solid States Ionics* **177** 779
[3] Yang J, Che Q, Zhou L, He R and Savinell R F 2011 *Electrochimica Acta* 56, 5940
[4] Schmidt C, Glück T and Schmidt-Naake G 2008 *Chem. Eng. Technol.* **31** 13
[5] Pham T, Cho C and Yun Y 2010 *Water Research* **44** 352
[6] Zicmanis A, Pavlovica S, Gzibovska E, Mekss P and Klavins M 2010 *Latvian J. Chem.* **3** 269
[7] Pavlovica S, Zicmanis A, Gzibovska E, Klavins M and Mekss P 2011 *Green and Sustainable Chemistry* **1** 103

Acknowledgments

Valeriy Garaev is grateful to the European Social Fund (ESF) for the financial support of this research.

AUTHOR INDEX

Abuova, F U ... 210
Akilbekov, A ... 205
Akilbekov, A T ... 210
Akylbekov, A T ... 199
Aleksejeva, J ... 146
Aleksejeva, L ... 190
Aleksiejunas, R ... 265
Anan'Ev, A ... 214
Andersone, A ... 169
Antisari, M V ... 1
Augustovs, P ... 68
Bagaturyants, A ... 184
Baitimirova, M ... 274
Bajars, G ... 107, 112, 117, 287
Bakulin, A ... 24, 86
Baluyan, T ... 150
Baranov, A N ... 43
Beganskiene, A ... 257
Berzina, B ... 265
Berzins, Dz ... 231, 235
Bikhert, Y V ... 199
Bocharov, D ... 283
Bochkov, I ... 261
Bogdanov, V ... 214
Bonetti, E ... 1
Borovanska, I ... 261
Boyko, V ... 49
Brighi, M ... 1
Butikova, J ... 55, 239
Chen, K H ... 265
Chen, L C ... 265
Chen, M W ... 265
Chen, Y T ... 265
Chernenko, K A ... 9
Chikvaidze, G ... 305
Churmanov, V N ... 43
Cvetkov, A ... 301
Dam, B ... 1
Dauletbekova, A ... 205
Dauletbekova, A K ... 199
Dizhbite, T ... 169
Dobele, G ... 169, 252
Doke, G ... 195
Dolgov, L ... 37, 49
Dovbeshko, G ... 49
D'Yachkov, P ... 16
Erts, D ... 243
Evarestov, R A ... 31
Fedotovs, A ... 231, 235
Fesenko, O ... 49
Fomins, S ... 130
Freidzon, A ... 184
Gabrusenoks, J ... 112, 117, 287
Galimov, D M ... 61
Galitskiy, A G ... 297
Garaev, V ... 309
Gavare, M ... 248
Gavars, E ... 218
Gerbreders, A ... 142
Gertners, U ... 74, 138
Gorelik, V ... 49
Gorokhova, E I ... 9
Grabis, J ... 261
Grigalovica, A ... 155, 261
Grigorjeva, J ... 265
Grigorjeva, L ... 94, 190, 226
Grinberga, L ... 287, 291
Gross, K A ... 164
Grossin, D ... 164
Grube, J ... 195
Gruduls, A ... 248
Gruzdev, N B ... 43
Gryaznov, D ... 283
Hanschmidt, K ... 126
Hocker, S ... 24, 86
Hussainov, M ... 126
Ivanov, V ... 190
Ivanov, V Yu ... 43
Jakovlevs, D ... 103
Janeliukštis, R ... 112
Jankovica, D ... 195, 248
Jankovskis, J ... 103
Jarašiunas, K ... 265
Jekabsone, L ... 117
Jeršova, A ... 164
Juršenas, S ... 177
Kalkis, V ... 155
Kampars, V ... 80, 173, 177
Kareiva, A ... 257
Katkevics, J ... 243, 274
Khenkin, L ... 150
Khoduk, I V ... 9
Kiisk, V ... 37, 49
Kink, I ... 126
Kiselova, O ... 231, 235
Kleperis, J ... 107, 287, 291, 309
Kokars, V ... 68, 173
Kolobjonoks, V ... 142
Kolosovska, G ... 305
Korepanov, V I ... 199
Kotomin, E A ... 90, 210, 283
Kotsilkova, R ... 261
Krasilnikova, J ... 169
Kreicberga, J ... 177
Krievs, O ... 270
Krumina, A ... 301
Kucinskis, G ... 107
Kukle, S ... 121
Kulikova, L ... 301
Kulis, P ... 239

AUTHOR INDEX

Kulish, V G 297
Kulkova, S 24, 86
Kundzins, K....94
Kuprin, A S....297
Kutlubaev, D 16
Lähderanta, E....61
Laipniece, L 173
Lashkul, A V....61
Lauberts, M 169
Lescinskis, A 243
Lesnicenoks, P 291
Liepina, I....287
Liepina, V 226
Liepins, J....248
Lisitsin, V....205
Lisitsyn, V M 199
Lisitsyna, L A 199
Lisovski, O 278
Lisunov, K G 61
Lobjakas, M....126
Lõhmus, A 126
Lomino, N S....297
Loot, A 37
Luse, K....130
Lusis, A 107, 112, 117, 287
Mäeorg, U 126
Maiorov, M....155
Makhaeva, E....150
Maksimov, L....214
Manika, I 98
Maniks, J 98, 205
Marcins, G....55, 239
Matiss, I 270
Meri, R M 261
Meri, R Merijs....155
Miasojedovas, S....177, 265
Mihailovs, I....177
Millers, D 190, 226
Mironova-Ulmane, N 103, 169, 222, 252
Moiseyenko, V 49
Montone, A 1
Monty, C 226
Morozov, O M 297
Moskvin, A S 43
Muhins, A 55
Muktepavela, F....94
Naidoo, Q-L 159
Naidoo, S....159
Nargelas, S....265
Ndungu, P 159
Nechaev, A 159
Neklyudov, I M....297
Nemilov, S....214
Novakova, A 150
Odinokov, A....184
Opalais, A....305
Ornicane, G....80
Osite, A 274, 305
Ovcharenko, V D 297
Ozolins, J 278
Ozolinsh, M 130
Ozols, A 68
Paalo, M 126
Palmisano, V 1
Pasquini, L 1
Pastors, P....80
Patmalnieks, A 248
Pavlovica, S 309
Pentjuss, E....112, 117
Pentjušs, E....270
Pervenecka, J 239
Petrik, L 159
Pikker, S....37
Pilipavicius, J 257
Pilmane, M 134
Piskunov, S....278
Polyakov, B 55, 239
Ponomarenko, J....169
Ponomarenko, N 103
Poplausks, R 218
Pudzs, K 239
Purans, J 90
Pustovarov, V A....43
Rebrin, I 150
Reinholds, I 155
Reiterov, V M 199
Reshetnyak, E N 297
Rey, C 164
Rodnyi, P A....9
Romanyuk, V....49
Rusakova, A 205
Rusan, V....214
Rutkis, M....173, 177
Saharov, D....68
Sakalauskas, D....257
Sarakovskis, A 195, 231, 235
Ščajev, P 265
Schmauder, S 24, 86
Schwartz, K....205
Serga, V 301
Shimane, O....142
Shkolnikov, E....252
Sildos, I....37, 49, 252
Sivars, A 291
Skagers, A 134
Skudra, A 218
Skuja, L 177
Skvortsova, V....222
Smits, J 107
Smits, K 226, 248

AUTHOR INDEX

Sokolov, P S 43
Sokolov, V I 43
Sorokin, A V 90
Sursajeva, V 94
Svagere, A 218
Tale, I 55, 239
Tätte, T 126
Telysheva, G 169
Teteris, J 74, 138, 142, 146
Tikhomirov, V 184
Titkov, A N 61
Tokmakovs, A 173
Traskovskis, K 68, 173
Trinkler, L 265
Uimin, M A 43
Vaivars, G 159, 309
Vamze, J 134
Vervikishko, D 252
Vihodceva, S 121
Viksna, A 164, 243, 274, 305
Voitkans, A 55
Volperts, A 252
Voss, M 195
Yermakov, A Ye 43
Zabels, R 94, 98
Zandersons, J 112
Zarins, E 173
Zdorovets, M 205
Zherebtsov, D A 61
Zhukovkii, Yu F 283
Zhukovskii, Y F 90, 278
Zhukovskii, Yu F 31
Zhurba, V I 297
Zicans, J 155, 261, 270
Zorina, N 218